Beneficial Chemical Elements of Plants

Beneficial Chemical Elements of Plants

Recent Developments and Future Prospects

Edited by

Sangeeta Pandey
Amity Institute of Organic Agriculture, Amity University
Uttar Pradesh, Noida, India

Durgesh Kumar Tripathi
Amity Institute of Organic Agriculture, Amity University
Uttar Pradesh, Noida, India

Vijay Pratap Singh
CMP Degree Collage, University of Allahabad, Prayagraj, India

Shivesh Sharma
Department of Biotechnology, Motilal Nehru National
Institute of Technology, Allahabad, Prayagraj, India

Devendra Kumar Chauhan
Department of Botany at the DD Pant Interdisciplinary Research Laboratory,
University of Allahabad, Allahabad, India

WILEY

This edition first published 2023

Registered Offices
John Wiley & Sons, Inc., 111 River Street, Hoboken, NJ 07030, USA
John Wiley & Sons Ltd, The Atrium, Southern Gate, Chichester, West Sussex, PO19 8SQ, UK

For details of our global editorial offices, customer services, and more information about Wiley products visit us at www.wiley.com.

Wiley also publishes its books in a variety of electronic formats and by print-on-demand. Some content that appears in standard print versions of this book may not be available in other formats.

Library of Congress Cataloging-in-Publication Data
Names: Pandey, Sangeeta, editor. | Tripathi, Durgesh Kumar, editor. | Singh, Vijay Pratap, editor. | Sharma, Shivesh, editor. | Chauhan, Devendra Kumar, editor.
Title: Beneficial chemical elements of plants : recent developments and future prospects / edited by Sangeeta Pandey, Durgesh Kumar Tripathi, Vijay Pratap Singh, Shivesh Sharma, and Devendra Kumar Chauhan.
Description: Hoboken, NJ : Wiley, 2023. | Includes bibliographical references and index.
Identifiers: LCCN 2022041344 (print) | LCCN 2022041345 (ebook) | ISBN 9781119688808 (hardback) | ISBN 9781119688815 (adobe pdf) | ISBN 9781119688839 (epub)
Subjects: LCSH: Plants–Effect of chemicals on. | Growth (Plants)
Classification: LCC QK746 .B36 2023 (print) | LCC QK746 (ebook) | DDC 575.9/7–dc23/eng/20220916
LC record available at https://lccn.loc.gov/2022041344
LC ebook record available at https://lccn.loc.gov/2022041345

Cover Design: Wiley
Cover Image: © New Africa/Shutterstock

Set in 9.5/12.5pt STIXTwoText by Straive, Pondicherry, India

Contents

Preface

Beneficial elements offer a new prospect as bio-stimulants in the field of plant biology due to their recent eminence as enhancers of plant productivity and stress tolerance. Aluminium (Al), cobalt (Co), sodium (Na), selenium (Se), and silicon (Si) are major members of this unique class of elements that are not essential for most plants but improve plant growth and nutritional quality when supplied at low concentrations. These elements stimulate various mechanisms, resulting in different phenotypical, physiological, biochemical, and molecular alterations in plants. Besides their growth-promoting effects, the elements trigger adaptive responses towards environmental stressors in some plants. When plants are exposed to challenging environmental stimuli including abiotic factors such as heat, drought, low temperatures, heavy metals, and salinity and biotic factors such as herbivory and pathogens, these beneficial elements boost their resistance and activate their defence responses. The acclimation against such stressors is achieved by the increase in nutrient uptake, synthesis, accumulation and activation of antioxidants, production of secondary metabolites and osmo-protectants, and stimulation of signalling cascades upon application of these beneficial elements at low doses. However, this low-dose growth stimulation by beneficial elements turns into inhibition at high dose. Thus, accurately applied beneficial elements can help in dealing with the most intimidating challenge of present times: that is, the production of food under unfavourable conditions to meet the food demands of incessantly growing human populations.

In this book, a total of 15 chapters address the current knowledge regarding the role of beneficial impacts on plants and the associated challenges. In its initial chapters, the book focuses on the interaction between plants and all beneficial elements. In later chapters, the book deals with individual beneficial elements and their respective interrelations with plant growth. The physiological roles of beneficial elements in plants and their potential to support plants under daunting stress conditions (Naz et al.) are compiled in an organized manner. Their capacity to induce epigenetic regulation of the plant genome in response to abiotic stressors (Kulak and Aydin) is also covered. The association of these beneficial elements with reactive oxygen and nitrogen species – the key signalling molecules in plants under stress conditions – and the related physiological impacts on plants are discussed by Pradhan et al. Trejo-Téllez et al. have efficiently summarized the concentration-dependent effects of these beneficial elements on plants and their potential as bio-stimulants. Furthermore, these beneficial elements are reported to positively impact the process of photosynthesis in plants growing under normal or stressed conditions as compiled and discussed by Ceccanti et al. Ulhassan et al. and Hu et al. have focused particularly on the uptake, translocation, and growth stimulatory effects of Co plants. The latter also addresses the problem of Co deficiency in plants and subsequent plant responses. Ulhassan et al. have also delineated the uptake, translocation, assimilation, and metabolism of Se in plants. Similarly, the uptake, transport, and deposition of Si in different forms inside plant tissues (Sharma et al.) are also covered. The protective role of different forms of Si on plant growth traits against heavy metal stress through induction of various physiological and molecular alterations to maintain their productivity (Rana et al.) has also been discussed. Hossain et al. have focused on the consequences of high concentrations of Al on plants and have highlighted the alleviation strategies for Al toxicity in plants. Furthermore, the function of Na in plants, the disturbance caused by its deficiency or hyperaccumulation in plants, mechanisms of plants to cope with Na stress (Kumlay et al.; Naz et al.), and the effect of Na content in the growing media on the biochemical content of some medicinal and aromatic plants (Kumlay et al.) are also explained. The sparsely available literature regarding the potential of lanthanides to act as beneficial elements with growth-enhancing impact on plants at low concentrations (Gómez-Merino et al.) is also discussed.

This edited book presents a systematic and well-organized compilation of the research related to the application of beneficial elements in the field of plant science and agriculture to allow convenient accessibility of this information to scientists, researchers, and students working in this domain. In summary, we firmly believe that this book will offer an easy understanding of concepts as a resource of crucial information regarding functions of beneficial elements in plants, their role as protective agents against stressful conditions, and other related aspects in plant life.

Dr. Sangeeta Pandey
Dr. Durgesh Kumar Tripathi
Dr. Vjay Pratap Singh
Prof. Shivesh Sharma
Prof. Devendra Kumar Chauhan

List of Contributors

Muhammad Rahil Afzal
Faculty of Life Sciences
Institute of Environmental and Agricultural Sciences
University of Okara
Okara, Pakistan

Zahoor Ahmad
Department of Botany
University of Central Punjab
Punjab Group of Colleges
Bahawalpur, Pakistan

M. Ashraful Alam
Plant Breeding Division
Spices Research Centre
Bangladesh Agricultural Research Institute
Bogura, Bangladesh

Md Jahangir Alam
On-Farm Research Division
Bangladesh Agricultural Research Institute
Gaibandha, Bangladesh

Maratab Ali
College of Food Science and Biotechnology
Key Laboratory of Fruits and Vegetables
Postharvest and Processing Technology
Research of Zhejiang Province
Zhejiang Gongshang University
Hangzhou, PR China

Mujahid Ali
Institute of Soil and Environmental Sciences
Faculty of Agriculture
University of Agriculture
Faisalabad, Pakistan

Mehmet Hakki Alma
Department of Forest Industry Engineering
Faculty of Forestry
Kahramanmaraş Sutcu Imam University
Kahramanmaraş, Türkiye

Muhammad Arif
Muhammad Nawaz Sharif
University of Agriculture, Multan

Muhammad Jawaad Atif
College of Horticulture
Northwest A&F University
Yangling, Shaanxi, China

Adnan Aydin
Department of Agricultural Biotechnology
Faculty of Agriculture
Igdir University
Igdir, Türkiye

Muhammad Ashar Ayub
Institute of Agro-Industry and Environment
Faculty of Agriculture and Environment
The Islamia University of Bahawalpur
Bahawalpur, Pakistan

Wardah Azhar
Institute of Crop Science
Ministry of Agriculture and Rural Affairs
Key Laboratory of Spectroscopy Sensing
Zhejiang University
Hangzhou, China

Ruchi Bansal
Department of Agri-Biotechnology
National Agri-Food Biotechnology Institute
Sahibzada Ajit Singh Nagar
Punjab, India

and

Department of Biotechnology
Panjab University
Chandigarh, India

Chhandashree Behera
P.G. Department of Botany
Berhampur University
Berhampur, Odisha, India

Preetha Bhadra
Department of Biotechnology
Centurion University of Technology and Management
Paralakhemundi
Odisha, India

Muhammad Adnan Bodlah
Fareed Biodiversity Conservation Centre
Department of Agricultural Engineering
Khwaja Fareed University of Engineering and Information Technology
Rahim Yar Khan
Punjab, Pakistan

Imran Bodlah
Department of Entomology
Pir Mehr Ali Shah Arid Agriculture University
Rawalpindi, Pakistan

Sarah Bouzroud
Laboratoire de Biotechnologie et Physiologie Végétales
Centre de Biotechnologie Végétale et Microbienne Biodiversité et Environnement
Faculté des Sciences
Université Mohammed V de Rabat
Rabat, Morocco

Marian Brestic
Department of Botany and Plant Physiology
Faculty of Agrobiology
Food and Natural Resources
Czech University of Life Sciences Prague
Prague, Czechia

and

Department of Plant Physiology
Slovak University of Agriculture
Nitra
Slovak Republic

Costanza Ceccanti
Department of Agriculture
Food and Environment
University of Pisa
Pisa, Italy

Ferdi Celikcan
Department of Organic Farming
College of Applied Science
Igdir University
Igdir, Türkiye

Jianjun Chen
Department of Environmental Horticulrture and Mid-Florida Research and Education Center
Institute of Food and Agricultural Sciences
University of Florida
Apopka, FL, USA

Soumya Ranjan Dash
P.G. Department of Botany
Berhampur University
Berhampur
Odisha, India

Rupesh Deshmukh
Department of Agri-Biotechnology
National Agri-Food Biotechnology
Institute, Sahibzada Ajit Singh Nagar
Punjab, India

and

Department of Biotechnology
Central University of Haryana
Mahendragarh
Haryana, India

Reza Mohammad Emon
Plant Breeding Division
Bangladesh Institute of Nuclear
Agriculture
Mymensingh, Bangladesh

Xiaorong Fan
State Key Laboratory of Crop Genetics and
Germplasm Enhancement
Nanjing Agricultural University
Nanjing, China

and

Key Laboratory of Plant Nutrition and
Fertilization in Lower-Middle Reaches of
the Yangtze River
Ministry of Agriculture
Nanjing Agricultural University
Nanjing, China

Muhammad Imran Ghani
College of Natural Resource and Environment
Northwest A&F University
Yangling
Shaanxi, China

Fernando C. Gómez-Merino
College of Postgraduates in Agricultural
Sciences
Laboratory of Plant Nutrition
Texcoco, State of Mexico, Mexico

Libia F. Gómez-Trejo
Department of Plant Protection
Chapingo Autonomous University
Texcoco, State of Mexico, Mexico

Lucia Guidi
Department of Agriculture
Food and Environment
University of Pisa, Pisa, Italy

and

CIRSEC
Centre for Climatic Change Impact
University of Pisa
Pisa, Italy

Yasir Hamid
Ministry of Education (MOE) Key Lab of
Environ. Remediation and Ecol. Health
College of Environmental and
Resources Science
Zhejiang University
Hangzhou, China

Akbar Hossain
Department of Agronomy
Bangladesh Wheat and Maize Research
Institute
Dinajpur, Bangladesh

Xiu Hu
College of Horticulture and Landscape
Architecture
Zhongkai University of Agriculture and
Engineering
Guangzhou
Guangdong, China

Nasrin Jahan
Plant Genetic Resources Centre
Bangladesh Agricultural Research Institute
Gazipur, Bangladesh

Asad Jamil
Institute of Soil and Environmental Sciences
Faculty of Agriculture
University of Agriculture
Faisalabad, Pakistan

Mrutyunjay Jena
P.G. Department of Botany
Berhampur University
Berhampur
Odisha, India

Ali Raza Khan
Institute of Crop Science
Ministry of Agriculture and Rural Affairs
Key Laboratory of Spectroscopy Sensing
Zhejiang University
Hangzhou, China

Mehmet Zeki Kocak
Department of Herbal and Animal Production
Vocational School of Technical Sciences
Igdir University,
Igdir, Türkiye

Muhittin Kulak
Department of Herbal and Animal Production
Vocational School of Technical Sciences
Igdir University
Igdir, Türkiye

Virender Kumar
Department of Agri-Biotechnology
National Agri-Food Biotechnology Institute
Sahibzada Ajit Singh Nagar
Punjab, India

Ahmet Metin Kumlay
Department of Field Crops
Faculty of Agriculture
Igdir University
Igdir, Türkiye

Marco Landi
Department of Agriculture
Food and Environment
University of Pisa
Pisa, Italy

and

CIRSEC
Centre for Climatic Change Impact
University of Pisa
Pisa, Italy

Jie Ling
He Xiangning College of Art and Design
Zhongkai University of Agriculture and Engineering
Guangzhou
Guangdong, China

Abdullah Al Mahmud
On-Farm Research Division
Bangladesh Agricultural Research Institute
Gaibandha, Bangladesh

Sagar Maitra
Department of Agronomy
Centurion University of Technology and Management
Odisha, India

Md Abdul Malek
Plant Breeding Division
Bangladesh Institute of Nuclear Agriculture
Mymensingh, Bangladesh

Debojyoti Moulick
Plant Stress Biology and Metabolomics Laboratory
Assam University
Silchar
Assam, India

Muntazir Mushtaq
School of Biotechnology
Sher-e-Kashmir University of Agricultural Sciences and Technology of Jammu
Chatha
J&K, India

Rabindra Nayak
P.G. Department of Botany
Berhampur University
Berhampur
Odisha, India

Misbah Naz
State Key Laboratory of Crop Genetics and Germplasm Enhancement
Nanjing Agricultural University,
Nanjing, China

Srimanta Patra
Department of Life Sciences
NIT, Rourkela
Odisha, India

Ermes Lo Piccolo
Department of Agriculture
Food and Environment
University of Pisa
Pisa, Italy

Biswajita Pradhan
P.G. Department of Botany
Berhampur University
Berhampur
Odisha, India

Md Atikur Rahman
Spices Research Center
Bangladesh Agricultural Research Institute (BARI)
Bogra, Bangladesh

Sana Rana
Institute of Soil and Environmental Sciences
Faculty of Agriculture
University of Agriculture
Faisalabad, Pakistan

Gaurav Raturi
Department of Agri-Biotechnology
National Agri-Food Biotechnology Institute
Sahibzada Ajit Singh Nagar
Punjab, India

and

Department of Biotechnology
Panjab University
Chandigarh, India

Muhammad Ammar Raza
College of Food Science and Biotechnology
Key Laboratory of Fruits and Vegetables Postharvest and Processing Technology Research of Zhejiang Province
Zhejiang Gongshang University
Hangzhou, China

Muhammad Zia Ur Rehman
Institute of Soil and Environmental Sciences
Faculty of Agriculture
University of Agriculture
Faisalabad, Pakistan

Muhammad Riaz
School of Agriculture and Biology
Shanghai Jiao Tong University
Minhang
Shanghai, China

Rubén Ruvalcaba-Ramírez
Department of Plant Protection
Chapingo Autonomous University
Texcoco
State of Mexico, Mexico

Saikat Saha
Nadia Krishi Vigyan Kendra
Bidhan Chandra Krishi Viswavidyalaya
Gayeshpur
West Bengal, India

Sukamal Sarker
School of Agriculture and Rural Development
Faculty Centre for IRDM
Ramakrishna Mission Vivekananda Educational and Research Institute
Ramakrishna Mission Ashrama
Narendrapur
Kolkata, India

Tariq Shah
Department of Agronomy
Faculty of Crop Production Sciences
University of Agriculture Peshawar
Peshawar, Pakistan

Aamir Mehmood Shah
State Key Joint Laboratory of Environment Simulation and Pollution Control
School of Environment
Beijing Normal University
Beijing, China

Shivani Sharma
Department of Agri-Biotechnology
National Agri-Food Biotechnology Institute
Sahibzada Ajit Singh Nagar
Punjab, India

Muhammad Tahir Shehzad
Institute of Soil and Environmental Sciences
Faculty of Agriculture
University of Agriculture
Faisalabad, Pakistan

S.M. Shivaraj
Department of Agri-Biotechnology
National Agri-Food Biotechnology Institute
Sahibzada Ajit Singh Nagar
Punjab, India

Milan Skalicky
Department of Botany and Plant Physiology
Faculty of Agrobiology
Food and Natural Resources
Czech University of Life Sciences Prague
Prague, Czechia

Sreeja Sudhakaran
Department of Agri-Biotechnology
National Agri-Food Biotechnology Institute
Sahibzada Ajit Singh Nagar
Punjab, India

Muhammad Tariq
Department of Pharmacology
Lahore Pharmacy Collage
Lahore, Pakistan

Vandana Thakral
Department of Agri-Biotechnology
National Agri-Food Biotechnology Institute
Sahibzada Ajit Singh Nagar
Punjab, India

and

Department of Biotechnology
Panjab University
Chandigarh, India

Libia I. Trejo-Téllez
College of Postgraduates in Agricultural Sciences
Laboratory of Plant Nutrition
Texcoco
State of Mexico, Mexico

Durgesh Kumar Tripathi
The Amity Institute of Organic Agriculture
Amity University Uttar Pradesh
Noida, India

Zaid Ulhassan
Institute of Crop Science
Ministry of Agriculture and Rural Affairs
Key Laboratory of Spectroscopy Sensing
Zhejiang University
Hangzhou, China

Muhammad Umair
Institute of Soil and Environmental Sciences
Faculty of Agriculture
University of Agriculture
Faisalabad, Pakistan

Sanskriti Vats
Department of Agri-Biotechnology
National Agri-Food Biotechnology Institute
Sahibzada Ajit Singh Nagar
Punjab, India

Hindu Vemuri
International Maize and Wheat Improvement Center
Hyderabad, India

Xiangying Wei
Fujian Key Laboratory on Conservation and Sustainable Utilization of Marine Biodiversity
Fuzhou Institute of Oceanography
College of Geography and Oceanography
Minjiang University, Fuzhou
Fujian, China

Weijun Zhou
Institute of Crop Science, Ministry of Agriculture and Rural Affairs Key Laboratory of Spectroscopy Sensing
Zhejiang University
Hangzhou, China

Akmal Zubair
Biochemistry Department
Quaid-i-Azam University Islamabad
Islamabad, Pakistan

1

Beneficial Elements in Plant Life Under A Changing Environment

Misbah Naz[1], Muhammad Ammar Raza[2], Muhammad Adnan Bodlah[3], Sarah Bouzroud[4], Muhammad Imran Ghani[5], Muhammad Riaz[6], Tariq Shah[7], Akmal Zubair[8], Imran Bodlah[9], and Xiaorong Fan[1, 10]

[1] *State Key Laboratory of Crop Genetics and Germplasm Enhancement, Nanjing Agricultural University, Nanjing, China*
[2] *College of Food Science and Biotechnology, Key Laboratory of Fruits and Vegetables Postharvest and Processing Technology Research of Zhejiang Province, Zhejiang Gongshang University, Hangzhou, China*
[3] *Fareed Biodiversity Conservation Centre, Department of Agricultural Engineering, Khwaja Fareed University of Engineering and Information Technology, Rahim Yar Khan, Punjab, Pakistan*
[4] *Laboratoire de Biotechnologie et Physiologie Végétales, Centre de Biotechnologie Végétale et Microbienne Biodiversité et Environnement, Faculté des Sciences, Université Mohammed V de Rabat, Rabat, Morocco*
[5] *College of Natural Resource and Environment, Northwest A&F University, Yangling, Shaanxi, China*
[6] *School of Agriculture and Biology, Shanghai Jiao Tong University, Minhang, Shanghai, China*
[7] *Department of Agronomy, Faculty of Crop Production Sciences, University of Agriculture Peshawar, Peshawar, Pakistan*
[8] *Biochemistry Department, Quaid-i-Azam University Islamabad, Islamabad, Pakistan*
[9] *Department of Entomology, Pir Mehr Ali Shah Arid Agriculture University, Rawalpindi, Pakistan*
[10] *Key Laboratory of Plant Nutrition and Fertilization in Lower-Middle Reaches of the Yangtze River, Ministry of Agriculture, Nanjing Agricultural University, Nanjing, China*

Introduction

The presence of many essential element nutrients affects plant growth and yield (Elser 2012). The availability of a single element, rather than the availability of all nutrients, limits plant absorption of all nutrients (Kirkby 2012). The scarcity of fertile land is a major concern as the world's population grows day by day (Niste et al. 2014). As a result, increasing food production and using fewer appropriate soils, and those with high salinity, poor nutrient availability, limited water holding capacity, and partially contaminated sites, would become critical. Plants suffer as a result of poor soil quality, leading to reduced food production (Niste et al. 2014). To improve this situation, the valuable elements are not considered important for all crops, but they may be critical for specific plant taxa. In the case of certain trace elements, distinguishing between useful and vital is often difficult (Pilon-Smits et al. 2009). Plants benefit from elements such as aluminium (Al), cobalt (Co), sodium (Na), silicon (Si), and selenium (Se). These elements are not needed for all plants, but they can enhance plant growth and yield (Kaur et al. 2016). At low levels, beneficial elements are said to improve resistance to abiotic stresses (drought, salinity, high

Beneficial Chemical Elements of Plants: Recent Developments and Future Prospects, First Edition. Edited by Sangeeta Pandey, Durgesh Kumar Tripathi, Vijay Pratap Singh, Shivesh Sharma, and Devendra Kumar Chauhan.

temperature, cold, UV stress, and nutrient toxicity or deficiency) as well as biotic stresses (pathogens and herbivores). The essential-to-lethal range for these elements, though, is quite narrow (Ahmad et al. 2012; Fahad et al. 2015). The effect of beneficial elements at lower doses merits more concern when it comes to using them to fertilize crops in order to boost crop production under stress and to improve plant nutritional value as a feed or food (Meena et al. 2014). A more proactive approach to plant nutrition might involve mineral elements at levels effective for best growth as well as nutrients necessary for survival (Coleman et al. 2014). We describe the mechanisms of absorption of various beneficial elements, their advantages, and the function of these elements in imparting tolerance to abiotic and biotic stresses in this paper. Sufficient intracellular doses of beneficial metal ions (in traces) are needed not only for optimal plant growth and development but also for pathogen infectivity and plant defences (Dighton and Krumins 2014).

Metal defences are primarily based on hyperaccumulators of important facets, anti-plant pathogen hypothesis stress signalling, and metal ion intermodulation is linked to plant responses to both abiotic and biotic stress factors and is an emerging research field in metal hyperaccumulator and non-hyperaccumulator plants (Bhardwaj et al. 2014). The impact of useful elements at low levels receives more respect in this period of research in order to fertilize crops with these nutrients to improve crop production in stressed environments as well as boost plant nutritional value as feed or food (Osorio Vega 2007). Unlike the toxic effects, the processes that accompany the beneficial effects (at low levels) to the plants have not been thoroughly investigated (at high levels). More research into how these elements protect against pathogens and abiotic stress factors is needed, particularly at the molecular level (Compant et al. 2010). It is necessary to investigate how these components have synergistic or antagonistic effects on plants developing in unstressed and stressed conditions. Foliar spray of these components must be checked in plants growing in stressful conditions (Ilangumaran and Smith 2017). Useful elements for agricultural crops, as well as their practical significance in stress defences, are a tool for increasing crop yield (van Boekel et al. 2010).

Beneficial Element Interaction with Environment

Plant behaviour is a strong demand for the supply of essential mineral nutrients, which affects many important functions (White and Brown 2010). Useful elements are not required for plant growth and development, but when they are present, they help to promote growth and development by stimulating resistance mechanisms against biological and abiotic stressors, promoting the use of other uses, and compensating for or alleviating the harmful effects of other elements (Broadley et al. 2012; Anjum et al. 2015). Plants' response mechanisms to environmental factors such as drought, heavy metal toxicity, high salt content soils, pests, or pathogens may contain useful elements. This analysis highlights the beneficial effects of aluminium (Al), methyl (Ce), cobalt (Co), iodine (I), lanthanum (la), sodium (Na), selenium (Se), silicon (Si), titanium (Ti), and sodium (N) in some plants that have observed major shifts, as well as the possible uses of novel ingredients in aluminium agricultural output (Paustenbach et al. 2013; Malagoli et al. 2015; Zhang et al. 2017; Muhammad et al. 2018). Aluminium controls flower colour, promotes plant growth and root production, extends the life of certain vases, and slows antioxidant mechanisms (Muhammad et al. 2018). Selenium can boost oxidative stress tolerance, slow the process of aging, enhance growth, and raise heavy metal consumption (Asemi et al. 2015). Silicon can counteract the toxic effects of heavy metals, drought, and salinity, leading to pest and disease tolerance, forming nanostructures, improving the multi-body and stiffness of plant tissue, stimulating antioxidant mechanisms, reducing ethylene synthesis, and extending vase existence (Zhu and Gong 2014). Sodium can act as a regulator, extend vase life, and induce the synthesis of amino acids like alanine. Titanium enhances N, P, K, Ca, and Mg experience, increases starch synthesis, reduces *Xanthomonas* damage, and produces better plant growth. Sodium is used as a secondary metabolism agent to boost plant growth (Vankova 2014). Plant nutrients are needed for plant growth and development. If there are not enough of them, it causes a particular deficient symptom. If a single plant nutrient is completely deficient, growth will halt and the plant cannot finish its life cycle. Recent scientific evidence indicates that there are 14 important plant nutrients based on these standards (Lambers et al. 2008). Plant nutrients are classified as macronutrients or micronutrients based on whether they are found in greater or lesser quantities in plants requiring several g/ha (Lambers et al. 2008). Present chapter emphasizes the role that useful elements such as Na, Si, Co, Se, and Al perform. Sodium could stimulate plant growth, especially in some C4 plants, by facilitating substrate movement between the mesophyll and the bundle sheath. It can also partially substitute K as an osmoticum, and applying Na fertilizers to sugar beet leads to an improvement in the leaf area index early in the growing season (Mahmoud et al. 2012), and under moderate drought stress, this increases light penetration and improves water usage quality of leaves (Elser 2012; Kirkby 2012). Silicon contributes to cell wall integrity by connecting polyuronides and promoting lignin synthesis. It may increase plant and leaf erectness, reduce water consumption, and safeguard plants from pests and diseases. Silicon is also useful in reducing the harmful effects of other metals such as Fe, Al, Cd, and Zn, that can be due to the presence of Si and metals in the apoplasm or symplasm (Cao et al. 2017). Cobalt is required for N2 fixing plants as it is a component of the coenzyme cobalamin (vitamin B12), that is required for nodule metabolism (Khan and Khan 2010). Se has a similar chemistry to S and can partially replace S in proteins, especially in Se hyperaccumulating plants. Since selenium is necessary for livestock, Se fertilization may also be helpful to human and animal health in areas with Se-deficient soils (Broadley et al. 2012). Aluminium is helpful to certain plants, such as tea, and can reduce proton toxicity and enhance antioxidant enzyme activity (Hayat et al. 2012, Chauhan et al. 2021) (Figures 1.1–1.5).

Aluminium (Al) in Plants

Aluminium (Al) is an amphoteric material that has no known biological role. Aluminium is the third most abundant chemical element in the lithosphere (Samac and Tesfaye 2003). Al is used in nature mostly as slightly soluble oxides and silicates, rather than as a free metal (Grzybkowski 2006). The supply of Al and, as a result, Al's ability to communicate with plants is largely limited to acidic environments (Neal 2008). Al $(H_2O)\ 6^{3+}$ is the leading monomeric Al species in aqueous media with a pH below 5. The key toxic Al species is thought to be this Al type, which is typically written as Al^{3+}. Different inorganic (e.g. fluoride, sulphate, silicon) and organic ligands (e.g. organic acids, phenolics, hydroxamates) are present in the chemically complex soil solutions. Aluminium (Al) is the third most common element in the Earth's crust and a key element hindering plant growth and reducing crop yield in acidic soil (Kopittke et al. 2005). While extensive research has been conducted on the phytotoxic effects and mechanisms underlying of Al when applied hydroponically, soil is a difficult medium containing various mineral elements that can associate with Al and other substances, as well as their bioavailability in plants (Yang and Watts 2005; Zheng 2010). We determine the process of Al in enhancing plant growth, increasing phosphorus supply and efficiency in plants, and reducing H+, iron, and manganese toxicity in acidic conditions in this study. Moreover, we explore the potential mechanisms of Al-induced increased abiotic stress tolerance (Chauhan et al. 2021).

Aluminium (Al) in Soil – Aluminium, a Friend or Foe of Higher Plants in Acidic Soils

Aluminium is even more accessible to the plants in acidic soils (soils with a very low pH), so acid-loving crops like blueberries and cranberries are also among the more aluminium-resistant varieties (Zheng 2010). Since calcium (Ca) cations in gypsum compete with aluminium (Al) cations, they become less soluble in water by plants. While aluminium (Al) seems to be the most affordable metal in the Earth's crust, its availability is affected by soil pH. Regardless its ample supply Al is not recognized as an important factor, and no experimental evidence for a biological function has been presented so far (Poschenrieder et al. 2008). Al may be beneficial or harmful to plants and other organisms, depends upon factors including metal concentration, chemical form of Al, growth conditions, and plant types (Watanabe and Osaki 2002). In this article, we review latest events in the study of Al in plants at the physiological, biochemical, and molecular levels, with an emphasis on the beneficial effect of Al in plants (stimulation of root growth, increased nutrient uptake, the increase in enzyme activity, and others) (Gupta and Huang 2014). Furthermore, we explore the potential mechanisms needed to enhance the growth of plants grown in acidic soils, along with mechanisms of tolerance to the toxic effect of Al. Acid soils, also known as ultisols or oxisols, have a pH of 5.5 or below and are commonly found in tropical and subtropical areas, accounting for about 30% of the total area of the planet and 50% of the world's agricultural land, as well as 25–80% of vegetable production (Gupta and Huang 2014; Silva 2017). Soil acidification may arise as a result of both natural and anthropogenic processes (Figure 1.1). The majority of acid soils are found in the tropics and subtropics, where acidification occurs naturally. This condition could be exacerbated by environmental pollution caused by the use of fertilizers and acidifying chemicals and the use of fossil energy sources (Iqbal 2012). For example, coal and oil, which emit nitrogen dioxide (NO_2) and sulphur dioxide (SO_2) into the atmosphere, and when combined with oxidizing agents, produce nitric acid (HNO_3) and sulphuric acid (H_2SO_4), rising acid rain accumulation and acidification of bodies of water and soil. Besides that, organic material decomposes; imbalances in the N, S, and C cycles; increased cation absorption over anions; and nutrients supplied by leguminous crops all increase the levels of H^+ and lower soil pH (Figure 1.1)

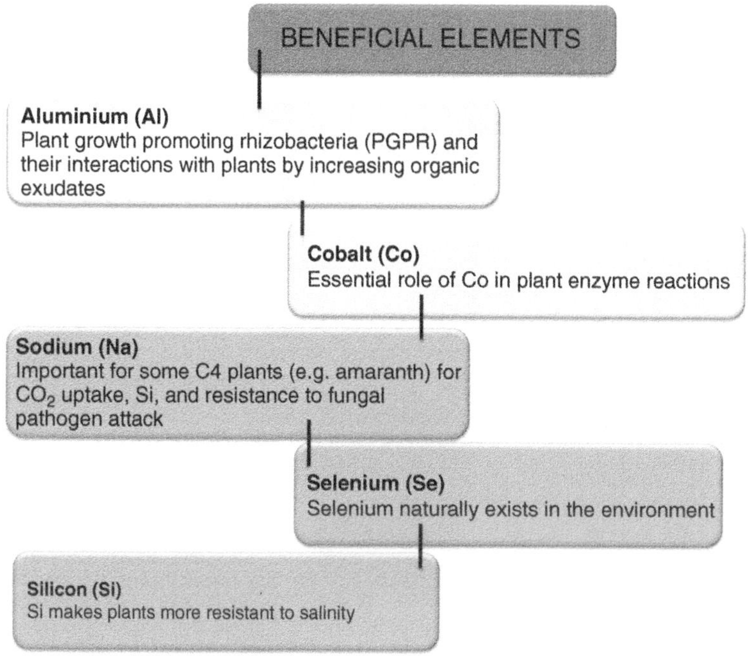

Figure 1.1 The schematic illustration of different beneficial trace elements necessary in plants' life. These natural elements play vital role in plant life cycle, i.e. defence, growth, and development.

(Nunes-Nesi et al. 2014; Sade et al. 2016; Singh et al. 2017). Acid soils are distinguished by poor nutrition and contamination by metals including manganese (Mn), iron (Fe), and aluminium (Al), with Al contamination being the primary factor affecting plant growth in acid soils (Kichigina et al. 2017).

Cobalt (Co) in Plants

Cobalt has previously been recognized as an important mineral for animals. After all, our awareness of the critical role in plant enzymatic reactions remains limited (Simonsen et al. 2012). Cobalt's most well-known role in plants is for N-fixing microorganisms like Rhizobia, that live symbiotically with legume plants (Gad et al. 2011). The overall effect of cobalt (Co) and copper (Cu) on plant toxicity are rarely reported, despite the fact that these two metals coexist frequently in soil. This study summarizes current knowledge of Cu-Co tolerance and deposition in plants (Nagajyoti et al. 2010; Lwalaba et al. 2019). Accretion of foliar Cu and Co to >300 g g-1 is pretty uncommon worldwide and is renowned from the Copperbelt of Central Africa. Cobalt deposition has also been reported in a small number of Ni hyperaccumulator plants found on ultramafic soils worldwide. Since foliar Cu-Co deposition is highly dose dependent, none of the alleged Cu or Co hyperaccumulator plants tend to follow the basic concept of hyperaccumulation (Faucon et al. 2018). Plant tissue Cu concentrations are unusually high only when plants are dealing with high soil Cu

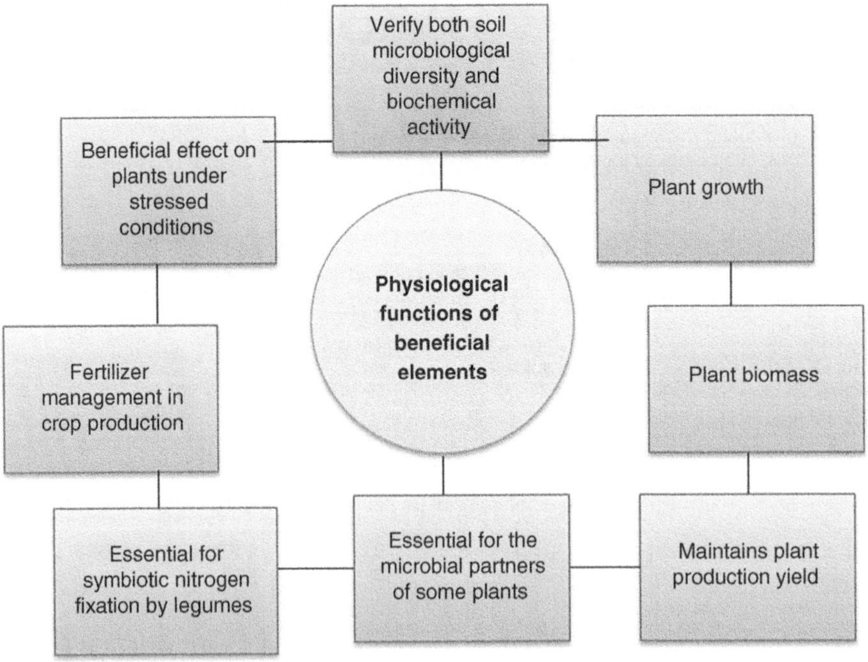

Figure 1.2 The role of different micro and macro essential nutrients involved in plant growth, development, and protection against different stress and disease at micro level.

concentration with a low shoot translocation factor. Although most Cu tolerant plants are Excluders sensu Baker and therefore setting Cu hyperaccumulation threshold values is useless (Lange et al. 2017). Cobalt deposition has also been found in a small number of nickel (Ni) hyperaccumulator plants found on ultramafic soils worldwide (Khan and Khan 2010). Due to their dose-dependent deposition features, the practical application of Cu–Co accumulator plants in phytomining is restricted; however, due to the extremely low metal content of Co, field trials on highly contaminated mineral wastes may be warranted (Sandrin and Hoffman 2007).

Cobalt (Co) in Soil

The European soil contains cobalt concentrations between 1 and 20 mg/kg on an average of the dry weight, whereas it has also been observed that these concentrations became higher in the areas those were geologically rich in cobalt including North Wales (Phoon et al. 2012). Similar levels of cobalt at over 2500 mg/kg dry weight were also observed by Paveley (1998). This element has been proved essential for the leguminous crops symbiotic nitrogen fixation by functioning as a coenzyme involved in nodule formation, growth, and N2 fixation that plays a critical role as cofactor of cobalamin (vitamin B12) (Weisany et al. 2013). The beneficial trace elements are those that are held essential for most of the crops but actually may be fixed vital towards the particular plant taxon (Sessitsch et al. 2013).

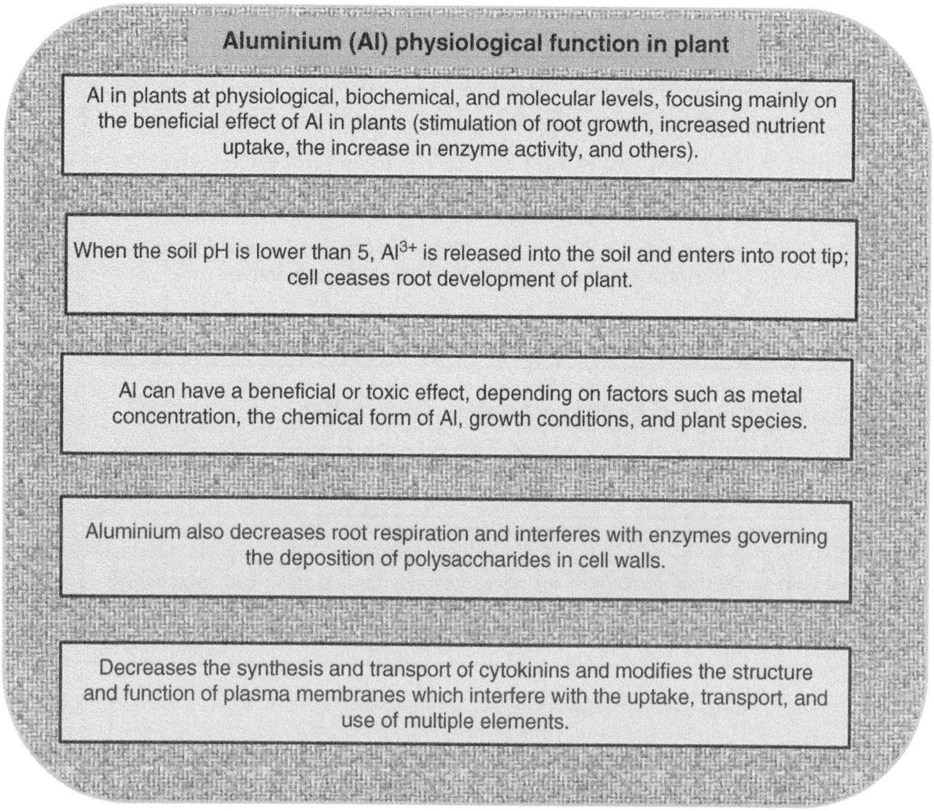

Figure 1.3 Role of aluminium for proper molecular functioning in life cycle of plant life. Aluminium takes part in disease control, nourishment, and other biochemical processes.

The pre-eminence between the essential and beneficial is most of the time difficult in the case of some trace elements. The elements including aluminium (Al), selenium (Se), silicon (Si), sodium (Na), and cobalt (Co) are believed to be beneficial for the plant growth. All these mentioned elements are not critically required for all the plants but may be supportive in growth and overall plant yield (Broadley et al. 2012). In fact, these beneficial elements evidently increase the plant resistance towards biotic stresses (herbivores, pathogens) and abiotic stress factors like high or low temperature, salinity, drought, UV stress, and nutrients deficiency or toxicity) at their low concentration levels (Ashraf and Foolad 2007).

Whereas their range being essential to lethal is somewhat too narrow, the low levels of the beneficial elements need more attention with respect to their use as fertilizer to increase crop nutritional value as food or animal feed and boosting up the crop production under the stressed field conditions. A more comprehensive perspective towards plant nutrition requirements would not be confined to essential nutrients towards survival but must also include other mineral elements at different levels beneficial for the magnificent plant growth (Pineda et al. 2010). Now, we discuss the uptake mechanism of different beneficial elements and their role in conferring tolerance against biotic as well as abiotic stress conditions with their favourable aspects (Vinocur and Altman 2005). The findings of the study

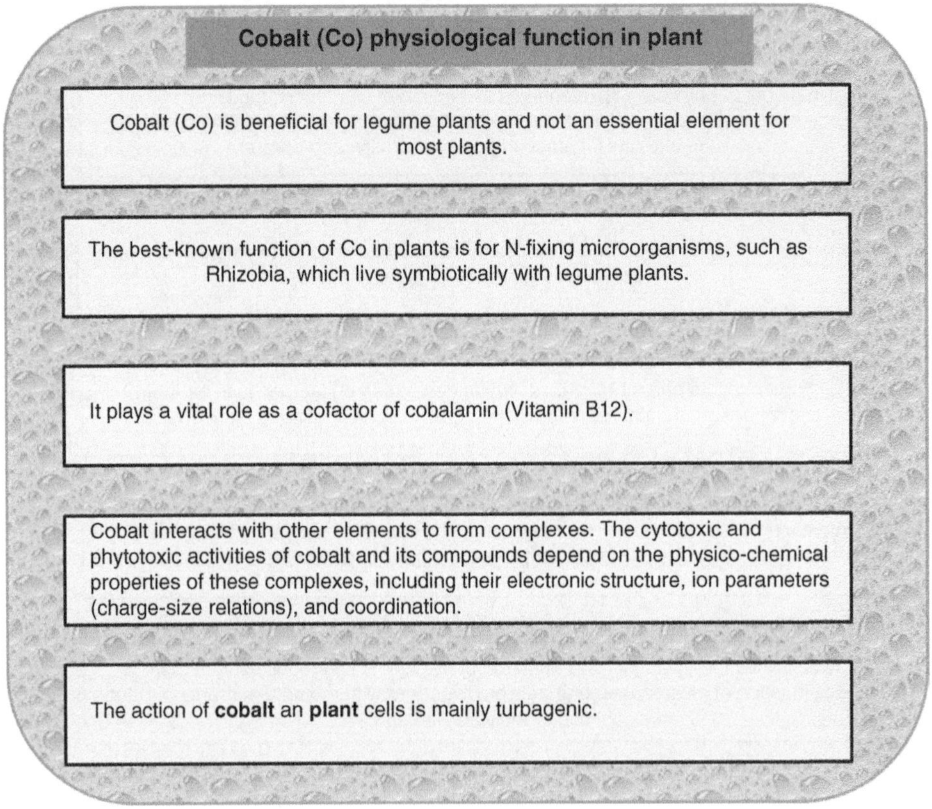

Figure 1.4 Explanation of cobalt's features involved in plant life cycle during different stages of growth and development.

suggest that the presence of trace elements in excess in the soil results in disturbing its homeostasis. They verify both biochemical activities as well as microbial biodiversity. If the recommended amount of an element is exceeded, it may hinder biological activity, which may limit plant development.

Cobalt has been proved to be the most toxic soil element. It not only induces very reduced resistant and enzymatic activities but also resulted into a substantial spring barley yield reduction. The soil reaction towards and Sn^{2+} excess was negative, but the problem scale was also not so alarming. The plants reacted extraordinarily positive to its increased doses in the soil. The barley crops grown during spring season did not face significant changes. In conclusion, this study is a crucial link in a sequence of studies on the perception of the quality of the environment where we function. It concedes the presence of the side effects to the soil contamination with the trace elements induced by the growing push towards an elevated standard of consumerism and living.

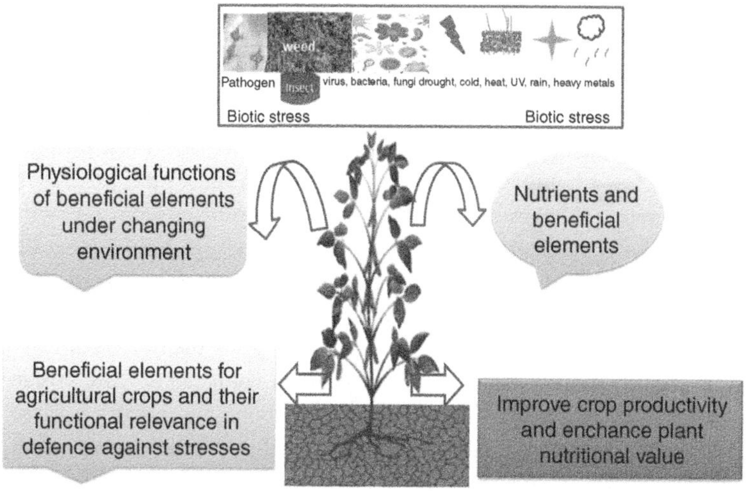

Figure 1.5 Schematic representation of relationship between plant and essential micro and macro elements that exists in the environment. These elements play vital role in plant and its development under different environmental conditions.

Silicon (Si)

The application of silicon (Si) has been discussed broadly in the recent years due to its effectiveness to increase plant resistance against the salinity (Guntzer et al. 2012; Kim et al. 2014, Dhiman et al. 2021) increase in the biomass of the crops and conversely decreases the uptake of different toxic elements. Silicon (Si) carries the properties to influence accumulation and uptake of various nutrients that are rarely investigated, especially the cereal crops like rice and cash crops like sugarcane those that are proved to be Si accumulator, and this element has been found to be beneficial for them (Tuna et al. 2008). Silicon (Si) is the most common element on the Earth surface. Whereas it is not completely available to the plants, as it remains locked in the minerals like recalcitrant silicate but a very trace fraction is available for the plants (Ahmad et al. 2007, Tripathi et al. 2020). The silicon soluble fractions are pH dependent and redox in nature (Diegoli et al. 2006).

The silicon configures towards the solid phase phytoliths once it is absorbed by the plants those that are recycled to the soil solution by the dead plants decay and will become available for the plants through soil again (Greger et al. 2018). The silicon is absorbed by the plants in the shape of undissociated silica acid (Ma and Yamaji 2008), which moves in the similar form through the plant xylem (Mitani et al. 2005). Silicon's uptake is considered to be passive (Su et al. 2010). In the past few years, different silicon transporters have been found in the plant roots' endo- and exodermis layers (Van Bockhaven et al. 2013). The silicon elements have been primarily found in the higher plant levels like monocotyledons with a higher content in rice up to 10% and the grasses with DW 0.3–1.2%. The plant tissues contain Si found as Si-organic, hydrogen bound complexes, and infuses the vessels and epidermis walls, where it serves as reducers of fungal infections, water transpiration, and provides strength towards the plant tissues (Allakhverdiev et al. 2010). The silicon elements are associated with the plant cell wall proteins, lignin, and polysaccharides (Guerriero et al. 2016). Most of the Si part is available in the plant in the form of narrowly dissolved phytoliths (Klotzbücher et al. 2016). The silicon impacts the distribution, uptake, and function of the different mineral nutrients present in the plants. In line with the literature, out of all the macronutrient elements, phosphorus (P), magnesium (Mg), nitrogen (N), potassium (K), calcium (Ca) are impacted in different ways (Ma and Yamaji 2008). Out of all micronutrients, manganese (Mn) and boron (B) look to be strongly impacted by the silicon (Alloway 2008). Uptake of other elements like Cl, Fe, and Zn by plants is also influenced by the silicon (Greger et al. 2011). Keeping in view the different Si effects, it looks that it might possible that various plant species nutrients acquisition and uptake modification differ within various plant taxons due to silicon presence. To study the non-uniformity of silicon elements on the plant nutrient uptake, it becomes valuable to differentiate between the studies made through hydroponic or directly through the Earth's surface. Through earth surfaces, the silicon impacts (Romero-Aranda et al. 2006) less or more availability and nutrient elements binding to the soil particles for the plant uptake. The silicon is well known in the reduction of P soil sorption, especially at lower pH levels, thus increasing the P portion availability in the soil surface for the plant (Hernandez-Apaolaza 2014). The phosphorous elements are sorbed mainly onto the Al, Fe, and Mn hydroxides in the soil. The silicon elements are associated with the Mn and Fe (III) and ultimately changes the availability of Mn and Fe and reduces the pool of hydroxides (Sommer et al. 2006). Up to now, a general silicon assembly impacts the plant's uptake of elements, status of different essential nutrients, and distribution is still missing (Greger et al. 2018). The main purpose of current study is to investigate the impacts of silicon on the plant's uptake and distribution of variable mineral nutrients in different five plant taxons and together with review of the literature obtained a common overview of silicon impacts on the plant elements status. We compared and probed different plant species like wheat and maize (monocotyledons), pea, lettuce, and carrot (dicotyledons) – wheat as a silicon accumulator, maize as C4 plants, wheat and maize as cereals, peas as nitrogen fixating plants, lettuce as a leafy vegetable, and carrot as root vegetable.

The plant cultivation media may impact both plant solution uptakes from the soil and elements availability as well as their distribution with the plant anatomy. Here in this study, we discussed and compared the silicon influence on: (i) plant nutrients presence in the different soil types and (ii) plants nutrients uptake from different soil nutrient solutions.

Function of Silicon

It seems that the silicon elements are beneficial for the plants when they are under stressful conditions (Ahmed et al. 2013, Shivaraj et al. 2021). This element has been observed to improve in the delay wilting and drought tolerance in various crops and increase plants' abilities to resist against micronutrients where the irrigation is withheld and other metals like copper, iron, zinc, manganese, and aluminium toxicities (Hasanuzzaman et al. 2017). Moreover, the silicon has been observed to support in the increase of the plants' stem strength. For instance, during a study rice and wheat plant were facing silicon deficiencies, their stems became weak and easily collapsed due to rain or wind pressures – a condition that is called as lodging, whereas poinsettias with silicon treatments have shown resistant and reduced stem collision and breakage. The silicon has been observed to increase plants' resistance against various attacks by fungal pathogens (Meharg and Meharg 2015). Contingent upon the phytophthora and powdery mildew, disease attacks were delayed in the silicon treated plants like gerbera (Phytophthora) and rose, sunflower zinnia and cucumber (powdery mildew), but the plant treated with silicon and untreated had the same amount of disease attack after three weeks (Moyer 2007). The modes of silicon actions are still ambiguous and uncertain and more research work is needed for the verification of these benefits.

Silicon in Soil

Silicon (Si) amount is ample in the lithosphere and is the second most abundant element in the Earth's crust (Ma 2010; Tripathi et al. 2020). Most of the soils contain 30% silicon, out of which majority has been found in rocks and minerals. Si is spotted and named as a beneficial quasi-essential nutrient. The Earth's surface layers are largely composed of Si that is observed primarily as secondary alumino silicates, silicate minerals, and multiple form of silicon dioxide (SiO_2) (Bhat et al. 2019). Whereas the riches of Si in the soils is not a sign that higher supplies of soluble Si are always available for the uptake of the plant (Sacala 2009). In the current chapter, the findings of multiple years of research work conducted in relation to Si are combined to create understanding regarding state of knowledge regarding Si fertilization for farmer's guidelines in the crop production process. Silicon is also used in the form of mono silicic acid (H_4SiO_4) by plants (Babu et al. 2016). That is found in both silicon's adsorbed and liquid phases in the soils. The total amount of mono silicic acid in the soil solution got affected by the pH of the soils, the amount of minerals, organic matter, clay, and Al/Fe hydroxides/oxides those are jointly in relation with the geological soil age (Tubaña and Heckman 2015). Fertilizer applications may cause rapid increase in the mono silicic acid concentration in the soil solutions. So, the Si concentration increases through fertilization, which has become a routine in areas with intensive crop production practices, especially for soils that are intrinsically low with the soluble silicon nutrients (Bhat et al. 2019).

Different procedures have been developed to estimate critical silicon levels in the soil and the available silicon in plants through the method of five-day Na_2CO_3-NH_4NO_3 extraction for the analysis of the silicon soluble fraction in the solid-state fertilizers that was among the most advanced methods in the agricultural research work in past few years (Sohail et al. 2020). These estimates were the key integrant essentially needed for the formation and execution of fruitful silicon management in the overall crop production system. However, multiple characteristics of the silicon remained understudied in the research work of the soil sciences, like the silicon impacts on the nutrient's status in plants is not well known and these features must be focused on in future studies (Lee et al. 2010). It is well known regarding silicon benefits for the plants under stressful conditions (Greger et al. 2018). The primary goal was to create a future analysis of the silicon impact: I The accumulation of diverse nutrients in hydroponically produced wheat, maize, peas, lettuce, and carrots (ii) The presence of nutrients in various soil types, including sand, clay, submerged soil, and alum shale soil (Greger et al. 2018). The silicon effects were homogenous in all the examined plant species and samples of soil types tested. According to the results, silicon increased the P, Mn, S, Zn, and Ca availability for the plants and availability of Fe and Cl also started to increase. Whereas availability of Mg and K were not significantly impacted by silicon (Sarwar et al. 2010). Moreover, the nutrients uptake from the soil solution like Mg, S, Ca Fe, Mn, and B increased; K, Zn, Cu, and N decreased, and P increased/decreased but Mo and Cl remained uninfluenced (Sharma and Chettri 2008). So, in conclusion we assumed that, during crop production, Si level in the soils maintained may be supportive in the availability of nutrients in the soil solution that would likely compensate for the decrease in nutrient elements in the plant tissue concentrations. The current finding indicates that silicon may also impact the plant nutrients uptake during a non-stressful condition.

Sodium in Plants

Sodium is not an important plant nutrient as others because a plant can complete its life cycle without it and can be replaced by other nutrients like potassium that plays a vital role during osmosis (Jones and Jacobsen 2005; Wakeel et al. 2011). However, sodium is important plant constituent as feed. Therefore, it is an important element for plants. For example, there are some exceptional cases of natrophilic crops from Chenopodioideae family, where some positive impacts on quality and yield were observed when fertilizers were applied with sodium combinations (Rodríguez-Navarro and Rubio 2006). An important industrial crop of this group like sugar beet (*Beta vulgaris*) is a famous example whose sodium requirements are very high. The glucose synthesis towards fructose is supported by the stored sodium of beet (Mengel et al. 2001). The efficiency of plant water use is also controlled by sodium by controlling the plants' osmotic cell pressure. Sometimes, sodium ions serve as substitute of K ions in various metabolic as well as osmoregulatory function. So, both of the nutrients remain interchangeable towards the various levels according to the group of the plants. Sodium plays a vital role for the uptake of carbon dioxide in some C4 plants like amaranth (Subbarao et al. 2003).

Sodium in Soil

Sodium presence in the soil is in the form of compounds, most commonly as salts (Warton et al. 2003). Clay minerals are adsorbed by the sodium, but the binding is mostly weaker than the K ions, therefore the sodium leach down propensity is always higher (Vimonses et al. 2009). In areas with the high rainfall, such as tropical to subtropical areas, soils are mostly depleted in sodium and washed down deeper into the lower soil layers (Kosmas et al. 2000). Conversely, sodium accumulation occurs mostly in the top soils of the arid to semi-arid climatic areas because of the exceeded evaporation levels of soil water (Rasouli et al. 2013). Ultimately deterioration of the soil structure occurs that has impacts on the air balance and the water contents of the soil. Moreover, the pH levels move towards more alkalinity with an increase in sodium contents (Liu et al. 2006).

Selenium (Se)

Selenium is a naturally occurring metalloid element that occurs in all kinds of climatic conditions like other elements. Moreover, it is believed to be a nonrenewable and limited resource of the Earth's crust (El-Ramady et al. 2016).

Selenium in Environment

More so than the silver elements, selenium is a backbone element of the Earth's surface. It is present in the atmosphere as a micron on the derivative. There are 40 well-known Se minerals, and the majority of them are found alone. In some cases, these minerals contain 30% of the selenium but rarely found in association with sulphide elements in zinc, lead, and copper metals (Purves 2012). Its main producers are United States, Russia, Bolivia, and Canada. About 1500 tons per year selenium is produced worldwide and 150 tons is recycled from industrial wastes like old photocopiers. Selenium is present in the climate and released through human activities and natural process. Over fertilization of agricultural soil is near about 400 mg ton^{-1}, as phosphorous fertilizers are also present naturally and added as micronutrients. Selenium is a natural element that cannot be destroyed or created easily, but it has capacities to change its form and increase water in the soil. Selenium waste from air ultimately ends up in the soil deposit sites (Kesler et al. 2015). Selenium remains fixed in the soil before reacting with oxygen. It cannot be soluble in the water and unmovable and low risky towards organisms. The soil acidity and oxygen levels increase the flow and movement of selenium that are caused usually through human activities like industrialization and other agricultural activities. Similar to that, the present chapter examines the several activities of selenium in plants, including acquisition, metabolism, and translocation. Plants utilize the released Se, which also meets the needs of humans and other animals in terms of Se. In the environment, selenium may be found in both organic forms like SeCys and SeMet and inorganic forms like cerel nitrate saline and character nitrate. Se is transferred to the plant roots' plasma membrane via a sulphate conveyor (El-Ramady et al. 2016).

Physiological Functions of Beneficial Elements Under A Changing Environment

The low-level benefits of nutrients Al, Co, Na, and Se are not well discussed in comparison with their harmful effects when applied at higher concentration levels. A clear understanding of beneficial part of these nutrients is essential to get increase in the plant nutritional values and overall productivity to feed the growing world populations. The climatic factors impact the plant growth by reducing their beneficial nutrients uptake through different ways like sunlight effects on the vegetative and flower completion stage and decreases the leaves colour shadow (Pilon-Smits et al. 2009; El-Ramady et al. 2015).

The photoperiod changes and control in different plants could affect its flowering stage. Similarly, temperature impacts the seed dormancy, germination, plant respiration, nutrients uptake, chlorophyll content, and plant organs transpiration. Due to the severe temperature impacts like heat or chilling factor on the photosynthesis process, the plants' overall health undergoes stressful conditions that result in the less availability of nutrients and make the plants nutrient deficient. In the end, the stressed plant is more vulnerable towards different infections caused by insects (El-Ramady et al. 2016).

5-Beneficial Elements Against Stresses

Healthy plants with no nutrient deficiencies are self-defenders and save themselves from different climatic stresses and severe losses. The elements with the recommended levels of concentrations support the plants against different stress conditions. The most dangerous abiotic stress condition is drought that causes severe molecular, physiological, and biochemical damage to the plants that eventually lead towards the significant losses in the yield. Drought has been described as the most significant threat towards world food security due to its limiting ability of crop productivity as well as the quality of crop production. The plant under drought stress loses its bud, root growth that causes reduction in photosynthesis, and its other important physiological activities (El-Ramady et al. 2015). Moreover, antioxidant plants' defence mechanism is also damaged due to severe drought stress. According to credible studies, the function and value of silicon towards its interactive network of plant drought stress resistance and other dynamic pathways are well documented. Silicon protects plants from the severe impacts and destructive activities of the drought stress conditions (Kong et al. 2005, Tripathi et al. 2020). In the Chapter 1 of this book, importance of silicon against drought stress crops has been discussed. It also discusses the role of silicon in the improvement of mineral distribution, physiological plant functions, enhancement of antioxidant defence system, oxidative markers release, and the future prospects in the counterproductive impacts of arid stress climatic conditions. In the current climatic change scenarios and overall increase in the food demand due to the increasing population levels, there is significant issue in the provision of so-called human's food security. The poor crop yields and nutritional values have become an essential worldwide issue by affecting billions of human populations. Whereas there is a belief that agronomic practices and new technologies can support by preventing the consequences of global warming and unpredictable climatic patterns (Lenz and Lens 2009). With increase in global temperature, many other abiotic stress factors have threatened agricultural sustainability and overall ecosystem. So, in the previous few decades the coercion of abiotic factors has become a serious concern for plant scientists. The survival through long-term or time-bound severe climatic changes, the plants must adapt the tolerance and resistance mechanism at organ level, cellular level, or the whole organism level (Hawrylak-Nowak et al. 2018). The plants' proper functioning depends at a higher extent on an adequate nutrient's levels. It is well described that the essential nutrients have a significant role in plant development, growth, and climatic stress resistance. According to the reports, the higher levels of nutrient concentrations in the plant's tissues may alter plants' responses towards stress conditions and cause increase in their nutritional value. Moreover, some beneficial nutrients like Se and Si at lower concentration levels can express positive impacts on the plant metabolic activities and contribute towards the increase in plant resistance against harmful climatic changes. So, as a result, different elements are discussed with various chemical compositions, examined as plants' abiotic stress reactive enhancers, including beneficial and essential for the plants. Most of them have potential to increase in crop quality and yield under stress conditions (El-Ramady et al. 2016).

Conclusion

In conclusion, the essential elements including Al, Co, Na, Se, and Si are recommended beneficial and important for the plants. The elements Na and Si are critical for different plant groups, whereas Se and Al are suggested as determining for some hyperaccumulator plant taxa.

The Co elements are essential for plant microbial partners in the trophic association rather than the plant themselves. Other elements like Cr and Fl are rarely explored and have beneficial functions towards plant biology. So, as humanity is facing multiple intimidating challenges including climate change and undemand-free growth, the discussed factors have significant importance. In the current scenario, more detailed research work on the benefits of the essential elements is recommended.

References

Ahmad, F., Aziz, T., Maqsood, M.A. et al. (2007). Effect of silicon application on wheat (Triticum aestivum L.) growth under water deficiency stress. *Emirates J. Food Agric.* 19 (2): 1–7.

Ahmad, P., Bhardwaj, R., and Tuteja, N. (2012). Plant signaling under abiotic stress environment. In: *Environmental Adaptations and Stress Tolerance of Plants in the Era of Climate Change* (ed. P. Ahmad and M. Prasad). Springer, New York, NY: https://doi.org/10.1007/978-1-4614-0815-4_14.

Ahmed, M., Kamran, A., Asif, M. et al. (2013). Silicon priming: a potential source to impart abiotic stress tolerance in wheat: a review. *Aust. J. Crop. Sci.* 7: 484–491.

Allakhverdiev, S.I., Thavasi, V., Kreslavski, V.D. et al. (2010). Photosynthetic hydrogen production. *J. Photochem. Photobiol. C Photochem. Rev.* 11: 101–113.

Alloway, B.J. (2008). Micronutrients and crop production: An introduction. In: *Micronutrient Deficiencies in Global Crop Production* (ed. B.J. Alloway), 1–39. Dordrecht: Springer https://doi.org/10.1007/978-1-4020-6860-7_1.

Anjum, N.A., Duarte, A.C., Pereira, E., and Ahmad, I. (2015). Plant-beneficial elements status assessment in soil-plant system in the vicinity of a chemical industry complex: shedding light on forage grass safety issues. *Environ. Sci. Pollut. Res.* 22: 2239–2246.

Asemi, Z., Jamilian, M., Mesdaghinia, E., and Esmaillzadeh, A. (2015). Effects of selenium supplementation on glucose homeostasis, inflammation, and oxidative stress in gestational diabetes: randomized, double-blind, placebo-controlled trial. *Nutrition* 31: 1235–1242.

Ashraf, M. and Foolad, M.R. (2007). Roles of glycine betaine and proline in improving plant abiotic stress resistance. *Environ. Exp. Bot.* 59: 206–216.

Babu, T., Tubana, B., Paye, W. et al. (2016). Establishing soil silicon test procedure and critical silicon level for rice in Louisiana soils. *Commun. Soil Sci. Plant Anal.* 47: 1578–1597.

Bhardwaj, D., Ansari, M.W., Sahoo, R.K., and Tuteja, N. (2014). Biofertilizers function as key player in sustainable agriculture by improving soil fertility, plant tolerance and crop productivity. *Microb. Cell Factories* 13: 1–10.

Bhat, J.A., Shivaraj, S.M., Singh, P. et al. (2019). Role of silicon in mitigation of heavy metal stresses in crop plants. *Plants* 8: 71.

van Boekel, M., Fogliano, V., Pellegrini, N. et al. (2010). A review on the beneficial aspects of food processing. *Mol. Nutr. Food Res.* 54: 1215–1247.

Broadley, M., Brown, P., Cakmak, I. et al. (2012). Chapter 8 – Beneficial elements. In: *Marschner's Mineral Nutrition of Higher Plants* (Third Edition) (ed. P. Marschner), 249–269. San Diego: Academic Press.

Cao, B., Wang, L., Gao, S. et al. (2017). Silicon-mediated changes in radial hydraulic conductivity and cell wall stability are involved in silicon-induced drought resistance in tomato. *Protoplasma* 254: 2295–2304.

Chauhan, D.K., Yadav, V., Vaculík, M. et al. (2021). Aluminum toxicity and aluminum stress-induced physiological tolerance responses in higher plants. *Crit. Rev. Biotechnol.* 41 (5): 715–730.

Coleman, D.C., Zhang, W., and Fu, S. (2014). Toward a holistic approach to soils and plant growth. In: *Interactions in Soil: Promoting Plant Growth. Biodiversity, Community and Ecosystems*, vol. 1 (ed. J. Dighton and J. Krumins), 211–223. Dordrecht: Springer https://doi.org/10.1007/978-94-017-8890-8_11.

Compant, S., Van Der Heijden, M.G.A., and Sessitsch, A. (2010). Climate change effects on beneficial plant–microorganism interactions. *FEMS Microbiol. Ecol.* 73: 197–214.

Dhiman, P., Rajora, N., Bhardwaj, S. et al. (2021). Fascinating role of silicon to combat salinity stress in plants: An updated overview. *Plant Physiol. Biochem.* 162: 110–123.

Diegoli, S., Mendes, P.M., Baguley, E.R. et al. (2006). pH-Dependent gold nanoparticle self-organization on functionalized Si/SiO_2 surfaces. *J. Exp. Nanosci.* 1: 333–353.

Dighton, J. and Krumins, J.A. (2014). *Interactions in Soil: Promoting Plant Growth*. Springer.

El-Ramady, H., Abdalla, N., Alshaal, T. et al. (2015). Selenium in soils under climate change, implication for human health. *Environ. Chem. Lett.* 13: 1–19.

El-Ramady, H., Abdalla, N., Taha, H.S. et al. (2016). Selenium and nano-selenium in plant nutrition. *Environ. Chem. Lett.* 14: 123–147.

Elser, J.J. (2012). Phosphorus: a limiting nutrient for humanity? *Curr. Opin. Biotechnol.* 23: 833–838.

Fahad, S., Hussain, S., Bano, A. et al. (2015). Potential role of phytohormones and plant growth-promoting rhizobacteria in abiotic stresses: consequences for changing environment. *Environ. Sci. Pollut. Res.* 22: 4907–4921.

Faucon, M.-P., Pourret, O., and Lange, B. (2018). Element case studies: cobalt and copper. In: *Agromining: Farming for Metals. Mineral Resource Reviews* (ed. A. van der Ent, A.J. Baker, G. Echevarria, et al.), 233–239. Cham: Springer https://doi.org/10.1007/978-3-030-58904-2_18.

Gad, N., El Zaher, A., Fatma, H. et al. (2011). Response of faba bean (Vicia faba L.) to cobalt amendements and nitrogen fertilization. *African J. Plant Sci. Biotechnol. Glob. Sci. Books* 41–45.

Greger, M., Landberg, T., Vaculik, M., and Lux, A. (2011). Silicon influences nutrient status in plants. In *Proceedings of the 5th International Conference on Silicon in Agriculture, Beijing, China* (pp. 57–58).

Greger, M., Landberg, T., and Vaculík, M. (2018). Silicon influences soil availability and accumulation of mineral nutrients in various plant species. *Plants* 7: 41.

Grzybkowski, W. (2006). Nature and properties of metal cations in aqueous solutions. *Pol. J. Environ. Stud.* 15 (4): 655–663.

Guerriero, G., Hausman, J.-F., and Legay, S. (2016). Silicon and the plant extracellular matrix. *Front. Plant Sci.* 7: 463.

Guntzer, F., Keller, C., and Meunier, J.-D. (2012). Benefits of plant silicon for crops: a review. *Agron. Sustain. Dev.* 32: 201–213.

Gupta, B. and Huang, B. (2014). Mechanism of salinity tolerance in plants: physiological, biochemical, and molecular characterization. *Int. J. Genomics* 2014.

Hasanuzzaman, M., Nahar, K., Rahman, A. et al. (2017). Actions of biological trace elements in plant abiotic stress tolerance. In: *Essential Plant Nutrients* (ed. M. Naeem, A. Ansari and S. Gill), 213–274. Cham: Springer https://doi.org/10.1007/978-3-319-58841-4_10.

Hawrylak-Nowak, B., Dresler, S., Rubinowska, K. et al. (2018). Selenium biofortification enhances the growth and alters the physiological response of lamb's lettuce grown under high temperature stress. *Plant Physiol. Biochem.* 127: 446–456.

Hayat, S., Hayat, Q., Alyemeni, M.N., et al. (2012). Role of proline under changing environments: a review. *Plant Signal. & Behav.* 7 (11), 1456–1466. doi: https://doi.org/10.4161/psb.21949.

Hernandez-Apaolaza, L. (2014). Can silicon partially alleviate micronutrient deficiency in plants? A review. *Planta* 240: 447–458.

Ilangumaran, G. and Smith, D.L. (2017). Plant growth promoting rhizobacteria in amelioration of salinity stress: a systems biology perspective. *Front. Plant Sci.* 8: 1768.

Iqbal, M.T. (2012). Acid tolerance mechanisms in soil grown plants. *Malaysian J. Soil Sci.* 16: 1–21.

Jones, C. and Jacobsen, J. (2005). Plant nutrition and soil fertility. *Nutr. Manag. Modul.* 2: 1–11.

Kaur, S., Kaur, N., Siddique, K.H.M., and Nayyar, H. (2016). Beneficial elements for agricultural crops and their functional relevance in defence against stresses. *Arch. Agron. Soil Sci.* 62: 905–920.

Kesler, S.E., Simon, A.C., and Simon, A.F. (2015). *Mineral Resources, Economics and the Environment*. Cambridge University Press.

Khan, M.R. and Khan, M.M. (2010). Effect of varying concentration of nickel and cobalt on the plant growth and yield of chickpea. *Aust. J. Basic Appl. Sci.* 4: 1036–1046.

Kichigina, N.E., Puhalsky, J.V., Shaposhnikov, A.I. et al. (2017). Aluminum exclusion from root zone and maintenance of nutrient uptake are principal mechanisms of Al tolerance in Pisum sativum L. *Physiol. Mol. Biol. Plants* 23: 851–863.

Kim, Y.H., Khan, A.L., Waqas, M. et al. (2014). Silicon application to rice root zone influenced the phytohormonal and antioxidant responses under salinity stress. *J. Plant Growth Regul.* 33: 137–149.

Kirkby, E. (2012). Chapter 1 – Introduction, definition and classification of nutrients. In: *Marschner's Mineral Nutrition of Higher Plants* (Third Edition) (ed. P. Marschner), 3–5. San Diego: Academic Press.

Klotzbücher, T., Marxen, A., Jahn, R., and Vetterlein, D. (2016). Silicon cycle in rice paddy fields: insights provided by relations between silicon forms in topsoils and plant silicon uptake. *Nutr. Cycl. Agroecosyst.* 105: 157–168.

Kong, L., Wang, M., and Bi, D. (2005). Selenium modulates the activities of antioxidant enzymes, osmotic homeostasis and promotes the growth of sorrel seedlings under salt stress. *Plant Growth Regul.* 45: 155–163.

Kopittke, P.M., Menzies, N.W., and Blamey, F.P.C. (2005). Rhizotoxicity of aluminate and polycationic aluminium at high pH. *Plant Soil* 266: 177–186.

Kosmas, C., Gerontidis, S., and Marathianou, M. (2000). The effect of land use change on soils and vegetation over various lithological formations on Lesvos (Greece). *Catena* 40: 51–68.

Lambers, H., Raven, J.A., Shaver, G.R., and Smith, S.E. (2008). Plant nutrient-acquisition strategies change with soil age. *Trends Ecol. Evol.* 23: 95–103.

Lange, B., van der Ent, A., Baker, A.J.M. et al. (2017). Copper and cobalt accumulation in plants: a critical assessment of the current state of knowledge. *New Phytol.* 213: 537–551.

Lee, C.W., Mahendra, S., Zodrow, K. et al. (2010). Developmental phytotoxicity of metal oxide nanoparticles to Arabidopsis thaliana. *Environ. Toxicol. Chem. An Int. J.* 29: 669–675.

Lenz, M. and Lens, P.N.L. (2009). The essential toxin: the changing perception of selenium in environmental sciences. *Sci. Total Environ.* 407: 3620–3633.

Liu, X., Herbert, S.J., Hashemi, A.M. et al. (2006). Effects of agricultural management on soil organic matter and carbon transformation-a review. *Plant Soil Environ.* 52: 531.

Lwalaba, J.L.W., Louis, L.T., Zvobgo, G. et al. (2019). Copper alleviates cobalt toxicity in barley by antagonistic interaction of the two metals. *Ecotoxicol. Environ. Saf.* 180: 234–241.

Ma, J.F. (2010). Silicon transporters in higher plants. In: *MIPs and Their Role in the Exchange of Metalloids. Advances in Experimental Medicine and Biology*, vol. 679 (ed. T.P. Jahn and G.P. Bienert), 99–109. New York, NY: Springer. https://doi.org/10.1007/978-1-4419-6315-4_8.

Ma, J.F. and Yamaji, N. (2008). Functions and transport of silicon in plants. *Cell. Mol. Life Sci.* 65: 3049–3057.

Mahmoud, E.A., Hassanin, M.A., and Emara, E.I.R. (2012). Effect of organic and mineral nitrogenous fertilizers and plant density on yield and quality of sugar beet (Beta vulgaris L.). *Egypt. J. Agron.* 34: 103–189.

Malagoli, M., Schiavon, M., and Pilon-Smits, E.A.H. (2015). Effects of selenium biofortification on crop nutritional quality. *Front. Plant Sci.* 6: 280.

Meena, V.D., Dotaniya, M.L., Coumar, V. et al. (2014). A case for silicon fertilization to improve crop yields in tropical soils. *Proc. Natl. Acad. Sci. India Sect. B Biol. Sci.* 84: 505–518.

Meharg, C. and Meharg, A.A. (2015). Silicon, the silver bullet for mitigating biotic and abiotic stress, and improving grain quality, in rice? *Environ. Exp. Bot.* 120: 8–17.

Mengel, K., Kirkby, E.A., Kosegarten, H., and Appel, T. (2001). Further elements of importance. In: *Principles of Plant Nutrition* (ed. K. Mengel, E.A. Kirkby, H. Kosegarten and T. Appel), 639–655. Dordrecht: Springer https://doi.org/10.1007/978-94-010-1009-2_19.

Mitani, N., Ma, J.F., and Iwashita, T. (2005). Identification of the silicon form in xylem sap of rice (Oryza sativa L.). *Plant Cell Physiol.* 46: 279–283.

Moyer, C. (2007). Evaluation of silicon and biofungicide products for managing powdery mildew caused by Podosphaera fusca in Gerbera Daisy (Gerbera jamesonii) (Doctoral dissertation, University of Florida).

Muhammad, N., Zvobgo, G., and Guo-ping, Z. (2018). A review: the beneficial effect of aluminum on plant growth in acid soil and the possible mechanisms. *J. Integr. Agric.* 17: 60345–60347.

Nagajyoti, P.C., Lee, K.D., and Sreekanth, T.V.M. (2010). Heavy metals, occurrence and toxicity for plants: a review. *Environ. Chem. Lett.* 8: 199–216.

Neal, A.L. (2008). What can be inferred from bacterium–nanoparticle interactions about the potential consequences of environmental exposure to nanoparticles? *Ecotoxicology* 17: 362–371.

Niste, M., Vidican, R., Rotar, I. et al. (2014). Plant nutrition affected by soil salinity and response of rhizobium regarding the nutrients accumulation. *ProEnvironment Promediu* 7 (18).

Nunes-Nesi, A., Brito, D.S., Inostroza-Blancheteau, C. et al. (2014). The complex role of mitochondrial metabolism in plant aluminum resistance. *Trends Plant Sci.* 19: 399–407.

Osorio Vega, N.W. (2007). A review on beneficial effects of rhizosphere bacteria on soil nutrient availability and plant nutrient uptake. *Rev. Fac. Nac. Agron. Medellín* 60: 3621–3643.

Paveley, N. D., Hims, M. J., Stevens, D. B., Clark, W. S., and Bryson, R. J. (1998). Appropriate fungicide doses for winter wheat and matching crop management to growth and yield potential. *HGCA Project Report (United Kingdom).*

Paustenbach, D.J., Tvermoes, B.E., Unice, K.M. et al. (2013). A review of the health hazards posed by cobalt. *Crit. Rev. Toxicol.* 43: 316–362.

Phoon, X., Ander, E. L., and Plant, J. A. (2012). Essential and beneficial trace elements, in Jane A. Plant, Nick Voulvoulis, K. Vala Ragnarsdottir *Pollutants, Human Health and the Environment: A Risk Based Approach*, 53 (47), 777–780.

Pilon-Smits, E.A.H., Quinn, C.F., Tapken, W. et al. (2009). Physiological functions of beneficial elements. *Curr. Opin. Plant Biol.* 12: 267–274.

Pineda, A., Zheng, S.-J., van Loon, J.J.A. et al. (2010). Helping plants to deal with insects: the role of beneficial soil-borne microbes. *Trends Plant Sci.* 15: 507–514.

Poschenrieder, C., Gunsé, B., Corrales, I., and Barceló, J. (2008). A glance into aluminum toxicity and resistance in plants. *Sci. Total Environ.* 400: 356–368.

Purves, D. (2012). *Trace-Element Contamination of the Environment*. Elsevier.

Rasouli, F., Pouya, A.K., and Karimian, N. (2013). Wheat yield and physico-chemical properties of a sodic soil from semi-arid area of Iran as affected by applied gypsum. *Geoderma* 193: 246–255.

Rodríguez-Navarro, A. and Rubio, F. (2006). High-affinity potassium and sodium transport systems in plants. *J. Exp. Bot.* 57: 1149–1160.

Romero-Aranda, M.R., Jurado, O., and Cuartero, J. (2006). Silicon alleviates the deleterious salt effect on tomato plant growth by improving plant water status. *J. Plant Physiol.* 163: 847–855.

Sacala, E. (2009). Role of silicon in plant resistance to water stress. *J. Elem.* 14: 619–630.

Sade, H., Meriga, B., Surapu, V. et al. (2016). Toxicity and tolerance of aluminum in plants: tailoring plants to suit to acid soils. *Biometals* 29: 187–210.

Samac, D.A. and Tesfaye, M. (2003). Plant improvement for tolerance to aluminum in acid soils–a review. *Plant Cell Tissue Organ Cult.* 75: 189–207.

Sandrin, T.R. and Hoffman, D.R. (2007). Bioremediation of organic and metal co-contaminated environments: effects of metal toxicity, speciation, and bioavailability on biodegradation. In:

Environmental Bioremediation Technologies (ed. S.N. Singh and R.D. Tripathi), 1–34. Berlin, Heidelberg: Springer https://doi.org/10.1007/978-3-540-34793-4_1.

Sarwar, N., Malhi, S.S., Zia, M.H. et al. (2010). Role of mineral nutrition in minimizing cadmium accumulation by plants. *J. Sci. Food Agric.* 90: 925–937.

Sessitsch, A., Kuffner, M., Kidd, P. et al. (2013). The role of plant-associated bacteria in the mobilization and phytoextraction of trace elements in contaminated soils. *Soil Biol. Biochem.* 60: 182–194.

Sharma, B. and Chettri, M.K. (2008). Impacts of heavy metals on accumulation of essential micronutrients and chlorophylls in some vegetables. *Pollut. Res.* 27: 355–363.

Shivaraj S.M., Mandlik R., Bhat J.A., et al. (2021). Outstanding questions on the beneficial role of silicon in crop plants. *Plant Cell Physiol.* doi: https://doi.org/10.1093/pcp/pcab145.

Silva, S.A. (2017). Aluminum concentration, nutritional status and metabolite profile in native Cerrado species with different resistance strategies to the metal.

Simonsen, L.O., Harbak, H., and Bennekou, P. (2012). Cobalt metabolism and toxicology—a brief update. *Sci. Total Environ.* 432: 210–215.

Singh, S., Tripathi, D.K., Singh, S. et al. (2017). Toxicity of aluminium on various levels of plant cells and organism: a review. *Environ. Exp. Bot.* 137: 177–193.

Sohail, M.I., ur Rehman, M.Z., Rizwan, M. et al. (2020). Efficiency of various silicon rich amendments on growth and cadmium accumulation in field grown cereals and health risk assessment. *Chemosphere* 244: 125481.

Sommer, M., Kaczorek, D., Kuzyakov, Y., and Breuer, J. (2006). Silicon pools and fluxes in soils and landscapes—a review. *J. Plant Nutr. Soil Sci.* 169: 310–329.

Su, Y.-H., McGrath, S.P., and Zhao, F.-J. (2010). Rice is more efficient in arsenite uptake and translocation than wheat and barley. *Plant Soil* 328: 27–34.

Subbarao, G.V., Ito, O., Berry, W.L., and Wheeler, R.M. (2003). Sodium—a functional plant nutrient. *CRC. Crit. Rev. Plant Sci.* 22: 391–416.

Tripathi, D.K., Singh, V.P., Lux, A., and Vaculik, M. (2020). Silicon in plant biology: from past to present, and future challenges. *Journal of Experimental Botany* 71 (21): 6699–6702. https://doi.org/10.1093/jxb/eraa448.

Tubaña, B.S. and Heckman, J.R. (2015). Silicon in soils and plants. In: *Silicon and Plant Diseases* (ed. F. Rodrigues and L. Datnoff), 7–51. Cham: Springer. https://doi.org/10.1007/978-3-319-22930-0_2.

Tuna, A.L., Kaya, C., Higgs, D. et al. (2008). Silicon improves salinity tolerance in wheat plants. *Environ. Exp. Bot.* 62: 10–16.

Van Bockhaven, J., De Vleesschauwer, D., and Höfte, M. (2013). Towards establishing broad-spectrum disease resistance in plants: silicon leads the way. *J. Exp. Bot.* 64: 1281–1293.

Vankova, R. (2014). Cytokinin regulation of plant growth and stress responses. In: *Phytohormones: A Window to Metabolism, Signaling and Biotechnological Applications*, 55–79. Springer.

Vimonses, V., Lei, S., Jin, B. et al. (2009). Kinetic study and equilibrium isotherm analysis of Congo Red adsorption by clay materials. *Chem. Eng. J.* 148: 354–364.

Vinocur, B. and Altman, A. (2005). Recent advances in engineering plant tolerance to abiotic stress: achievements and limitations. *Curr. Opin. Biotechnol.* 16: 123–132.

Wakeel, A., Farooq, M., Qadir, M., and Schubert, S. (2011). Potassium substitution by sodium in plants. *CRC. Crit. Rev. Plant Sci.* 30: 401–413.

Warton, B., Matthiessen, J.N., and Shackleton, M.A. (2003). Cross-enhancement: enhanced biodegradation of isothiocyanates in soils previously treated with metham sodium. *Soil Biol. Biochem.* 35: 1123–1127.

Watanabe, T. and Osaki, M. (2002). Mechanisms of adaptation to high aluminum condition in native plant species growing in acid soils: a review. *Commun. Soil Sci. Plant Anal.* 33: 1247–1260.

Weisany, W., Raei, Y., and Allahverdipoor, K.H. (2013). Role of some of mineral nutrients in biological nitrogen fixation. *Bull. Environ. Pharmacol. Life Sci.* 2: 77–84.

White, P.J. and Brown, P.H. (2010). Plant nutrition for sustainable development and global health. *Ann. Bot.* 105: 1073–1080.

Yang, L. and Watts, D.J. (2005). Particle surface characteristics may play an important role in phytotoxicity of alumina nanoparticles. *Toxicol. Lett.* 158: 122–132.

Zhang, W., Xie, Z., Lang, D. et al. (2017). Beneficial effects of silicon on abiotic stress tolerance in legumes. *J. Plant Nutr.* 40: 2224–2236.

Zheng, S.J. (2010). Crop production on acidic soils: overcoming aluminium toxicity and phosphorus deficiency. *Ann. Bot.* 106: 183–184.

Zhu, Y. and Gong, H. (2014). Beneficial effects of silicon on salt and drought tolerance in plants. *Agron. Sustain. Dev.* 34: 455–472.

2

Role of Beneficial Elements in Epigenetic Regulation of Plants in Response to Abiotic Stress Factors

Muhittin Kulak[1,] and Adnan Aydin[2,*]*

[1] *Department of Herbal and Animal Production, Vocational School of Technical Sciences, Igdir University, Igdir, Türkiye*
[2] *Department of Agricultural Biotechnology, Faculty of Agriculture, Igdir University, Igdir, Türkiye*

Introduction

Regarding essentiality, Arnon and Stout (1939) postulated that without an essential element a plant cannot complete its normal growth and development and consequently reproduction, but beneficial elements at an appropriate/low concentration are needed for particular taxa of the plants, not all plants. Of those elements, Al, Co, Na, Se, and Si are the relevant elements in this regard (Pilon-Smits et al. 2009). These elements exhibit various roles through induction, activation, or inhibition of activities of enzymes or molecules, as signal molecules like plant growth regulators under stress conditions.

Beneficial Elements for Crop and Non-Crop Plants

Selenium

Selenium (Se) is not grouped into the essentiality for plant kingdom but for many animals, bacteria, and green algae (*Chlamydomonas reinhardtii*) (Fu et al. 2002; Pilon-Smits et al. 2009). In spite of different forms of selenium available in soils, selenite and selenate are of the forms that can be absorbed and sequestered by the plants (Kaur et al. 2016). Many and various functions have been attributed to Se but in a concentration-dependent manner. The impacts on growth and development (Mroczek-Zdyrska and Wójcik 2012), transpiration and photosynthesis (Germ et al. 2007; Zhang et al. 2014), modulation of antioxidant system (Kong et al. 2005), and biosynthesis of secondary metabolites (Robbins et al. 2005; Elguera et al. 2013; Astaneh et al. 2018) have been reported.

* Both authors equally contributed to the chapter.

Beneficial Chemical Elements of Plants: Recent Developments and Future Prospects, First Edition. Edited by Sangeeta Pandey, Durgesh Kumar Tripathi, Vijay Pratap Singh, Shivesh Sharma, and Devendra Kumar Chauhan.

Silicon

Silicon (Si) is one of the beneficial elements with positive impacts on some iconic plants, viz. rice, wheat, and barley. In the case of deficit of this element, retarding in plant growth, development, and reproduction has been observed (Pilon-Smits et al. 2009). Silicon is taken up as monosilicic (Si $(OH)_4$) and deposited in amorphous silica (SiO_2-nH_2O) chemical structure in the plant cell walls. Here, interaction of silicon with pectins and polyphenols available in cell wall makes the cell wall rigid (Banerjee and Roychoudhury 2018; Shivaraj et al. 2021). Furthermore, Si exhibits significant roles in combating with abiotic stresses including heavy metal stress, water stress, and salinity and high and low temperature (Kaur et al. 2016; Dhiman et al. 2021). Of all these relevant fundamental roles, this element has been revealed and characterized with beneficial roles in biosynthesis of phenolics (Goto et al. 2003; Malčovská et al. 2014; Vega et al. 2019), regulation of antioxidant responses (Tripathi et al. 2012a, 2012b), and morphology and growth of plants (Fan et al. 2016; Mattson and Leatherwood 2010).

Aluminium

Aluminium (Al) is one of the most widely available elements, being beneficial but confined to specific plant taxa at a low concentration. Enhancement herbivore defence, prevention of Fe-toxicity, and promotion of P-uptake are some of the functions for plant system (Pilon-Smits et al. 2009). In *Camelia sinensis*, activities of antioxidant enzymes were enhanced with Al-induction, which in turn increased membrane integrity and consequently stimulated growth of the plant (Ghanati et al. 2005; Chauhan et al. 2021).

Sodium

Sodium (Na) functioning and its essential role are still much-discussed, but low concentrations of Na^+ might be needed for proper growth and development of some plant species (Subbarao et al. 2003). Of all the essential roles of Na, Na^+ exhibits significant role in conversion of pyruvate to phosphoenolpyruvate for C4 plants (Brownell and Bielig 1996). A role regarding replacement functions of potassium is given for Na^+ (Subbarao et al. 2003).

Cobalt

Cobalt (Co) acts as a component of several enzymes and co-enzymes, being effective on growth and metabolism of plants but in a concentration-dependent manner (Palit et al. 1994). A major role with respect to the nitrogen fixation through being component of cobalamin required for enzymes responsible for N-fixation has been attributed to Co^{2+} (Gad 2006). Furthermore, leaf senescence through suppression of ethylene biosynthesis was retarded by Co. Moreover, accumulation of isoquinoline, an alkaloid, has been stimulated with Co-stimulated biosynthesis regulation alkaloid precursors (Palit et al. 1994).

Abiotic Stress Factors

Biotic and abiotic stress factors are the main environmental factors influencing the geographical distribution of plants in nature, limiting plant productivity in agriculture and threatening food security. The negative effects of these stressors are expected to worsen with global warming and sudden climate change (Fedoroff et al. 2010; Li et al. 2019). Especially, abiotic stress conditions, viz. salinity, temperature, cold, drought, and heavy metal, are of serious environmental problems that disrupt plant germination, plant growth, geographical distribution, and agricultural products (Liang et al. 2018; Dilnur et al. 2019; Deshmukh et al. 2020). Plants' response to stress factors is through upregulation or downregulation of genes for common and specific pathways that enable the plants to cope with and tolerate the stress conditions (Mahajan and Tuteja 2005; Roychoudhury et al. 2021).

Epigenetic Modifications Under Stressful Conditions

Regarding genetics of the plants, genome hosts the vital information required for/affecting the distribution of the plants in a certain ecosystem. Corresponding to the relevant distribution of the plants under environmental fluctuations, genetic recombination might result in new gene combinations that can expand or enhance the tolerance ranges of the plants. However, as clearly reported and uttered by Peng and Zhang (2009), the relevant modification regarding new gene combinations might not be as fast as the environmental fluctuations rate. In this context, in order to cope with the stress, plants have to exhibit changes in their genomes.

Since plants are sessile organisms, they constantly struggle with biotic and abiotic stresses from the environment. Plants undergo both physiological and indirect genetic changes in their cells in order to adapt to the environment with the signals they receive from the environment. Subsequently, that improves adaptation ability by making changes in the plant's phenotype (Liu et al. 2015). Continuous and frequent changes at molecular and morphological levels are required to adapt to this environment. The adaptation processes herein are provided with epigenetic modifications that ensure the survival of plants by increasing stress tolerance (Saraswat et al. 2017).

Although the concept of epigenetics has been known for long time, its emergence as a field of science covers a short period of 20–30 years. Along with the concept of epigenetics, the concepts of episome, epigenomics, and epiallel have become widespread recently. Although there is no complete harmony between different scientific circles, Holliday (1990) defines it as a branch of science that studies variations in expression of hereditary genes that are not linked or consequence of the relevant changes on the DNA sequence of the plant. The concept of epiallel refers to alleles that generate regular and inherited phenotypical or physiological variation without occurrences of differences in DNA sequence. The main difference that distinguishes epigenetics from modifications is the presence of epialleles.

Considering the relevant regulatory mechanisms of epigenetics, Saraswat et al. (2017) reported three mechanisms such as RNA interaction, DNA methylation, and histone modification. Of these modifications for epigenetic regulations, DNA methylation involves significant modifications at fifth carbon position of the cytosine ring, which is primarily methylated in the context of cytosine-guanine dinucleotides in both animals and plants.

In plant system, DNA methylation is observed in the context of all three cytosines – methylations at CG-, CHG, and CHH regions. Herein, H replaces adenine, cytosine, or thymine (Chen et al. 2010; Zemach et al. 2010), being catalyzed with DNA methyltransferase 1

(DNMT1) family. Furthermore, DNMT3 family is responsible for de novo methylation. Regarding CHH, DRM2 (domain rearranged methyltransferase 2), a homologue of *DNMT3*, catalyzes, whilst CG methylation and CHG methylation are provided by *DNA methyltransferase 1* (*MET1* or *DMT1*) and chromethylase *3* (*CMT3*), respectively (Gehring and Henikoff 2007; Chen et al. 2010; Law and Jacobsen 2010). Considering removal of DNA methylation, methylated cytosines are replaced with the unmodified cytosines after DNA replication, and the process concerned with cytosine methyl groups is coupled with the activities of relevant DNA glycosylases (Tariq and Paszkowski 2004).

Also, it was revealed that methylation in the 5′ and 3′ region can result in inhibition of gene expression (Zhang et al. 2006; Gehring and Henikoff 2007; Zilberman et al. 2007). However, the effects of DNA methylation on the level of transcription of the relevant were not still clearly revealed. Zhang et al. (2010) reported that cytosine methylation roles were equally diversified and confined for each gene group. The relevant methylated cytosines affect the methyl binding proteins, which then hold histone modifier and chromatin modelling proteins for constructing a new complex that can retard the transcription factors binding (Fransz and de Jong 2002), resulting in an epigenetic memory (Zhang 2008). Particularly heterochromatin regions in the genome contain highly methylated cytosine. This makes transcription inactive. Cytosine methylation has been found to be lower in euchromatin regions, but still at significant levels (Santos et al. 2011).

Abiotic stresses can lead to substantial alterations concerned with the levels of DNA methylation that are hypothesized to be linked with chromatin remodelling and stress-responsive transcriptional regulation (as clearly and nicely reviewed by Santos et al. 2011; Table 2.1). According to the findings of genome-wide analysis, clear DNA methylation arrangements in response to adverse environmental fluctuations, mostly corresponding to the demethylation, have been reported (Labra et al. 2002; Aina et al. 2004; Zhong et al. 2009; Wang et al. 2011). Transcriptional activation of specific loci concerned with changes in DNA methylation were described for transposons and protein coding sequences corresponding to the environmental perturbations (Steward et al. 2002; Hashida et al. 2006; Choi and Sano 2007).

The increments in relevant genomic DNA methylation can modify/modulate the expression of the transcriptome, which then decelerate the plant's metabolism. However, the reduction in the levels of tolerance-linked genes promotes activation of chromatin as well as expression of new gene combinations, which subsequently lead to long-term tolerance against the stress factors.

As the case of water deficit for crassulacean acid metabolism (CAM) plants, the plants modify their photosynthetic C3 cycle to CAM pathway. Along with the modifications or transitions, the water loss is reduced and subsequently an enhanced tolerance level is provided. This transition is coupled and correlated with the genomic methylation and satellite DNA hypermethylation (Dyachenko et al. 2006). For that reason, satellite DNA hypermethylation must be in accordance with the synthesis of specified chromatin structure. Herewith the relevant process, expression levels of a number genes contribute to the required transition to CAM for maintaining the adaptation under restricted water supply. A clear hypermethylation was noted at the root tip DNA of *Pisum sativum* under water deficit. Also, analysis of methylation-sensitive amplified polymorphism (MSAP) verified a substantial increment concerned with the level of methylation of cytosine residues in CCGG, in particular for internal cytosine (Labra et al. 2002).

Table 2.1 Epigenetic modifications in various plants under abiotic stress factors. Santos et al. 2011 / Mary Ann Liebert, Inc.[a]

Stress	Plant species	Chromatin mark	Enrichment (+) Depletion (−) No change (0)	Authors
Cold	*Zea mays*	5-mC	−	Steward et al. (2002)
	Arabidopsis thaliana	H3K27me3	−	Kwon et al. (2009)
		H3ac	+	Pavangadkar et al. (2010)
	Cell lines:	H3Ser10P	+	Sokol et al. (2007)
	Nicotiana (By2)	H3Ser10PK14ac	+	
	Arabidopsis thaliana (T87)	H4ac	+	
Drought	*Pisum sativum*	5-mC	+	Labra et al. (2002)
	Oryza sativa	5-mC	−	Wang et al. (2011)
	Arabidopsis thaliana	H3K23ac	+	Kim et al. (2008)
		H3K27ac	+	
		H3K4me3	+	
	Arabidopsis thaliana	H3K4me3	+	Van Dijk et al. (2010)
		H3K4me2	+	
		H3K4me	+	
Salinity	*Laguncularia racemosa*	5-mC	−	Lira-Medeiros et al. 2010
	Triticum aestivum	5-mC	−	Zhong et al. (2009)
	Nicotiana (By2)	H3Ser10P;	+	Sokol et al. (2007)
		H3K14ac	+	
	Arabidopsis thaliana	H3K9ac	+	Chen et al. (2010a)
		H3K14ac	+	
		H3K4me3	+	
		H3K9me2	−	
Heat	*Arabidopsis thaliana*	5-mC	0	Pecinka et al. (2010)
		H3K9me2	−	
		H3K4me3	0/-	
Hypoxia	*Oryza sativa*	H3K4me2	−	Tsuji et al. (2006)
		H3ac	+	
		H3K4me3	+	
Aluminium	*Nicotiana tabacum*	5-mC	−	Choi and Sano (2007)
UV	*Arabidopsis thaliana*	H3ac	+	Lang-Mladek et al. (2010)

[a] The current table was obtained from the study Santos et al. 2011 / Mary Ann Liebert, Inc.

In addition to the DNA methylation, histone modifications are of clearly revealed epigenetic mechanisms, constituting a fundamental topic of epigenetic regulations. Post-translational modification for each histone includes acetylation, methylation, phosphorylation, ubiquitination, ribosylation, and biotinylation (Tariq and Paszkowski 2004; Liu et al. 2010). Amino acids localized at N-terminal tails of H3 and H4 histones are easily modified relative to other relevant histone amino acids (Chen et al. 2010). Ha et al. (2011) reported that histone modifications might be interrelated with the alterations in gene transcription. The relevant transcription increases through acetylation, phosphorylation, monobichitination, and trimethylation in histone H3 lysine 4 (H3K4), whilst sumoylation and dimethylation in histone regions decrease the transcription (Zhang et al. 2007; Chinnusamy and Zhu 2009; Veiseth et al. 2011). Considering the major enzymes of histone modifications involved in response to plants under stress conditions, *HDAC, ATX, SUVH,* and *UBC* are the revealed enzymes. Former findings have clearly documented that N-terminal histone modifications are involved in expression levels of genes, but the studies regarding understanding how the relevant modifications are linked with responses of plants against stress are of great interest for researchers (Zhu et al. 2008; Kim et al. 2009; Van Dijk et al. 2010). In this regard, Vermaak (2003) documented that two chromation states, viz. euchromatin and heterochromatin, have been revealed to be correlated with histone modifications that might be responsible for activating or silencing chromatin and cause substantial alterations in nucleosomes mobility. Regarding silent chromation, H3K9Me is one of the documented markers, binding to methylated H3K9. Also, it might recruit *heterochromatin protein 1 (LHP1),* which has been associated with the contributions in spread of heterochromatin to adjacent regions of the chromosome (Lachner et al. 2001; Grewal and Moazed 2003; Boyko and Kovalchuk 2008). However, in both animals and plants, active chromatin is related to H3 acetylation, which might be responsive to reduce the relevant charge interactions between DNA and nucleosomes and also might facilitate the passage of RNA polymerases (Workman and Kingston 1998; Vermaak 2003).

RNA interaction is also of epigenetic mechanisms. In this regard, small RNA mechanism has recently emerged as a new epigenetic editing system, being groups of RNA but not encoding proteins (Ben Amor et al. 2009). Small RNAs, splitting of translation, repression, or methylation of DNA can alter gene expression (Ramachandran and Chen 2008). Small RNAs are classified according to their mode of action and biogenesis pathway. Small interfering RNAs (siRNAs) and micro RNAs (miRNAs) are the known key small RNAs (Phillips et al. 2007), ranging from 20 to 27 nucleotides for miRNA and siRNA families. miRNAs are localized on a universally conserved sequence in eukaryotes, suggesting their substantial editing roles corresponding to an ancient and evolutionary based approach/mechanism (Ramachandran and Chen 2008). They are arranged with the enzyme *dicer-like 1* (*DCL1*), being bound by *Argonaute* family (*AGO1*) members and forming a new structure, namely, RNA-induced silencing complex (RISC) for triggering the relevant division or repressing the translation of an mRNA target (Bartel 2004). siRNAs are derivative of long double-stranded RNA (dsRNA) molecules that are catalyzed by RNA-dependent RNA polymerases (RDRs), being reported to play substantial roles in regulation of gene expression (Ramachandran and Chen 2008). Endogenous siRNA classes are dsRNAs composed of mRNAs encoded by nat-siRNAs (natural antisense siRNAs) and natural cis-antisense gene pairs (Borsani et al. 2005; Sunkar et al. 2007); Trans-active siRNAs (ta-siRNAs) are miRNA-directed cleavage products, being then converted by RDRs (Hunter et al. 2006; Sunkar et al. 2007). Heterochromatic siRNAs (hc-siRNAs), heterochromatin, and dsRNAs are product of repeats of DNA (Mallory and Vaucheret 2006; Sunkar et al. 2007). Despite the reported structural differences, both miRNAs and siRNAs exhibited interrelations concerned with their sequence-specific inhibitory functions (Carthew and Sontheimer 2009), being suggested to have significant roles regarding response regulation of plants against biotic and abiotic stress conditions. Considering the epigenetic mechanisms involved in response of plants against stress, small RNAs are of interest for molecular scientists. Gene expression regulations are core points of epigenetic modifications as the case of plants exposed to stress (Chinnusamy and Zhu 2009; Zhang et al. 2009; Lv et al. 2010; Grativol et al. 2012).

Studies Regarding the Effect of Beneficial Elements on Epigenetic Changes in the Genome of Plants

Selenium

Activator roles of selenium regarding particular classes of genes in combating abiotic stress factors have been demonstrated in *Zea mays* L. Along with the study, significant roles have been attributed to Se in prevention or counteraction of alterations in methylation status (Bocchini et al. 2018). Kumar et al. (2012) has also reported the protective role of selenium in amending the Cd-induced toxicity in *Gracilaria dura* through regulation of antioxidants and stabilization of DNA methylation. In *Capsicum annuum*, low concentration of Se nanoparticles (0.5 and 1 mg L^{-1}) exhibited growth-promoting effect, whilst high concentration such as 10 and 30 mg L^{-1} caused severe toxicity and abnormalities, which were responses to epigenetic modifications of toxic doses of Se particles (Sotoodehnia-Korani et al. 2020). The positive impacts of low doses were observed in higher biomass accumulation but as expected higher doses of Se nanoparticles induced DNA hypermethylations in *Momordica charantia*. Epigenetic modifications regarding DNA cytosine methylation, chromatin conformation, and cellular transcription program were hypothesized to be associated with high doses of Se nanoparticles (Rajaee Behbahani et al. 2020). These findings are consistent with the fundamental principle 'sola dosis facit venemum' (only the dose makes the poison) (attributed to Paracelsus) (Kronzucker et al. 2013), suggesting that dose moves the elements from being beneficial to toxic status.

Cobalt

Erturk et al. (2015) assayed an array of Co concentration (5–40 mmol L^{-1}) impacts on epigenetic modifications in maize genome, reporting that 40 mmol L^{-1} concentration of Co was notably effective in altering the methylation status.

Yigider et al. (2020) investigated the retrotransposon polymorphism and genomic template stability in *Z. mays* L. exposed to Co. Along with the study, 10, 20, and 40 mM concentration of Co were used, reporting that increasing concentration of Co caused increases in retrotransposon polymorphism but decreased the genomic template stability.

Rancelis et al. (2012) observed treatment-timing-dependent effect of Co on DNA methylation, chlorophyll morphosis induction, and superoxide dismutase (SOD) polymorphism in *Vicia faba* seeds. Similarly, Čėsnienė et al. (2006) also revealed individual plant polymorphisms after Co stress.

Sodium

In addition to the beneficial effect of low concentration of sodium in growing media, high levels of sodium cause osmotic and ionic stress that induced epigenetic modifications on the genome of the plants, which in turn limits the plant growth and development. The impact of high level of sodium as NaCl stress was observed on DNA methylation patterns in pepper cultivars, of which DNA methylation ratio increased by 5.45, 10.00, and 11.11% according to the cultivars (Shams et al. 2020). Furthermore, NaCl alone and its combination with putrescine caused significant changes in DNA methylation classes in cabbage (*Brassica oleraceae* L. cv Yalova-F1) (Orhan et al. 2020). In *Triticum aestivum* under 50, 100, and 150 mM NaCl concentration, 50 and 150 mM NaCl caused 76.9 and 46.4% DNA methylation value, respectively. However, in the same study, the putrescine treatment improved the DNA hypermethylation damage resulting from high concentration of NaCl (Taspinar et al. 2017). In a similar study on *T. aestivum* exposed to the various NaCl concentrations (50–300 mM NaCl) combined with treatment of putrescine concentrations (0.01–1 mM), Sigmaz et al. (2015) revealed that 300 mM NaCl concentration caused highest retrotransposon polymorphism ratio but this retrotransposon was significantly amended with the highest dose of putrescine (1 mM).

Aluminium

In *Z. mays* L., the increasing concentration of aluminium caused reductions in genomic template stability and increased DNA-damage-related band frequencies coupled with increases in DNA methylation ratio. Taspinar et al. (2018) explained changes with respect to mobilization and methylation status as a response regarding defence mechanisms against stress. Furthermore, Al has been reported to alter cytosine methylation (Lukens and Zhan 2007). In order to combat with stress, plants exhibit some epigenetic regulations for adaptation but without change in their DNA sequence (Causevic et al. 2005; Tan 2010). Al stress resulted in occurrences of hypermethylation and hypomethylation and DNA methylation (Pour et al. 2019).

Regarding adaptation to the new conditions, the substantial roles of histone methylation have been revealed and have been correlated with induction and repression of gene via methylations in four histone proteins (Ezaki et al. 2016). In *Arabidopsis thaliana*, Ezaki et al. (2016) observed changes of H3 methylation status as a consequence of Al stress.

Kashino-Fujii et al. (2018) revealed a novel strategy in barley (*Hordeum vulgare*) accessions in order to combat with Al stress. Corresponding to the Al treatments, retrotransposon-like sequences in the *HvAACT1* upstream genomic region and demethylation additions have been observed and subsequently an increase in *HvAACT1* expression was noted. These alterations were positively correlated with enhanced tolerance against Al stress (Kashino-Fujii et al. 2018).

Kimatu et al. (2011) highlighted that Al^{3+} acts as a signal in triggering secondary stress signals, as a situation of plant growth regulators, which subsequently cause important modifications in epigenetic regulators, viz. like DNA methyltransferases, histone modification enzymes, and chromatin remodelling factors. These changes bring about new changes in cytosine DNA methylation and histone modifications.

Silicon

With the uptake by plants, Si exhibits significant roles against biotic and abiotic stress factors. For example, resistant plants of *Arabidopsis* contained higher levels of salicylic acid coupled with higher levels of transcript encoding defence-related genes. These responses were attributed to the possible roles of Si (Vivancos et al. 2015).

Conclusion

Alterations in regular climatic conditions cause serious impacts, namely stress factors, on plants and their natural distribution in the nature. As a response against stress factors, plants have developed various multiple mechanisms to adapt to the stress conditions for their survival. Of those mechanisms, epigenetic modifications of plant genome are the fundamental regulatory mechanisms including DNA methylation, histone modifications, and RNA interactions. In the three decades, significant efforts have been undertaken in order to reveal the epigenetic responses of plants against stress conditions. Epigenetic interactions are of great importance in adapting to environmental changes that require complex regulation of the genome's transcriptional output. These interactions can be inherited and reversible, which do not alter the original DNA sequence. Euchromatin and heterochromatin domains in plant genomes play an important role in the expression of genes. Although euchromatin areas are regions where gene expression is high, heterochromatin areas are known as the region where genes are suppressed. The remodelling of these areas in the genome is the result of epigenetic interactions with the signals the plant receives from the environment. As a result of the full understanding of this mechanism in plants under stress, a crop with desired characteristics based on both yield and quality can be obtained especially in cultivated plants under abiotic stress conditions. Corresponding to understanding the relevant epigenetic mechanism, it might be possible to meet the nutritional and food demands of world's population.

References

Aina, R., Sgorbati, S., Santagostino, A. et al. (2004). Specific hypomethylation of DNA is induced by heavy metals in white clover and industrial hemp. *Physiol. Plant.* 121: 472–480.

Arnon, D.I. and Stout, P.R. (1939). The essentiality of certain elements in minute quantity for plants with special reference to copper. *Plant Physiol.* 14: 371–375.

Astaneh, R.K., Bolandnazar, S., Nahandi, F.Z., and Oustan, S. (2018). Effect of selenium application on phenylalanine ammonia-lyase (PAL) activity, phenol leakage and total phenolic content in garlic (*Allium sativum* L.) under NaCl stress. *Inf. Process Agric.* 5 (3): 339–344.

Banerjee, A. and Roychoudhury, A. (2018). Role of beneficial trace elements in salt stress tolerance of plants. In: *Plant Nutrients and Abiotic Stress Tolerance* (ed. M. Hasanuzzaman, M. Fujita, H. Oku, et al.), 377–390. Singapore: Springer https://doi.org/10.1007/978-981-10-9044-8_16.

Bartel, D.P. (2004). MicroRNAs: genomics, biogenesis, mechanism, and function genomics. *Cell* 116: 281–297.

Ben Amor, B., Wirth, S., Merchan, F. et al. (2009). Novel long non-protein coding RNAs involved in Arabidopsis differentiation and stress responses. *Genome Res.* 19: 57–69.

Bocchini, M., D'Amato, R., Ciancaleoni, S. et al. (2018). Soil selenium (Se) biofortification changes the physiological, biochemical and epigenetic responses to water stress in Zea mays L. by inducing a higher drought tolerance. *Front. Plant Sci.* 9: 389.

Borsani, O., Zhu, J., Verslues, P.E. et al. (2005). Endogenous siRNAs derived from a pair of natural cis-antisense transcripts regulate salt tolerance in Arabidopsis. *Cell* 123: 1279–1291.

Boyko, A. and Kovalchuk, I. (2008). Epigenetic control of plant stress response. *Environ. Mol. Mutagen.* 49: 61–72.

Brownell, P.F. and Bielig, L.M. (1996). The role of sodium in the conversion of pyruvate to phosphoenolpyruvate in mesophyll chloroplasts of C4 plants. *Funct. Plant Biol.* 23 (2): 171–177.

Carthew, R.W. and Sontheimer, E.J. (2009). Origins and mechanisms of miRNAs and siRNAs. *Cell* 136: 642–655.

Causevic, A., Delaunay, A., Ounnar, S. et al. (2005). DNA methylating and demethylating treatments modify phenotype and cell wall differentiation state in sugarbeet cell lines. *Plant Physiol. Biochem.* 43 (7): 681–691.

Česnienė, T., Kleizaitė, V., and Rančelis, V. (2006). Polymorphic response of Vicia faba plants to cobalt excess: comparison of intact plants and callus cultures. *Biologija* 1: 73–76.

Chauhan, D.K., Yadav, V., Vaculík, M. et al. (2021). Aluminum toxicity and aluminum stress-induced physiological tolerance responses in higher plants. *Crit. Rev. Biotechnol.* 41 (5): 715–730.

Chen, L.T., Luo, M., Wang, Y.Y., and Wu, K. (2010). Involvement of Arabidopsis histone deacetylase HDA6 in ABA and salt stress response. *J. Exp. Bot.* 61: 3345–3353.

Chinnusamy, V. and Zhu, J.-K. (2009). Epigenetic regulation of stress responses in plants. *Curr. Opin. Plant Biol.* 12: 133–139.

Choi, C.-S. and Sano, H. (2007). Abiotic-stress induces demethylation and transcriptional activation of a gene encoding a glycerophosphodiesterase-like protein in tobacco plants. *Mol. Gen. Genomics.* 277: 589–600.

Deshmukh, R., Tripathi, D.K., and Guerriero, G. (2020). *Metalloids in Plants: Advances and Future Prospects*, vol. 1, 1–520. John Wiley & Sons.

Dhiman, P., Rajora, N., Bhardwaj, S. et al. (2021). Fascinating role of silicon to combat salinity stress in plants: An updated overview. *Plant Physiol. Biochem.* 162: 110–123.

Dilnur, T., Peng, Z., Pan, Z. et al. (2019). Association Analysis of Salt Tolerance in Asiatic cotton (*Gossypium arboretum*) with SNP Markers. *Int. J. Mol. Sci.* https://doi.org/10.3390/ijms20092168.

Dyachenko, O.V., Zakharchenko, N.S., Shevchuk, T.V. et al. (2006). Effect of hypermethylation of CCWGG sequences in DNA of *Mesembryanthemum crystallinum* plants on their adaptation to salt stress. *Biochem Biokhimiia* 71: 461–465.

Elguera, J.C.T., Barrientos, E.Y., Wrobel, K., and Wrobel, K. (2013). Effect of cadmium (Cd (II)), selenium (Se (IV)) and their mixtures on phenolic compounds and antioxidant capacity in *Lepidium sativum*. *Acta Physiol. Plant.* 35 (2): 431–441.

Erturk, F.A., Agar, G., Nardemir, G., and Arslan, E. (2015). Cytogenetic and epigenetic alterations by cobalt and nickel on *Zea mays* L. *Toxicol. Environ. Chem.* 97 (10): 1350–1362.

Ezaki, B., Higashi, A., Nanba, N., and Nishiuchi, T. (2016). An S-adenosyl Methionine Synthetase (SAMS) gene from Andropogon virginicus L. confers aluminum stress tolerance and facilitates epigenetic gene regulation in Arabidopsis thaliana. *Front. Plant Sci.* 7: 1627.

Fan, X., Wen, X., Huang, F. et al. (2016). Effects of silicon on morphology, ultrastructure and exudates of rice root under heavy metal stress. *Acta Physiol. Plant.* 38 (8): 197.

Fedoroff, N.V., Battisti, D.S., Beachy, R.N. et al. (2010). Radically rethinking agriculture for the 21st century. *Science* 327: 833–834.

Fransz, P.F. and De Jong, J.H. (2002). Chromatin dynamics in plants. *Curr. Opin. Plant Biol.* 5: 560–567.

Fu, L.-H., Wang, X.-F., Eyal, Y. et al. (2002). A selenoprotein in the plant kingdom: Mass spectrometry confirms that an opal codon (UGA) encodes selenocysteine in *Chlamydomonas reinhardtii* glutathione peroxidase. *J. Biol. Chem.* 277: 25983–25991.

Gad, N. (2006). Increasing the efficiency of nitrogen fertilization through cobalt application to pea plant. *Res. J. Agric. Biol. Sci.* 2: 433–442.

Gehring, M. and Henikoff, S. (2007). DNA methylation dynamics in plant genomes. *Biochim. Biophys. Acta* 1769: 276–286.

Germ, M., Kreft, I., Stibilj, V., and Urbanc-Berčič, O. (2007). Combined effects of selenium and drought on photosynthesis and mitochondrial respiration in potato. *Plant Physiol. Biochem.* 45 (2): 162–167.

Ghanati, F., Morita, A., and Yokota, H. (2005). Effects of aluminum on the growth of tea plant and activation of antioxidant system. *Plant Soil* 276: 133–141.

Goto, M., Ehara, H., Karita, S. et al. (2003). Protective effect of silicon on phenolic biosynthesis and ultraviolet spectral stress in rice crop. *Plant Sci.* 164 (3): 349–356.

Grativol, C., Hemerly, A.S., and Ferreira, P.C.G. (2012). Genetic and epigenetic regulation of stress responses in natural plant populations. *Biochim. Biophys. Acta* 1819: 176–185.

Grewal, S.I.S. and Moazed, D. (2003). Heterochromatin and epigenetic control of gene expression. *Science* 301: 798–802.

Ha, M., Ng, D.W., Li, W.-H., and Chen, Z.J. (2011). Coordinated histone modifications are associated with gene expression variation within and between species. *Genome Res.* 21 (4): 590–598. https://doi.org/10.1101/gr.116467.110.

Hashida, S.-N., Uchiyama, T., Martin, C. et al. (2006). The temperature-dependent change in methylation of the Antirrhinum transposon Tam3 is controlled by the activity of its transposase. *Plant Cell* 18: 104–118.

Holliday, R. (1990). Mechanisms for the control of gene activity during development. *Biol. Rev. Camb. Philos. Soc.* 65: 431–471.

Hunter, C., Willmann, M.R., Wu, G. et al. (2006). Trans-acting siRNA-mediated repression of ETTIN and ARF4 regulates heteroblasty in Arabidopsis. *Development* 133: 2973–2981.

Kashino-Fujii, M., Yokosho, K., Yamaji, N. et al. (2018). Retrotransposon insertion and DNA methylation regulate aluminum tolerance in European barley accessions. *Plant Physiol.* 178 (2): 716–727.

Kaur, S., Kaur, N., Siddique, K.H., and Nayyar, H. (2016). Beneficial elements for agricultural crops and their functional relevance in defence against stresses. *Arch. Agron. Soil Sci.* 62 (7): 905–920.

Kim, K.C., Lai, Z., Fan, B., and Chen, Z. (2008). Arabidopsis WRKY38 and WRKY62 transcription factors interact with histone deacetylase 19 in basal defense. *Plant Cell* 20: 2357–2371.

Kim, J., To, T., Ishida, J. et al. (2009). Alterations of lysine modifications on the histone H3 N-tail under drought stress conditions in *Arabidopsis thaliana*. *Plant Cell Physiol.* 50: 1856–1864.

Kimatu, J.N., Moussa, D., Song, C. et al. (2011). DNA cytosine methylation alterations associated with aluminium toxicity and low pH in Sorghum bicolor. *Afr. J. Agric. Res.* 6 (19): 4579–4593.

Kong, L., Wang, M., and Bi, D. (2005). Selenium modulates the activities of antioxidant enzymes, osmotic homeostasis and promotes the growth of sorrel seedlings under salt stress. *Plant Growth Regul.* 45 (2): 155–163.

Kronzucker, H.J., Coskun, D., Schulze, L.M. et al. (2013). Sodium as nutrient and toxicant. *Plant Soil* 369 (1–2): 1–23.

Kumar, M., Bijo, A.J., Baghel, R.S. et al. (2012). Selenium and spermine alleviate cadmium induced toxicity in the red seaweed Gracilaria dura by regulating antioxidants and DNA methylation. *Plant Physiol. Biochem.* 51: 129–138.

Kwon, C.S., Lee, D., Choi, G., and Chung, W.I. (2009). Histone occupancy-dependent and -independent removal of H3K27 trimethylation at cold-responsive genes in Arabidopsis. *Plant J.* 60: 112–121.

Labra, M., Ghiani, A., Citterio, S. et al. (2002). Analysis of cytosine methylation pattern in response to water deficit in pea root tips. *Plant Biol.* 4: 694–699.

Lachner, M., O'Carroll, D., Rea, S. et al. (2001). Methylation of histone H3 lysine 9 creates a binding site for HP1 proteins. *Nature* 410: 116–120.

Lang-Mladek, C., Popova, O., Kiok, K. et al. (2010). Transgenerational inheritance and resetting of stress-induced loss of epigenetic gene silencing in Arabidopsis. *Mol. Plant* 3: 594–602. https://doi.org/10.1093/mp/ssq014.

Law, J.A. and Jacobsen, S.E. (2010). Establishing, maintaining and modifying DNA methylation patterns in plants and animals. *Nat. Rev. Genet.* 11: 204–220.

Li, H.M., Liu, S.D., Ge, C.W. et al. (2019). Analysis of drought tolerance and associated traits in upland cotton at the seedling stage. *Int. J. Mol. Sci.* https://doi.org/10.3390/ijms20163888.

Liang, W., Ma, X., Wan, P., and Liu, L. (2018). Plant salt-tolerance mechanism: a review. *Biochem. Biophys. Res. Commun.* https://doi.org/10.1016/j.bbrc.2017.11.043.

Lira-Medeiros, C.F., Parisod, C., Fernandes, R.A. et al. (2010). Epigenetic variation in mangrove plants occurring in contrasting natural environment. *PLoS ONE* 5: e10326.

Liu, C., Lu, F., Cui, X., and Cao, X. (2010). Histone methylation in higher plants. *Annu. Rev. Plant Biol.* 61: 395–420.

Liu, J., Feng, L., Li, J., and He, Z. (2015). Genetic and epigenetic control of plant heat responses. *Front. Plant Sci.* 6: 267.

Lukens, L.N. and Zhan, S.H. (2007). The plant genome's methylation status and response to stress, implications for plant improvement. *Curr. Opin. Plant Biol.* 10: 317–322.

Lv, D.K., Bai, X., Li, Y. et al. (2010). Profiling of cold-stress-responsive miRNAs in rice by microarrays. *Gene* 459 (1–2): 39–47.

Mahajan, S. and Tuteja, N. (2005). Cold, salinity and drought stresses: an overview. *Arch. Biochem. Biophys.* 444: 139–158.

Malčovská, S.M., Dučaiová, Z., Maslaňáková, I., and Bačkor, M. (2014). Effect of silicon on growth, photosynthesis, oxidative status and phenolic compounds of maize (*Zea mays* L.) grown in cadmium excess. *Water Air Soil Pollut.* 225 (8): 2056.

Mallory, A.C. and Vaucheret, H. (2006). Functions of microRNAs and related small RNAs in plants. *Nat. Genet.* 38 (6): S31–S36.

Mattson, N.S. and Leatherwood, W.R. (2010). Potassium silicate drenches increase leaf silicon content and affect morphological traits of several floriculture crops grown in a peat-based substrate. *HortScience* 45 (1): 43–47.

Mroczek-Zdyrska, M. and Wójcik, M. (2012). The influence of selenium on root growth and oxidative stress induced by lead in *Vicia faba* L. minor plants. *Biol. Trace Elem. Res.* 147 (1–3): 320–328.

Orhan, E., Uzundumlu, F., Yiğider, E., and Aydın, M. (2020). The effect of putrescine on DNA methylation on cabbage plant under salt stress conditions. *Turk. J. Agric. For.* 44 (3): 301–311.

Palit, S., Sharma, A., and Talukder, G. (1994). Effects of cobalt on plants. *Bot. Rev.* 60 (2): 149–181.

Pavangadkar, K., Thomashow, M.F., and Triezenberg, S.J. (2010). Histone dynamics and roles of histone acetyltransferases during cold-induced gene regulation in Arabidopsis. *Plant Mol. Biol.* 74: 183–200.

Pecinka, A., Dinh, H.Q., Baubec, T. et al. (2010). Epigenetic regulation of repetitive elements is attenuated by prolonged heat stress in Arabidopsis. *Plant Cell* 22: 3118–3129.

Peng, H. and Zhang, J. (2009). Plant genomic DNA methylation in response to stresses: potential applications and challenges in plant breeding. *Prog. Nat. Sci.* 19: 1037–1045.

Phillips, J.R., Dalmay, T., and Bartels, D. (2007). The role of small RNAs in abiotic stress. *FEBS Lett.* 581: 3592–3597.

Pilon-Smits, E.A., Quinn, C.F., Tapken, W. et al. (2009). Physiological functions of beneficial elements. *Curr. Opin. Plant Biol.* 12 (3): 267–274.

Pour, A.H., Özkan, G., Nalcı, Ö., and Haliloğlu, K. (2019). Estimation of genomic instability and DNA methylation due to aluminum (Al) stress in wheat (*Triticum aestivum* L.) using iPBS and CRED-iPBS analyses. *Turk. J. Bot.* 43 (1): 27–37.

Rajaee Behbahani, S., Iranbakhsh, A., Ebadi, M. et al. (2020). Red elemental selenium nanoparticles mediated substantial variations in growth, tissue differentiation, metabolism, gene transcription, epigenetic cytosine DNA methylation, and callogenesis in bittermelon (*Momordica charantia*); an in vitro experiment. *PLoS One* 15 (7): e0235556.

Ramachandran, V. and Chen, X. (2008). Small RNA metabolism in Arabidopsis. *Trends Plant Sci.* 13: 368–374.

Rancelis, V., Cesniene, T., Kleizaite, V. et al. (2012). Influence of cobalt uptake by Vicia faba seeds on chlorophyll morphosis induction, SOD polymorphism, and DNA methylation. *Environ. Toxicol.* 27 (1): 32–41.

Robbins, R.J., Keck, A.S., Banuelos, G., and Finley, J.W. (2005). Cultivation conditions and selenium fertilization alter the phenolic profile, glucosinolate, and sulforaphane content of broccoli. *J. Med. Food* 8 (2): 204–214.

Roychoudhury, S., Chakraborty, S., Choudhury, A.P. et al. (2021). Environmental factors-induced oxidative stress: hormonal and molecular pathway disruptions in hypogonadism and erectile dysfunction. *Antioxidants 10*: 837. https://doi.org/10.3390/antiox10060837.

Santos, A.P., Serra, T., Figueiredo, D.D. et al. (2011). Transcription regulation of abiotic stress responses in rice: a combined action of transcription factors and epigenetic mechanisms. *Omics J. Integr. Biol.* 15 (12): 839–857.

Saraswat, S., Yadav, A.K., Sirohi, P., and Singh, N.K. (2017). Role of epigenetics in crop improvement: water and heat stress. *J. Plant Biol.* 60: 231–240.

Shams, M., Yildirim, E., Arslan, E., and Agar, G. (2020). Salinity induced alteration in DNA methylation pattern, enzyme activity, nutrient uptake and H. *Acta Physiol. Plant.* 42 (59): 59.

Shivaraj, S.M., Mandlik, R., Bhat, J.A. et al. (2021). Outstanding questions on the beneficial role of silicon in crop plants. *Plant Cell Physiol.* https://doi.org/10.1093/pcp/pcab145.

Sigmaz, B., Agar, G., Arslan, E. et al. (2015). The role of putrescine against the long terminal repeat (LTR) retrotransposon polymorphisms induced by salinity stress in Triticum aestivum. *Acta Physiol. Plant.* 37 (11): 251.

Sokol, A., Kwiatkowska, A., Jerzmanowski, A., and Prymakowska-Bosak, M. (2007). Up-regulation of stress-inducible genes in tobacco and Arabidopsis cells in response to abiotic stresses and ABA treatment correlates with dynamic changes in histone H3 and H4 modifications. *Planta* 227: 245–254.

Sotoodehnia-Korani, S., Iranbakhsh, A., Ebadi, M. et al. (2020). Selenium nanoparticles induced variations in growth, morphology, anatomy, biochemistry, gene expression, and epigenetic DNA methylation in *Capsicum annuum*; an in vitro study. *Environ. Pollut.* 114727.

Steward, N., Ito, M., Yamaguchi, Y. et al. (2002). Periodic DNA methylation in maize nucleosomes and demethylation by environmental stress. *J. Biol. Chem.* 277: 37741–37746.

Subbarao, G.V., Ito, O., Berry, W.L., and Wheeler, R.M. (2003). Sodium-a functional plant nutrient. *Crit. Rev. Plant Sci.* 22: 391–416.

Sunkar, R., Chinnusamy, V., Zhu, J., and Zhu, J.-K. (2007). Small RNAs as big players in plant abiotic stress responses and nutrient deprivation. *Trends Plant Sci.* 12: 301–309.

Tan, M.P. (2010). Analysis of DNA methylation of maize in response to osmotic and salt stress based on methylation-sensitive amplified polymorphism. *Plant Physiol. Biochem.* 48 (1): 21–26.

Tariq, M. and Paszkowski, J. (2004). DNA and histone methylation in plants. *Trends Genet.* 20 (6): 244–251.

Taspinar, M.S., Aydin, M., Arslan, E. et al. (2017). Salinity and putrescine effects on DNA methylation changes in *Triticum aestivum*. *J. Biotechnol. 256*: S101.

Taspinar, M.S., Aydin, M., Sigmaz, B. et al. (2018). Aluminum-induced changes on DNA damage, DNA methylation and LTR retrotransposon polymorphism in maize. *Arab. J. Sci. Eng.* 43 (1): 123–131.

Tripathi, D.K., Singh, V.P., Kumar, D., and Chauhan, D.K. (2012a). Rice seedlings under cadmium stress: effect of silicon on growth, cadmium uptake, oxidative stress, antioxidant capacity and root and leaf structures. *Chem. Ecol.* 28 (3): 281–291.

Tripathi, D.K., Singh, V.P., Kumar, D., and Chauhan, D.K. (2012b). Impact of exogenous silicon addition on chromium uptake, growth, mineral elements, oxidative stress, antioxidant capacity, and leaf and root structures in rice seedlings exposed to hexavalent chromium. *Acta Physiol. Plant.* 34 (1): 279–289.

Tsuji, H., Saika, H., Tsutsumi, N. et al. (2006). Dynamic and reversible changes in histone H3-Lys4 methylation and H3 acetylation occurring at submergenceinducible genes in rice. *Plant Cell Physiol.* 47: 995–1003.

Van Dijk, K., Ding, Y., Malkaram, S. et al. (2010). Dynamic changes in genome-wide histone H3 lysine 4 methylation patterns in response to dehydration stress in *Arabidopsis thaliana*. *BMC Plant Biol.* 10: 238.

Vega, I., Nikolic, M., Pontigo, S. et al. (2019). Silicon improves the production of high antioxidant or structural phenolic compounds in barley cultivars under aluminum stress. *Agronomy* 9 (7): 388.

Veiseth, S.V., Rahman, M.A., Yap, K.L. et al. (2011). The SUVR4 histone lysine methyltransferase binds ubiquitin and converts H3K9me1 to H3K9me3 on transposon chromatin in Arabidopsis. *PLoS Genet.* 7: e1001325.

Vermaak, D. (2003). Maintenance of chromatin states: an open-and-shut case. *Curr. Opin. Cell Biol.* 15: 266–274.

Vivancos, J., Labbé, C., Menzies, J.G., and Bélanger, R.R. (2015). Silicon-mediated resistance of A rabidopsis against powdery mildew involves mechanisms other than the salicylic acid (SA)-dependent defence pathway. *Mol. Plant Pathol.* 16 (6): 572–582.

Wang, W.-S., Pan, Y.-J., Zhao, X.-Q. et al. (2011). Drought-induced site-specific DNA methylation and its association with drought tolerance in rice (*Oryza sativa* L.). *J. Exp. Bot.* 62: 1951–1960.

Workman, J.L. and Kingston, R.E. (1998). Alteration of nucleosome structure as a mechanism of transcriptional regulation. *Annu. Rev. Biochem.* 67: 545–579.

Yigider, E., Taspinar, M.S., Aydin, M., and Agar, G. (2020). Cobalt-induced retrotransposon polymorphism and humic acid protection on maize genome. *Biol. Futura* 1–8.

Zemach, A., McDaniel, I.E., Silva, P., and Zilberman, D. (2010). Genome-wide evolutionary analysis of eucaryotic DNA methylation. *Science* 328 (5980): 916–919.

Zhang, X. (2008). The epigenetic landscape of plants. *Science* 320: 489–492.

Zhang, X., Yazaki, J., Sundaresan, A. et al. (2006). Genomewide high-resolution mapping and functional analysis of DNA methylation in Arabidopsis. *Cell* 126: 1189–1201.

Zhang, K., Sridhar, V.V., Zhu, J. et al. (2007). Distinctive core histone posttranslational modification patterns in *Arabidopsis thaliana*. *PLoS ONE* 2: e1210.

Zhang, J., Xu, Y., Huan, Q., and Chong, K. (2009). Deep sequencing of Brachypodium small RNAs at the global genome level identifies microRNAs involved in cold stress response. *BMC Genomics* 10: 449.

Zhang, M., Kimatu, J.N., Xu, K., and Liu, B. (2010). DNA cytosine methylation in plant development. *J. Gen. Genom.* 37: 1–12.

Zhang, M., Tang, S., Huang, X. et al. (2014). Selenium uptake, dynamic changes in selenium content and its influence on photosynthesis and chlorophyll fluorescence in rice (*Oryza sativa* L.). *Environ. Exp. Bot.* 107: 39–45.

Zhong, L., Xu, Y.-H., and Wang, J.-B. (2009). DNA-methylation changes induced by salt stress in wheat *Triticum aestivum*. *Afr. J. Biotechnol.* 8: 6201–6207.

Zhu, J., Jeong, J., Zhu, S.I. et al. (2008). Involvement of Arabidopsis HOS15 in histone deacetylation and cold tolerance. *Proc. Natl. Acad. Sci.* 105: 4945–4950.

Zilberman, D., Gehring, M., Tran, R.K. et al. (2007). Genome-wide analysis of *Arabidopsis thaliana* DNA methylation uncovers an interdependence between methylation and transcription. *Nat. Genet.* 39: 61–69.

3

Beneficial Elements and Status of ROS and RNS in Plants: Current Evidence and Future Prospects

Biswajita Pradhan[1], Rabindra Nayak[1], Srimanta Patra[2], Chhandashree Behera[1], Soumya Ranjan Dash[1], and Mrutyunjay Jena[1]

[1] *P.G. Department of Botany, Berhampur University, Berhampur, Odisha, India*
[2] *Department of Life Sciences, NIT, Rourkela, Odisha, India*

Introduction

The required essential elements are indispensable for plant growth and development (Arnon and Stout 1939). There are well-known 17 essential elements of plants that include both micronutrients and macronutrients. The micronutrients are called trace elements that are established in plants at lower concentrations ($\leq$100 mg^{-1} kg^{-1} dry $weight^{-1}$) like boron (B), iron (Fe), chlorine (Cl), molybdenum (Mo), copper (Cu), zinc (Zn), manganese (Mn), and nickel (NI). Some reports suggest Silicon (Si) as a micronutrient, but few mention it as a beneficial element (Römheld and Marschner 1991, Shivaraj et al. 2021). Micronutrients are involved in several metabolic processes such as primary and secondary metabolism, signal transduction, cell defence mechanism, gene regulation, energy metabolism, and hormonal perception (Hänsch and Mendel 2009; Vatansever et al. 2017a). Additionally, the deficiencies or over supplementation of these elements causes toxic effects to plant life cycle and may lead to plant death (Mou et al. 2000; Soetan et al. 2010; Alam 1999). In this context, the foremost question remains to be concluded as the classification of micronutrients either as essential or as beneficial elements. Additionally, to drive a conclusion, physiological and molecular investigations are required for the inclusion or exclusion of these controversial elements from the micronutrient list. The macronutrients include oxygen (O), carbon (C), hydrogen (H), calcium (Ca), potassium (K), nitrogen (N), magnesium (Mg), phosphorous (P), and sulphur (S) (Hawkesford et al. 2012; Waraich et al. 2011; Sarwar et al. 2010). Among these macronutrients, C, H, and O constitute 95% of plant dry material and the other elements contribute very low in dry weight (Pilon-Smits et al. 2009). The beneficial elements such as growth-promoting elements are essential to specific taxa but not essential to all plant taxa (Pilon-Smits et al. 2009; Kaur et al. 2016). There are five beneficial elements implicated for the plant growth and regulation that are Al, Na, Co, Si, and Se. These elements are the growth promoters for various

Beneficial Chemical Elements of Plants: Recent Developments and Future Prospects, First Edition. Edited by Sangeeta Pandey, Durgesh Kumar Tripathi, Vijay Pratap Singh, Shivesh Sharma, and Devendra Kumar Chauhan.

plant species under convinced environmental situations (Pilon-Smits et al. 2009; Kaur et al. 2016). However, the concentration and functions are distinct in plant species. These essential elements have multiple effects such as growth promotion, phenotypic differences, and physiological changes at cellular and tissue levels (Hasanuzzaman et al. 2017; Hajiboland 2012). At high tissue concentration, these advocate a crucial role as structural or osmotic regulator (Zhu et al. 1997; Prashanth et al. 2015; Pilon-Smits et al. 2009). However, at low tissue concentration, they act as specific cofactors on behalf of several enzymes (Prashanth et al. 2015; Hänsch and Mendel 2009).

Many times, the induction of nutrients is accountable for the manufacture of ROS and reactive nitrogen species (RNS) are due to defective physiological function and disrupted metabolism (Ashley et al. 2006). ROS is formed in different compartments of cell in plants due to oxidative metabolism (Janků et al. 2019; Rodríguez-Serrano et al. 2006; Pradhan et al. 2020a). ROS includes both oxygen radicals and other non-radicals. The oxygen radicals comprise superoxide (O_2^-), peroxyl (ROO˙), hydroxyl (·OH), while the non-radicals include $^1O_2$ (singlet oxygen) and H_2O_2 (hydrogen peroxide) (Del Río 2015). Similarly, RNS includes radicals like nitric dioxide ($NO_2^{\cdot}$) and nitric oxide (NO·) and non-radicals include N_2O_4 (dinitrogen tetroxide) and HNO_2 (nitrous acid) (Del Río 2015). Both RNS and ROS play a key role in plant growth, development, and physiological adaptiveness under extreme environmental conditions (Del Río 2015; Halliwell 2006). The oxidizing species produces ROS (hydroxyl radicals and singlet oxygen) that causes cellular damages to the biological systems due to oxidative stress. However, to sustain such oxidative stress, plants deploy several antioxidants and their precursor stress-responsive enzymes that can scavenge the excess oxidants to abide by the toxic effects (Del Río 2015; Halliwell 2006; Halliwell and Gutteridge 2007). The main notion of 'oxidative stress' was reconsidered in modern years, and the term 'oxidative signalling' has emerged as a booming field in stress biology. Hence, considering the present bankable knowledge of ROS and RNS and their physiological significance, this article summarizes the key aspects of some essential elements and their physiological role in response to RNS and ROS (Rinalducci et al. 2008; Foyer and Noctor 2016).

Essential and Beneficial Elements in Plant Physiology: A Pleasant Dilemma

The beneficial elements have not been regarded as essential chief element for all plants but can be attributed as necessary for some plants (Kaur et al. 2016). A plant is incapable to complete its life cycle due to the lack of essential elements, and its function cannot mimic the function of another element. The beneficial elements are directly involved in plant metabolism (Kaur et al. 2016; Broadley et al. 2012; Asher 1991). The beneficial elements can reimburse for the noxious properties of other elements or may substitute few nutrients contributing to a fewer specific physiological function (Pirson 1955). The omission of beneficial elements can hinder the growth of plants to their ideal genetic potential as these are merely formed at an existence level. In this section, we have discussed some chief beneficial essential elements like Al, Co, Na, Se, and Si and their physiological role in plant nutrition, growth, and survival as well as in reproduction (Fageria and Moreira 2011) (Figure 3.1).

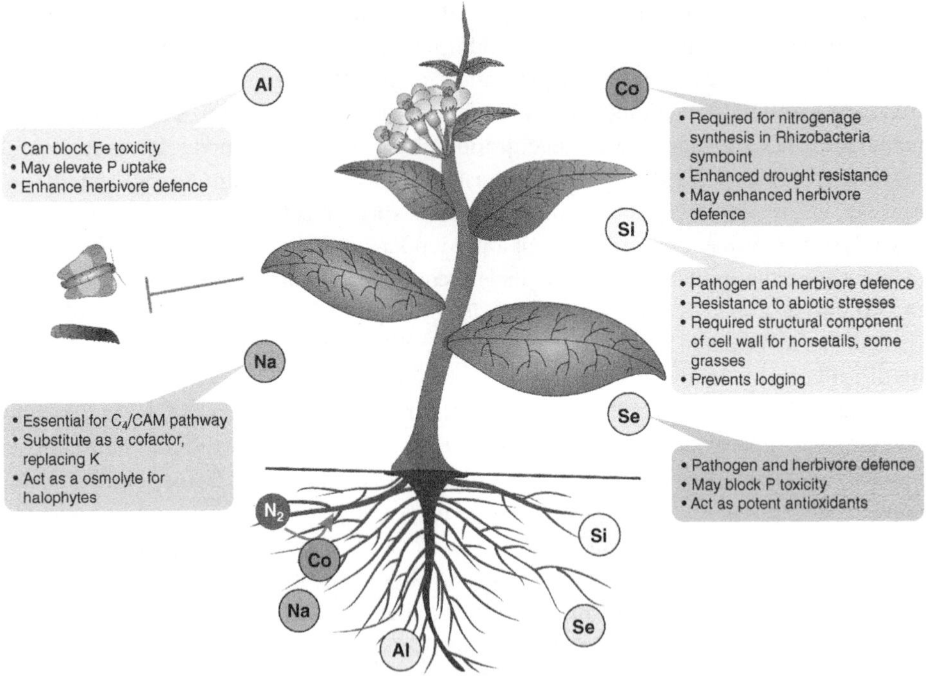

Figure 3.1 Mechanism of beneficial elements for plant growth and their role in plant metabolism.

Aluminium

Aluminium (Al) is considered as the third foremost plentiful essential element on the surface of Earth's crust (Pradhan et al. 2020b). Al concentration increased in the environment by anthropogenic sources such as industrialization and acid rainfall. Al elevated toxicity for both plants and animals (Kochian et al. 2004; Pradhan et al. 2020b). Aluminium toxicity is a chief problem for the crops growing in acidic soil (Pradhan et al. 2020b, Chauhan et al. 2021). The bioavailability of aluminium (Al) is maximum in the acidic soils at lower pH range of < 5.5. Aluminium toxicity resulted in a decrease of growth of roots and hinders photosynthesis in plants (Kochian et al. 2004). Nevertheless, it also triggers adverse effects in plant growth through accumulation in plant cells (Pradhan et al. 2020b). Hence, to sustain under Al-mediated toxic environment, plants develop antioxidant defence mechanism that sustains cellular viability and tolerance to promote growth simulation (Pradhan et al. 2020b). To warrant such adverse modalities, understanding the molecular mechanism must be investigated. In this notion, more physiological and molecular investigations are required for Al accumulation in plants to validate the contrary role of Al in plants (Pradhan et al. 2020b).

Mechanistically, plants develop Al tolerance through symplastic or apoplastic detoxification mechanisms. Apoplastic mechanisms confront Al binding to the cell wall and prevent its transformation into the symplasm (Wang et al. 2014; Rengel 1997). The secretions in the root raised the proximal in soil pH and made less Al bioavailability, and the mucilage reduces Al mobility (Kisnierienė and Lapeikaitė 2015; Horst et al. 1982). Some of the plant species tolerate Al in symplast and metabolically converts and stores as it is in lesser toxic forms and in complexion with organic acids (Taylor 1988). Moreover, a high concentration of Al causes metal toxicity in plants. On the other hand, at low concentrations, Al is beneficial for some plants (Foy et al. 1978). In *Miscanthus sinensis*, a low concentration of Al exhibits a growth-promoting effect. Al demonstrates higher growth of plants in acidic soils (Yoshii 1937). In *Camellia sinensis*, Al increases antioxidant enzymes that support plant growth (Ghanati et al. 2005, Chauhan et al. 2021). In addition to this, high Al accumulation from the soil by plant-like *Melastoma malabathricum* also upregulates the shoot and root growth that can avoid Fe toxicity (Watanabe et al. 2005, 2006; Watanabe et al. 2008). The hyperaccumulator plant species accumulates more than 1 g kg^{-1} of Al/dry weight in their tissues of leaves of plants. The hyperaccumulator plant species such as *Festuca arundinacea* accumulate Al in their tissues to prevent herbivorous (Potter et al. 1996).

Cobalt

Cobalt is well-known as well as a chief essential element for prokaryotes and animals, while in plants the abundance is very less (Wells et al. 2017; Schlegel and Jannasch 2006). Cobalt is not considered an essential element but can act as a key beneficial element for plants (Palit et al. 1994). Like heavy metals, cobalt (Co) is also the reason of toxicity in plants at higher concentrations (Kaur et al. 2016). Cobalt concentrations in plants are generally very low, but at higher concentrations, it is toxic to plants that block metabolic functions. Higher concentrations of Co hamper plant growth, photosynthesis, and seed germination (Nagajyoti et al. 2010; Palit et al. 1994). Mechanistically, Co was transported as divalent cation (Co^{2+}) and persistent in the cells by transporters and maintain homeostasis in plants. Co has beneficial effects in leguminous plants at low concentrations (Palit et al. 1994). At 8 ppm Co increased the growth, nodule weight, number, levels of nutrition, increased seed quality, and seedpod yield in *Pisum sativum* L. (Abdel-Aziz and Ismail 2016). These effects of cobalt (Co) invite the symbiotic rhizobia to live in the root nodules of leguminous plants. Cobalt is a constituent of cobalamin (vitamin B12) and essential for numerous enzymes for nitrogen fixers like *Cyanobacteria* and *Rhizobium* (Knack et al. 2015). Moreover, Co regulates the delay in senescence of leaves via modulation of biosynthesis of ethylene. It also regulates the drought resistance in plant seeds (Pilon-Smits et al. 2009). Cobalt also acts as a strong stimulator of isoquinoline (alkaloids) accumulation in medicinal plants via the upregulation of aromatic amino acid biosynthesis as precursors of alkaloids. Cobalt hyperaccumulations at tissue level protect the plants from herbivores or pathogens. Cobalt also has a supplementary effect and enhanced nutritional quality (Solaiman et al. 2007; Palit et al. 1994).

Cobalt is not very rich and abundantly present like Al with the concentration ranges of 15–25 ppm in soils and 0.04 ppm in natural waters. The concentration of Cobalt in plants is 0.1–10 ppm (Baker 2000). Recent studies have revealed that out of 670 experimented plant species, the cobalt concentration in leaf was less than 0.2 ppm. In *Euasterids ericales* and *Asparagales* the concentration ranges from 0.3 to 0.5 ppm (Baker 2000). There are 26 cobalt hyperaccumulator plant species reported containing more than 1000 ppm in their leaf tissues. These plant species belong to the families of Asteraceae, Lamiaceae, Fabaceae, and Scrophulariaceae (Baker 2000). In higher plants, cobalt has been absorbed from the soil solution and strongly binds to roots. Cobalt and nickel enter into cells by specific plasma membrane carriers because of their similarity affinity. Baxter et al. (2008) reported that *Arabidopsis* was grown under abridged concentration of Fe in the solution of fertilizer and the cobalt concentration was inclined in the shoot (Baxter et al. 2008). In addition, the simultaneous inclination of Fe transporter IRT1 expression in the shoot confirmed that IRT1 transport is responsible for cobalt transport (Baxter et al. 2008).

Sodium

Sodium is significantly associated with soil salinity. A higher concentration of Na^+ is toxic to plants (Davenport and Tester 2000; Demidchik et al. 2002). Na^+ accumulation in the stems and roots are harmful to plants in woody plant species like grapevine and citrus. In some of the higher plants, Na^+ causes ion-specific damage (Ohnishi et al. 1990). Nevertheless, high Na^+ treatments exhibit substantial growth in dicotyledonous halophytes and some C_4 plants. Sodium is an essential and chief element for C4 or CAM plants and is used in photosynthetic pathways (Ohnishi et al. 1990; Marschner 2011). These CAM/C4 plants used phosphoenolpyruvate (PEP) to fix the atmospheric carbon for the photosynthesis. In this context, Na is required for revival of PEP from the pyruvate (Ohnishi et al. 1990; Marschner 2011). In certain halophytic plants, e.g. *Atriplex*, Na accumulation in vacuoles at high concentration contributes towards the exhibition of osmotic potential in the plant and allows to take up water from the saline environment. Additionally, halophytes have conveyed that the use of Na facilitates nitrate uptake through a Na^+/NO_3^- cotransporter (García-Sánchez Ma et al. 2000). Additionally, Na also boosts crop productivity and enhances the nutritional value of the plant (Kelly and Horning 1999). Sodium has a negative effect at higher concentrations rather than having beneficial effects. Na^+ is similar to potassium (K^+) that enters into plants through K^+ channels (Ohnishi et al. 1990; Marschner 2011). However, plants also have transporters of Na^+. The main mechanism deployed in salt tolerance is the transport of sodium (Na) out of the root hooked on the vacuole or the shoot of the phloem (Ren et al. 2005; Hasegawa 2013). Moreover, it has been evidenced that Na acts as a pivotal beneficial as well as essential element for the growth of plant and acts as an osmoregulator in stomatal movement (Pilon-Smits et al. 2009). Further, a little concentration of Na acts as a beneficial effect on growth under K-deficient conditions. In addition to this, Na^+ can be the substitute of K^+ that acts as a cofactor for specific enzymes (Bassil et al. 2011).

Selenium

Selenium is non-essential but acts as a chief beneficial element for plants. Selenium is present in a different forms in different soils mostly as selenate in the alkaline form and in well-oxidized soils. The type of selenite found in well-drained mineral soils is also found in selenide in vigorously reduced soils (Hasanuzzaman et al. 2014). Selenium not only is beneficial for crop fertilization for plant productivity but also enhances its nutritional value. As both selenate and sulphate possess the same chemical structure, plants could transport the selenate by sulphate through the specific membrane transporters to the cell (Zhu et al. 2009; Lima 2016) and both play a role in the construction of chloroplast and cell membranes (Terry et al. 2000). Selenium is available in selenite form in the soils. It can be taken up by plants via sulphated transporters and assimilated into selenocysteine (SeCys) and also selenomethionine (SeMet) (Fu et al. 2002). Selenium is needed for numerous animals, bacteria, and the green alga *Chlamydomonas reinhardtii* (Novoselov et al. 2002), while it has not been a crucial element for higher plants (Brown and Shrift 1982). SeCys in selenoproteins is encrypted through the opal stop codon when existing in place of an explicit secondary mRNA structure (SeCys insertion sequence). In higher plants, the SeCys supplement sequences have not been found yet, and the plants' homologs of selenoproteins are like glutathione peroxidase (GPX) that were found to contain Cys in their place of SeCys in active site (Novoselov et al. 2002). It takes central role in antioxidant and regulation of ROS, heavy metal curiosity, and also photosynthetic enhancement. Besides this, it also helps plants to ameliorate the abiotic stress like drought, temperature, water, cold, desiccation, heavy metals, salinity, senescence, and UV (Arif et al. 2016).

Reports on selenium as a beneficial element of plants and its effects on plant growth due to accumulating have been observed in selenium hyperaccumulation in plants (El Mehdawi and Pilon-Smits 2012). The beneficial properties of selenium on the hyperaccumulator of growth of the plant were lesser than the plants grown under lesser phosphate conditions (Pilon-Smits and Quinn 2010). Moreover, selenium also acts as an antagonist against the phosphate toxicity in hyperaccumulation (Feng et al. 2013). Se triggers the growth at trace amounts in the non-hyperaccumulator type of species such as lettuce, ryegrass, duckweed, and potato (El-Ramady et al. 2016). The Se accumulated plants exhibit lipid peroxidation at lower concentration and GPX activity at higher concentrations. Additionally, it also demonstrates more resistance to UV radiation (Feng et al. 2013). Se induces antioxidant capacity and became stress resistant (Feng et al. 2013). Se protects and defends plants from harmful biotic stresses (Feng et al. 2013; Broyer et al. 1972). Se on non-hyperaccumulator and hyperaccumulator plants protected them from herbivores and fungal infections. Se treatment in plant displayed the upregulation of ethylene and jasmonic acid (JA) production (Tamaoki et al. 2008; Van Hoewyk et al. 2008) and also displayed the upregulated selenite/sulphate assimilation (Daly et al. 2017). Moreover, fertilization through lower doses of selenium can endorse the growth of plant and increases resistance potential to pests (Malagoli et al. 2015). Se hyperaccumulator species such as *Astragalus bisulcatus* and *Brassica oleracea* (broccoli) consume a selenium explicit selenocysteine methyltransferase, leading to the addition of selenium as a relatively harmless methylselenocysteine (Lyi et al. 2005; Sors et al. 2005). Moreover, *Arabidopsis thaliana* is Se tolerant because it has Se-binding protein (Agalou et al. 2005).

At low concentrations (1 mg L^{-1}), it increases plant growth and development and also improves the fruit quality in pears and peaches (Zhu et al. 2017). At lower concentrations, selenium alleviates the stress factors, but at higher concentrations, it is toxic in ryegrass (1 mg kg^{-1} H_2SeO_4 addition to soil) (Hartikainen and Xue 1999). Besides this, in some selenium accumulators like *red seaweed*, *Spirulina platensis*, and *Pteris vittata*, it has exhibited beneficial effect at up to 5 mg L^{-1}. In most soils, the selenium concentration usually differs from 0.01 to 2.0 mg kg^{-1} (Kaur et al. 2014). So, it is necessary to illustrate that the sulphated transporter homologs are chiefly in Se hyperaccumulators. Moreover, the plant having a selenite accumulator is lesser known and its uptake is metabolically not connected. Phosphate and carbonyl cyanide chlorophenyl hydrazone (CCCP) hinder selenite uptake in the plant of wheat but lack of phosphate improves the selenite uptake (Vatansever et al. 2017a; El-Ramady et al. 2015). Selenium is classically below 1 ppm mg kg^{-1} soil but 4–100 ppm in seleniferous soils (Zhao et al. 2005). Most of the vegetation in the soils have less than 1 ppm mg/kg plant dry weight of selenium (Arvy 1992). On the other hand, plants of seleniferous soils contain 1–10 ppm of selenium and are called selenium hyperaccumulator plants (El Mehdawi et al. 2011). The example of hyperaccumulator plants of genera *Stanleya* and *Astragalus* accumulates 1000–15 000 ppm of selenium (Mehdawi 2013; He et al. 2018). Selenium is similar to sulphur in its chemical structure and is metabolized through the same pathways (Kopsell and Randle 1997; Pilon-Smits et al. 2002).

Silicon

Among the prevalent elements on the Earth, silicon (Si) constitutes 25% of the surface of Earth's crust (Sommer et al. 2006, Tripathi et al. 2020). Silicon is present in soil water as a monosilicic acid, $Si(OH)_4$, with concentrations ranging from 0.1 to 0.6 mM. Silicon is considered as a beneficial element among plant micronutrients (Epstein and Bloom 2005). It triggers photosynthesis, decreases loss of transpiration, and also increases resistance against various abiotic/biotic stresses like drought, cold, temperature, salinity, and several fungal and bacterial diseases (Shafi and Zahoor 2020). Silicon deficiency causes several abnormalities in plant growth, improvement, development, and reproduction (Miyake and Takahashi 1978, Tripathi et al. 2020). Silicon supplement exerts several beneficial properties on plant growth and yield (Miyake and Takahashi 1978, Shivaraj et al. 2021). It improves leaf introduction to light, resistance to lodging, declined vulnerability to pathogens and also to root parasites, and also enhances the abiotic condition of stresses (Ma 2004). Additionally, Si has numerous beneficial properties for plants and is essential for certain plant taxa such as sedges, horsetails, and grasses (Epstein 1994). Hyperaccumulation of Si in the cell walls endorses plant sturdiness and resistance under several biotic and abiotic stresses (Pandey 2015; Pilon-Smits et al. 2009, Tripathi et al. 2020). Si is also considered as having beneficial nutrient properties for plants and animals, and therefore Si crops fertilization may enhance their nutritive value (Kaur et al. 2016).

Silicon is accumulated in plants as amorphous silica (SiO_2^- nH_2O) (Currie and Perry 2007). It is deposited in cell walls of plants, and it interacts with polyphenols and pectins giving strength and rigidity to the cell wall (Currie and Perry 2007). The concentration of Si in plants is about (0.1–15)% of dry weight (Hodson et al. 2005). Monocotyledonous plants accumulate higher amount of Si (10–15%) than dicotyledonous plants (0.5%) (Ma and Yamaji 2008). Though several studies have conveyed the beneficial effects of silicon (Si) in tolerance to plant stress, but its rate of accumulation in various plants and their physiological and biochemical role are yet to be accessed. The beneficial properties of silicon depend on the accumulation potency of plants. Silicon is accumulated and translocated to the shoots as silicic acid ($Si(OH)_4$) and also converted to hydrated amorphous silica. These are stored in cell walls (Epstein and Bloom 2005). It also forms silica cuticle and silica cellulose of double layers on the leaf, stem, and surfaces. In this context, two transporters of silicon such as Lsi1 and Lsi2 are involved in efficient silicon uptake in rice (Ma and Yamaji 2015; Zhao et al. 2010). The knockout of either of the transporters causes the reduction in uptake of silicon. Plants without Lsi1 or Lsi2 remained either lesser accumulators or incompetently accumulation of silicon (Vatansever et al. 2017b; Ma and Yamaji 2015). Silicon is the major essential constituent of some members of plants such as Equisetaceae (*Equisetum arvense*) and Poaceae (*Oryza sativa*) (Kaufman et al. 1981).

Silicon can also avert abiotic stress in plants. Silicon alleviates heavy metal toxicity symptoms in plants (Ma 2004; Liang et al. 2007, Tripathi et al. 2020). Several antioxidants such as both enzymatic such as ascorbate peroxidase (APX), superoxide dismutase (SOD), glutathione reductase (GR), and dehydroascorbate reductase and non-enzymatic such as glutathione and ascorbate (Adrees et al. 2015; Liang et al. 2007). By the elevation of this, toxic metal ions reacted with silicon in the cytoplasm (Cd, Zn, and Al silicates) and that hinders more toxicity. This mechanism maintains cellular homeostasis by confiscation of metals in the vacuoles. Recent studies have demonstrated that silicon mediates moderate the salinity stress via increasing sodium prohibiting and decreased lipid membrane peroxidation by stimulating both non-enzymatic and enzymatic antioxidant molecules (Adrees et al. 2015).

Silicon reduces the plant vulnerability to fungal diseases (Datnoff et al. 1997). There are two probable mechanisms of silicon-enhanced resistance to pathogens in plants. The first mechanism is Si polymerization that supports the cell walls via physical inhibition of fungal germ tube insertion into epidermis and then-impending infections. The second one is silicon mediated signal that triggers the natural defence response to both monocots and dicots by eliciting enzymes such as peroxidases, chitinases, polyphenol oxidases, and increased phenolic compounds generation, antimicrobial compounds production, phytoalexins, and the systemic stress signals (jasmonic acid salicylic acid and also ethylene). However, silicon induces resistance to pathogens on plants (Datnoff et al. 1997).

ROS and RNS Production Sites in Plant Cells: Cellular Redox Compartments with Regards to Essential Elements

Both ROS and RNS are formed in specific cellular organelles. ROS is formed via electron transportation actions of plasma membranes, mitochondria, and chloroplasts or as the by-product of several metabolism pathways (Del Río 2015; Sandalio and Romero-Puertas 2015). In contrast, the generation of nitric oxide (NO) is more complicated and uncertain in plant cells. NO can be produced with or without enzymatic catalysis. Plant NO production can be originated from oxidative catabolism of polyamines or the oxidation of hydroxylamines (Astier et al. 2018). NO can be produced by the action of nitrate reductase activity with the involvement of the nitrate reductase (NR) enzyme (Astier et al. 2018). The cytosolic enzyme nitrate reductase endures a regulatory switch from its special substrate, nitrate to nitrite, and generates NO by one-electron reduction of nitrite (Kolbert and Feigl 2018). Mitochondria is signified as a major source for intracellular ROS production. In plants, excessive accumulation of ROS in the mitochondria can be ameliorated via an alternative mechanism for electron transport that is carried out by alternative oxidases (Saini et al. 2018; Asada 2006). Moreover, in the reductive mechanism, nitrite acts as a terminal electron acceptor for mitochondrial cytochrome C oxidase/reductase in low oxygen availability (Halliwell and Gutteridge 2007). The superoxide radicals are mainly produced in the electron transportation chain by NADH dehydrogenase (complex I) and ubiquinone cytochrome C reductase (complex III). Furthermore, superoxide dismutase and other antioxidants are accountable for the production of H_2O_2 from O_2^- in the mitochondrion (Han et al. 2001). In certain physiological circumstances, NO and O_2^- may be created at the same time in the electron transportation chain and result in the formation of mitochondrial peroxynitrite (Rasmusson et al. 2008). The plasma membrane plays a key role in intracellular ROS production as it binds to the major superoxide-generating system such as the NADPH oxidase enzyme complex (Gupta et al. 2018). The plasma membrane also provides space for NO production. The cell wall is the chief site of non-enzymatic production of NO (Das and Roychoudhury 2014). Altogether, chloroplast, mitochondrion, and peroxisome are the organelles that are responsible for intracellular ROS and RNS production (Luis et al. 2006; Foyer and Noctor 2003). Furthermore, the plasma membrane and the plant cell wall are also clarified as the additional contributor to both ROS and RNS production (O'Brien et al. 2012; Del Río 2015) and status of ROS production in diverse cellular compartments of the plant cell (Figure 3.2).

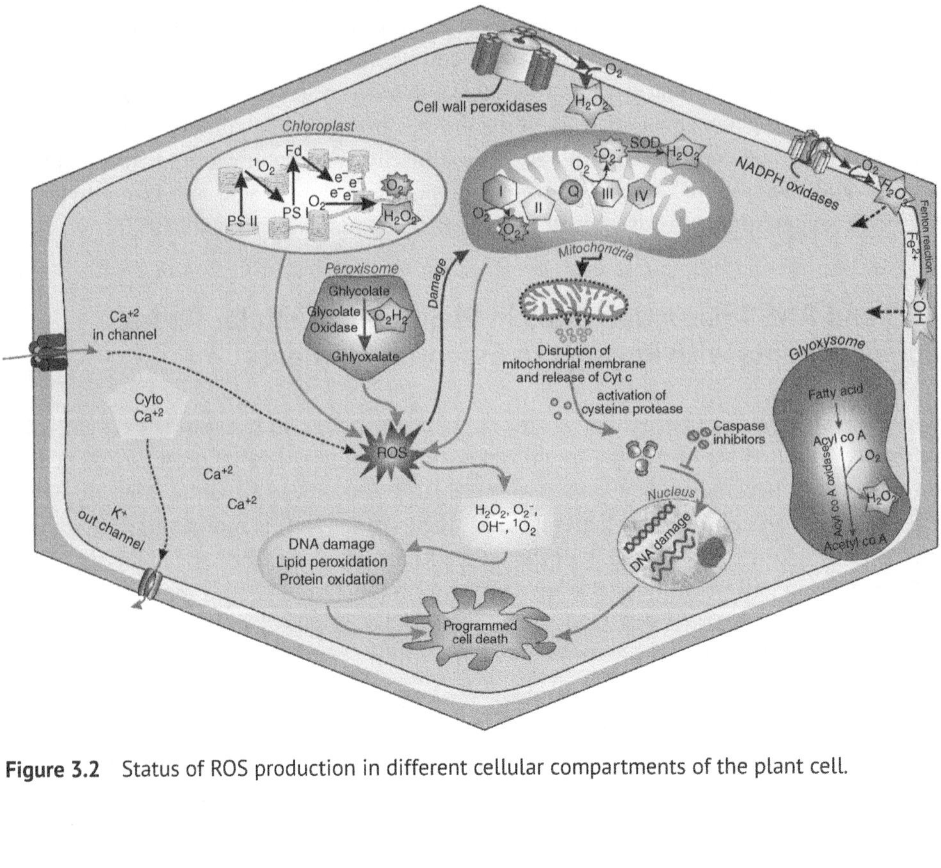

Figure 3.2 Status of ROS production in different cellular compartments of the plant cell.

ROS and RNS Production and Their Function in Plants: Connecting Physiology to Stress Physiology

It is also well recognized that the main source of O_2^- is the plasma membrane-localized NADPH oxidase (NOX), but the extra sources of superoxide and H_2O_2 are produced in mitochondria, chloroplasts, and peroxisomes by electron transport reactions and during oxidative stress (Heyno et al. 2008). H_2O_2 is produced as the by-product of catalytic reactions in plants (Giorgio et al. 2007). The chief mechanism of synthesis of H_2O_2 and the catabolism is diverse in several cellular structures (Warm and Laties 1982). The H_2O_2 is produced by peroxisomal oxidases with the involvement of the enzymes glycolate oxidase and acyl-CoA (Corpas et al. 2001). In addition, there is the involvement of fatty acid β-oxidation and photo-respiratory pathways in H_2O_2 production (Impa et al. 2012). Another main source of ROS is the cell wall-bounded peroxidases that produce hydrogen peroxide (H_2O_2). Moreover, chloroplasts are the chief manufacturer sites of non-radical ROS like $^1O_2$ (singlet oxygen) via several photodynamic reactions (Edreva 2005). Several ROS like singlet oxygen ($^1O_2$) and hydroxyl radicals (·OH) are oxidizing species that can be used as commanding oxidants (McKersie and Lesheim 2013). These are ROS components that react with all constituents of the living cells and bring injury to lipids, proteins, and nucleic acids during the oxidative upset (Das and Roychoudhury 2014). However, to avoid such oxidative imbalance, plant cells deploy both non-enzymatic and enzymatic antioxidants to evade the toxic effects of these reactive biomolecules (Pradhan et al. 2020a, 2020b). To overcome ROS production, the management of oxidative stress and deployment of 'oxidative signalling' or 'redox signalling' is important in plants (Foyer 2018). Moreover, modulation of signalling network in response to environmental challenges to organizing signal transducers is important in oxidative stress management. It is normally accepted that ROS plays a key signalling role in plants' stress response (Foyer 2018). Extreme ROS are produced as a result of several environmental conditions and oxidative imbalance, which resulting in cell damage (Sharma et al. 2012). For the prevention of ROS-mediated cellular damage in plants, reinforce antioxidant mechanisms to detoxify ROS concurrently (Sharma et al. 2012; Scandalios 2005). The ROS generation in mitochondria, chloroplasts, and peroxisomes are vital for the initiation of signalling cascades (Foyer and Noctor 2003). In plants, ROS generation by external stimuli elicits signal transduction for the onset of particular cellular responses (Foyer and Noctor 2003).

In plants, potential sources of NO production include both non-enzymatic and enzymatic systems. However, a little evidence about subcellular NO production has been reported. The activity of NOS in the peroxisomes was first observed in the tissues of plants. Besides peroxisomes, mitochondria and chloroplasts also generate NO (Foyer and Noctor 2003). Nitric oxide is an intracellular and intercellular signalling molecule that has a potential role in the growth of plant and plant development (del Rıo et al. 2004). Besides this, NO also controls different processes such as induction of gene transcription or triggering secondary messengers (del Rıo et al. 2004). In addition to this, NO has multiple functions in plants such as pollen tube growth, seed germination, root organogenesis, cell wall lignification, the establishment of legume–*Rhizobium* symbiosis, fruit ripening, flowering, and senescence (Neill et al. 2003).

Like ROS, RNS also plays a major role under abiotic stress in plants and acts as signalling molecule. Generally, ROS and RNS production take place under oxidative upset conditions in plants (Foyer et al. 2016). The NO and ROS production play a key role in programmed cell death (PCD). PCD regulates the growth and development of environmental upsets and pathogen attacks (De Pinto et al. 2012).

In the existence of O_2, NO reacts with the reduced glutathione (GSH) to form S-nitrosoglutathione (GSNO) through S-nitrosylation reaction (Leterrier et al. 2011; Hasanuzzaman et al. 2019). Additionally, RNS peroxynitrite ($ONOO^-$) acts as a commanding nitrating/oxidant species that are produced via the rapid reaction between NO and O_2^- in the peroxisomes (Corpas et al. 2018; Luis et al. 2014). In peroxisomes, catalase (CAT) and glycolate oxidase activity are repressed by S-nitrosylation and control the cellular level of vital redox signalling molecules such as hydrogen peroxide (H_2O_2) (Piacentini et al. 2020). Moreover, the $ONOO^-$ production produces tyrosine nitration of plant proteins and mediates nitrosative injury in the plant cells (Chaki et al. 2011). Recently, studies in pea plants by electron microscopy (EM) techniques immune gold-labelling have displayed that the presence of nitrated proteins in diverse subcellular compartments of the leaf cells, which include mitochondria, chloroplast, peroxisomes, and the cytosol (Yemets et al. 2019). Moreover, the proteomic exploration of isolation of pea leaf peroxisomes has displayed that the peroxisomal NADH-dependent hydroxypyruvate reductase is the chief target for nitration. This type of reaction of peroxynitrite facilitates the loss of purpose in the enzyme NADH-dependent hydroxypyruvate reductase (Hasanuzzaman et al. 2019).

Conclusion and Future Perspectives

There are few minerals and essential elements (B, Fe, Cu, Mo, Ni, Mn, Na, Zn, Al, Cl, Se, Co, and Si) that play implicate role in the plant growth, development, PCD, and host defence to biotic and abiotic stresses. These elements even in trace amounts mediate cellular, physiological, and biochemical responses in several plants. Although these are not essential for some groups of plants, these elements have a significant effect on plant growth that might improve the production yield and nutritional value. These beneficial elements with low concentration are more desired as they exhibit significant effects on improving plant nutrition enhanced crop production. The essentialities of some micronutrients have been well recognized, while some others have not been explored to a greater extend. Therefore, it is challenging to clarify whether these elements are essential or beneficial for the plants. To conclude this debate, physiological and molecular studies are highly needed to elucidate their effective role in the growth, development, and survival of plants. These essential elements such as Al, Co, Na, Se, and Si give positive properties in plant growth, development, and stress resistance. These elements are essential for all plants. In this context, Si and Na are essential for specific plants, while Se and Al are essential for the hyperaccumulator species. Cobalt (Co) not only is necessary for microbial partners of plants but also is required by plants themselves. In addition to this, the effects of beneficial elements including several other elements such as cerium (Ce), silver (Ag), fluorine (F), chromium (Cr), lanthanum (La), iodine (I), tin (Sn), rubidium (Rb), titanium (Ti), tungsten (W), serium (Sr), and vanadium (V) have also been described in recent studies, but more studies are needed to warrant a conclusion about the role and involvement of these elements in plants.

Although ROS and RNS have a significant function in intracellular, intercellular, and molecular level of communication, it is vital to know the regulatory functions of ROS and RNS signalling in cellular redox changes. Moreover, understanding the critical functions of ROS- and RNS-mediated regulation of hormones like auxin, salicylic acid, cytokinin, jasmonic acid, and also ethylene responsible for plant growth, nutrition, and development needs to be accessed. The physiological role of NO production and biosynthesis is needed to be investigated to delineate further molecular association in redox management. Therefore, it is highly essential to illustrate the gene(s) or protein(s) accountable for L-arginine dependent nitric oxide synthase (NOS) activity under several physiological conditions in plants. Although few genes involved in RNS and ROS production, regulation, and signal transduction have also been well recognized, it also remains a great contest to recognize other regulatory gene networks associated with these genes. Moreover, a critical assessment is required to decrypt their mode of action in ROS- and RNS-mediated transduction and in epigenetic character transformation. Hence, understanding the physiological role of essential elements and their dysfunction in metabolic pathways leading to ROS and RNS production in association with the ROS and RNS regulation under oxidative stress is highly desired to decrypt the plant growth, development, and stress response in plants.

Acknowledgments

The authors kindly thank MoEF & CC, Govt. of India, for carrying out this research work. The authors are thankful to Berhampur University for providing the necessary facilities.

Conflicts of Interest

The authors have no conflicts of interest to disclose.

References

Abdel-Aziz, M. and Ismail, A. (2016). Response of pea plant (Pisum sativum L.) for levels of nitrogen, *Rhizobium* inoculation and spraying of molybdenum on growth, green pods, dry seed yield and its components. *Journal of Plant Production* 7 (9): 991–1000.

Adrees, M., Ali, S., Rizwan, M. et al. (2015). Mechanisms of silicon-mediated alleviation of heavy metal toxicity in plants: a review. *Ecotoxicology and Environmental Safety* 119: 186–197.

Agalou, A., Roussis, A., and Spaink, H.P. (2005). The Arabidopsis selenium-binding protein confers tolerance to toxic levels of selenium. *Functional Plant Biology* 32 (10): 881–890.

Alam, S.M. (1999). Nutrient uptake by plants under stress conditions. *Handbook of Plant and Crop Stress* 2: 285–313.

Arif, N., Yadav, V., Singh, S. et al. (2016). Assessment of antioxidant potential of plants in response to heavy metals. In: *Plant Responses to Xenobiotics* (ed. A. Singh, S.M. Prasad and R.P. Singh), 97–125. Springer.

Arnon, D.I. and Stout, P. (1939). The essentiality of certain elements in minute quantity for plants with special reference to copper. *Plant Physiology* 14 (2): 371.

Arvy, M. (1992). Some aspects of selenium relationships in soils and plants. *Communications in Soil Science and Plant Analysis* 23 (13-14): 1397–1407.

Asada, K. (2006). Production and scavenging of reactive oxygen species in chloroplasts and their functions. *Plant Physiology* 141 (2): 391–396.

Asher, C.J. (1991). Beneficial elements, functional nutrients, and possible new essential elements. *Micronutrients in Agriculture* 4: 703–723.

Ashley, M.K., Grant, M., and Grabov, A. (2006). Plant responses to potassium deficiencies: a role for potassium transport proteins. *Journal of Experimental Botany* 57 (2): 425–436.

Astier, J., Gross, I., and Durner, J. (2018). Nitric oxide production in plants: an update. *Journal of Experimental Botany* 69 (14): 3401–3411.

Baker, A. (2000). Metal hyperaccumulator plants: a review of the ecology and physiology of a biological resource for phytoremediation of metal-polluted soils. In: *Phytoremediation of Contaminated Soil and Water*, 85–107. CRC Press.

Bassil, E., Tajima, H., Liang, Y.-C. et al. (2011). The Arabidopsis Na+/H+ antiporters NHX1 and NHX2 control vacuolar pH and K+ homeostasis to regulate growth, flower development, and reproduction. *The Plant Cell* 23 (9): 3482–3497.

Baxter, I.R., Vitek, O., Lahner, B. et al. (2008). The leaf ionome as a multivariable system to detect a plant's physiological status. *Proceedings of the National Academy of Sciences* 105 (33): 12081–12086.

Broadley, M., Brown, P., Cakmak, I. et al. (2012). Beneficial elements. In: *Marschner's Mineral Nutrition of Higher Plants* (ed. P. Marschner), 249–269. Elsevier.

Brown, T.A. and Shrift, A. (1982). Selenium: toxicity and tolerance in higher plants. *Biological Reviews* 57 (1): 59–84.

Broyer, T., Johnson, C., and Huston, R. (1972). Selenium and nutrition of Astragalus. *Plant and Soil* 36 (1-3): 651–669.

Chaki, M., Valderrama, R., Fernández-Ocaña, A.M. et al. (2011). High temperature triggers the metabolism of S-nitrosothiols in sunflower mediating a process of nitrosative stress which provokes the inhibition of ferredoxin–NADP reductase by tyrosine nitration. *Plant, Cell & Environment* 34 (11): 1803–1818.

Chauhan, D.K., Yadav, V., Vaculík, M. et al. (2021). Aluminum toxicity and aluminum stress-induced physiological tolerance responses in higher plants. *Critical Reviews in Biotechnology* 41 (5): 715–730.

Corpas, F.J., Barroso, J.B., and del Río LA. (2001). Peroxisomes as a source of reactive oxygen species and nitric oxide signal molecules in plant cells. *Trends in Plant Science* 6 (4): 145–150.

Corpas, F.J., Luis, A., and Palma, J.M. (2018). A role for RNS in the communication of plant peroxisomes with other cell organelles? In: *Proteomics of Peroxisomes* (ed. L.A. del Río and M. Schrader), 473–493. Springer.

Currie, H.A. and Perry, C.C. (2007). Silica in plants: biological, biochemical and chemical studies. *Annals of Botany* 100 (7): 1383–1389.

Daly, D.H., Velivelli, S.L., and Prestwich, B.D. (2017). The role of soil microbes in crop biofortification. In: *Agriculturally Important Microbes for Sustainable Agriculture* (ed. V.S. Meena, P.K. Mishra, J.K. Bisht and A. Pattanayak), 333–356. Springer.

Das, K. and Roychoudhury, A. (2014). Reactive oxygen species (ROS) and response of antioxidants as ROS-scavengers during environmental stress in plants. *Frontiers in Environmental Science* 2: 53.

Datnoff, L., Deren, C., and Snyder, G. (1997). Silicon fertilization for disease management of rice in Florida. *Crop Protection* 16 (6): 525–531.

Davenport, R.J. and Tester, M. (2000). A weakly voltage-dependent, nonselective cation channel mediates toxic sodium influx in wheat. *Plant Physiology* 122 (3): 823–834.

De Pinto, M., Locato, V., and De Gara, L. (2012). Redox regulation in plant programmed cell death. *Plant, Cell & Environment* 35 (2): 234–244.

Del Río, L.A. (2015). ROS and RNS in plant physiology: an overview. *Journal of Experimental Botany* 66 (10): 2827–2837.

Demidchik, V., Davenport, R.J., and Tester, M. (2002). Nonselective cation channels in plants. *Annual Review of Plant Biology* 53 (1): 67–107.

Edreva, A. (2005). Generation and scavenging of reactive oxygen species in chloroplasts: a submolecular approach. *Agriculture, Ecosystems & Environment* 106 (2-3): 119–133.

El Mehdawi, A. and Pilon-Smits, E. (2012). Ecological aspects of plant selenium hyperaccumulation. *Plant Biology* 14 (1): 1–10.

El Mehdawi, A.F., Quinn, C.F., and Pilon-Smits, E.A. (2011). Effects of selenium hyperaccumulation on plant–plant interactions: evidence for elemental allelopathy? *New Phytologist* 191 (1): 120–131.

El-Ramady, H., Abdalla, N., Alshaal, T. et al. (2015). Selenium and its role in higher plants. In: *Pollutants in Buildings, Water and Living Organisms* (ed. E. Lichtfouse, J. Schwarzbauer and D. Robert), 235–296. Springer.

El-Ramady, H., Abdalla, N., Taha, H.S. et al. (2016). Selenium and nano-selenium in plant nutrition. *Environmental Chemistry Letters* 14 (1): 123–147.

Epstein, E. (1994). The anomaly of silicon in plant biology. *Proceedings of the National Academy of Sciences* 91 (1): 11–17.

Epstein, E. and Bloom, A.J. (2005). *Mineral Nutrition of Plants: Principles and Perspectives*. Sunderland, Mass: Sinauer Associates. Inc.

Fageria, N.K. and Moreira, A. (2011). The role of mineral nutrition on root growth of crop plants. In: *Advances in Agronomy*, vol. 110, 251–331. Elsevier.

Feng, R., Wei, C., and Tu, S. (2013). The roles of selenium in protecting plants against abiotic stresses. *Environmental and Experimental Botany* 87: 58–68.

Foy, C., Chaney, R.T., and White, M. (1978). The physiology of metal toxicity in plants. *Annual Review of Plant Physiology* 29 (1): 511–566.

Foyer, C.H. (2018). Reactive oxygen species, oxidative signaling and the regulation of photosynthesis. *Environmental and Experimental Botany* 154: 134–142.

Foyer, C.H. and Noctor, G. (2003). Redox sensing and signalling associated with reactive oxygen in chloroplasts, peroxisomes and mitochondria. *Physiologia Plantarum* 119 (3): 355–364.

Foyer, C.H. and Noctor, G. (2016). Stress-triggered redox signalling: what's in pROSpect? *Plant, Cell & Environment* 39 (5): 951–964.

Foyer, C.H., Rasool, B., Davey, J.W., and Hancock, R.D. (2016). Cross-tolerance to biotic and abiotic stresses in plants: a focus on resistance to aphid infestation. *Journal of Experimental Botany* 67 (7): 2025–2037.

Fu, L.-H., Wang, X.-F., Eyal, Y. et al. (2002). A selenoprotein in the plant kingdom mass spectrometry confirms that an opal codon (UGA) encodes selenocysteine in *Chlamydomonas reinhardtii* glutathione peroxidase. *Journal of Biological Chemistry* 277 (29): 25983–25991.

García-Sánchez Ma, J., Jaime, M.P., Ramos, A. et al. (2000). Sodium-dependent nitrate transport at the plasma membrane of leaf cells of the marine higher plant *Zostera marina* L. *Plant Physiology* 122 (3): 879–886.

Ghanati, F., Morita, A., and Yokota, H. (2005). Effects of aluminum on the growth of tea plant and activation of antioxidant system. *Plant and soil* 276 (1-2): 133–141.

Giorgio, M., Trinei, M., Migliaccio, E., and Pelicci, P.G. (2007). Hydrogen peroxide: a metabolic by-product or a common mediator of ageing signals? *Nature Reviews Molecular Cell Biology* 8 (9): 722–728.

Gupta, K.J., Kumari, A., Florez-Sarasa, I. et al. (2018). Interaction of nitric oxide with the components of the plant mitochondrial electron transport chain. *Journal of Experimental Botany* 69 (14): 3413–3424.

Hajiboland, R. (2012). Effect of micronutrient deficiencies on plants stress responses. In: *Abiotic Stress Responses in Plants* (ed. P. Ahmad and M.N.V. Prasad), 283–329. Springer.

Halliwell, B. (2006). Reactive species and antioxidants. Redox biology is a fundamental theme of aerobic life. *Plant Physiology* 141 (2): 312–322.

Halliwell, B. and Gutteridge, J. (2007). Antioxidant defences: endogenous and diet derived. *Free Radicals in Biology and Medicine* 4: 79–186.

Han, D., Williams, E., and Cadenas, E. (2001). Mitochondrial respiratory chain-dependent generation of superoxide anion and its release into the intermembrane space. *Biochemical Journal* 353 (2): 411–416.

Hänsch, R. and Mendel, R.R. (2009). Physiological functions of mineral micronutrients (cu, Zn, Mn, Fe, Ni, Mo, B, cl). *Current Opinion in Plant Biology* 12 (3): 259–266.

Hartikainen, H. and Xue, T. (1999). The promotive effect of selenium on plant growth as triggered by ultraviolet irradiation. *Journal of Environmental Quality* 28 (4): 1372–1375.

Hasanuzzaman, M., Nahar, K., and Fujita, M. (2014). Silicon and selenium: two vital trace elements that confer abiotic stress tolerance to plants. In: *Emerging Technologies and Management of Crop Stress Tolerance* (ed. P. Ahmad and S. Rasool), 377–422. Elsevier.

Hasanuzzaman, M., Nahar, K., Rahman, A. et al. (2017). Actions of biological trace elements in plant abiotic stress tolerance. In: *Essential Plant Nutrients* (ed. M. Naeem, A.A. Ansari and S.S. Gill), 213–274. Springer.

Hasanuzzaman, M., Fotopoulos, V., Nahar, K., and Fujita, M. (2019). *Reactive Oxygen, Nitrogen and Sulfur Species in Plants*, 1–945. Wiley.

Hasegawa, P.M. (2013). Sodium (Na+) homeostasis and salt tolerance of plants. *Environmental and Experimental Botany* 92: 19–31.

Hawkesford, M., Horst, W., Kichey, T. et al. (2012). Functions of macronutrients. In: *Marschner's Mineral Nutrition of Higher Plants* (ed. P. Marschner), 135–189. Elsevier.

He, Y., Xiang, Y., Zhou, Y. et al. (2018). Selenium contamination, consequences and remediation techniques in water and soils: a review. *Environmental Research* 164: 288–301.

Heyno, E., Klose, C., and Krieger-Liszkay, A. (2008). Origin of cadmium-induced reactive oxygen species production: mitochondrial electron transfer versus plasma membrane NADPH oxidase. *New Phytologist* 179 (3): 687–699.

Hodson, M., White, P., Mead, A., and Broadley, M. (2005). Phylogenetic variation in the silicon composition of plants. *Annals of Botany* 96 (6): 1027–1046.

Horst, W., Wagner, A., and Marschner, H. (1982). Mucilage protects root meristems from aluminium injury. *Zeitschrift für Pflanzenphysiologie* 105 (5): 435–444.

Impa, S., Nadaradjan, S., and Jagadish, S. (2012). *Drought stress induced reactive oxygen species and anti-oxidants in plants*, 131–147. Abiotic Stress Responses in Plants. Springer.

Janků, M., Luhová, L., and Petřivalský, M. (2019). On the origin and fate of reactive oxygen species in plant cell compartments. *Antioxidants* 8 (4): 105.

Kaufman, P., Dayanandan, P., Takeoka, Y. et al. (1981). Silica in shoots of higher plants. In: *Silicon and Siliceous Structures in Biological Systems*, 409–449. Springer.

Kaur, N., Sharma, S., Kaur, S., and Nayyar, H. (2014). Selenium in agriculture: a nutrient or contaminant for crops? *Archives of Agronomy and Soil Science* 60 (12): 1593–1624.

Kaur, S., Kaur, N., Siddique, K.H., and Nayyar, H. (2016). Beneficial elements for agricultural crops and their functional relevance in defence against stresses. *Archives of Agronomy and Soil Science* 62 (7): 905–920.

Kelly, C.K. and Horning, K. (1999). Acquisition order and resource value in Cuscuta attenuata. *Proceedings of the National Academy of Sciences* 96 (23): 13219–13222.

Kisnierienė, V. and Lapeikaitė, I. (2015). When chemistry meets biology: the case of aluminium–a review. *Chemija* 26 (3): 148–158.

Knack, J., Wilcox, L., Delaux, P.-M. et al. (2015). Microbiomes of streptophyte algae and bryophytes suggest that a functional suite of microbiota fostered plant colonization of land. *International Journal of Plant Sciences* 176 (5): 405–420.

Kochian, L.V., Hoekenga, O.A., and Pineros, M.A. (2004). How do crop plants tolerate acid soils? Mechanisms of aluminum tolerance and phosphorous efficiency. *Annual Review of Plant Biology* 55: 459–493.

Kolbert Z, Feigl G. (2018) Cross-talk of Reactive Oxygen Species and Nitric Oxide in Various Processes of Plant Development: Past and Present. Reactive Oxygen Species in Plants:261.

Kopsell, D.A. and Randle, W.M. (1997). Selenate concentration affects selenium and sulfur uptake and accumulation byGranex 33'onions. *Journal of the American Society for Horticultural Science* 122 (5): 721–726.

Leterrier, M., Chaki, M., Airaki, M. et al. (2011). Function of S-nitrosoglutathione reductase (GSNOR) in plant development and under biotic/abiotic stress. *Plant Signaling & Behavior* 6 (6): 789–793.

Liang, Y., Sun, W., Zhu, Y.-G., and Christie, P. (2007). Mechanisms of silicon-mediated alleviation of abiotic stresses in higher plants: a review. *Environmental Pollution* 147 (2): 422–428.

Lima LW. (2016) Selenium and sulfur: mitigation in plant stresses.

Luis, A., Sandalio, L.M., Corpas, F.J. et al. (2006). Reactive oxygen species and reactive nitrogen species in peroxisomes. Production, scavenging, and role in cell signaling. *Plant Physiology* 141 (2):330-335.

Luis, A., Corpas, F.J., Barroso, J.B. et al. (2014). Function of peroxisomes as a cellular source of nitric oxide and other reactive nitrogen species. In: *Nitric Oxide in Plants: Metabolism and Role in Stress Physiology*, 33–55. Springer.

Lyi, S.M., Heller, L.I., Rutzke, M. et al. (2005). Molecular and biochemical characterization of the selenocysteine Se-methyltransferase gene and Se-methylselenocysteine synthesis in broccoli. *Plant Physiology* 138 (1): 409–420.

Ma, J.F. (2004). Role of silicon in enhancing the resistance of plants to biotic and abiotic stresses. *Soil science and plant nutrition* 50 (1): 11–18.

Ma, J.F. and Yamaji, N. (2008). Functions and transport of silicon in plants. *Cellular and Molecular Life Sciences* 65 (19): 3049–3057.

Ma, J.F. and Yamaji, N. (2015). A cooperative system of silicon transport in plants. *Trends in Plant Science* 20 (7): 435–442.

Malagoli, M., Schiavon, M., and Pilon-Smits, E.A. (2015). Effects of selenium biofortification on crop nutritional quality. *Frontiers in Plant Science* 6: 280.

Marschner, H. (2011). *Marschner's mineral nutrition of higher plants.* Academic press.

McKersie, B.D. and Lesheim, Y. (2013). *Stress and stress coping in cultivated plants.* Springer Science & Business Media.

Mehdawi, A.F.E. (2013). *Ecological aspects of plant selenium hyperaccumulation: effects of selenium hyperaccumulation on plant-plant interactions.* Libraries: Colorado State University.

Miyake, Y. and Takahashi, E. (1978). Silicon deficiency of tomato plant. *Soil Science and Plant Nutrition* 24 (2): 175–189.

Mou, Z., He, Y., Dai, Y. et al. (2000). Deficiency in fatty acid synthase leads to premature cell death and dramatic alterations in plant morphology. *The Plant Cell* 12 (3): 405–417.

Nagajyoti, P.C., Lee, K.D., and Sreekanth, T. (2010). Heavy metals, occurrence and toxicity for plants: a review. *Environmental Chemistry Letters* 8 (3): 199–216.

Neill, S.J., Desikan, R., and Hancock, J.T. (2003). Nitric oxide signalling in plants. *New Phytologist* 159 (1): 11–35.

Novoselov, S.V., Rao, M., Onoshko, N.V. et al. (2002). Selenoproteins and selenocysteine insertion system in the model plant cell system. *Chlamydomonas reinhardtii. The EMBO Journal* 21 (14): 3681–3693.

O'Brien, J.A., Daudi, A., Butt, V.S., and Bolwell, G.P. (2012). Reactive oxygen species and their role in plant defence and cell wall metabolism. *Planta* 236 (3): 765–779.

Ohnishi, J.-i., Flügge, U.-I., Heldt, H.W., and Kanai, R. (1990). Involvement of Na+ in active uptake of pyruvate in mesophyll chloroplasts of some C4 plants: Na+/pyruvate cotransport. *Plant Physiology* 94 (3): 950–959.

Palit, S., Sharma, A., and Talukder, G. (1994). Effects of cobalt on plants. *The botanical review* 60 (2): 149–181.

Pandey, R. (2015). *Mineral nutrition of plants*, 499–538. In: Plant biology and biotechnology. Springer.

Piacentini, D., Corpas, F.J., D'Angeli, S. et al. (2020). Cadmium and arsenic-induced-stress differentially modulates Arabidopsis root architecture, peroxisome distribution, enzymatic activities and their nitric oxide content. *Plant Physiology and Biochemistry* 148: 312–323.

Pilon-Smits, E.A. and Quinn, C.F. (2010). *Selenium metabolism in plants*, 225–241. Cell biology of metals and nutrients. Springer.

Pilon-Smits, E.A., Garifullina, G.F., Abdel-Ghany, S. et al. (2002). Characterization of a NifS-like chloroplast protein from Arabidopsis. Implications for its role in sulfur and selenium metabolism. *Plant Physiology* 130 (3): 1309–1318.

Pilon-Smits, E.A., Quinn, C.F., Tapken, W. et al. (2009). Physiological functions of beneficial elements. *Current Opinion in Plant Biology* 12 (3): 267–274.

Pirson, A. (1955). Functional aspects in mineral nutrition of green plants. *Annual Review of Plant Physiology* 6 (1): 71–114.

Potter, D.A., Powell, A.J., Spicer, P.G., and Williams, D.W. (1996). Cultural practices affect root-feeding white grubs (Coleoptera: Scarabaeidae) in turfgrass. *Journal of Economic Entomology* 89 (1): 156–164.

Pradhan, B., Baral, S., Patra, S. et al. (2020a). Delineation of gamma irradiation (^{60}Co) induced oxidative stress by decrypting antioxidants and biochemical responses of microalga, *Chlorella sp. Biocatalysis and Agricultural Biotechnology* 25:101595. doi:https://doi.org/10.1016/j.bcab.2020.101595.

Pradhan, B., Patra, S., Dash, S.R. et al. (2020b) Antioxidant responses against aluminum metal stress in Geitlerinema amphibium. *SN Applied Sciences* 2 (5):800. doi:https://doi.org/10.1007/s42452-020-2599-1.

Prashanth, L., Kattapagari, K.K., Chitturi, R.T. et al. (2015). A review on role of essential trace elements in health and disease. *Journal of Dr. NTR University of Health Sciences* 4 (2):75.

Rasmusson, A.G., Geisler, D.A., and Møller, I.M. (2008). The multiplicity of dehydrogenases in the electron transport chain of plant mitochondria. *Mitochondrion* 8 (1): 47–60.

Ren, Z.-H., Gao, J.-P., Li, L.-G. et al. (2005). A rice quantitative trait locus for salt tolerance encodes a sodium transporter. *Nature Genetics* 37 (10): 1141–1146.

Rengel Z. (1997) 11. Mechanisms of plant resistance to toxicity of aluminium and heavy metals. Mechanisms of environmental stress resistance in plants:241.

Rinalducci, S., Murgiano, L., and Zolla, L. (2008). Redox proteomics: basic principles and future perspectives for the detection of protein oxidation in plants. *Journal of Experimental Botany* 59 (14): 3781–3801.

del Rıo, L.A., Corpas, F.J., and Barroso, J.B. (2004). Nitric oxide and nitric oxide synthase activity in plants. *Phytochemistry* 65 (7): 783–792.

Rodríguez-Serrano, M., ROMERO-PUERTAS, M.C., Zabalza, A. et al. (2006). Cadmium effect on oxidative metabolism of pea (Pisum sativum L.) roots. Imaging of reactive oxygen species and nitric oxide accumulation in vivo. *Plant, Cell & Environment* 29 (8): 1532–1544.

Römheld, V. and Marschner, H. (1991). Function of micronutrients in plants. *Micronutrients in Agriculture* 4: 297–328.

Saini P, Gani M, Kaur JJ, Godara LC, Singh C, Chauhan S, Francies RM, Bhardwaj A, Kumar NB, Ghosh M. (2018). Reactive oxygen species (ROS): a way to stress survival in plants. In: Abiotic Stress-Mediated Sensing and Signaling in Plants: An Omics Perspective. Springer, pp 127–153.

Sandalio, L. and Romero-Puertas, M. (2015). Peroxisomes sense and respond to environmental cues by regulating ROS and RNS signalling networks. *Annals of Botany* 116 (4): 475–485.

Sarwar, N., Malhi, S.S., Zia, M.H. et al. (2010). Role of mineral nutrition in minimizing cadmium accumulation by plants. *Journal of the Science of Food and Agriculture* 90 (6): 925–937.

Scandalios, J. (2005). Oxidative stress: molecular perception and transduction of signals triggering antioxidant gene defenses. *Brazilian Journal of Medical and Biological Research* 38 (7): 995–1014.

Schlegel, H.G. and Jannasch, H.W. (2006). Prokaryotes and their habitats. *The prokaryotes* 1: 137–184.

Shafi, A. and Zahoor, I. (2020). *Abiotic and Biotic Stress-Induced Alterations in the Micronutrient Status of Plants*, 285–309. In: Plant Micronutrients. Springer.

Sharma, P., Jha, A.B., Dubey, R.S., and Pessarakli, M. (2012). Reactive oxygen species, oxidative damage, and antioxidative defense mechanism in plants under stressful conditions. *Journal of Botany* 2012: 1–26.

Shivaraj, S.M., Mandlik, R., Bhat, J.A. et al. (2021). Outstanding Questions on the Beneficial Role of Silicon in Crop Plants. *Plant and Cell Physiology* doi.org/https://doi.org/10.1093/pcp/pcab145.

Soetan, K., Olaiya, C., and Oyewole, O. (2010). The importance of mineral elements for humans, domestic animals and plants-a review. *African Journal of Food Science* 4 (5): 200–222.

Solaiman, A., Hossain, D., and Rabbani, M.G. (2007). Influence of Rhizobium inoculant and mineral nitrogen on some chickpea varieties. *Bangladesh Journal of Microbiology* 24 (2): 146–150.

Sommer, M., Kaczorek, D., Kuzyakov, Y., and Breuer, J. (2006). Silicon pools and fluxes in soils and landscapes—a review. *Journal of Plant Nutrition and Soil Science* 169 (3): 310–329.

Sors, T.G., Ellis, D.R., and Salt, D.E. (2005). Selenium uptake, translocation, assimilation and metabolic fate in plants. *Photosynthesis Research* 86 (3): 373–389.

Tamaoki, M., Freeman, J., Marquès, L., and Pilon-Smits, E. (2008). New insights into the roles of ethylene and jasmonic acid in the acquisition of selenium resistance in plants. *Plant Signaling & Behavior* 3 (10): 865–867.

Taylor, G.J. (1988). The physiology of aluminum phytotoxicity. *Metal Ions in Biological Systems: Aluminum and its Role in Biology* 24: 123–163.

Terry, N., Zayed, A., De Souza, M., and Tarun, A. (2000). Selenium in higher plants. *Annual Review of Plant Biology* 51 (1): 401–432.

Tripathi, D.K., Singh, V.P., Lux, A., and Vaculik, M. (2020). Silicon in plant biology: from past to present, and future challenges. *Journal of Experimental Botany* 71 (21): 6699–6702.

Van Hoewyk, D., Takahashi, H., Inoue, E. et al. (2008). Transcriptome analyses give insights into selenium-stress responses and selenium tolerance mechanisms in *Arabidopsis*. *Physiologia Plantarum* 132 (2): 236–253.

Vatansever, R., Ozyigit, I.I., and Filiz, E. (2017a). Essential and beneficial trace elements in plants, and their transport in roots: a review. *Applied Biochemistry and Biotechnology* 181 (1): 464–482.

Vatansever, R., Ozyigit, I.I., Filiz, E., and Gozukara, N. (2017b). Genome-wide exploration of silicon (Si) transporter genes, Lsi1 and Lsi2 in plants; insights into Si-accumulation status/capacity of plants. *BioMetals* 30 (2): 185–200.

Wang, Z.Q., Xu, X.Y., Gong, Q.Q. et al. (2014). Root proteome of rice studied by iTRAQ provides integrated insight into aluminum stress tolerance mechanisms in plants. *Journal of Proteomics* 98: 189–205.

Waraich, E.A., Ahmad, R., and Ashraf, M. (2011). Role of mineral nutrition in alleviation of drought stress in plants. *Australian Journal of Crop Science* 5 (6): 764.

Warm, E. and Laties, G.G. (1982). Quantification of hydrogen peroxide in plant extracts by the chemiluminescence reaction with luminol. *Phytochemistry* 21 (4): 827–831.

Watanabe, T., Jansen, S., and Osaki, M. (2005). The beneficial effect of aluminium and the role of citrate in Al accumulation in *Melastoma malabathricum*. *New Phytologist* 165: 773–780.

Watanabe, T., Jansen, S., and Osaki, M. (2006). Al–Fe interactions and growth enhancement in *Melastoma malabathricum* and *Miscanthus sinensis* dominating acid sulphate soils. *Plant, Cell & Environment* 29 (12): 2124–2132.

Watanabe, T., Misawa, S., Hiradate, S., and Osaki, M. (2008). Characterization of root mucilage from *Melastoma malabathricum*, with emphasis on its roles in aluminum accumulation. *New Phytologist* 178 (3): 581–589.

Wells, M.L., Potin, P., Craigie, J.S. et al. (2017). Algae as nutritional and functional food sources: revisiting our understanding. *Journal of Applied Phycology* 29 (2): 949–982.

Yemets, A.I., Karpets, Y.V., Kolupaev, Y.E., and Blume, Y.B. (2019). Emerging technologies for enhancing ROS/RNS homeostasis. In: *Reactive Oxygen, Nitrogen and Sulfur Species in Plants: Production, Metabolism, Signaling and Defense Mechanisms*, 873–922. Wiley.

Yoshii, Y. (1937). Aluminium requirements of solfatara-plants. *Shokubutsugaku Zasshi* 51 (605): 262–270.

Zhao, C., Ren, J., Xue, C., and Lin, E. (2005). Study on the relationship between soil selenium and plant selenium uptake. *Plant and Soil* 277 (1-2): 197–206.

Zhao, X.Q., Mitani, N., Yamaji, N. et al. (2010). Involvement of silicon influx transporter OsNIP2; 1 in selenite uptake in rice. *Plant Physiology* 153 (4): 1871–1877.

Zhu, J.-K., Hasegawa, P.M., Bressan, R.A., and Bohnert, H.J. (1997). Molecular aspects of osmotic stress in plants. *Critical Reviews in Plant Sciences* 16 (3): 253–277.

Zhu, Y.-G., Pilon-Smits, E.A., Zhao, F.-J. et al. (2009). Selenium in higher plants: understanding mechanisms for biofortification and phytoremediation. *Trends in Plant Science* 14 (8): 436–442.

Zhu, Z., Chen, Y., Shi, G., and Zhang, X. (2017). Selenium delays tomato fruit ripening by inhibiting ethylene biosynthesis and enhancing the antioxidant defense system. *Food Chemistry* 219: 179–184.

4

Biostimulant Effects and Concentration Patterns of Beneficial Elements in Plants

Libia I. Trejo-Téllez[1], *Libia F. Gómez-Trejo*[2], *and Fernando C. Gómez-Merino*[1]

[1] *College of Postgraduates in Agricultural Sciences, Laboratory of Plant Nutrition, Texcoco, State of Mexico, Mexico*
[2] *Department of Plant Protection, Chapingo Autonomous University, Texcoco, State of Mexico, Mexico*

Introduction

Beneficial elements have drawn great attention because of several positive effects they may have on plants. The current accepted list of beneficial elements includes aluminium (Al), cerium (Ce), cobalt (Co), iodine (I), lanthanum (La), sodium (Na), selenium (Se), silicon (Si), titanium (Ti), and vanadium (V) (Gómez-Merino and Trejo-Téllez 2018). Moreover, other non-essential trace elements, such as fluorine (F^{9}), chromium (Cr^{24}), arsenic (As^{33}), silver (Ag^{47}), cadmium (Cd^{48}), tungsten or wolfram (W^{74}), mercury (Hg^{80}), and lead (Pb^{82}), may also have beneficial effects on plants (Poschenrieder et al. 2013; Escobar-Sepúlveda et al. 2017; Tripathi et al. 2015), but they have not been sufficiently studied to arrive at a firm determination in this regard.

Despite their lack of essentiality for plants, these elements can stimulate adaptive responses to environmental stimuli and stressors, while enhancing numerous traits related to germination, biomass production, flower development, yield, and crop quality in a dose-response relationship. Concentration, rate, and method of application to match the needs of the crop are key factors in determining the success of beneficial elements supply in agriculture. In turn, the needs of the crops for the application of any beneficial element would depend on the genotype and the phenological stage of the cultivated species, agricultural management practices, and environment prevailing, among other critical factors. Thus, when properly managed, these elements may stimulate tolerance and resistance to biotic (e.g. insect pests, pathogenic fungi, oomycetes, bacteria, or viruses) and abiotic (e.g. drought, saline soils, nutrient deficiency, toxicity of heavy metals, or low temperatures) stresses. Such adaptive responses can be achieved because beneficial elements may induce changes in the biochemistry, metabolism, physiology, anatomy, and morphology of the plant, so that it can cope with challenging environments. Consequently, production, productivity, and quality traits can be enhanced in crops properly treated with these elements (Chatzistathis 2018; Gómez-Merino and Trejo-Téllez 2018).

Beneficial Chemical Elements of Plants: Recent Developments and Future Prospects, First Edition. Edited by Sangeeta Pandey, Durgesh Kumar Tripathi, Vijay Pratap Singh, Shivesh Sharma, and Devendra Kumar Chauhan.

Notably, beneficial elements can prompt hormesis, a natural phenomenon observed in a wide range of living organisms including bacteria, fungi, plants, animals, and humans. During hormesis, the application of a low or moderate dose of an environmental agent or condition results in an adaptive stimulatory/beneficial outcome, whereas the application of a high dose results in an inhibitory/detrimental outcome (Mattson and Calabrese 2010; Agathokleous et al. 2020). Consequently, it is critically important to determine the exact range of doses at which these elements exhibit positive effects on plants (Gómez-Merino and Trejo-Téllez 2018). Furthermore, uptake, transport, and accumulation of these elements in plant tissues are processes of primary importance to design better application approaches for beneficial elements in agriculture (Vatansever et al. 2016).

Herewith, we outline some of the most recent and salient advances in research concerning the positive effects of 10 chemical elements classified as beneficial for plants: Al, Ce, Co, I, La, Na, Se, Si, Ti, and V. In addition, we describe the concentration patterns of these elements in different tissues, analyzing the levels at which they trigger beneficial effects in plants and depicting the mechanisms explaining such effects with an extensive range of applications in crop production. In Table 4.1 we include a summary of the most prominent findings on the concentration patterns of beneficial elements found in different plant species published in recent years. This table includes the list of the species and cultivars used, the treatments applied, the positive effects found, the concentration of the beneficial elements recorded in different tissues (when available), and the corresponding references.

Aluminium

Just after oxygen (O) and silicon (Si), aluminium (Al) is the third most abundant element and the most abundant metal on the Earth's surface, accounting for approximately 8.23% of its mass (Lide 2017). Despite its abundance, Al does not show any essential functions in plants, but rather may exert stimulatory or detrimental effects, depending on factors such as its concentration and chemical form in the growth medium, the prevalent growth conditions, and genotypes of plants (Vatansever et al. 2016; Bojórquez-Quintal et al. 2017). Among the beneficial effects Al can trigger in plants are delayed senescence, increased nutrient absorption and utilization, higher enzymatic activity, enhanced root growth, and improved resistance to some environmental stressors (Moreno-Alvarado et al. 2017; Bojórquez-Quintal et al. 2017). Nevertheless, when the levels of Al in soils or nutrient media surpass the capacity of plants to overcome the challenge, Al may exhibit toxic effects, with limiting outcomes on plant productivity causing yield losses of 25 to 80%, depending on the crop (Sade et al. 2016); Al is present in about 40% of the areas with agricultural potential (de Jesus et al. 2017).

From the total absorbed Al of root cells, a range of 30–80% of it is mainly accumulated in root apoplast (Sade et al. 2016). Furthermore, from the total Al accumulated in the cell, 70–90% is found in the cell wall, making it more rigid. After entering the cell wall, Al may bind to the components of the cell membrane, moving into the symplast and giving rise to changes in the nutrient absorption, structure, and metabolism of the whole plant (Silva et al. 2020a).

Table 4.1 Concentration patterns of beneficial elements in different plant tissues.

Aluminium

Species and cultivar	Treatment	Positive effects	Al concentration (mg kg^{-1})	References
Camellia sinensis L.	50, 100, and 300 μM Al (hydroponic). Source: $AlCl_3$	Growth stimulation, higher photosynthesis rate, and increased antioxidant defence. Greater root surface area	Shoot: 489–635 Root: 2059–4416	Hajiboland et al. 2013
Drimia elata Jacq.	18.53 mM Al (500 mg L^{-1}) (hydroponic and quartz sand as substrate). Source: Al_3NO_3	Slight positive effect on total phenolic and flavonoid content. Good antibacterial activity against *Staphylococcus aureus*. Growth was not affected	Shoot: 639 Bulb: 419	Okem et al. 2015
Oryza sativa L. cv. Cotaxtla, Tres Ríos, Huimanguillo and Temporalero	200 μM Al (hydroponic). Source: $AlCl_3$	140% more root dry biomass, 30% taller than control plants. Sugar concentration increased. Synergic relationships with P, K, and Mg	Shoot: 38.3–55.1 Root: 581–810.4	Moreno-Alvarado et al. 2017
Vaccinium corymbosum L. cv. Cargo	200 μM Al (48 h Al exposure, hydroponic). Source: $AlCl_3$	Increased total chlorophyll, carotenoids, and protein concentration. Enhancement of antioxidant activity	Root: 3944 Leaves: 128.5	Cárcamo et al. 2019
Camellia sinensis cv. 'Longjing43'	0.2 to 1 mM Al (hydroponic)	Improved growth. Increased accumulation of P and K	Lateral root: 1356.5 to 2056.5 Main root: 360.9 to 478.3 Mature leaves: 1569.6 to 2165.2	Fan et al. 2019
Camellia japonicum	0.5 and 1 mM Al (hydroponic). Source: $AlCl_3$	Increased root biomass, chlorophyll, soluble sugars, and total soluble protein contents, as well as photosynthesis rate	1st/2nd mature leaves means: 956 and 1587 3rd/4th mature leaves means: 5859 and 7189 white fine roots: 8598 and 7731	Liu et al. 2020

Cerium

Species and cultivar	Treatment	Positive effects	Ce concentration (mg kg^{-1})	References
Zea mays L.	50 mg dm^{-3} (soil). Source: $Ce(NO_3)_3 \times 6H_2O$	Promotion of plant growth, chlorophyll index, and mycorrhizal colonization in tropical soil	Shoot: 0.52 Root: 28.32	Ferreira-Vilela et al. 2018
Oryza sativa L. cv. Morelos A-98	25, 50, and 100 μM (hydroponic). Source: $CeCl_3 \times 7H_2O$	Improved biomass production, root volume, number of tillers, and plant height	Shoot: 14.3–46.0 Root: 1952.7–4923.1	Ramírez-Olvera et al. 2018
Triticum aestivum L. cv. Bainong 207	10 mM under salt stress with 80 mM NaCl (hydroponic). Source: $Ce(NO_3)_3$	Protection of chloroplasts in seedlings against oxidative stress. Improved photosynthesis and plant growth under salt stress	Shoot: 2.2 (control without salt stress: 3.8) Root: 15.0 (control without salt stress: 18.5)	Chen and Shan 2019

Cobalt

Species and cultivar	Treatment	Positive effects	Co concentration (mg kg^{-1})	Reference
Brassica napus L. var. Pactol	12.5 mg Co kg^{-1} (soil). Source: $CoSO_4$	Stimulated growth, seed yield, oil yield, and quality	Seeds: 3.0	Gad 2010
Hordeum vulgare cv. Mill Giza 2000	169.7 μM (10 mg L^{-1}, soil watering)	Stimulated growth, dry matter content, yield, and quality	Shoot: 5.44	Gad et al. 2011
Glycine max L. Merr	12.0 mg Co kg^{-1} (soil with different N fertilizers). Source: $CoSO_4$	Enhanced growth, yield, and nutrient status, oil, and protein contents in seeds	Leaf mean: 7.56	Kandil et al. 2013

(*Continued*)

Table 4.1 (Continued)

Cobalt

Species and cultivar	Treatment	Positive effects	Co concentration (mg kg^{-1})	Reference
Zea mays L. cv. Giza 310 and Hi tick 2030	20 mg Co kg^{-1} (soil). Source: $CoSO_4$	Enhanced growth, yield, and its quality. Promotive effect on endogenous hormones, leaf water potential, proline content, yield, quantity, and quality under salinity condition. Promotion of tolerance to high soil salinity. Reduction of the harmful effect of salinity on maize plants	Corn grain: 3.66	Gad and El–Metwally 2015
Moringa oleifera Lam	127.3 to 212.1 μM Co (7.5-12.5 mg L^{-1}, irrigation once to soil with three organic fertilizers. Source: $CoSO_4$	Higher levels of proteins; total carbohydrates; total soluble solids; total phenolics; carotenoids; vitamin C; and N, P, K, Mn, Zn, Cu, Fe, and Co concentration in leaves	Leaf mean: 7.59	Gad et al. 2019
Paulownia trees	678.7 μM Co (40 mg L^{-1} soil watering). Source: $CoSO_4$	Increased vegetative growth parameters, photosynthetic pigments (chlorophyll a, b, and carotenoid), total carbohydrates, nitrogen, phosphorus, and potassium concentration in leaves	Seedling: 3.09	Hashish et al. 2019

Iodine

Species and cultivar	Treatment	Positive effects	I concentration (mg kg^{-1})	Reference
Lactuca sativa L. var. Phillipus	20 and 40 μM I (hydroponic, under salt stress 100 Mm NaCl). Source: KIO_3	Higher biosynthesis of phenolic compounds and lower oxidation under salt stress. Improved lettuce plant growth. Improved nutritional value for human diet	Leaves: 238.3	Blasco et al. 2013

Fragaria × *ananassa*	1.97 and 3.94 μM I 0.25 and 0.50 mg L^{-1} (hydroponic). Source: KI	Enhanced plant growth, higher contents of vitamin C and soluble sugar contents, reduced NO_3^- contents, and improved acidity of the fruits	Leaves: 77.4 and 108.5 Roots: 280.5 and 670.0 Fruits: 28.6 and 37.0	Li et al. 2016
Capsicum annuum L. var. Zidenka	10 and 15 μM I (foliar spraying, seedlings established in peat moss) Source: KI	Enhanced growth of seedlings and nutrient uptake	Seedling shoot: 93.4 and 106.3	Cortés-Flores et al. 2016
Ocimum basilicum L. cv. Superbo	1 mM I (soil watering). Source: KI	Biofortification. Moderate enrichment of antioxidant compounds in leaves	Leaves: 68.3	Kiferle et al. 2019
Ocimum basilicum L. cv. Tigullio and Red Rubin	0.1, 10, 50, 100, and 200 μM, supplied as KI or KIO_3 in a hydroponic system	Biofortification	Leaves: 4250 (KI) or 687 (KIO_3) in Tigullio, and 5433 (KI) or 717 (KIO_3) in Red Rubin	Incrocci et al. 2019

Lanthanum

Species and cultivar	Treatment	Positive effects	La concentration (mg kg^{-1})	Reference
Oryza sativa L. cv. Shengdao 16	0.1 mM La (agar medium). Source: La $(NO_3)_3$	Enhancement of seed germination and biomass accumulation	Shoot: 11.05	Liu et al. 2012
Oryza sativa L. cv. Shengdao 16	0.05 mM La (agar medium). Source: $La(NO_3)_3$	Promotion of root growth	Root: 566.52	Liu et al. 2013
Zea mays hybrid 2B604Hx	25 μM La (hydroponic)	Slight increase in growth, accompanied by a stimulated photosynthetic rate and chlorophyll index	Shoot: 47.3 Root: 3185.1	Duarte et al. 2018

(*Continued*)

Table 4.1 (Continued)

Lanthanum

Species and cultivar	Treatment	Positive effects	La concentration (mg kg^{-1})	Reference
Vigna angularis var. 'Jin No. 5'	0.4 mM La (sprayed twice). Source: $LaCl_3$	Promotion of growth. P acquisition in adzuki beans under P limitation by inducing changes in root morphology, acid phosphatase activity, and hydraulic conductivity	Seedling: 85.5	Lian et al. 2019

Selenium

Species and cultivar	Treatment	Positive effects	Se concentration (mg kg^{-1})	Reference
Lactuca sativa L. cv. Vera	2, 4, and 8 μM Se (hydroponic, nutrient solution). Source: Na_2SeO_4	Increased growth, shoot biomass production, Se content, and antioxidant activity	Shoot: 0.34 to 1.82	Ramos et al. 2010
Beta vulgaris L. cv. Fordhook Giant Swiss Chard	126.6 and 253.3 μM Se (four foliar applications). Source: Na_2SeO_4	Greater accumulation of proteins. Foliar applications did not affect the accumulation of nitrate	922 and 1408 mg per shoot	Hernández-Castro et al. 2015
Ocimum basilicum var. 'Red Rubin' and 'Dark Green'	316.6 and 633.2 μM Se m^{-2} (foliar application). Source: Na_2SeO_4	Selenium content increased after foliar fortification, without influencing essential oils content	2.31 and 7.86 (1st harvest, Red Rubin) 1.71 and 4.08 (1st harvest Dark Green)	Mezeyová et al. 2016
Medicago sativa	3.17 to 126.64 μM Se (hydroponic, nutrient solution). Source: Na_2SeO_3	Improved Se content and uptake of important minerals: K, Ca, Cu, Mn, and Mo	Shoot: 20.7 to 75.0	Guevara-Moreno et al. 2018

Valerianella locusta L.	633.2 μM Se (foliar application of 10 cm^3) 633.2 μM (soil application of 50 cm^3). Source: Na_2SeO_4	Stimulated plant growth, thermo-tolerance, and antioxidant enzymatic activity	Shoot (foliar Se application under normal temperature): 19.5 Shoot (soil Se application under heat stress): 28.0 Shoot (soil Se application under normal temperature): 56.5 Shoot (foliar Se application under heat stress): 61.0	Hawrylak-Nowak et al. 2018

Silicon

Species and cultivar	Treatment	Positive effects	Si concentration (g kg^{-1})	Reference
Oryza sativa 'Oochikara' (accumulator)	1.07 mM Si (hydroponic). Source: silica gel	Promoted Casparian band formation	Shoot: 29.66 Root: 1.754	Fleck et al. 2015
Zea mays cv. Helix (accumulator)			Shoot: 7.77 Root: 1.038	
Allium cepa Hercules I			Shoot: 0.339 Root: 0.819	
Zea mays L. cv. Reduta (accumulator)	5 mM Si (7 days, hydroponic). Source: K_2SiO_3	Increased the net accumulation of N, K, P, Mg, Ca, S, Cl, Mn, Fe, Mo, and B	Shoot: 10.294 Root: 14.702	Greger et al. 2018
Lactuca sativa L. cv. Amerikanischer brauner	1 mM Si (3 weeks, hydroponic). Source K_2SiO_3		Shoot: 0.887 Root: 1.844	
Triticum aestivum cv. Tjalve (accumulator)		Increased the net accumulation of Mg, Ca, S, Cl, Mn, Fe, and Mo	Shoot: 4.008 Root: 8.439	

(*Continued*)

Table 4.1 (Continued)

Sodium

Species and cultivar	Treatment	Positive effects	Na concentration (g kg^{-1})	Reference
Cynara cardunculus L. var. altilis DC cv. 'Bianco Avorio' and 'Gigante di Romagna'	30 mM Na (hydroponic). Source: NaCl	Increased the antioxidant activity as well as contents of cynarin, chlorogenic acid, luteolin, and polyphenols	Leaf means: 26.2	Colla et al. 2013
Cynara cardunculus L. subsp. *scolymus* cv. 'Romolo', 'Violetto di Provenza' and 'Violetto di Romagna'			Leaf means: 26.5	
Beta vulgaris L. cv. 'Gantang7'	50 mM Na (hydroponic under osmotic stress by 240 mM sorbitol). Source: NaCl	Strengthened resistance to osmotic stress by osmotic adjustment, Na^+, and proline accumulation	Shoot: 43.5 Root: 28.5	Wu et al. 2015
Salicornia bigelovii Torr.	100 mM Na (hydroponic)	Promoted growth	Shoot: 77.7	Yamada et al. 2015
Beta vulgaris L. subsp. cicla cv. Seiyou Shirokuki	Source: NaCl (96 mM) and $NaNO_3$ (4 mM)		Shoot: 74.3	
Beta vulgaris L. subsp. vulgaris cv. Detroit Dark Red			Shoot: 61.2	

Titanium

Species and cultivar	Treatment	Positive effects	Ti concentration (mg kg^{-1})	Reference
Brassica napus L cv. Chagal	5 g Ti ha^{-1} (doses divided into three foliar applications) (soil). Source: Mg-Tytanit (MgTi) containing 8.5 g dm^{-3} Ti. Ti is in the form of titanium ascorbate, whereas S (4%) and Mg (3%) contained in the biostimulant are in the form of $MgSO_4$	Increased seed and straw yield and oil production	Seed: 1.69 Straw: 11.27 Aerial phytomass (growing season): 53.6 Underground phytomass (growing season): 210.5	Kováčik et al. 2016

Lactuca sativa L.	25 mg Ti L^{-1} (watering to soil). Source: TiO_2	Stimulated plant growth and leaf quality of lettuce plant. Reduction in the content of nitrate in lettuce leaves	Shoot: 0.19	El-Ghamry et al. 2018
Solanum lycopersicum cv. 'Foria'	20.89 and 41.78 μM Ti (1 and 2 mg Ti L^{-1}) (hydroponic in peat)	Increased photosynthesis rate and ion uptake by increasing root volume	Leaves: 0.71 and 0.69	Haghighi and Daneshmand 2018

Vanadium

Species and cultivar	Treatment	Positive effects	V concentration (μg kg^{-1})	Reference
Equisetum arvense	No treatments were applied. Medicinal plants were collected from several areas in the Aninei Mountains of Romania. The collection sites believed to be unpolluted were located far away from roads and kilometres outside villages or towns	Positive effects were not determined	Aerial parts: 45 to 410	Antal et al. 2009
Hypericum perforatum			Aerial parts: 92 to 2060	
Origanum vulgare			Aerial parts: 61 to 1015	
Thymus pulegioides			Aerial parts: 263 to 76300	
Allium ursinum			Leaves: 43 to 1890	
Urtica dioica			Leaves: 133 to 14500	
Geum urbanum			Subterranean parts: 225 to 28700	
Valeriana officinalis			Subterranean parts: 72 to 1846	
Beta vulgaris	No treatments were applied. Plants were collected in four typical land-use districts in Panzhihua region, southwestern China		Leaf: 6500 to 42800	Teng et al. 2011
Vigna radiata			Leaf: 3000 to 22700	
Brassica rapa			Leaf: 26600 Root: 12600 Seed: 18800	

Tropical species such as buckwheat (*Fagopyrum esculentum*), Indian rhododendron (*Melastoma malabathricum*), teaplant (*Camellia sinensis*), Chinese silver grass (*Miscanthus sinensis*), cranberry (*Vaccinium macrocarpon*), or jolcham oak (*Quercus serrate*) are considered Al hyperaccumulators and detoxify Al internally by forming complexes with organic acids (Jansen et al. 2002; Vondráčková et al. 2015; Bojórquez-Quintal et al. 2017). In these species, the uptake of Al is a passive process stimulated by transpiration, which allows accumulation levels from 1 to 20 g Al kg^{-1} on a dry basis (Jansen et al. 2002; Poschenrieder et al. 2015).

Most plant species, however, are not Al hyperaccumulators but may exhibit stimulatory effects of Al in a hormetic manner. Thus, when applied at low doses, the stimulatory effects of Al may include enhanced plant growth (in both roots and shoots), improved nutrient uptake, alleviation of abiotic stress, and increased metabolism and antioxidant mechanisms of both enzymatic and non-enzymatic nature (Bojórquez-Quintal et al. 2017; Chauhan et al. 2021). A typical beneficial application of Al in horticulture is the modulation of sepal colours in hortensia (*Hydrangea macrophylla*). If the concentration of Al (expressed as μg Al g^{-1} fresh biomass) is between the range of 0 and 10, it produces red sepals; if it is increased to 10-40, then purple sepals may be observed; and finally, if Al concentration is higher than 40, blue sepals are brought about (Schreiber et al. 2011). The blue sepal colour is the result of a supramolecular metal-complex pigment in aqueous solution at pH 4.0, which consists of delphinidin 3-*O*-glucoside; 1, 5-*O*-caffeoylquinic acid; and 2 and/or 5-*O*-*p*-coumaroylquinic acid bound to trivalent aluminium ions (Al^{3+}) in a ratio of 1:1:1 (Oyama et al. 2015; Ito et al. 2018).

In growth media where availability of P is low, both divalent/trivalent iron ($Fe^{2/3+}$) and trivalent aluminium (Al^{3+}) ions may comparably contribute to stabilize the protein STOP1 (*S*ensitive *to P*roton rhizotoxicity *1*), which renders proton tolerance in plants grown in acid soils (Sawaki et al. 2009). STOP1 is a transcription factor whose accumulation in the nucleus results in the transcriptional activation of ALMT1 (*Al*uminium-activated *M*alate *T*ransporter *1*), which is involved in detoxification and nutrient security in plant cells (Sharma et al. 2016; Godon et al. 2019).

In the Brazilian cagaiteira tree (*Eugenia dysenterica*), the application of 200 μmol Al L^{-1} increased root growth, while the supply of 800 μmol Al L^{-1} stimulated the activation of enzymatic antioxidant mechanisms and the accumulation of phenolic compounds and maintained some photosynthetic activity attributes unchangeable (Almeida Rodrigues et al. 2019). In rice (*Oryza sativa*), the expression of *HtNHX1* (a $Na^+(K^+)/H^+$ antiporter of the Na^+-H^+ exchangers family) increased citrate acid secretion with a concomitant acidification of the rhizosphere and decreased K^+ efflux from the root, while the co-expression of *HtNHX1* and *HtNHX2* further improved rice growth under toxic levels of trivalent aluminium (Li et al. 2020b). Previously, in rice, Moreno-Alvarado et al. (2017) reported that Al (200 μM) enhanced growth and chlorophyll and sugar concentrations, with an evident increase in the contents of P, K, and Mg, while the expression of *NAC* transcription factors was differentially regulated by Al.

Since Al is the most prevalent metal in the Earth's surface, it is ubiquitous in all ecosystems, thus plants are usually exposed to a certain concentration of this metal, especially if they grow in tropical acidic soils. In such ecosystems, there is a plethora of evidence demonstrating that under certain circumstances, Al may elicit beneficial responses in plants, which will depend on factors related to the plant itself, the metal, and the environment (Pax et al. 2015).

Usually, Al mainly accumulates in underground tissues as compared to aboveground tissues. Thus, in tea plants, Al concentrations in root tissues fluctuate between 2059 and 4416 mg kg^{-1}, while in shoots its concentrations fluctuate from 489 to 635 mg Al kg^{-1} (Hajiboland et al. 2013). In satin squill (*Drimia elata*), bulbs concentrate 419 mg Al kg^{-1} and shoots 639 mg Al kg^{-1} (Okem et al. 2015). In rice, Al concentrations vary in the range of 581-810 mg kg^{-1} in roots; instead, in shoots, its concentrations fluctuate from 38 to 55 mg kg^{-1} (Moreno-Alvarado et al. 2017). In blueberry (*Vaccinium corymbosum*), there is an average of 3944 mg Al kg^{-1} in roots, while in shoots the Al concentration is 128.5 mg kg^{-1} (Cárcamo et al. 2019). In camellia (*Camellia japonicum*), roots concentrate between 7700 and 8500 mg Al kg^{-1}, while in leaves, Al concentrations fluctuate between 956 and 7189 mg Al kg^{-1}, and such concentrations are dependent on the position of the leaves in the plant (Liu et al. 2020) (Table 4.1).

Cerium

Among the lanthanide series, Cerium (Ce) is the most abundant element in the Earth's crust (Kotelnikova et al. 2019). It constitutes about 6.65×10^{1} mg kg^{-1} of the Earth's crust and 1.2×10^{-6} mg L^{-1} of the seawater (Lide 2017).

Light lanthanides have recently been shown to strongly impact the metabolism of methylotrophic bacteria, which may play pivotal roles in plant biology, affecting numerous processes and metabolic functions (Skovran et al. 2019). Methylotrophic bacteria are ubiquitous in nature, promoting the growth of various crop species since they may help fix atmospheric nitrogen (Masuda et al. 2018). For instance, the aerobic, facultative Ln^{3+}-utilizing methylotrophic and diazotrophic bacterium *Oharaeibacter diazotrophicus* strain SM30 (isolated from rice rhizosphere) harbours alpha-proteobacterial methylotrophy genes having La^{3+}, Ce^{3+}, Pr^{3+}, and Nd^{3+} as cofactors for the corresponding enzymatic activity (Lv and Tani 2018). Importantly, plant growth has a major impact on both the diversity and activity of methanol-utilizing methylotrophic bacteria in the soil system (Macey et al. 2020). Moreover, this diverse group of functional bacteria plays a crucial role in mitigating greenhouse gases that are causal agents of global climate change, which in turn contributes to improving ecological conditions and agricultural sustainability (Kumar et al. 2018).

As a beneficial element for plants, Ce has been proven to enhance various attributes in numerous plant species. In common bean (*Phaseolus vulgaris*), the application of 80 and 160 mg Ce^{3+} L^{-1} enhanced the content of Zn in pods and the content of malondialdehyde (MDA), as compared to the control (Xie et al. 2019). In legumes such as common bean, soya bean, and chickpea, the natural endosymbiotic bacterium *Bradyrhizobium diazoefficiens* is a crucial component of N fixation (Salas et al. 2020). This bacterium can utilize CH_3OH to grow in environments where light lanthanides such as La^{3+}, Ce^{3+}, praseodymium (Pr^{3+}), or neodymium (Nd^{3+}) are abundant, exhibiting enhanced enzymatic activity of methanol dehydrogenase (MDH) in the presence of CH_3OH/Ln^{3+} (Wang et al. 2019b). In tomato (*Solanum lycopersicum*), significant increases in root growth and biomass production were observed when applying 20-100 mg $CeCl_3$ L^{-1} (Singh et al. 2019). Also in tomato, the content of lycopene and the weight of fruits were significantly increased when applying cerium acetate, as compared to the control (Adisa et al. 2020). In common duckweed (*Lemna minor*), Ce ions affected growth and metabolism following a biphasic trend, with stimulatory effects at lower concentrations (i.e. 0.1 and 0.5 mM Ce^{3+}) and inhibitory effects at higher concentrations (1.0 mM Ce^{3+}) (Zicari et al. 2018).

Apart from the ionic form of Ce, there is plenty of data recently published concerning the effects of nanoceria or CeO_2 nanoparticles ($nCeO_2$) on plant biology (i.e. Hayder et al. 2020; Hu et al. 2020; Iftikhar et al. 2020). Nevertheless, we did not consider such data in this chapter since the effects of $nCeO_2$ have been proven different from those of conventional ionic Ce. Importantly, the significant reduction of size in nanomaterials renders a increase in their surface mass ratio, with a corresponding painful amount of particles moving gradually and constantly for more time (Subramanian et al. 2015; Monreal et al. 2016). Key factors determining the metabolism of nanoparticles within the plant cell (i.e. absorption, transport, and utilization) include genotypic variations both among and within species, phenological stage of the plant, environmental conditions where plants are established, as well as particular attributes of the nanoparticles themselves, such as physical, chemical, functional, and delivery properties and characteristics (Solanki et al. 2015).

As compared to Al, the concentrations of Ce found in plants are much lower. For instance, in maize (*Zea mays*), Ce concentration in roots is 28.32 mg kg^{-1}, whereas shoots concentrate 0.52 mg Ce kg^{-1} (Ferreira-Vilela et al. 2018). In rice plants, Ce concentrations are between 1952.7 and 4923.1 mg kg^{-1} in roots and between 14.3 and 46.0 mg kg^{-1} in shoots (Ramírez-Olvera et al. 2018). In wheat, root tissues have between 15 and 18.5 mg Ce kg^{-1}, whereas shoots have 2.2-3.8 mg Ce kg^{-1} (Chen and Shan 2019) (Table 4.1).

Cobalt

Cobalt (Co) is a relatively rare element in nature (Okamoto and Eltis 2011). According to Lide (2017), it constitutes 2.5×10^{1} mg kg^{-1} of the Earth's crust and 2×10^{-5} mg^{-1} of the seawater.

The beneficial effects of Co in plants are widely recognized. Indeed, for some groups of plants like legumes as well as marine protists such as crysophytes and dinoflagellates, Co is also considered as an essential element (Pérez-Espinosa et al. 2015). In legume root nodules, Co is utilized by symbiotic bacteria for the fixation of N_2 (Lange et al. 2016; Chatzistathis 2018). For instance, in pea (*Pisum sativum*), Co applications increased number and weight of nodules, which also improve nutrient use efficiency (Singh et al. 2012). In fenugreek (*Trigonella foenum-graecum*) inoculated with *Rhizobium tibeticum*, nodulation, glutamine synthetase, and nitrogenase activities, as well as biomass production, were dramatically enhanced when supplying 50 mg Co kg^{-1} (Abd-Alla et al. 2014). In the root nodules of alfalfa (*Medicago sativa*), another symbiotic nitrogen-fixing bacterium *Sinorhizobium meliloti* harbours an ABC-type cobalt importer, CbtJKL, which is of paramount importance for bacterial growth at trace Co concentrations in the growth medium (Cheng et al. 2011). In other non-legume species such as alder (*Alnus glutinosa*) (Callender et al. 2016), river oak (*Casuarina cunninghamiana*) (Echbab et al. 2004), and weeping willow (*Salix babylonica*) (Wang et al. 2020b), which also exhibit symbiotic associations with different beneficial bacteria, positive effects of Co have been recorded.

Cobalt may differentially affect plant responses, which depends on the levels and availability of the metal in the soil and the rhizosphere. Cobalt is usually absorbed by the plant's root system as a divalent cation (Co^{2+}) or as part of organic molecules (Chatzistathis 2018). Normal concentrations of Co in plants are from 0.1 to 10 μg g^{-1} on dry weight basis (Bakkaus et al. 2005).

In summer squash (*Cucurbita pepo*), growth, reproductive traits, and yield were significantly stimulated when soaking seeds in 0.25 ppm Co^{2+}, which was attributed to a potential increase in ethylene concentrations within the plant (Atta-Aly 1998). Nevertheless, Co functions as an inhibitor of ethylene too (Alarcón et al. 2009; Abdelsamad et al. 2019), since it hampers the activity of ACC synthase, which plays a pivotal role in ethylene biosynthesis (Schaller and Binder 2017). Indeed, in date (*Phoenix dactylifera*), applications of 1–3% Co to fruits induced them into entering the rutab stage (26-28 weeks turning brown) at a slow rate, consequently extending the marketing season and preserving good taste during handling (Abd El-Naby and Gad 2014). Furthermore, in mango (*Mangifera indica*), foliar applications of Co significantly reduced fruitlet abscission, while fruit weight, volume, and pulp weight, as well as fruit firmness, soluble solid concentration, and total sugars of fruits were increased (Wahdan 2011).

In tomato (*Solanum lycopersycum*), Co increased growth and yield (Gad and Kandil 2010), while also stimulating the enzymatic activity of superoxide dismutase (SOD) and catalase (CAT) (Li et al. 2007; Hasan et al. 2011). In cowpea (*Vigna unguiculata*), Co supply improved biomass production and productivity, while increasing the concentrations of the macronutrients N, P, K, and Mn, as well as N use efficiency (Siam et al. 2012). In cucumber (*Cucumis sativus*), applying 10-20 ppm Co significantly increased growth and yield, with 15 ppm Co being more effective than other treatments, thus indicating a hormetic dose-response effect (Gad et al. 2018). In onion (*Allium cepa*) grown in saline soils (3.3-5.5 dS m^{-1}), Co (10-20 ppm) significantly increased growth, yield, mineral composition, and bulb quality, especially when plants were treated with 12.5 ppm Co (Gad et al. 2020).

However, Co may also cause toxic effects to plants, and thus its applications must be tightly regulated, taking into consideration the principles of hormesis. Inhibition of PSII, impaired chlorophyll biosynthesis, decreased starch content, obstruction of CO_2 fixation, as well as mineral balance disruption are among the negative effects Co can cause when applied at doses beyond the homeostatic capacity of plants (Chatzistathis 2018). Importantly, plants exhibit different mechanisms to cope with toxic levels of Co. For instance, the ferroprotein IREG2/FPN2 transports Co^{2+} inside the vacuoles of root epidermal and cortical cells (Morrissey et al. 2009), while P_{1B}-ATPases are central to Co^{2+} homeostasis in all kingdoms of life, coupling ATP hydrolysis to the translocation of Co^{2+} and other metal ions across membranes (Smith et al. 2015). Since this element has a low mobility, the transport is limited from stems to leaves, while there is evidence of a high retention of Co in roots without xylem load (Chatzistathis 2018). The ways in which Co is accumulated and tolerated by plants is still not completely understood. However, since some species known as Co accumulators are also Cu accumulators, it has been deduced that there are some shared mechanisms (Lange et al. 2016).

Concentration patterns of Co in plants have been more explored as compared to other beneficial elements. In rape, Co concentrations reached 3 mg kg^{-1} in seeds (Gad 2010). In barley (*Hordeum vulgare*), an average concentration of 5.44 mg kg^{-1} was found in shoots

(Gad et al. 2011). In soya bean, Co concentration in leaves was 7.56 mg kg^{-1} (Kandil et al. 2013), while in moringa (*Moringa oleifera*) it was 7.59 mg kg^{-1} (Gad et al. 2019). In maize, a total of 3.66 mg Co kg^{-1} was reported (Gad and El-Metwally 2015). In paulownia (*Paulownia* spp.), Co concentration in seedlings was 3.09 mg kg^{-1} (Hashish et al. 2019).

Iodine

Iodine (I) is ranked 63rd based on its abundance in the Earth's crust, with 4.5×10^{-1} mg kg^{-1}, making it less plentiful than thulium (Tm), the least abundant rare-earth element (REE). Its concentration in the sea is 6×10^{-2} mg L^{-1} (Lide 2017; Zhang et al. 2018).

Though iodine is not considered a mineral nutrient for land plants, it plays a critical role for some aquatic plants and is indeed an essential trace element in the human diet, whose deficiency represents a significant health problem related with cognitive disabilities and certain types of cancer (Jha and Warkentin 2020). Hence, global efforts are being carried out to address I malnutrition and thus guaranty the adequate ingestion of this mineral in most affected and vulnerable groups, for example, by using different iodine fertilization techniques to biofortify crops (Medrano-Macías et al. 2016). Concomitantly, I is being used as a beneficial element to biostimulate plant physiology and metabolism, whose application causes hormetic dose-response curves.

Like any other beneficial element, I may improve growth and stimulate stress tolerance and antioxidant capacity in plants. However, it may also elicit neutral or even detrimental outcomes, depending on the plant species, environmental conditions, farming systems, concentrations applied, chemical forms used, and application methods (Medrano-Macías et al. 2016). Henceforth, we will focus on describing the beneficial effects iodine may cause in selected crop plants.

In celery (*Apium graveolens*) and Chinese cabbage (*Brassica rapa* ssp. *pekinensis*), I applied to soil increased leaf biomass (Dai et al. 2004). Furthermore, the latter species takes up I more effectively if the element is supplied as IO_3^- as compared to I^- if the concentration is low (<0.5 mg L^{-1}), but the tendency is reversed when the I concentration is 0.5 mg L^{-1} or higher (Weng et al. 2008a). In water spinach (*Ipomoea aquatica*), positive effects on plant growth were observed when applying low doses of iodine, with evident increases in vitamin C when using I^-, and negative effects with IO_3^-, while both forms of I increased nitrates content in leaves (Weng et al. 2008b). In radish (*Raphanus sativus*), higher concentrations of free amino acids were found when applying foliar KI as well as soil fertilization with KIO_3, while dry biomass, nitrites, nitrates, ammonium, total soluble sugars, and photosynthetic pigments were not affected by the treatments tested (Strzetelski et al. 2010). In lettuce (*Lactuca sativa*), concentrations ≤ 40 μM I^- did not affect but rather decreased NO_3^- concentrations in leaves, which substantially enhances the nutritional quality of this leafy vegetable (Blasco et al. 2010). Moreover, soil applications of KI (0.5, 1.0, and 2.0 kg I ha^{-1}) and foliar application of KIO_3 (0.0005, 0.005, and 0.05%) increased concentrations of K, Mg, Ca, Mn, and Cd in lettuce too (Smoleń et al. 2011). Synergic interaction between I and selenium (Se) for absorption through leaves via foliar spray were observed in lettuce (Smoleń et al. 2014a). In carrot (*Daucus carota*), I increased the concentrations of soluble solids and total sugars, as well as those of sucrose, fructose, and glucose, but did not change biomass production (Smoleń et al. 2014b). In soya bean (*Glycine max*), the application of 1 kg KIO_3 ha^{-1} significantly increased biomass production (Mao et al. 2014), as well as the concentrations of the enzymes SOD, glutathione reductase (GR), and ascorbate peroxidase (APX), in response to Cd exposure (Gupta et al. 2015). In tomato, dipping its fruits into an iodide-doped chitosan solution extended their shelf life (Limchoowong et al. 2016), while applying KIO_3 increased the concentrations of phenols, ascorbic acid, glucose, fructose, and total soluble solids (Smoleń et al. 2015). In melon (*Cucumis melo*), foliar applications of 1 mM KI L^{-1} once a week enhanced plant growth and gas exchange, without modifying the production of fruit, while foliar application of 2 mM KI L^{-1} resulted in higher antioxidant contents in leaves, no changes in growth and gas exchange, but decreased fruit production (Gordillo-Melgoza et al. 2016). Likewise, in sweet pepper (*Capsicum annuum*) seedlings, iodine stimulated growth and mineral concentrations, especially those of N and Ca (Cortés-Flores et al. 2016). In sweet basil (*Ocimum basilicum*), supply of 10 μM KI or 100 μM KIO_3 did not affect plant growth, whereas applying ≥ 50 μM KI reduced plant height and leaf area and dry matter production drastically decreased. In this study, the 'Red Rubin' variety exhibited greater tolerance to I, which was attributed to a higher capacity of this genotype to cope with higher concentrations of this element in leaves, instead of reducing the concentration of the element in the leaves (Incrocci et al. 2019).

According to Medrano-Macías et al. (2016), iodine effects on plants may be affected by interaction with other microelements, as well as by different physical, chemical, and biogeochemical environments that result in divergent dynamics of I in the environment (i.e. deposition, precipitation, absorption, or volatilization). Importantly, microbial communities may critically influence such dynamics, which in turn may affect carbon and nitrogen metabolism as well as vital functions in plants including photosynthesis, respiration, and photorespiration. In agronomic approaches, the source, ionic form, concentration, rate and technique of application, crop management, growth condition, and genotype are key factors affecting the final effect of iodine on crop plants. Therefore, negative, neutral, or positive responses can be observed given the hormetic effect it causes in plants.

Since I is widely used in biofortification approaches in order to enrich its content in different plant species, the concentration patterns it exhibits is of pivotal importance to delimit threshold levels for consumption of agricultural products such as roots, shoots, leaves, fruits, or seeds. Nevertheless, just a few studies have reported these data. In lettuce, leaves concentrated 238.3 mg kg^{-1} (Blasco et al. 2013). In strawberry (*Fragaria × ananassa*), I concentrations were as follows: 77.4 mg kg^{-1} in leaves, 280.5 mg kg^{-1} in roots, and 28.6 mg kg^{-1} in fruits (Li et al. 2016). In pepper, seedling shoots concentrated 93.4-106.3 mg I kg^{-1} (Cortés-Flores et al. 2016), while sweet basil leaves concentrated 68.3 mg I kg^{-1} (Kiferle et al. 2019) when supplied as KI in a fertigation system, and up to 5433 mg I kg^{-1} when supplied as KIO_3 in a hydroponic system with constant supply of iodine (Incrocci et al. 2019).

Lanthanum

Lanthanum (La) is the third most abundant lanthanide on Earth, after cerium (Ce) and neodymium (Nd) (Thornton and Burdette 2019). It constitutes 3.9×10^1 mg kg^{-1} of the Earth's crust and 3.4×10^{-6} mg L^{-1} of the seawater (Lide 2017).

La is one of the most studied beneficial elements, and its hormetic effects have been well characterized in plants. After analyzing more than 700 hormetic assays elicited by La, the geometric mean of biological response was found to be 142% of the control response, when applying 56 μM La, while the geometric mean concentration of the no-observed-adverse-effect-level (NOAEL) was observed at 249 μM La (Agathokleous et al. 2018, 2019). In consequence, no deleterious effects of La are to be expected if supplied in doses ranging from c. 50 to 250 μM. The ultimate response will depend on factors such as the genotypic variation of the plant species evaluated, pH of the growth media, source and ionic form, and time of exposure to La (García-Jiménez et al. 2017).

Depending on the ionic form in which La is supplied and metabolic changes it triggers in plant cells, this lanthanide may exhibit analogous functions to those played by Ca^{2+} and K^+ (d'Aquino et al. 2009). In Himalayan yew (*Taxus yunnanensis*), cell growth rate and taxol production were significantly increased when applying ≤46.2 μM La^{3+} (Wu et al. 2001). In desert broomrape (*Cistanche deserticola*), enhanced biomass and phenylpropanoid glycosides accumulation were obtained at ≤0.1 mmol La L^{-1} (Ouyang et al. 2003). In common bean, the application of 3.5-14.4 μmol $LaCl_3$ kg^{-1} increased yield by up to 10% as compared to the untreated control, while in spinach (*Spinacea oleracea*) treated with 3.6 μmol $LaCl_3$ kg^1, yield increased by 35.4% compared to the control (von Tucher and Schmidhalter 2005). In durum wheat (*Triticum durum*), the maximum biological response regarding root growth was recorded at 10 μM La (d'Aquino et al. 2009). In horseradish (*Armoracia rusticana*), chlorophyll content, photosynthetic rate, yield, and antioxidant activity were all enhanced with 35 μM La^{3+} (Wang et al. 2014).

Significant stimulation in wet weight was observed in rangpur lime (*Citrus limonia*) when treated with 50 mg $LaCl_3 \times 7H_2O$ (Turra et al. 2015). In rice, leaf area; fresh and dry weight of leaves; relative growth rate; contents of P, Ca, K, Mg, Mn, Fe, and Ni; and expression of ATPase genes displayed maximum biological response at 80 μM $LaCl_3 \times 6H_2O$ (Hu et al. 2016). In bell pepper (*C. annuum*), supplying 10 μM $LaCl_3$ enhanced seedling growth and metabolism, while stimulating flower differentiation (García-Jiménez et al. 2017). Flower differentiation is associated with an enhanced biosynthesis of cytokinins, which are key components of floral stimuli in plants (He and Loh 2002), while both La and Ce have been found to promote reproductive growth and flowering by stimulating cytokinins concentration in *Arabidopsis thaliana* (He and Loh, 2000). In ornamental species such as Easter lily (*Lolium longiflorum*) and tulip (*Tulipa gesneriana*), La has been proven to delay senescence by fostering antioxidant mechanisms and maintaining high water contents in cut flowers (Shan and Zhao 2015), while also influencing the geotropism of cut stems by reducing stem elongation but increasing stem diameter, thus preventing stem bending (Kim et al. 2005; Ramírez-Martínez et al. 2009). La also prolongs lifespan of cut tulip flowers by stimulating water consumption and concentrations of vital biomolecules such as chlorophylls, proteins, and sugars (Gómez-Merino et al. 2020a). Furthermore, La may ameliorate copper (Cu^{2+}) stressed rice by decreasing Cu^{2+} content in roots, stems, and leaves by 26.29, 59.46, and 55.56%, as compared to control plants, while maintaining the ultrastructure of mesophyll cells (Zhong and Chen 2020). Moreover, La and abscisic acid jointly regulate chlorophyll biosynthesis in switchgrass (*Panicum virgatum*) seedlings (He et al. 2020). However, La may also trigger toxic effects in plants, interfering with vital physiological processes including photosynthesis and growth (Hu et al. 2016; Siddiqui et al. 2019; Sun et al. 2019). In order to strengthen tolerance to La and other REE, plants encode vitronectin-like proteins (VN), which are localized outside plant plasma membranes. VN harbour active domains that combine strongly with REE, acting defence agents to cope with toxic La^{3+} concentrations (Yang et al. 2016). Low La^{3+} concentrations may modulate the VN molecular conformation, while high La^{3+} doses are fairly bound to VN and conform steady La-VN structures, with a consequent disruption of the molecular conformation of VN (Wang et al. 2017). In hormetic terms, with 30 μM La^{3+}, initial endocytosis increases extracellular VN contents by 5.46 times, keeping plasma membranes unchangeable; with 80 μM La^{3+} enhanced endocytosis increases extracellular VN contents by 119 times (Cheng et al. 2018b). Furthermore, clathrin-mediated endocytosis (CME) of extracellular La^{3+} cargoes requires extracellular arabinogalactan proteins (AGPs) that are anchored on the outer face of the plasma membrane and colocalized with La^{3+} on the cell surface and in La^{3+}-induced endocytic vesicles in *Arabidopsis* leaf cells (Wang et al. 2019a).

As a transition metal, La exhibits typical hormetic dose-response curves, while plants are capable of metabolizing significant amounts of this element before manifesting negative effects. From an agronomic perspective, the capacity of La to trigger hormesis represents an advantageous property to increase metabolic pathways involved in enhancing crop production, productivity, and quality, as well as stimulating internal mechanisms to cope with environmental stressors of an abiotic nature. All the aforementioned issues, together with detailed data on the impact of the application of this element on the environment will be of prime importance in forthcoming research.

Lanthanum concentration patterns also differ among plant species and tissues analyzed. In rice, Liu et al. (2012) reported 11.05 mg La kg^{-1} in shoot tissues, while in roots, Liu et al. (2013) found 556.5 mg La kg^{-1}. In maize, Duarte et al. (2018) reported 47.3 mg La kg^{-1} in shoots and 3185.1 mg La kg^{-1} in roots. In azuki bean (*Vigna angularis*), Lian et al. (2019) observed 85.5 mg La kg^{-1} in seedlings. The corresponding details of all these experiments are described in Table 4.1.

Selenium

Selenium (Se) is the 30th most bountiful metalloid in the Earth's crust with 0.05–0.5 mg kg^{-1} (Tan et al. 2016; Lide 2017). In seawater, it has a concentration of 2×10^{-4} mg L^{-1} (Lide 2017). Se is located in group VIa of the periodic table, between sulphur (S) and tellurium (Te), and thus displays chemical attributes analogous to them both (Pilbeam et al. 2015).

In diverse organisms including bacteria, archaea, and eukaryotes, Se has been proven to be essential (Araie and Shiraiwa 2016). Among eukaryotes, Se is essential for algae, animals, and humans, though its essentiality in higher plants has not been demonstrated yet (Malagoli et al. 2015; Garousi 2017). However, it is estimated that anywhere from 500 million to 1 billion people worldwide are affected by Se deficiency, due to inadequate intake (Shreenath et al. 2020). Selenium deficiencies in animals and humans result in health affections such as oxidative damage, fertility disorders, autoimmune thyroiditis, and carcinogenesis, among others (Hatfield et al. 2014; Qazi et al. 2018). Since plants are capable of accumulating Se in different tissues, agronomic biofortification (i.e. biofortification to augment Se contents in food crops) with the application of Se-containing fertilizers has been the focus of great attention worldwide. Furthermore, genetic biofortification through the selection of outstanding varieties and breeding approaches may also bring about enhanced Se-accumulation capacity in food crops (Broadley et al. 2006; Schiavon et al. 2020).

Thanks to its chemical similarity to S, Se uptake and translocation with plant cells is under the control of specialized S transporters located on the membranes, while its incorporation into cysteine (Cys) and methionine (Met) takes place via sulphur assimilatory metabolism, and its volatilization into the atmosphere takes place via the soil-plant system (Pilbeam et al. 2015; Gupta and Gupta 2017). However, since Se and S compete for the same transporters and metabolic pathways in order to be incorporated into molecules such as Cys and Met, a higher risk of Se toxicity is to be expected if S availability is not efficient (White et al. 2004). Indeed, high amounts of Se cause toxic symptoms in plants, with a consequent formation of selenoproteins, oxyproteins, and nitroproteins due to Se stress (Kolbert et al. 2019).

Once absorbed and while maintaining homeostasis, Se can be translocated from vegetative organs (i.e. roots, stems, leaves) to reproductive ones (flowers, fruits, seeds) (Galeas et al. 2007), while remobilization within plant tissues may also take place (Keskinen et al. 2010; Shinmachi et al. 2010).

Selenium essentiality has not been proven for plants, albeit it can improve growth and other productive attributes if properly managed (Pilbeam et al. 2015). Furthermore, when applied at low concentrations, Se elicits improved responses in plants to cope with environmental cues causing abiotic stress, such as dehydration, low temperatures, and heavy metals, among others (Gupta and Gupta 2017). In wheat (*Triticum aestivum*), Se has been proven to protect the plant against cold stress (Chu et al. 2010). In rye grass (*Lolium perenne*), Se mediated the reduction of the superoxide anion ($\bullet O_2^-$) to peroxide (H_2O_2) induced by Al stress (Cartes et al. 2010). In Cd-stressed rape (*Brassica napus*) (Filek et al. 2008), sorghum (*Sorghum bicolor*) exposed to heat stress (Djanaguiraman et al. 2010), and mung bean (*Phaseolus aureus* [synonym *Vigna radiata*]) stressed with As (Malik et al. 2012), reductions in ROS accumulation were observed after application of Se. In rapeseed (*B. napus* subsp. *napus*) seedlings, Se pretreatment upregulated antioxidant and detoxification systems and conferred improved tolerance to dehydration (Hasanuzzaman and Fujita 2011). In recalcitrant seeds of sugar maple (*Acer saccharium*), Se played a protective role during desiccation (Pukacka et al. 2011). In the intertidal red macroalga *Gracilaria dura* (Gracilariaceae, Rhodophyta), the application of Se in combination with spermine elicited DNA methylation and antioxidant mechanisms and thus alleviated toxic effects induced by Cd toxicity (Kumar et al. 2012). In rice seedlings exposed to As stress, Se and auxin improved growth; concentrations of MDA, proteins, and chlorophylls (stress molecular signals); and the contents of Cys and proline (Pro) (stress molecular modulators) (Pandey and Gupta 2015). In cucumber exposed to Cd stress, Se improved dry weight of roots and shoots, photosynthesis, stomatal conductance, and reduced Cd uptake (Haghighi et al. 2016). In maize exposed to oxidative stress induced by H_2O_2, Se kept homeostatic levels of cytosolic Ca^{2+}, while maintaining pollen grain germination parameters unchangeable (Del Pino et al. 2019).

Thus, Se biofortification and its beneficial effects in plants are of paramount importance in current and future agricultural approaches, albeit the exact mechanism behind the effect; the linkage between S and Se biochemistry and metabolism; the differential capacities of plants to accumulate, tolerate, and detoxify Se; and the screening of safer limits of Se for human consumption need to be further investigated.

Se deficiency as a major global health issue has been widely addressed over the years through different biofortification approaches aimed at alleviating such deficiency. In lettuce, Ramos et al. (2010) found a maximum Se concentration of 1.82 mg kg^{-1} in shoot tissues. In Swiss chard (*Beta vulgaris*), Se concentrations varied between 922 and 1408 mg kg^{-1} in shoots (Hernández-Castro et al. 2015). In basil leaves, Se concentrations reached 7.86 mg kg^{-1} in the cultivar Red Robin and 4.08 mg kg^{-1} in the cultivar Dark Green (Mezeyová et al. 2016). In alfalfa, Guevara-Moreno et al. (2018) reported Se concentrations between 20.7 and 75.0 mg kg^{-1}. In salad (*Valerianella locusta*), Se concentrations in shoots ranged from 19.5 to 61.0 mg kg^{-1}, which was dependent on the kind of stress plants were subjected to (Hawrylak-Nowak et al. 2018).

Silicon

Silicon (Si) is the second most bountiful element on Earth, just after oxygen (O) (Jurkowska and Świerczewska-Gładysz 2020). Its concentration is 2.82×10^5 mg kg^{-1} in the Earth's crust and 2.2 mg L^{-1} in seawater (Lide 2017).

Si is considered a beneficial and a *quasi*-essential element for plants, and by far, it is the beneficial element most studied until now. Since silicon dioxide (SiO_2) comprises more than 50% of soils, all plants that are in contact with soil through their root systems contain a certain amount of Si in their tissues (Ma 2015). Together with silicon dioxide, silicate minerals and aluminosilicates make up most silicon compounds found in the soil system. Such minerals are inert and therefore, they are unavailable for plant uptake. Indeed, plants can take up and transport silicon within the cells only in the form of monosilicic/orthosilicic acid [$Si(OH)_4$ or H_4SiO_4], whose concentration is low (i.e. 100–600 μM) in the soil solution of most agricultural fields (Gropper and Smith 2018; Laane 2018). Once absorbed by plant roots through different Si channels located at the plasma membrane, Si is transported inside the plant, with the apoplast space being the main nonprotoplasmic component where Si is largely deposited outside the plasma membrane. In plant tissues, Si concentrations may vary from 0.1 to 10% on a dry weight basis (Epstein 1994). Such variation range in Si contents in different plant tissues is dependent on the genotypic diversity of plants, as well as on the properties of the soil from which Si is taken up by root cells (Coskun et al. 2019). Importantly, rice, wheat, barley, sugarcane (*Saccharum* spp. hybrids), soya bean, and sugar beet (*B. vulgaris* subsp. *vulgaris*) are among the most cultivated crop species worldwide, which have been categorized as Si accumulators (Guntzer et al. 2012; Elsokkary 2018; Artyszak et al. 2019). Interestingly, exogenous Si applications elicit improved production, productivity, and quality traits in the vast majority of these species (Tubana et al. 2016).

In sugarcane, increased yield, biomass, sugar content, and resistance against stress have been reported when using stabilized orthosilicic acid (Solomon et al. 2016). Furthermore, Si fertilization reduces the deleterious effects of water deficit in sugarcane (Camargo et al. 2017). Indeed, most monocots such as wheat, rice, and barley significantly benefit from Si applications, and they may contain 10-100 g Si kg^{-1} on a dry weight basis (Tubana et al. 2016). In particular, Si increased cell wall thickness and lignification of root tips in rice plants exposed to excessive Fe supply (Mehrabanjoubani et al. 2019). Moreover, Si contents in rice grains were significantly reduced when plants were treated with a combination of biochar and ash obtained from silicon-rich rice husk burning (Leksungnoen et al. 2019). In wheat, organosilicon fertilizers stimulated Si absorption in roots and shoots, while decreasing Cd and Pb concentrations in both vegetative and reproductive tissues (Huang et al. 2019). In barley, monosilicic acid stimulated a more effective denitrification and thus contributed to reduce the emission of greenhouse gases in the atmosphere (Włodarczyk et al. 2019). In barley exposed to osmotic stress, the expression of the Si transporter *HvLsi1* and *HvLsi2* increased Si concentration in shoots and contributed to the homeostatic regulation of starch, abscisic acid (ABA), and potassium metabolism (Hosseini et al. 2017). Furthermore, in barley exposed to Cr stress, Si alleviated anatomical and functional disruptions in both root and leaf tissues, demonstrating that Si is a good candidate for Cr detoxification (Ali et al. 2013). In dicots such as soya bean, Si applied as foliar spray under normal light and shading significantly improved photosynthetic efficiency and growth parameters, as well as the contents of molecules such as chlorophylls, soluble sugars, and lignin, enhancing stem strength and yield (Hussain et al. 2020). By efficiently regulating reactive O and reactive N species metabolism, Si may also result in enhanced resistance to salt stress in soya bean (Chung et al. 2020; Dhiman et al. 2021). In sugar beet, Ca and Si foliar fertilization increased root yield, biological sugar yield, and technological yield of sugar, compared with the control (Artyszak et al. 2015). Furthermore, Si foliar fertilization induced higher concentrations of P in leaves and those of N and K in roots, though sugar yield is not affected by Si (Artyszak et al. 2019). According to the 'apoplastic obstruction hypothesis' proposed by Coskun et al. (2019), Si plays a key role in the induction of plant prophylaxis aimed at conferring tolerance or resistance against diverse biotic and abiotic stress factors. Recently, numerous reviews have evidenced the beneficial effects of Si in plants, with novel models explaining its action (Ahanger et al. 2020; Gaur et al. 2020; Gómez-Merino et al. 2020a; Kreszies et al. 2020; Mandlik et al. 2020; Sheng and Chen 2020; Trejo-Téllez et al. 2020; Vaculík et al. 2020; Katz et al. 2021; Ramírez-Olvera et al. 2021; Schaller et al. 2021; Shivaraj et al. 2021).

By far, Si is the most studied beneficial element in plants, and thus it is referred to as a *quasi*-essential element. Nonetheless, in technical terms, determining the essentiality of Si in plant biology has become a challenging task. This is because Si has a crustal abundance of 28.5%, making it the second most bountiful element on the Earth's surface, and thus it is a common ubiquitous contaminant for laboratory analyses. Consequently, Si is hard to be completely remove from plant growth media and laboratory instruments (Epstein 1994; Coskun et al. 2019). In our search, we found two studies reporting Si concentrations in plant tissues in different species. For instance, Si concentrations in roots ranged from approximately 0.82 g kg^{-1} in onion to c. 1.8 g kg^{-1} in rice, with maize exhibiting 1.04 g Si kg^{-1}; in shoot tissues, onion had roughly 0.34 g Si kg^{-1}, rice 29.66 g Si kg^{-1}, and maize 7.77 g Si kg^{-1} (Fleck et al. 2015). In shoots of maize, lettuce, and wheat, the concentrations were 10.29, 0.88, and 4.01 g Si kg^{-1}, respectively, whereas root tissues exhibited 14.70, 1.84, and 8.44 g Si kg^{-1} in the same corresponding species, on a dry basis, respectively (Greger et al. 2018).

Sodium

Sodium (Na) is one of the most bountiful elements on Earth. It has an even distribution and can be considered as practically infinite in the ocean (Chen et al. 2019). It has a concentration of 2.36×10^4 mg kg^{-1} in the Earth's crust and of 1.08×10^4 mg L^{-1} in seawater (Lide 2017).

Like potassium (K), Na is classified as an alkali metal that belongs to group V in the periodic table of the chemical elements, though they both display different significance in the biosphere. Potassium is a major macronutrient for plants, while sodium is considered a beneficial element for some taxa at low levels of supply under certain conditions, but a toxic element for other plants if applied at high concentrations (Schubert 2015). Importantly, in some plant species, the role of K^+ in the osmotic mechanism of stomata aperture modulation can be replaced by Na^+ (Subbarao et al. 1999; Shabala and Mackay 2011; Gattward et al. 2012). Moreover, for some C4 species (i.e. *Atriplex portulacoides*, *Kochia childsii*, and *Amaranthus tricolor*), Na is considered essential (Redondo-Gómez et al. 2007; Kronzucker et al. 2013). When those C4 species face Na deficiency, part of the photosynthetic activity is disrupted (Grof et al. 1989; Johnston et al. 1989), while NO_3^- uptake and assimilation can be enhanced when Na is applied (Ohta et al. 1989). Furthermore, in the mesophyll cells of C4 species, Na facilitates the synthesis of phosphoenolpyruvate (PEP) from pyruvate, prior to the Calvin–Benson cycle to fix carbon during photosynthesis (Johnston et al. 1988; Kaur et al. 2015), and Na also promotes the translocation of pyruvate into chloroplasts by the action of bile acid sodium symporters (BASS) localized at the chloroplast envelope membrane (Nieves-Cordones et al. 2016). In Na^+-deficient soils, C4 species belonging to the families Cyperaceae, Amaranthaceae, and Chenopodiaceae exhibit symptoms of Na deficiency such as necrosis and chlorosis of leaves, poor growth, and decreased flowering, which can be corrected by supplementing 100 μM Na^+ (Johnston et al. 1988). For instance, in some species of the family Amaranthaceae, maximum growth and increased biomass production were observed between the ranges of 80 and 200 mol Na m^{-3}. Furthermore, Na also promoted a more balanced water content because of the osmotic pressure caused by this element (Yamada et al. 2016). In CAM species, the enzyme PEP carboxylase kinase that regulates the activity of PEP carboxylase is activated by Na (Kronzucker et al. 2013). In the halophytic C3 shrub glaucous glasswort (*Arthrocnemum macrostachyum*), the application of 171-510 mm NaCl elicited significant growth and increase in the specific shoot area (Redondo-Gómez et al. 2010).

Plants have evolved and developed unique mechanisms to absorb, metabolize, perceive, and transduce signals triggered by Na^+ ions. In root cells, Na^+ is absorbed by K^+ transporters and subsequently translocated, accumulated in shoot tissues, and loaded into the xylem in a NO_3^- dependent manner (Raddatz et al. 2020). Hence, an adequate balance among K^+-Na^+, Na^+-NO_3^-, and K^+-NO_3^- transport, uptake, and allocation is crucial to maintain plant homeostatic growth and sustain crop productivity. Sodium contributes to the maintenance of osmotic balance in plants by stimulating the synthesis of osmoprotectant like proline and glycine-betaine, which in turn leads to amides and polyamines accumulation (Pilon-Smits et al. 2009). Polyamides are key factors affecting diverse vital processes including growth and tolerance or resistance mechanisms upon environmental stressors, stimulating adaptive responses such inhibition of oxidative stress; membrane stabilization;

binding to polysaccharides, phospholipids, proteins, and DNA; protein phosphorylation; and post-transcriptional modifications (Khare et al. 2018).

In tomato, potato (*Solanum tuberosum*), carrot (*D. carota*), cacao (*Theobroma cacao*), and numerous cereals, Na has also demonstrated beneficial effects on plant growth and metabolism, which are observed within concentrations higher than those found in macronutrient ions including nitrate, ammonium, and potassium, with maximum levels found in halophytes (Flowers and Colmer 2008). For instance, Na can positively affect metabolism and sensorial/nutritional value of tomato fruits (D'Amico et al. 2003), including total soluble content (°Brix) and titratable acidity, while stimulating the defence system of the plant through an enhanced accumulation of secondary metabolites (Moles et al. 2019). Indeed, Na can accelerate lycopene biosynthesis of tomato fruits in hydroponics (Wu and Kubota 2008).

In the succulent coastal groundcover plant karkalla (*Carpobrotus rossii*) exposed to Cd stress, NaCl supply enhanced plant growth and decreased Cd uptake and accumulation in shoots, while the synthesis of organic acids increased in shoots (Cheng et al. 2018a). In sea lavender (*Limonium bicolor*), appropriated supply of Na_2SO_4, KCl, or NaCl substantially increased root elongation, with Na^+ playing a critical role in halotropism and gravitropic setpoint angle, while endogenous concentrations of indoleacetic acid (IAA) were dramatically higher in seedlings exposed to NaCl as compared to the untreated controls (Leng et al. 2019). In sword-leaf dogbane (*Apocynum venetum*), supplying 25 mM NaCl notably enhanced Na^+ accumulation in leaves and improved the relative water content, with a concomitant higher photosynthesis and increased plant growth (Cui et al. 2019).

In black locust (*Robinia pseudoacacia*) grown in Pb–Zn polluted soils, the application of sodium sulphide (Na_2S) increased biomass production of shoots, while decreasing Pb and Zn bioavailability and uptake by plants (Zhang et al. 2020b).

In summary, Na is not essential for most plant taxa, though it can be beneficial in numerous conditions, particularly in K-deficient environments, and thus it is regarded as a 'functional' element. The mechanisms of uptake and distribution of Na play key roles in mitigating salt stress and in optimizing its uses as a functional element. Therefore, further research is needed to identify the optimal concentration of Na in plant tissues, while exploring the exact mechanisms that regulate transcriptional and transport activities.

Concentration patterns of Na in different plant species have also been explored. For instance, in cardoon (*Cynara cardunculus*), leaves had 26.2 g Na kg^{-1} (Colla et al. 2013). In sugar beet, Na concentration in roots was 28.5 and in shoots, it was 43.5 g kg^{-1} (Wu et al. 2015). In dwarf saltwort (*Salicornia bigelovii*) shoots, Yamada et al. (2015) reported 77.7 g Na kg^{-1}, while in chard shoots, Na concentration was 74.3 and in beetroot, it was 61.2 mg Na kg^{-1}.

Titanium

Since titanium (Ti) is the ninth most bountiful element in the Earth's crust, it is considered as almost inexhaustible. Ti is also the fourth most bountiful metallic element just after Al, Fe, and Mg (Zheng et al. 2019). It constitutes 5.65×10^3 mg kg^{-1} of the Earth's crust and 1×10^{-3} mg L^{-1} of the seawater (Lide 2017).

Titanium has been categorized as a beneficial element for plants. When applied at low concentrations, Ti may enhance productivity and quality attributes of crop plants by improving the enzymatic activity of various proteins, stimulating the biosynthesis of certain biomolecules involved in photosynthesis and other processes, conferring better nutrient uptake and metabolism, and eliciting improved responses to different types of stress (Lyu et al. 2017). In grape (*Vitis vinifera*), foliar sprayed Ti improved the accumulation of macronutrients (i.e. N, P, K, Ca, and Mg) and micronutrients (Fe, Zn, and Cu) and Ti in leaf tissues, while flesh and skin of fruits exhibited increased concentrations of Ca, Fe, Zn, and Cu, demonstrating that Ti affects mineral absorption, translocation, and assimilation processes (Alcaraz-López et al. 2005). In barley, wheat and potato cultivated in field conditions under two regimes of N supply (reduced or optimum N supply), the N content of plants significantly affected the responses to Ti foliar sprays, while in N-deficient plants, the application of Ti had no effect (Tlustoš et al. 2005). In geranium (*Pelargonium* x *hortorum*), application of 0–100 mg Ti-ascorbate L^{-1} as foliar sprays increased plant growth and quality (Whitted-Haag et al. 2014). In Timothy grass (*Phleum pratense*), foliar application of 0.2–0.8 L Tytanit (0.85% Ti complex) ha^{-1} increased seed yield, grain weight, and seed germination (Radkowski et al. 2015). In tomato, applications of up to 960 g ha^{-1} a year of Ti significantly increased Fe, Mn, and Zn uptake and lycopene content (Markiewicz and Kleiber 2014). In wheat, foliar application of Mg-Titanit (containing 8.5 g Ti L^{-1}) increased chlorophyll content and crop yield (Kováčik et al. 2014). In common duckweed, the application of TiO_2 (both as bulk and nano-sized particles) displayed no concentration-dependent effects and phytotoxic symptoms were not observed, demonstrating that Ti is harmless for this plant (Dolenc Koce 2017). Titanium salts (i.e. titanium tetrachloride and polytitanium tetrachloride) not only provide a cost-effective alternative to conventional coagulants for the removal of algal organic matter in seawater by producing a valuable by-product but also minimize the environmental impact of sludge production (Chekli et al. 2017). In soya bean, the application of a commercial titanium complex positively affected plant growth, seed yield, and the contents of cellulose, lignin, and fiber, as well as the antioxidative activity of seeds, while lipid and protein concentrations were significantly decreased in seeds (Szparaga et al. 2018). Also in soya bean, when supplying 2.5 mg Ti per plant, chlorophyll and carotenoid concentrations, plant biomass, and some photosynthetic attributes were significantly improved, though when applying 5-10 mg Ti per plant, anatomical and structural anomalies in leaves and chloroplasts were detected (Hussain et al. 2019). In tomato grown hydroponically at 50 mM NaCl in solution, Ti enhanced quality attributes of fruits, including electrical conductivity, titratable acidity, and total soluble solids (Carbajal-Vázquez et al. 2020). In rice grown in paddy soils containing toxic levels of As, Cd, and Pb, the application of 0.15 and 0.30% titanium gypsum (TG) reduced the uptake of the toxic metals (i.e. As, Cd, and Pb), thus avoiding serious health problems for humans and animals after grain consumption (Zhai et al. 2020).

According to Lyu et al. (2017), the interaction of Ti with essential elements such as Fe represents one of the primary mechanisms triggering beneficial effects of Ti in plants. Both iron and titanium are classified as transition metals, thus displaying analogous chemical and physical attributes but also exhibiting antagonism and synergism (Cigler et al. 2010). Under Fe-deficient conditions, the transcriptional activity of genes involved in iron uptake is upregulated, which consequently improves plant growth. Nonetheless, if plants accumulate high levels of Ti, it may cause serious toxic effects as a result of its competition with iron ions in plant metabolism (Lyu et al. 2017). Thus, Ti applications are recommended in Fe-deficient conditions, especially if soil pH is higher than 6.5.

Similar to Ce, a plethora of publications on the effect of TiO_2 nanoparticles are recently appearing (i.e. Daryabeigi-Zand et al. 2020; Silva et al. 2020b; Zhang et al. 2020a), though such nanotechnological approaches are not the focus of our attention in this chapter.

Like most of the beneficial elements reported herein, Ti concentrations in plant tissues are reported in the range of mg kg^{-1} (on a dry basis). In winter oilseed rape, Ti concentrations were found as follows: seeds, 1.69; straw, 11.27; aboveground phytomass: 53.6; underground phytomass: 210.5 mg kg^{-1} (Kováčik et al. 2016). In lettuce, El-Ghamry et al. (2018) reported 0.19 mg Ti kg^{-1} in shoots, while in tomato, a range of 0.69 to 0.71 mg Ti kg^{-1} was found in leaves (Haghighi and Daneshmand, 2018).

Vanadium

In the Earth's crust vanadium is a bountiful element with a concentration of 1.2×10^2 mg kg^{-1}. In seawater, its concentration is nearly 2.5×10^{-3} mg L^{-1} (Petranikova et al. 2020; Lide 2017).

As the rest of the beneficial elements, V may trigger hormetic dose-response effects in plants.

In sweet basil, V supply at concentrations ≤40 mg V L^{-1} did not retard plant growth, though root dry matter increased linearly with increasing addition of V and prevented Pb absorption and accumulation in plant tissues (Akoumianaki-Ioannidou et al. 2016). Similarly, in cucumber, V displayed antagonist effects on Cd and Pb uptake and accumulation (Osu and Onyema 2016). In sunflower (*Helianthus annuus*), the application of V at either 3.25, 7.5, or 15 mg L^{-1} significantly decreased plant growth, total protein, and carotenoid contents, but increased the total chlorophyll and soluble sugars contents (Abedini and Mohammadian 2018). In maize, V applied at a low concentration (20 ppm) resulted in improved morphological, biochemical, and yield attributes (Magar et al. 2018). Furthermore, in sugarcane (*Sachharum* spp. Hybrids), application of 20 μM V enhanced plant height, whereas flowering and pith formation were significantly reduced in V-treated plants (Sentíes-Herrera et al. 2018). In pepper, applying 5 μM V resulted in significant increases of plant growth and enhanced metabolism, whereas applying 10-15 μM V severe damages were detected in plant physiology and biochemistry (García-Jiménez et al. 2018). Furthermore, in dog tail's grass (*Setaria viridis*), vanadium improved P and Fe accumulation and translocation, with a simultaneous stimulation of plant growth (Aihemaiti et al. 2019). In general, trace amounts of V (usually less than 2 ppm) increase chlorophyll biosynthesis, nutrient acquisition, N assimilation, P utilization, and plant growth (Aihemaiti et al. 2020).

Furthermore, in several species of green algae, fungi, and nitrogen-fixing microorganisms, V is an essential element. In green algae, V plays functional roles in chlorophyll biosynthesis and Fe metabolism, thus improving growth (Pilbeam et al. 2015). Indeed, the inactive apoenzymes bromoperoxidase, iodoperoxidase, and chloroperoxidase can be reconstituted by V to active haloenzymes (Anke 2004).

Given the beneficial outcomes of V on the biology of different plant species, it is evident that V can function as potent inorganic biostimulant and that it can be potentially used to improve plant performance under limiting environmental conditions and increase the production and quality of crops of agricultural importance. Significantly, since the ion vanadate resembles the structure of the ion phosphate, it may interfere in metabolic processes depending on phosphate, like phosphorylation and dephosphorylation, interaction with DNA molecules, and modulation of reactive oxygen species. Therapeutic approaches point to crucial roles of V as an insulin enhancer (antidiabetic properties) and in the treatment of hearth and neuron anomalies, tumors development prevention, tropical diseases, as well as viral and bacterial infections (Rehder 2013). Therefore, novel approaches utilizing new and emerging technologies in the fields of modern imaging techniques (Li et al. 2020a), spectroscopy (Yang 2020), or genomics (Wang et al. 2020a), to cite some, would further shed light and uncover unveiled physiological and metabolic effects of V in plants to be used in biostimulation and biofortification approaches. These possibilities are vast and await additional and more in-depth research.

Among the beneficial elements addressed herein, V is by far the least abundant element in plant tissues. In different medicinal plants grown in conventional conditions (i.e. not exposed to V applications), V concentrations can range from 45-410 μg kg^{-1} in field horsetail (*Equisetum arvense*) to 255-28700 μg kg^{-1} in wood avens (*Geum avens*) and to even more than 76000 μg V kg^{-1} in wild thyme (*Thymus pulegioides*), which has a unique ability to accumulate V (Antal et al. 2009). Other V-accumulator species are common nettle (*Urtica dioica*), St. John's wort (*Hypericum perforatum*), and garden heliotrope (*Valeriana officinalis*). In beet leaves, Teng et al. (2011) reported V concentrations ranging from 6500 to 42800 μg kg^{-1}, while in mung bean leaves V concentrations fluctuated from 3000 to 22700 μg kg^{-1}, and in rape, V concentrations were found as follows: in leaves, 26600; in roots, 12600; and in seeds, 18800 μg kg^{-1}.

Conclusions and Perspectives

To date, aluminium (Al), cobalt (Co), cerium (Ce), iodine (I), lanthanum (La), selenium (Se), silicon (Si), sodium (Na), titanium (Ti), and vanadium (V) have been recognized as beneficial elements for different plant species. Though they are not considered essential elements from a pure plant nutrition perspective, they can enhance plant growth and even play pivotal roles in structure and function of particular taxa within the plant kingdom. Furthermore, beneficial elements may trigger different mechanisms that help enhance crop yield and quality, as well as elicit tolerance or resistance to biotic and abiotic stressors. Regarding the concentration patterns of these elements in different plant tissues, Si and Na display the highest concentrations (g kg^{-1} on a dry basis), while Al, Ce, Co, I, La, Se, and Ti exhibit moderate concentrations (mg kg^{-1} on a dry basis), and V the lowest (μg kg^{-1} on a dry basis). Although past research has mainly focused on the toxic or negative effects of this group of elements on plant biology, in the past two decades significant advances have been made demonstrating the potential biostimulant effects of these elements. As biostimulants, beneficial elements may trigger hormesis. Hence, much experimental work is needed to determine the optimal dose-response curves in each case. Finally, given their undeniable value as evidenced in this chapter, beneficial elements could play a key role in stepped-up efforts to manage several of the current challenges facing humanity, such as climate change, world population growth resulting in increased food demand, and pandemic disease outbreaks.

References

Abd El-Naby, S.K.M. and Gad, N. (2014). Effect of cobalt as postharvest treatment on fruit quality of fresh Zaghloul and Samani dates. *World Applied Sciences Journal* 31: 1321–1332. https://doi.org/10.5829/idosi.wasj.2014.31.07.14420.

Abd-Alla, M.H., Bagy, M.K., El-enany, A.W., and Bashandy, S.R. (2014). Activation of *Rhizobium tibeticum* with flavonoids enhances nodulation, nitrogen fixation, and growth of fenugreek (*Trigonella foenum-graecum* L.) grown in cobalt-polluted soil. *Archives of Environmental Contamination and Toxicology* 66: 303–315. https://doi.org/10.1007/s00244-013-9980-7.

Abdelsamad, N.A., MacIntosh, G.C., and Leandro, L.F.S. (2019). Induction of ethylene inhibits development of soybean sudden death syndrome by inducing defense-related genes and reducing *Fusarium virguliforme* growth. *PLoS ONE* 14: e0215653. https://doi.org/10.1371/journal.pone.0215653.

Abedini, M. and Mohammadian, F. (2018). Vanadium effects on phenolic content and photosynthetic pigments of sunflower. *South Western Journal of Horticulture, Biology and Environment* 9: 77–86.

Adisa, I.O., Rawat, S., Pullagurala, V.L.R. et al. (2020). Nutritional status of tomato (*Solanum lycopersicum*) fruit grown in *Fusarium*-infested soil: Impact of cerium oxide nanoparticles. *Journal of Agricultural and Food Chemistry* 68: 1986–1997. https://doi.org/10.1021/acs.jafc.9b06840.

Agathokleous, E., Kitao, M., and Calabrese, E.J. (2018). The rare earth element (REE) lanthanum (La) induces hormesis in plants. *Environmental Pollution* 238: 1044–1047. https://doi.org/10.1016/j.envpol.2018.02.068.

Agathokleous, E., Kitao, M., and Calabrese, E.J. (2019). Hormetic dose responses induced by lanthanum in plants. *Environmental Pollution* 244: 332–341. https://doi.org/10.1016/j.envpol.2018.10.007.

Agathokleous, E., Kitao, M., and Calabrese, E.J. (2020). Hormesis: Highly generalizable and beyond laboratory. *Trends in Plant Science* S1360-1385 (20): 30188–30186. https://doi.org/10.1016/j.tplants.2020.05.006.

Ahanger, M.A., Bhat, J.A., Siddiqui M.H. et al. (2020). Silicon and secondary metabolites integration in plants: A significant association in stress tolerance. *Journal of Experimental Botany* 2020: eraa291. https://doi.org/10.1093/jxb/eraa291.

Aihemaiti, A., Jiang, J., Gao, Y. et al. (2019). The effect of vanadium on essential element uptake of *Setaria viridis*' seedlings. *Journal of Environmental Management* 237: 399–407. https://doi.org/10.1016/j.jenvman.2019.02.054.

Aihemaiti, A., Gao, Y., Meng, Y. et al. (2020). Review of plant-vanadium physiological interactions, bioaccumulation, and bioremediation of vanadium-contaminated sites. *Science of The Total Environment* 712: 135637. https://doi.org/10.1016/j.scitotenv.2019.135637.

Akoumianaki-Ioannidou, A., Barouchas, P.E., Ilia, E. et al. (2016). Effect of vanadium on dry matter and nutrient concentration in sweet basil (*Ocimum basilicum* L.). *Australian Journal of Crop Science* 10: 199–206.

Alarcón, M.V., Lloret-Salamanca, A., Lloret, P.G. et al. (2009). Effects of antagonists and inhibitors of ethylene biosynthesis on maize root elongation. *Plant Signaling & Behavior* 4: 1154–1156. https://doi.org/10.4161/psb.4.12.9948.

Alcaraz-López, C., Botia, M., Alcaraz, C.F., and Riquelme, F. (2005). Induction of fruit calcium assimilation and its influence on the quality of table grapes. *Spanish Journal of Agricultural Research* 3: 335–343. https://doi.org/10.5424/sjar/2005033-156.

Ali, S., Farooq, M.A., Yasmeen, T. et al. (2013). The influence of silicon on barley growth, photosynthesis and ultra-structure under chromium stress. *Ecotoxicology and Environmental Safety* 89: 66–72. https://doi.org/10.1016/j.ecoenv.2012.11.015.

Almeida, R.A., Carvalho, V.F.S., Müller, C. et al. (2019). Tolerance of *Eugenia dysenterica* to aluminum: Germination and plant growth. *Plants (Basel)* 8: 317. https://doi.org/10.3390/plants8090317.

Anke, M. (2004). Vanadium. An element both essential and toxic to plants, animals and humans? *Anales de la Real Academia Nacional de Farmacia* 70: 961–999.

Antal, D.S., Dehelean, C.A., Canciu, C.M., and Anke, M. (2009). *Vanadium in medicinal plants: New data on the occurrence of an element both essential and toxic to plants and man*, 5–10. Fascicula de Biologie XVI: *Analele Universității din Oradea*.

Araie, H. and Shiraiwa, Y. (2016). Selenium in algae. In: *The Physiology of Microalgae, Developments in Applied Phycology* (ed. M.A. Borowitzka, J. Beardall and J.A. Raven), 281–288. Cham: Springer https://doi.org/10.1007/978-3-319-24945-2_12.

Artyszak, A., Gozdowski, D., and Kucińska, K. (2015). The effect of calcium and silicon foliar fertilization in sugar beet. *Sugar Tech* 18: 109–114. https://doi.org/10.1007/s12355-015-0371-4.

Artyszak, A., Gozdowski, D., and Kucińska, K. (2019). Impact of foliar fertilization on the content of silicon and macronutrients in sugar beet. *Plants (Basel)* 8: 136. https://doi.org/10.3390/plants8050136.

Atta-Aly, M.A. (1998). Soaking summer squash seeds in low concentrations of cobalt solution before sowing increased plant growth, femaleness, and fruit yield via increasing plant ethylene level. *Journal of Plant Growth Regulation* 17: 25–32. https://doi.org/10.1007/PL00007008.

Bakkaus, E., Gouget, B., Gallien, J.P. et al. (2005). Concentration and distribution of cobalt in higher plants: The use of micro-PIXE spectroscopy. *Nuclear Instruments and Methods in Physics Research Section B: Beam Interactions with Materials and Atoms* 231: 350–356. https://doi.org/10.1016/j.nimb.2005.01.082.

Blasco, B., Rios, J.J., Cervilla, L.M. et al. (2010). Photorespiration process and nitrogen metabolism in lettuce plants (*Lactuca sativa* L.): induced changes in response to iodine biofortification. *Journal of Plant Growth Regulation* 29: 477–486. https://doi.org/10.1007/s00344010-9159-7.

Blasco, B., Leyva, R., Romero, L., and Ruiz, J.M. (2013). Iodine effects on phenolic metabolism in lettuce plants under salt stress. *Journal of Agricultural and Food Chemistry* 61: 2591–2596. https://doi.org/10.1021/jf303917n.

Bojórquez-Quintal, E., Escalante-Magaña, C., Echevarría-Machado, I., and Martínez-Estévez, M. (2017). Aluminum, a friend or foe of higher plants in acid soils. *Frontiers in Plant Science* 8: 1767. https://doi.org/10.3389/fpls.2017.01767.

Broadley, M.R., White, P.J., Bryson, R.J. et al. (2006). Biofortification of UK food crops with selenium. *Proceedings of the Nutrition Society* 65: 169–181. https://doi.org/10.1079/PNS2006490.

Callender, K.L., Roy, S., Khasa, D.P. et al. (2016). Actinorhizal alder phytostabilization alters microbial community dynamics in gold mine waste rock from Northern Quebec: A greenhouse study. *PLoS ONE* 11: e0150181. https://doi.org/10.1371/journal.pone.0150181.

Camargo, M.S., de Bezerra, B.K.L., Vitti, A.C. et al. (2017). Silicon fertilization reduces the deleterious effects of water deficit in sugarcane. *Journal of Soil Science and Plant Nutrition* 17: 99–111. http://dx.doi.org/10.4067/S0718-95162017005000008.

Carbajal-Vázquez, V.H., Gómez-Merino, F.C., Herrera-Corredor, J.A. et al. (2020). Effect of titanium foliar applications on tomato fruits from plants grown under salt stress conditions. *Notulae Botanicae Horti Agrobotanici Cluj-Napoca* 48: 924–937. https://doi.org/10.15835/nbha48211904.

Cárcamo, M.P., Reyes-Díaz, M., Rengel, Z. et al. (2019). Aluminum stress differentially affects physiological performance and metabolic compounds in cultivars of highbush blueberry. *Scientific Reports* 9: 11275. https://doi.org/10.1038/s41598-019-47569-8.

Cartes, P., Jara, A.A., Pinilla, L. et al. (2010). Selenium improves the antioxidant ability against aluminium-induced oxidative stress in ryegrass roots. *Annals of Applied Biology* 156: 297–307. https://doi.org/10.1111/j.1744-7348.2010.00387.x.

Chatzistathis, T. (2018). Physiological importance of manganese, cobalt and nickel and the improvement of their uptake and utilization by plants. In: *Plant Micronutrient Use Efficiency* (ed. M.A. Hossain, T. Kamiya, D.J. Burritt, et al.), 129–131. London: Academic Press https://doi.org/10.1016/B978-0-12-812104-7.00008-3.

Chauhan, D.K., Yadav, V., Vaculík, M. et al. (2021). Aluminum toxicity and aluminum stress-induced physiological tolerance responses in higher plants. *Critical Reviews in Biotechnology.* 41 (5): 715–730.

Chekli, L., Corjon, E., Tabatabai, S.A.A. et al. (2017). Performance of titanium salts compared to conventional $FeCl_3$ for the removal of algal organic matter (AOM) in synthetic seawater: Coagulation performance, organic fraction removal and floc characteristics. *Journal of Environmental Management* 201: 28–36. https://doi.org/10.1016/j.jenvman.2017.06.025.

Chen, X.D. and Shan, C.J. (2019). Cerium nitrate improves salt tolerance of wheat seedlings by regulating the antioxidant capacity of chloroplasts. *Biologia Plantarum* 63: 323–327. https://doi.org/10.32615/bp.2019.037.

Chen, M., Liu, Q., Wang, S.W. et al. (2019). High-abundance and low-cost metal-based cathode materials for sodium-ion batteries: problems, progress, and key technologies. *Energy Materials* 9: 1803609. https://doi.org/10.1002/aenm.201803609.

Cheng, J., Poduska, B., Morton, R.A., and Finan, T.M. (2011). An ABC-type cobalt transport system is essential for growth of *Sinorhizobium meliloti* at trace metal concentrations. *Journal of Bacteriology* 193: 4405–4416. https://doi.org/10.1128/JB.05045-11.

Cheng, M., Wang, A., Liu, Z. et al. (2018a). Sodium chloride decreases cadmium accumulation and changes the response of metabolites to cadmium stress in the halophyte *Carpobrotus rossii*. *Annals of Botany* 122: 373–385. https://doi.org/10.1093/aob/mcy077.

Cheng, M., Wang, L., Yang, Q., and Huang, X. (2018b). A detection method in living plant cells for rapidly monitoring the response of plants to exogenous lanthanum. *Ecotoxicology and Environmental Safety* 158: 94–99. https://doi.org/10.1016/j.ecoenv.2018.04.021.

Chu, J., Yao, X., and Zhang, Z. (2010). Responses of wheat seedlings to exogenous selenium supply under cold stress. *Biological Trace Element Research* 136: 355–363. https://doi.org/10.1007/s12011-009-8542-3.

Chung, Y.S., Kim, K.S., Hamayun, M., and Kim, Y. (2020). Silicon confers soybean resistance to salinity stress through regulation of reactive oxygen and reactive nitrogen species. *Frontiers in Plant Science* 10: 1725. https://doi.org/10.3389/fpls.2019.01725.

Cigler, P., Olejnickova, J., Hruby, M. et al. (2010). Interactions between iron and titanium metabolism in spinach: a chlorophyll fluorescence study in hydropony. *Journal of Plant Physiology* 167: 1592–1597. https://doi.org/10.1016/j.jplph.2010.06.021.

Colla, G., Rouphael, Y., Cardarelli, M. et al. (2013). Effects of saline stress on mineral composition, phenolic acids and flavonoids in leaves of artichoke and cardoon genotypes grown in floating system. *Journal of Science of Food and Agriculture* 93: 1119–1127. https://doi.org/10.1002/jsfa.5861.

Cortés-Flores, C., Rodríguez-Mendoza, M.N., Benavides-Mendoza, A. et al. (2016). Iodine increases the growth and mineral concentration in sweet pepper seedlings. *Agrociencia* 50: 747–758.

Coskun, D., Deshmukh, R., Sonah, H. et al. (2019). The controversies of silicon's role in plant biology. *New Phytology* 221: 67–85. https://doi.org/10.1111/nph.15343.

Cui, Y.N., Xia, Z.R., Ma, Q. et al. (2019). The synergistic effects of sodium and potassium on the xerophyte *Apocynum venetum* in response to drought stress. *Plant Physiology and Biochemistry* 135: 489–498. https://doi.org/10.1016/j.plaphy.2018.11.011.

D'Amico, M.L., Izzo, R., Tognoni, F. et al. (2003). Application of diluted sea water to soilless culture of tomato (*Lycopersicon esculenttum* Mill.): effects on plant growth, yield, fruit quality and antioxidant capacity. *Journal of Food, Agriculture and Environment* 1: 112–116. https://doi.org/10.1234/4.2003.350.

d'Aquino, L., de Pinto, M.C., Nardi, L. et al. (2009). Effect of some light rare earth elements on seed germination, seedling growth and antioxidant metabolism in *Triticum durum*. *Chemosphere* 75: 900–905. https://doi.org/10.1016/j.chemosphere.2009.01.026.

Dai, J.L., Zhu, Y.G., Zhang, M., and Huang, Y.Z. (2004). Selecting iodine enriched vegetables and the residual effect of iodate application to soil. *Biological Trace Element Research* 101: 265–276. https://doi.org/10.1385/BTER:101:3:265.

Daryabeigi-Zand, A., Tabrizi, A.M., and Heir, A.V. (2020). Co-application of biochar and titanium dioxide nanoparticles to promote remediation of antimony from soil by *Sorghum bicolor*: metal uptake and plant response. *Heliyon* 6: e04669. https://doi.org/10.1016/j.heliyon.2020.e04669.

Del Pino, A.M., Guiducci, M., D'Amato, R. et al. (2019). Selenium maintains cytosolic Ca^{2+} homeostasis and preserves germination rates of maize pollen under H_2O_2-induced oxidative stress. *Scientific Reports* 9: 13502. https://doi.org/10.1038/s41598-019-49760-3.

Dhiman, P., Rajora, N., Bhardwaj, S. et al. (2021). Fascinating role of silicon to combat salinity stress in plants: An updated overview. *Plant Physiology and Biochemistry* 162: 110–123.

Djanaguiraman, M., Prasad, P.V., and Seppanen, M. (2010). Selenium protects sorghum leaves from oxidative damage under high temperature stress by enhancing antioxidant defense system. *Plant Physiology and Biochemistry* 48: 999–1007. https://doi.org/10.1016/j.plaphy.2010.09.009.

Dolenc Koce, J. (2017). Effects of exposure to nano and bulk sized TiO_2 and CuO in *Lemna minor. Plant Physiology and Biochemistry* 119: 43–49. https://doi.org/10.1016/j.plaphy.2017.08.014.

Duarte, O.A.C., de Oliveira, C., Ramos, S.J. et al. (2018). Lanthanum content and effects on growth, gas exchanges, and chlorophyll index in maize plants. *Acta Scientiarum. Biological Sciences* 40: e38469. https://doi.org/10.4025/actascibiolsci.v40i1.38469.

Echbab, H., Prin, Y., Ducousso, M. et al. (2004). Helper bacteria associated with *Casuarina cunninghamiana-Frankia* symbiosis: selection of isolates for their effects on plant growth in axenic conditions. *Acta Botanica Gallica* 151: 429–440. https://doi.org/10.1080/12538078.2004.10515445.

El-Ghamry, A.M., Ghazi, D.A., and Mousa, Z.R. (2018). Effect of titanium dioxide on lettuce plants grown on sandy soil. *Journal of Soil Science and Agricultural Engineering* 9: 461–466. https://doi.org/10.21608/JSSAE.2018.36290.

Elsokkary, I. (2018). Silicon as a beneficial element and as an essential plant nutrient: An outlook (Review). *Alexandria Science Exchange Journal* 39: 534–550. https://doi.org/10.21608/asejaiqjsae.2018.16920.

Epstein, E. (1994). The anomaly of silicon in plant biology. *Proceedings of the National Academy of Sciences of the United States of America* 91: 11–17. https://doi.org/10.1073/pnas.91.1.11.

Escobar-Sepúlveda, H.F., Trejo-Téllez, L.I., Pérez-Rodríguez, P. et al. (2017). Diacylglycerol kinases are widespread in higher plants and display inducible gene expression in response to beneficial elements, metal, and metalloid ions. *Frontiers in Plant Science* 8: 129. https://doi.org/10.3389/fpls.2017.00129.

Fan, K., Wang, M., Gao, Y. et al. (2019). Transcriptomic and ionomic analysis provides new insight into the beneficial effect of Al on tea roots' growth and nutrient uptake. *Plant Cell Reports* 38: 715–729. https://doi.org/10.1007/s00299-019-02401-5.

Ferreira-Vilela, L., Ramos, S.J., Carneiro, M.A.C. et al. (2018). Cerium (Ce) and lanthanum (La) promoted plant growth and mycorrhizal colonization of maize in tropical soil. Australian. *Journal of Crop Sciences* 12: 704–710. https://doi.org/10.21475/ajcs.18.12.05.PNE754.

Filek, M., Keskinen, R., Hartikainen, H. et al. (2008). The protective role of selenium in rape seedlings subjected to cadmium stress. *Journal of Plant Physiology* 165: 833–844. https://doi.org/10.1016/j.jplph.2007.06.006.

Fleck, A.T., Schulze, S., Hinrichs, M. et al. (2015). Silicon promotes exodermal Casparian band formation in Si-accumulating and Si-excluding species by forming phenol complexes. *PLoS ONE* 10: e0138555. https://doi.org/10.1371/journal.pone.0138555.

Flowers, T.J. and Colmer, T.D. (2008). *Salinity tolerance in halophytes. New Phytology* 179: 945–963. https://doi.org/10.1111/j.1469-8137.2008.02531.x.

Gad, N. (2010). Improving quantity and quality of canola oil yield through cobalt nutrition. *Agriculture and Biology Journal of North America* 1: 1090–1097.

Gad, N. and El-Metwally, I.M. (2015). Chemical and physiological response of maize to salinity using cobalt supplement. *International Journal of ChemTech Research* 8: 45–52.

Gad, N. and Kandil, H. (2010). Influence of cobalt on phosphorus uptake, growth and yield of tomato. *Agriculture and Biology Journal of North America* 1: 1069–1075. https://doi.org/10.5251/abjna.2010.1.5.1069.1075.

Gad, N., El-Moez, M.R.A., and Kandil, H. (2011). Barley response to salt stress at varied levels of cobalt I. Growth and mineral composition. *Journal of Applied Sciences Research* 7: 1454–1459.

Gad, N., Abdel-Moez, M.R., Fekry Ali, M.E., and Abou-Hussein, S.D. (2018). Increasing salt tolerance in cucumber by using cobalt. *Middle East Journal of Applied Sciences* 8: 345–354.

Gad, N., Sekara, A., and Abdelhamid, M.T. (2019). The potential role of cobalt and/or organic fertilizers in improving the growth, yield, and nutritional composition of *Moringa oleifera*. *Agronomy* 9: 862. https://doi.org/10.3390/agronomy9120862.

Gad, N., Sayed, S., and Hassan, N.M.K. (2020). Influence of cobalt on tolerating climatic change (salinity) in onion plant with reference to growth, yield, and mineral composition. *Plant Archives* 20: 1491–1495.

Galeas, M.L., Zhang, L.H., Freeman, J.L. et al. (2007). Seasonal fluctuations of selenium and sulfur accumulation in selenium hyperaccumulators and related nonaccumulators. *New Phytology* 173: 517–525. https://doi.org/10.1111/j.1469-8137.2006.01943.x.

García-Jiménez, A., Gómez-Merino, F.C., Tejeda-Sartorius, O., and Trejo-Téllez, L.I. (2017). Lanthanum affects bell pepper seedling quality depending on the genotype and time of exposure by differentially modifying plant height, stem diameter and concentrations of chlorophylls, sugars, amino acids, and proteins. *Frontiers in Plant Science* 8: 308. https://doi.org/10.3389/fpls.2017.00308.

García-Jiménez, A., Trejo-Téllez, L.I., Guillén-Sánchez, D., and Gómez-Merino, F.C. (2018). Vanadium stimulates pepper plant growth and flowering, increases concentrations of amino acids, sugars and chlorophylls, and modifies nutrient concentrations. *PLoS ONE* 13: e0201908. https://doi.org/10.1371/journal.pone.0201908.

Garousi, F. (2017). The essentiality of selenium for humans, animals, and plants, and the role of selenium in plant metabolism and physiology. *Acta Universitatis Sapientiae, Alimentaria* 10: 75–90. https://doi.org/10.1515/ausal-2017-0005.

Gattward, J.N., Almeida, A.A.F., Souza, J.J.O. et al. (2012). Sodium-potassium synergism in *Theobroma cacao*: stimulation of photosynthesis, water-use efficiency and mineral nutrition. *Physiologia Plantarum* 146: 350–362. https://doi.org/10.1111/j.1399-3054.2012.01621.x.

Gaur, S., Kumar, J., Kumar, D. et al. (2020). Fascinating impact of silicon and silicon transporters in plants: A review. *Ecotoxicology and Environmental Safety* 202: 110885. https://doi.org/10.1016/j.ecoenv.2020.110885.

Godon, C., Mercier, C., Wang, X. et al. (2019). Under phosphate starvation conditions, Fe and Al trigger accumulation of the transcription factor STOP1 in the nucleus of Arabidopsis root cells. *The Plant Journal* 99: 937–949. https://doi.org/10.1111/tpj.14374.

Gómez-Merino, F.C. and Trejo-Téllez, L.I. (2018). The role of beneficial elements in triggering adaptive responses to environmental stressors and improving plant performance. In: *Biotic and Abiotic Stress Tolerance in Plants* (ed. S. Vats), 137–172. Singapore: Springer Nature https://doi.org/10.1007/978-981-10-9029-5_6.

Gómez-Merino, F.C., Ramírez-Martínez, M., Castillo-González, A.M., and Trejo-Téllez, L.I. (2020a). Lanthanum prolongs vase life of cut tulip flowers by increasing water consumption, and concentrations of sugars, proteins, and chlorophylls. *Scientific Reports* 10: 4209. https://doi.org/10.1038/s41598-020-61200-1.

Gómez-Merino, F.C., Trejo-Téllez, L.I., García-Jiménez, A. et al. (2020b). Silicon flow from root to shoot in pepper: a comprehensive *in silico* analysis reveals a potential linkage between gene expression and hormone signaling that stimulates plant growth and metabolism. *PeerJ* 8: e10053. https://doi.org/10.7717/peerj.10053.

Gordillo-Melgoza, F.A., Borrego-Escalante, F., Lozano-Cavazos, C.J. et al. (2016). Melon plant response to applications potassium iodine. *Revista Mexicana de Ciencias Agrícolas* 17: 3465–3476.

Greger, M., Landberg, T., and Vaulík, M. (2018). Silicon influences soil availability and accumulation of mineral nutrients in various plant species. *Plants* 7: 41. http://dx.doi.org/10.3390/plants7020041.

Grof, C.P.L., Johnston, M., and Brownell, P.F. (1989). Effect of sodium nutrition on the ultrastructure of chloroplasts of C4 plants. *Plant Physiology* 89: 539–543. https://doi.org/10.1104/pp.89.2.539.

Gropper, S.S. and Smith, J.L. (2018). Nonessential trace and ultratrace elements. In: *Advanced Nutrition and Human Metabolism* (ed. S.S. Gropper and J.L. Smith), 543–556. Belmont: Wadsworth/Cengage Learning.

Guevara-Moreno, O.D., Acevedo-Aguilar, F.J., and Yanez-Barrientos, E. (2018). Selenium uptake and biotransformation and effect of selenium exposure on the essential and trace elements status: comparative evaluation of four edible plants. *Journal in the Mexican Chemical Society* 62: 247–258. http://dx.doi.org/10.29356/jmcs.v62i2.388.

Guntzer, F., Keller, C., and Meunier, J.D. (2012). Benefits of plant silicon for crops: a review. *Agronomy for Sustainable Development* 32: 201–213. https://doi.org/10.1007/s13593-011-0039-8.

Gupta, M. and Gupta, S. (2017). An overview of selenium uptake, metabolism, and toxicity in plants. *Frontiers in Plant Science* 7: 2074. https://doi.org/10.3389/fpls.2016.02074.

Gupta, N., Bajpai, M.S., Majumdar, R.S., and Mishra, P.K. (2015). Response of iodine on antioxidant levels of *Glycine max* L. grown under Cd^{2+} stress. *Advances in Biological Research* 9: 40–48. https://doi.org/10.5829/idosi.abr.2015.9.1.9183.

Haghighi, M. and Daneshmand, B. (2018). Beneficial effect of titanium on plant growth, photosynthesis and nutrient trait of tomato cv. *Foria. Iran Journal Research* 37: 83–88. http://dx.doi.org/10.22099/IAR.2017.4804.

Haghighi, M., Sheibanirad, A., and Pessarakli, M. (2016). Effects of selenium as a beneficial element on growth and photosynthetic attributes of greenhouse cucumber. *Journal of Plant Nutrition* 39: 1493–1498. https://doi.org/10.1080/01904167.2015.1109116.

Hajiboland, R., Rad, S.B., Barceló, J., and Poschenrieder, C. (2013). Mechanisms of aluminum-induced growth stimulation in tea (*Camellia sinensis*). *Journal of Plant Nutrition and Soil Science* 176: 616–625. https://doi.org/10.1002/jpln.201200311.

Hasan, S.A., Hayat, S., Wani, A.S., and Ahmad, A. (2011). Establishment of sensitive and resistant variety of tomato on the basis of photosynthesis and antioxidative enzymes in the presence of cobalt applied as shotgun approach. *Brazilian Journal of Plant Physiology* 23: 175–185. http://dx.doi.org/10.1590/S1677-04202011000300001.

Hasanuzzaman, M. and Fujita, M. (2011). Selenium pretreatment upregulates the antioxidant defense and methylglyoxal detoxification system and confers enhanced tolerance to drought stress in rapeseed seedlings. *Biological Trace Element Research* 143: 1758–1776. https://doi.org/10.1007/s12011-011-8998-9.

Hashish, K.I., Youssef, N.M., and Mazhar, A.M. (2019). Effect of different concentration of cobalt on growth and chemical constituents of paulownia seedlings. *Current Science International* 8: 518–522.

Hatfield, D.L., Tsuji, P.A., Carlson, B.A., and Gladyshev, V.N. (2014). Selenium and selenocysteine: roles in cancer, health, and development. *Trends in Biochemical Science* 39: 112–120. https://doi.org/10.1016/j.tibs.2013.12.007.

Hawrylak-Nowak, B., Dresler, S., Rubinowska, K. et al. (2018). Selenium biofortification enhances the growth and alters the physiological response of lamb's lettuce grown under high temperature stress. *Plant Physiology and Biochemistry* 127: 446–456. https://doi.org/10.1016/j.plaphy.2018.04.018.

Hayder, M., Wojcieszek, J., Asztemborska, M. et al. (2020). Analysis of cerium oxide and copper oxide nanoparticles bioaccessibility from radish using SP-ICP-MS. *Journal of the Science of Food and Agriculture* 100: 4950–4958. https://doi.org/10.1002/jsfa.10558.

He, Y.W. and Loh, C.S. (2000). Cerium and lanthanum promote floral initiation and reproductive growth in *Arabidopsis thaliana*. *Plant Science* 159: 117–124. https://doi.org/10.1016/s0168-9452(00)00338-1.

He, Y.W. and Loh, C.S. (2002). Induction of early bolting in *Arabidopsis thaliana* by triacontanol, cerium and lanthanum is correlated with increased endogenous concentration of isopentenyl adenosine (iPAdos). *Journal of Experimental Botany* 53: 505–512. https://doi.org/10.1093/jexbot/53.368.505.

He, X., You, P., and Sun, Y. (2020). Lanthanum and abscisic acid coregulate chlorophyll production of seedling in switchgrass. *PLoS ONE* 15: e0232750. https://doi.org/10.1371/journal.pone.0232750.

Hernández-Castro, E., Trejo-Téllez, L.I., Gómez-Merino, F.C. et al. (2015). Bioaccumulation of iron, selenium, nitrate, and proteins in chard shoots. *Journal of Soil Science and Plant Nutrition* 15: 694–710.

Hosseini, S.A., Maillard, A., Hajirezaei, M.R. et al. (2017). Induction of barley silicon transporter HvLsi1 and HvLsi2, increased silicon concentration in the shoot and regulated Starch and ABA homeostasis under osmotic stress and concomitant potassium deficiency. *Frontiers in Plant Science* 8: 1359. https://doi.org/10.3389/fpls.2017.01359.

Hu, H., Wang, L., Li, Y. et al. (2016). Insight into mechanism of lanthanum (III) induced damage to plant photosynthesis. *Ecotoxicology and Environmental Safety* 127: 43–50. https://doi.org/10.1016/j.ecoenv.2016.01.008.

Hu, P., An, J., Faulkner, M.M. et al. (2020). Nanoparticle charge and size control foliar delivery efficiency to plant cells and organelles. *ACS Nano* 14: 7970–7986. https://doi.org/10.1021/acsnano.9b09178.

Huang, H., Rizwan, M., Li, M. et al. (2019). Comparative efficacy of organic and inorganic silicon fertilizers on antioxidant response, Cd/Pb accumulation and health risk assessment in wheat (*Triticum aestivum* L.). *Environmental Pollution* 255 (Pt1): 113146. https://doi.org/10.1016/j.envpol.2019.113146.

Hussain, S., Iqbal, N., Brestic, M. et al. (2019). Changes in morphology, chlorophyll fluorescence performance and Rubisco activity of soybean in response to foliar application of ionic titanium under normal light and shade environment. *Science of The Total Environment* 658: 626–637. https://doi.org/0.1016/j.scitotenv.2018.12.182.

Hussain, S., Shuxian, L., Mumtaz, M. et al. (2020). Foliar application of silicon improves stem strength under low light stress by regulating lignin biosynthesis genes in soybean (*Glycine max* (L.) Merr.). *Journal of Hazardous Materials* 401: 123256. https://doi.org/10.1016/j.jhazmat.2020.123256.

Iftikhar, A., Rizwan, M., Adrees, M. et al. (2020). Effect of gibberellic acid on growth, biomass, and antioxidant defense system of wheat (*Triticum aestivum* L.) under cerium oxide nanoparticle stress. *Environmental Science and Pollution Research International* 27: 33809–33820. https://doi.org/10.1007/s11356-020-09661-9.

Incrocci, L., Carmassi, G., Maggini, R. et al. (2019). Iodine accumulation and tolerance in sweet basil (*Ocimum basilicum* L.) with green or purple leaves grown in floating system technique. *Frontiers in Plant Science* 10: 1494. https://doi.org/10.3389/fpls.2019.01494.

Ito, T., Oyama, K.I., and Yoshida, K. (2018). Direct observation of hydrangea blue-complex composed of 3-*o*-glucosyldelphinidin, Al^{3+} and 5-*O*-acylquinic acid by ESI-Mass Spectrometry. *Molecules* 23: 1424. https://doi.org/10.3390/molecules23061424.

Jansen, S., Broadley, M.R., Robbrecht, E., and Smets, E. (2002). Aluminum hyperaccumulation in angiosperms: A review of its phylogenetic significance. *The Botanical Review* 68: 235–269. https://doi.org/10.1663/0006-8101(2002)068[0235:AHIAAR]2.0.CO;2.

de Jesus, L.R., Batista, B.L., and da Silva Lobato, A.K. (2017). Silicon reduces aluminum accumulation and mitigates toxic effects in cowpea plants. *Acta Physiologiae Plantarum* 39: 138. https://doi.org/10.1007/s11738-017-2435-4.

Jha, A.B. and Warkentin, T.D. (2020). Biofortification of pulse crops: Status and future perspectives. *Plants (Basel)* 9: 73. https://doi.org/10.3390/plants9010073.

Johnston, M., Grof, C.P.L., and Brownell, P.F. (1988). The effect of sodium nutrition on the pool sizes of intermediates of the C4 photosynthetic pathway. *Australian Journal of Plant Physiology* 15: 749–760. https://doi.org/10.1071/pp9880749.

Johnston, M., Grof, C.P.L., and Brownell, P.F. (1989). Chlorophyll *a*/*b* ratios and photosystem activity of mesophyll and bundle sheath fractions from sodium-deficient C4 plants. *Australian Journal of Plant Physiology* 16: 449–457. https://doi.org/10.1071/PP9890449.

Jurkowska, A. and Świerczewska-Gładysz, E. (2020). New model of Si balance in the Late Cretaceous epicontinental European Basin. *Global and Planetary Change* 186: 103108. https://doi.org/10.1016/j.gloplacha.2019.103108.

Kandil, H., Farid, I.M., and El-Maghraby, A. (2013). Effect of cobalt level and nitrogen source on quantity and quality of soybean plant. *Journal of Basic and Applied Scientific Research* 3: 185–192.

Katz, O., Puppe, D., Kaczorek, D. et al. (2021). Silicon in the soil–plant continuum: Intricate feedback mechanisms within ecosystems. *Plants 10*: 652. https://doi.org/10.3390/plants10040652.

Kaur, S., Kaur, N., Siddique, K.H.M., and Nayyar, H. (2015). Beneficial elements for agricultural crops and their functional relevance in defence against stresses. *Archives of Agronomy and Soil Science* 62: 905–920. https://doi.org/10.1080/03650340.2015.1101070.

Keskinen, R., Turakainen, M., and Hartikainen, H. (2010). Plant availability of soil selenate additions and selenium distribution within wheat and ryegrass. *Plant and Soil* 333: 301–313. https://doi.org/10.1007/s11104-010-0345-y.

Khare, T., Srivastav, A., Shaikh, S., and Kumar, V. (2018). Polyamines and their metabolic engineering for plant salinity stress tolerance. In: *Salinity Responses and Tolerance in Plants*, vol. 1 (ed. V. Kumar, S. Wani, P. Suprasanna and L.S. Tran), 339–358. Cham: Springer https://doi.org/10.1007/978-3-319-75671-4_13.

Kiferle, C., Ascrizzi, R., Martinelli, M. et al. (2019). Effect of Iodine treatments on *Ocimum basilicum* L.: Biofortification, phenolics production and essential oil composition. *PLoS ONE* 14: e0226559. https://doi.org/10.1371/journal.pone.0226559.

Kim, H.J., Holcomb, E.J., and Brown, K.M. (2005). Lanthanum effects on gravitropic responses of cut tulip flowers. *Acta Horticulturae* 669: 417–423. https://doi.org/10.17660/ActaHortic.2005.669.55.

Kolbert, Z., Molnár, Á., Feigl, G., and Van Hoewyk, D. (2019). Plant selenium toxicity: Proteome in the crosshairs. *Journal of Plant Physiology* 232: 291–300. https://doi.org/10.1016/j.jplph.2018.11.003.

Kotelnikova, A., Fastovets, I., Rogova, O. et al. (2019). Toxicity assay of lanthanum and cerium in solutions and soil. *Ecotoxicology and Environmental Safety* 167: 20–28. https://doi.org/10.1016/j.ecoenv.2018.09.117.

Kováčik, P., Hudec, J., Ondrišík, P., and Poliaková, N. (2014). The effect of liquid Mg-Titanit on creation of winter wheat phytomass. *Research Journal of Agricultural Science* 46: 125–131.

Kováčik, P., Šimanský, V., Wierzbowska, J., and Renčo, M. (2016). Impact of foliar application of the biostimulator Mg-Tytanit on the formation of winter oilseed rape phytomass and titanium content. *Journal of Elementology* 21: 1235–1251. https://doi.org/10.5601/jelem.2016.21.2.1155.

Kreszies, T., Kreszies, V., Ly, F. et al. (2020). Suberized transport barriers in plant roots: the effect of silicon. *Journal of Experimental Botany* 2020: eraa203. https://doi.org/10.1093/jxb/eraa203.

Kronzucker, H.J., Coskun, D., Schulze, L.M. et al. (2013). Sodium as nutrient and toxicant. *Plant and Soil* 369: 1–23. https://doi.org/10.1007/s11104-013-1801-2.

Kumar, M., Bijo, A.J., Baghel, R.S. et al. (2012). Selenium and spermine alleviate cadmium induced toxicity in the red seaweed *Gracilaria dura* by regulating antioxidants and DNA methylation. *Plant Physiology and Biochemistry* 51: 129–138. https://doi.org/10.1016/j.plaphy.2011.10.016.

Kumar, M., Saxena, R., Tomar, R.S. et al. (2018). Role of methylotrophic bacteria in climate change mitigation. In: *Microbes for Climate Resilient Agriculture* (ed. P.L. Kashyap, A.K. Srivastava, S.P. Tiwari and S. Kumar), 149–164. Hoboken: Wiley https://doi.org/10.1002/9781119276050.ch7.

Laane, H.M. (2018). The effects of foliar sprays with different silicon compounds. *Plants (Basel)* 7: 45. https://doi.org/10.3390/plants7020045.

Lange, B., van der Ent, A., Baker, A.J.M. et al. (2016). Copper and cobalt accumulation in plants: a critical assessment of the current state of knowledge. *New Phytologist* 213: 537–551. https://doi.org/10.1111/nph.14175.

Leksungnoen, P., Wisawapipat, W., Ketrot, D. et al. (2019). Biochar and ash derived from silicon-rich rice husk decrease inorganic arsenic species in rice grain. *Science of The Total Environment* 684: 360–370. https://doi.org/10.1016/j.scitotenv.2019.05.247.

Leng, B., Geng, F., Dong, X. et al. (2019). Sodium is the critical factor leading to the positive halotropism of the halophyte *Limonium bicolor*. Plant Biosystems – An International

Journal Dealing with all Aspects of. Plant Biology 153: 544–551. https://doi.org/10.1080/11263504.2018.1508085.

Li, M., Zhu, Q., Hu, C.W. et al. (2007). Cobalt and manganese stress in the microalga *Pavlova viridis* (Prymnesiophyceae): effects on lipid peroxidation and antioxidant enzymes. *Journal of Environmental Sciences* 19: 1330–1335. https://doi.org/10.1016/s1001-0742(07)60217-4.

Li, R., Liu, H.P., Hong, C.L. et al. (2016). Iodide and iodate effects on the growth and fruit quality of strawberry. *Journal of Science of Food and Agriculture* 97: https://doi.org/10.1002/jsfa.7719.

Li, D., Li, C.H., Yao, Y. et al. (2020a). Modern imaging techniques in plant nutrition analysis: A review. *Computers and Electronics in Agriculture* 174: 105459. https://doi.org/10.1016/j.compag.2020.105459.

Li, W., Du, J., Feng, H. et al. (2020b). Function of NHX-type transporters in improving rice tolerance to aluminum stress and soil acidity. *Planta* 251: 71. https://doi.org/10.1007/s00425-020-03361-x.

Lian, H., Qin, C., Zhang, C. et al. (2019). Foliar-applied lanthanum chloride increases growth and phosphorus acquisition in phosphorus-limited adzuki bean seedlings. *Plant and Soil* 442: 385–399. https://doi.org/10.1007/s11104-019-04197-5.

Lide, R.D. (2017). Abundance of elements in the earth's crust and in the sea. In: *CRC Handbook of Chemistry and Physics* (ed. R.D. Lide), 14–17. Boca Raton: CRC Press.

Limchoowong, N., Sricharoen, P., Techawongstien, S., and Chanthai, S. (2016). An iodine supplementation of tomato fruits coated with an edible film of the iodide-doped chitosan. *Food Chemistry* 200: 223–229. https://doi.org/10.1016/j.foodchem.2016.01.042.

Liu, D., Lin, Y., and Wang, X. (2012). Effects of lanthanum on growth, element uptake, and oxidative stress in rice seedlings. *Journal of Plant Nutrition of Soil Science* 175: 907–911. https://doi.org/10.1002/jpln.201200016.

Liu, D., Wang, X., Zhang, X., and Gao, Z. (2013). Effects of lanthanum on growth and accumulation in roots of rice seedlings. *Plant, Soil and Environment* 59: 196–200. https://doi.org/10.17221/760/2012-PSE.

Liu, Y., Tao, J., Cao, J. et al. (2020). The beneficial effects of aluminum on the plant growth in *Camellia japonica*. *Journal of Soil Science and Plant Nutrition.* 20: 1799–1809. https://doi.org/10.1007/s42729-020-00251-9.

Lv, H. and Tani, A. (2018). Genomic characterization of methylotrophy of *Oharaeibacter diazotrophicus* strain $SM30^T$. *Journal of Bioscience and Bioengineering* 126: 667–675. https://doi.org/10.1016/j.jbiosc.2018.05.023.

Lyu, S., Wei, X., Chen, J. et al. (2017). Titanium as a Beneficial Element for Crop Production. *Frontiers in Plant Science* 8: 597. https://doi.org/10.3389/fpls.2017.00597.

Ma, J.M. (2015). Silicon. In: *Handbook of Plant Nutrition* (ed. A.V. Barker and D.J. Oilbeam), 681–695. Boca Raton: CRC Press.

Macey, M.C., Pratscher, J., Crombie, A.T., and Murrell, J.C. (2020). Impact of plants on the diversity and activity of methylotrophs in soil. *Microbiome* 8: 31. https://doi.org/10.1186/s40168-020-00801-4.

Magar, A.S., Misra, P., Latke, M.B. et al. (2018). Effect of different doses of vanadium on morphological, biochemical and yield attributing characters of sweet corn (*Zea mays* L.). *Journal of Pharmacognosy and Phytochemistry* 7: 1672–1675.

Malagoli, M., Schiavon, M., dall'Acqua, S., and Pilon-Smits, E.A.H. (2015). Effects of selenium biofortification on crop nutritional quality. *Frontiers in Plant Science* 6: 280. https://doi.org/10.3389/fpls.2015.00280.

Malik, J.A., Goel, S., Kaur, N. et al. (2012). Selenium antagonizes the toxic effects of arsenic on mungbean (*Phaseolus aureus* Roxb.) plants by restricting its uptake and enhancing the antioxidative and detoxification mechanisms. *Environmental and Experimental Botany* 77: 242–248. https://doi.org/10.1016/j.envexpbot.2011.12.001.

Mandlik, R., Thakral, V., Raturi, G. et al. (2020). Significance of silicon uptake, transport, and deposition in plants. *Journal of Experimental Botany* 2020: eraa301. https://doi.org/10.1093/jxb/eraa301.

Mao, H., Wang, J., Wang, Z. et al. (2014). Using agronomic biofortification to boost zinc, selenium, and iodine concentrations of food crops grown on the loess plateau in China. *Journal of Soil Science and Plant Nutrition* 14: 459–470. https://doi.org/10.4067/s0718-95162014005000036.

Markiewicz, B. and Kleiber, T. (2014). The effect of Tytanit application on the content of selected microelements and the biological value of tomato fruits. *Journal of Elementology* 19: 1065–1072. https://doi.org/10.5601/jelem.2014.19.3.486.

Masuda, S., Suzuki, Y., Fujitani, Y. et al. (2018). Lanthanide-Dependent Regulation of Methylotrophy in *Methylobacteriumaquaticum* strain 22A. *mSphere* 3: e00462–e00417. https://doi.org/10.1128/mSphere.00462-17.

Mattson, M.P. and Calabrese, E.J. (2010). Hormesis: What it is and why it matters. In: *A Revolution in Biology, Toxicology and Medicine* (ed. M.P. Mattson and E.J. Calabrese), 1–13. New York: Humana Press https://doi.org/10.1007/978-1-60761-495-1_1.

Medrano-Macías, J., Leija-Martínez, P., González-Morales, S. et al. (2016). Use of iodine to biofortify and promote growth and stress tolerance in crops. *Frontiers in Plant Science* 7: 1146. https://doi.org/10.3389/fpls.2016.01146.

Mehrabanjoubani, P., Abdolzadeh, A., Sadeghipour, H.R. et al. (2019). Silicon increases cell wall thickening and lignification in rice (*Oryza sativa*) root tip under excess Fe nutrition. *Plant Physiology and Biochemistry* 144: 264–273. https://doi.org/10.1016/j.plaphy.2019.09.047.

Mezeyová, I., Hegedűsová, A., Andrejiová, A. et al. (2016). Phytomass and content of essential oils in *Ocimum basilicum* after foliar treatment with selenium. *Agriculture & Food* 4: 19–27.

Moles, T.M., de Brito, F.R., Mariotti, L. et al. (2019). Salinity in autumn-winter season and fruit quality of tomato landraces. *Frontiers in Plant Science* 10: 1078. https://doi.org/10.3389/fpls.2019.01078.

Monreal, C.M., DeRosa, M., Mallubhotla, S.C. et al. (2016). Nanotechnologies for increasing the crop use efficiency of fertilizer-micronutrients. *Biology and Fertility of Soils* 52: 423–437. https://doi.org/10.1007/s00374-015-1073-5.

Moreno-Alvarado, M., García-Morales, S., Trejo-Téllez, L.I. et al. (2017). Aluminum enhances growth and sugar concentration, alters macronutrient status and regulates the expression of NAC transcription factors in rice. *Frontiers in Plant Science* 8: 73. https://doi.org/10.3389/fpls.2017.00073.

Morrissey, J., Baxter, I.R., Lee, J. et al. (2009). The ferroportin metal efflux proteins function in iron and cobalt homeostasis in Arabidopsis. *The Plant Cell* 21: 3326–3338. https://doi.org/10.1105/tpc.109.069401.

Nieves-Cordones, M., Al Shiblawi, F.R., and Sentenac, H. (2016). Roles and transport of sodium and potassium in plants. In: *The Alkali Metal Ions: Their Role for Life* (ed. A. Sigel, H. Sigel and R. Sigel), 291–324. Cham: Springer https://doi.org/10.1007/978-3-319-21756-7_9.

Ohta, D., Yasuoka, S., Matoh, T., and Takahashi, E. (1989). Sodium stimulates growth of *Amaranthus tricolor* L. plants through enhanced nitrate assimilation. *Plant Physiology* 89: 1102–1105. https://doi.org/10.1104/pp.89.4.1102.

Okamoto, S. and Eltis, L.D. (2011). The biological occurrence and trafficking of cobalt. *Metallomics* 3: 963–970. https://doi.org/10.1039/C1MT00056J.

Okem, A., Southway, C., Stirk, W.A. et al. (2015). Effect of cadmium and aluminum on growth, metabolite content and biological activity in *Drimia elata* (Jacq.) Hyacinthaceae. *South African Journal of Botany* 98: 142–147. http://dx.doi.org/10.1016/j.sajb.2015.02.013.

Osu, C. and Onyema, M. (2016). Vanadium inhibition capacity on nutrients and heavy metal uptake by *Cucumis sativus*. *Journal of American Science* 12: 63–66. https://doi.org/10.7537/marsjas121016.09.

Ouyang, J., Wang, X., Zao, B. et al. (2003). Effects of rare earth elements on the growth of *Cistanche deserticola* cells and the production of phenylethanoid glycosides. *Journal of Biotechnology* 102: 129–134. https://doi.org/10.1016/S0168-1656(03)00019-1.

Oyama, K., Yamada, T., Ito, D. et al. (2015). Metal complex pigment involved in the blue sepal color development of *Hydrangea*. *Journal of Agricultural and Food Chemistry* 63: 7630–7635. https://doi.org/10.1021/acs.jafc.5b02368.

Pandey, C. and Gupta, M. (2015). Selenium and auxin mitigates arsenic stress in rice (*Oryza sativa* L.) by combining the role of stress indicators, modulators and genotoxicity assays. *Journal of Hazardous Materials* 287: 384–391. https://doi.org/10.1016/j.jhazmat.2015.01.044.

Pax, F., Blamey, C., Kopittke, P.M. et al. (2015). Aluminum. In: *Handbook of Plant Nutrition* (ed. A.V. Barker and D.J. Pilbeam), 567–605. Boca Raton: CRC Press.

Pérez-Espinosa, A., Moral-Herrero, R., Pérez-Murcia, M.D. et al. (2015). Cobalt. In: *Handbook of Plant Nutrition* (ed. A.V. Barker and D.J. Oilbeam), 607–623. Boca Raton: CRC Press.

Petranikova, M., Tkaczyk, A.H., Bartl, A. et al. (2020). Vanadium sustainability in the context of innovative recycling and sourcing development. *Waste Management* 113: 521–544. https://doi.org/10.1016/j.wasman.2020.04.007.

Pilbeam, D.J., Grethead, H.M.R., and Drihem, K. (2015). Selenium. In: *Handbook of Plant Nutrition* (ed. A.V. Barker and D.J. Oilbeam), 651–679. Boca Raton: CRC Press.

Pilon-Smits, E.A.H., Quinn, C.F., Tapken, W. et al. (2009). Physiological functions of beneficial elements. *Current Opinion in Plant Biology* 12: 267–274. https://doi.org/10.1016/j.pbi.2009.04.009.

Poschenrieder, C., Cabot, C., Martos, S. et al. (2013). Do toxic ions induce hormesis in plants? *Plant Science* 212: 15–25. https://doi.org/10.1016/j.plantsci.2013.07.012.

Poschenrieder, C., Tolrà, R., Hajiboland, R. et al. (2015) Mechanisms of hyper-resistance and hyper-tolerance to aluminum in plants. In: *Aluminum Stress Adaptation in Plants. Signaling and Communication in Plants* (eds. S. Panda, and F. Baluška), 81–98. Cham: Springer.

Pukacka, S., Ratajczak, E., and Kalemba, E. (2011). The protective role of selenium in recalcitrant *Acer saccharium* L. seeds subjected to desiccation. *Journal of Plant Physiology* 168: 220–225. https://doi.org/10.1016/j.jplph.2010.07.021.

Qazi, I.H., Angel, C., Yang, H. et al. (2018). Selenium, selenoproteins, and female reproduction: A Review. *Molecules* 23 (12): 3053. https://doi.org/10.3390/molecules23123053.

Raddatz, N., Morales de los Ríos, L., Lindahl, M. et al. (2020). Coordinated transport of nitrate, potassium, and sodium. *Frontiers in Plant Science* 11: 247. https://doi.org/10.3389/fpls.2020.00247.

Radkowski, A., Radkowska, I., and Lemek, T. (2015). Effects of foliar application of titanium on seed yield in timothy (*Phleum pratense* L.). *Ecology Chemistry and Engineering* 22: 691–701. https://doi.org/10.1515/eces-2015-0042.

Ramírez-Martínez, M., Gómez-Merino, F.C., Trejo-Téllez, L.I. et al. (2009). Effect of lanthane on quality of tulip flower 'Ile de France'. *Acta Horticulturae* 847: 295–300. https://doi.org/10.17660/ActaHortic.2009.847.39.

Ramírez-Olvera, S.M., Trejo-Téllez, L.I., García-Morales, S. et al. (2018). Cerium enhances germination and shoot growth, and alters mineral nutrient concentration in rice. *PLoS ONE* 13: e0194691. https://doi.org/10.1371/journal.pone.0194691.

Ramírez-Olvera, S.M., Trejo-Téllez, L.I., Gómez-Merino, F.C. et al. (2021). Silicon stimulates plant growth and metabolism in rice plants under conventional and osmotic stress conditions. *Plants* 10: 777. https://doi.org/10.3390/plants10040777.

Ramos, S.J., Faquin, V., Guilherme, L.R.G. et al. (2010). Selenium biofortification and antioxidant activity in lettuce plants fed with selenate and selenite. *Plant Soil and Environment* 56: 584–588. https://doi.org/10.17221/113/2010-PSE.

Redondo-Gómez, S., Mateos-Naranjo, E., Davy, A.J. et al. (2007). Growth and photosynthetic responses to salinity of the salt-marsh shrub *Atriplex portulacoides*. *Annals of Botany* 100: 555–563. https://doi.org/10.1093/aob/mcm119.

Redondo-Gómez, S., Mateos-Naranjo, E., Figueroa, M.E., and Davy, A.J. (2010). Salt stimulation of growth and photosynthesis in an extreme halophyte, *Arthrocnemum macrostachyum*. *Plant Biology* 12: 79–87. https://doi.org/10.1111/j.1438-8677.2009.00207.x.

Rehder, D. (2013). Vanadium. Its role for humans. *Metal Ions in Life Sciences* 13: 139–169. https://doi.org/10.1007/978-94-007-7500-8_5.

Sade, H., Meriga, B., Surapu, V. et al. (2016). Toxicity and tolerance of aluminum in plants: tailoring plants to suit to acid soils. *Biometals* 29: 187–210. https://doi.org/10.1007/s10534-016-9910-z.

Salas, A., Tortosa, G., Hidalgo-García et al. (2020). The hemoglobin Bjgb from *Bradyrhizobium diazoefficien* controls NO homeostasis in soybean nodules to protect symbiotic nitrogen fixation. *Frontiers in Microbiology* 10: 2915. https://doi.org/10.3389/fmicb.2019.02915.

Sawaki, Y., Iuchi, S., Kobayashi et al. (2009). STOP1 regulates multiple genes that protect arabidopsis from proton and aluminum toxicities. *Plant Physiology* 150: 281–294. https://doi.org/10.1104/pp.108.134700.

Schaller, G.E. and Binder, B.M. (2017). Inhibitors of ethylene biosynthesis and signaling. *Methods in Molecular Biology* 1573: 223–235. https://doi.org/10.1007/978-1-4939-6854-1_15.

Schaller, J., Puppe, D., Kaczorek, D. et al. (2021). *Silicon cycling in soils revisited. Plants* 10: 295. https://doi.org/10.3390/plants10020295.

Schiavon, M., Nardi, S., dalla Vecchia, F. et al. (2020). Selenium biofortification in the 21st century: status and challenges for healthy human nutrition. *Plant and Soil* 453: 245–270. https://doi.org/10.1007/s11104-020-04635-9.

Schreiber, H.D., Jones, A.H., Lariviere, C.M. et al. (2011). Role of aluminum in red-to-blue color changes in *Hydrangea macrophylla* sepals. *Biometals* 24: 1005–1015. https://doi.org/10.1007/s10534-011-9458-x.

Schubert, S. (2015). Sodium. In: *Handbook of Plant Nutrition* (ed. A.V. Barker and D.J. Oilbeam), 697–710. Boca Raton: CRC Press.

Senties-Herrera, H.E., Trejo-Téllez, L.I., Volke-Haller, V.H. et al. (2018). Iodine, silicon, and vanadium differentially affect growth, flowering, and quality components of stalks in sugarcane. *Sugar Tech* 20: 518–533. https://doi.org/10.1007/s12355-017-0572-0.

Shabala, S. and Mackay, A.S. (2011). Ion transport in halophytes. *Advances in Botanical Research* 57: 151–199. https://doi.org/10.1016/B978-0-12-387692-8.00005-9.

Shan, C. and Zhao, X. (2015). Lanthanum delays the senescence of *Lilium longiflorum* cut flowers by improving antioxidant defense system and water retaining capacity. *Scientia Horticulturae* 197: 516–520. https://doi.org/10.1016/j.scienta.2015.10.012.

Sharma, T., Dreyer, I., Kochian, L., and Piñeros, M.A. (2016). The ALMT family of organic acid transporters in plants and their involvement in detoxification and nutrient security. *Frontiers in Plant Science* 7: 1488. https://doi.org/10.3389/fpls.2016.01488.

Sheng, H. and Chen, S. (2020). Plant silicon-cell wall complexes: Identification, model of covalent bond formation and biofunction. *Plant Physiology and Biochemistry* 155: 13–19. https://doi.org/10.1016/j.plaphy.2020.07.020.

Shinmachi, F., Buchner, P., Stroud, J.L. et al. (2010). Influence of sulfur deficiency on the expression of specific sulfate transporters and the distribution of sulfur, selenium, and molybdenum in wheat. *Plant Physiology* 153: 327–336. https://doi.org/10.1104/pp.110.153759.

Shivaraj, S.M., Mandlik, R., Bhat, J.A. et al. (2022). Outstanding questions on the beneficial role of silicon in crop plants. *Plant and Cell Physiology.* 63: 4–16. https://doi.org/10.1093/pcp/pcab145.

Shreenath, A.P., Ameer, M.A., and Dooley, J. (2020). Selenium deficiency. In: *StatPearls*. Treasure Island: StatPearls Publishing.

Siam, H.S., Saffaa, S.M., and Abdel Hady, M. (2012). Increasing the utilization efficiency of nitrogenous fertilizers by cobalt addition and its effect on growth, yield parameters and nutrients contents of cowpea plants. *Journal of Applied Sciences Research* 8: 915–929.

Siddiqui, M.H., Alamri, S., Alsubaie, Q.D. et al. (2019). Potential roles of melatonin and sulfur in alleviation of lanthanum toxicity in tomato seedlings. *Ecotoxicology and Environmental Safety* 180: 656–667. https://doi.org/10.1016/j.ecoenv.2019.05.043.

Silva, C.O., Brito, D.S., da Silva, A.A. et al. (2020a). Differential accumulation of aluminum in root tips of soybean seedlings. *Brazilian Journal of Botany* 43: 99–107. https://doi.org/10.1007/s40415-020-00593-9.

Silva, S., Ribeiro, T.P., Santos, C. et al. (2020b). TiO_2 nanoparticles induced sugar impairments and metabolic pathway shift towards amino acid metabolism in wheat. *Journal of Hazardous Materials* 399: 122982. https://doi.org/10.1016/j.jhazmat.2020.122982.

Singh, D.K., Singh, A.K., Singh, M. et al. (2012). Production potential and nutrient uptake efficiency of pea (*Pisum sativum* L) as influenced by different fertility levels and micronutrients. *Journal of the Indian Society of Soil Science* 60: 1–7.

Singh, A., Hussain, I., Singh, N.B., and Singh, H. (2019). Uptake, translocation and impact of green synthesized nanoceria on growth and antioxidant enzymes activity of *Solanum lycopersicum* L. *Ecotoxicology and Environmental Safety* 182: 109410. https://doi.org/10.1016/j.ecoenv.2019.109410.

Skovran, E., Raghuraman, C., and Martinez-Gomez, N.C. (2019). Lanthanides in Methylotrophy. *Current Issues in Molecular Biology* 33: 101–116. https://doi.org/10.21775/cimb.033.101.

Smith, A.T., Barupala, D., Stemmler, T.L., and Rosenzweig, A.C. (2015). A new metal binding domain involved in cadmium, cobalt and zinc transport. *Nature Chemical Biology* 11: 678–684. https://doi.org/10.1038/nchembio.1863.

Smoleń, S., Rozek, R., Ledwozyw-Smoleń, I., and Strzetelski, P. (2011). Preliminary evaluation of the influence of soil fertilization and foliar nutrition with iodine on the efficiency of iodine biofortification and chemical composition of lettuce. *Journal of Elementology* 16: 613–622. https://doi.org/10.5601/jelem.2011.16.4.10.

Smoleń, S., Kowalska, I., and Sady, W. (2014a). Assessment of biofortification with iodine and selenium of lettuce cultivated in the NFT hydroponic system. *Scientia Horticulturae* 166: 9–16. https://doi.org/10.1016/j.scienta.2013.11.011.

Smoleń, S., Sady, W., Ledwozyw-Smolen, I. et al. (2014b). Quality of fresh and stored carrots depending on iodine and nitrogen fertilization. *Food Chemistry* 159: 316–322. https://doi.org/10.1016/j.foodchem.2014.03.024.

Smoleń, S., Wierzbinska, J., Sady, W. et al. (2015). Iodine biofortification with additional application of salicylic acid affects yield and selected parameters of chemical composition of tomato fruits (*Solanum lycopersicum* L.). *Scientia Horticulturae* 188: 89–96. https://doi.org/10.1016/j.scienta.2015.03.023.

Solanki, P., Bhargava, A., Chhipa, H. et al. (2015). Nano-fertilizers and their smart delivery system. In: *Nanotechnologies in Food and Agriculture* (ed. M. Rai, C. Ribeiro, L. Mattoso and N. Duran), 81–101. Cham: Springer https://doi.org/10.1007/978-3-319-14024-7_4.

Solomon, S., Jain, N., Chaab, A. et al. (2016). Improvement of the sucrose content of early harvested sugarcane using stabilised orthosilicic acid. *Proceedings of the International Society of Sugar Cane Technologists* 29: 1–3.

Strzetelski, P., Smoleń, S., Rozek, S., and Sady, W. (2010). The effect of diverse iodine fertilization on nitrate accumulation and content of selected compounds in radish plants (*Raphanus sativus* L.). *Acta Scientiarum Polonorum. Hortorum Cultus* 9: 65–73.

Subbarao, G.V., Wheeler, R.M., Stutte, G.W., and Levine, L.H. (1999). How far can sodium substitute for potassium in red beet? *Journal of Plant Nutrition* 22: 1745–1761. https://doi.org/10.1080/01904169909365751.

Subramanian, K.S., Manikandan, A., Thirunavukkarasu, M., and Rahale, C.S. (2015). Nano-fertilizers for balanced crop nutrition. In: *Nanotechnologies in Food and Agriculture* (ed. M. Rai, C. Ribeiro, L. Mattoso and N. Duran), 69–80. Cham: Springer https://doi.org/10.1007/978-3-319-14024-7_4.

Sun, D., He, N., Chen, Q., and Duan, S. (2019). Effects of lanthanum on the photosystem II energy fluxes and antioxidant system of *Chlorella vulgaris* and *Phaeodactylum tricornutum*. *International Journal of Environmental Research Public Health* 16: 2242. https://doi.org/10.3390/ijerph16122242.

Szparaga, A., Kocira, S., Kocira, A. et al. (2018). Modification of growth, yield, and the nutraceutical and antioxidative potential of soybean through the use of synthetic biostimulants. *Frontiers in Plant Science* 9: 1401. https://doi.org/10.3389/fpls.2018.01401.

Tan, L.C., Nancharaiah, Y.V., van Hullebusch, E.D., and Lens, P.N.L. (2016). Selenium: environmental significance, pollution, and biological treatment technologies. *Biotechnology Advances* 34: 886–907. https://doi.org/10.1016/j.biotechadv.2016.05.005.

Teng, Y., Yang, J., Sun, Z. et al. (2011). Environmental vanadium distribution, mobility and bioaccumulation in different land-use Districts in Panzhihua Region, SW China.

Environmental Monitoring and Assessment 176: 605–620. https://doi.org/10.1007/s10661-010-1607-0.

Thornton, B.F. and Burdette, S.C. (2019). Seekers of the lost lanthanum. *Nature Chemistry* 11: 188. https://doi.org/10.1038/s41557-018-0208-3.

Tlustoš, P., Cigler, P., Hrubý, M. et al. (2005). The role of titanium in biomass production and its influence on essential elements' contents in field growing crops. *Plant, Soil and Environment* 51: 19–25. https://doi.org/10.17221/3551-PSE.

Trejo-Téllez, L.I., García-Jiménez, A., Escobar-Sepúlveda, H.F. et al. (2020). Silicon induces hormetic dose-response effects on growth and concentrations of chlorophylls, amino acids and sugars in pepper plants during the early developmental stage. *PeerJ* 8: e9224. https://doi.org/10.7717/peerj.9224.

Tripathi, D.K., Singh, S., Singh, S. et al. (2015). Micronutrients and their diverse role in agricultural crops: advances and future prospective. *Acta Physiologiae Plantarum.* 37 (7): 1–4.

Tubana, B.S., Babu, T., and Datnoff, L.E. (2016). A review of silicon in soils and plants and its role in US agriculture. *History and future perspectives. Soil Science* 181: 393–411. https://doi.org/10.1097/SS.0000000000000179.

Turra, C., Fernandes, E.A.N., Bacchi, M.A. et al. (2015). Effects of lanthanum on *Citrus* plant. *International Journal of New Technology and Research* 1: 48–50.

Vaculík, M., Lukačová, Z., Bokor, B. et al. (2020). Alleviation mechanisms of metal(loid) stress in plants by silicon: A review. *Journal of Experimental Botany* 2020: eraa288. https://doi.org/10.1093/jxb/eraa288.

Vatansever, R., Ozyigit, I.I., and Fili, E. (2016). Essential and beneficial trace elements in plants, and their transport in roots: a review. *Applied Biochemistry and Biotechnology* 181: 464–482. https://doi.org/10.1007/s12010-016-2224-3.

Von Tucher, S. and Schmidhalter, U. (2005). Lanthanum uptake from soil and nutrient solution and its effects on plant growth. *Journal of Plant Nutrition and Soil Science* 168: 574–580. https://doi.org/10.1002/jpln.200520506.

Vondráčková, S., Száková, J., Drábek, O. et al. (2015). Aluminium uptake and translocation in Al-hyperaccumulator *Rumex obtusifolius* is affected by low-molecular-weight organic acids content and soil pH. *PLoS ONE* 10: e0123351. https://doi.org/10.1371/journal.pone.0123351.

Wahdan, M.T. (2011). Effects of cobalt sulfate and choline chloride on fruiting and fruit quality of mango cv. *Succary Abiad. Life Science Journal* 8: 337–343.

Wang, L., Li, J., Zhou, Q. et al. (2014). Rare earth elements activate endocytosis in plant cells. *Proceedings of the National Academy of Sciences of the United States of America* 111: 12936–12941. https://doi.org/10.1073/pnas.1413376111.

Wang, L., He, J., Yang, Q. et al. (2017). A preliminary study on the effects of lanthanum (III) on plant vitronectin-like protein and its toxicological basis. *Ecotoxicology and Environmental Safety* 145: 227–234. https://doi.org/10.1016/j.ecoenv.2017.07.039.

Wang, L., Cheng, M., Yang, Q. et al. (2019a). Arabinogalactan protein-rare earth element complexes activate plant endocytosis. *Proceedings of the National Academy of Sciences of the United States of America* 116: 14349–14357. https://doi.org/10.1073/pnas.1902532116.

Wang, L., Suganuma, S., Hibino, A. et al. (2019b). Lanthanide-dependent methanol dehydrogenase from the legume symbiotic nitrogen-fixing bacterium *Bradyrhizobium diazoefficiens* strain USDA110. *Enzyme and Microbial Technology* 130: 109371. https://doi.org/10.1016/j.enzmictec.2019.10937.

Wang, H., Cimen, E., Singh, N., and Buckler, E. (2020a). Deep learning for plant genomics and crop improvement. *Current Opinion in Plant Biology* 54: 34–41. https://doi.org/10.1016/j.pbi.2019.12.010.

Wang, Y.M., Yang, Q., Xu, H. et al. (2020b). Physiological and transcriptomic analysis provide novel insight into cobalt stress responses in willow. *Scientific Reports* 10: 2308. https://doi.org/10.1038/s41598-020-59177-y.

Weng, H.X., Hong, C.L., Yan, A.L. et al. (2008a). Mechanism of iodine uptake by cabbage: effects of iodine species and where it is stored. *Biological Trace Element Research* 125: 59–71. https://doi.org/10.1007/s12011-008-8155-2.

Weng, H.X., Yan, A.L., Hong, C.L. et al. (2008b). Uptake of different species of iodine by water spinach and its effect to growth. *Biological Trace Element Research* 124: 184–194. https://doi.org/10.1007/s12011-008-8137-4.

White, P.J., Bowen, H.C., Parmaguru, P. et al. (2004). Interactions between selenium and sulphur nutrition in *Arabidopsis thaliana*. *Journal of Experimental Botany* 55: 1927–1937. https://doi.org/10.1093/jxb/erh192.

Whitted-Haag, B., Kopsell, D.E., Kopsell, D.A., and Rhykerd, R.L. (2014). Foliar silicon and titanium applications influence growth and quality characteristics of annual bedding plants. *The Open Horticulture Journal* 7: 6–15. https://doi.org/10.2174/1874840601407010006.

Włodarczyk, T., Balakhnina, T., Matichenkov, V. et al. (2019). Effect of silicon on barley growth and N_2O emission under flooding. *Science of The Total Environment* 685: 1–9. https://doi.org/10.1016/j.scitotenv.2019.05.410.

Wu, M. and Kubota, C. (2008). Effects of high electrical conductivity of nutrient solution and its application timing on lycopene, chlorophyll and sugar concentrations of hydroponic tomatoes during ripening. *Scientia Horticulturae* 116: 122–129. https://doi.org/10.1016/j.scienta.2007.11.014.

Wu, J., Wang, C., and Mei, X. (2001). Stimulation of taxol production and excretion in *Taxus* spp. cell cultures by rare earth chemical lanthanum. *Journal of Biotechnology* 85: 67–73. https://doi.org/10.1016/s0168-1656(00)00383-7.

Wu, G.Q., Feng, R.J., Liang, N. et al. (2015). Sodium chloride stimulates growth and alleviates sorbitol-induced osmotic stress in sugar beet seedlings. *Plant Growth Regulation* 75: 307–316. http://dx.doi.org/10.1007/s10725-014-9954-4.

Xie, C., Ma, Y., Yang, J. et al. (2019). Effects of foliar applications of ceria nanoparticles and $CeCl_3$ on common bean (*Phaseolus vulgaris*). *Environmental Pollution* 250: 530–536. https://doi.org/10.1016/j.envpol.2019.04.042.

Yamada, M., Kuroda, C., and Fujiyama, H. (2015). Function of sodium and potassium in growth of sodium-loving Amaranthaceae species. *Soil Science and Plant Nutrition* 62: 20–26. http://dx.doi.org/10.1080/00380768.2015.1075365.

Yamada, M., Kuroda, C., and Fujiyama, H. (2016). Growth promotion by sodium in amaranthaceous plants. *Journal of Plant Nutrition* 39: 1186–1193. https://doi.org/10.1080/01904167.2015.1069341.

Yang, J. (2020). *Novel Imaging and Spectroscopy*. Rijeka: Intech.

Yang, Q., Wang, L., He, J. et al. (2016). Vitronectin-like protein is a first line of defense against lanthanum (III) stress in Arabidopsis leaf cells. *Environmental and Experimental Botany* 130: 86–94. https://doi.org/10.1016/j.envexpbot.2016.05.011.

Zhai, W., Dai, Y., Zhao, W. et al. (2020). Simultaneous immobilization of the cadmium, lead and arsenic in paddy soils amended with titanium gypsum. *Environmental Pollution* 258: 113790. https://doi.org/10.1016/j.envpol.2019.113790.

Zhang, L., Goncalves, A.A.S., Jiang, B., and Jaroniec, M. (2018). Capture of iodide by $BiVO_4$ and Bi_2O_3: an insightful look at the process and its aftermath. *ChemSusChem* 11: 1486–1493. https://doi.org/10.1002/cssc.201800327.

Zhang, W., Long, J., Geng, J. et al. (2020a). Impact of titanium dioxide nanoparticles on Cd phytotoxicity and bioaccumulation in rice (*Oryza sativa* L.). *International Journal of Environmental Research and Public Health* 17: 2979. https://doi.org/10.3390/ijerph17092979.

Zhang, X., Lou, X., Zhang, H. et al. (2020b). Effects of sodium sulfide application on the growth of *Robinia pseudoacacia*, heavy metal immobilization, and soil microbial activity in Pb-Zn polluted soil. *Ecotoxicology and Environmental Safety* 197: 110563. https://doi.org/10.1016/j.ecoenv.2020.110563.

Zheng, C., Ouchi, T., Iizuka, A. et al. (2019). Deoxidation of titanium using Mg as deoxidant in $MgCl_2$-YCl_3 flux. *Metallurgical and Materials Transactions B* 50: 622–631. https://doi.org/10.1007/s11663-018-1494-2.

Zhong, Y. and Chen, J. (2020). Ameliorative effects of lanthanum(III) on copper(II) stressed rice (*Oryza sativa*) and its molecular mechanism revealed by transcriptome profiling. *Plant Physiology and Biochemistry* 152: 184–193. https://doi.org/10.1016/j.plaphy.2020.05.004.

Zicari, M.A., d'Aquino, L., Paradiso, A. et al. (2018). Effect of cerium on growth and antioxidant metabolism of *Lemna minor* L. *Ecotoxicology and Environmental Safety* 163: 536–543. https://doi.org/10.1016/j.ecoenv.2018.07.113.

5

Targeted Effects of Beneficial Elements in Plant Photosynthetic Process

Costanza Ceccanti[1,*], *Ermes Lo Piccolo*[1,*], *Lucia Guidi*[1,2], *and Marco Landi*[1,2]

[1] *Department of Agriculture, Food and Environment, University of Pisa, Pisa, Italy*
[2] *CIRSEC, Centre for Climatic Change Impact, University of Pisa, Pisa, Italy*

Introduction

The issue of essential elements is a complex and dynamic field of research in plant biology (Marschner 2012), and plant mineral nutrients are commonly classified by plant researchers based on the concept of 'essentiality' defined by Arnon and Stout (1939) and modified by Epstein (1965) as the necessary elements to complete the life cycle of a plant. Nowadays, there are no doubts about the number of macroelements (K, Ca, Mg, S, N, and P) whose concentrations are usually >1000 mg kg^{-1} dry weight in plant tissues in addition, obviously, to carbon, oxygen, and hydrogen. Conversely, the exact number of micronutrients, because of their lower concentrations in plants (≤100 mg kg^{-1} dry weight), still represents a debated topic since different authors may consider some microelements as essential or beneficial (Gupta and Gupta 2017; Vatansever et al. 2017). Indeed, due to the low levels of their requirement, additional micronutrients could be identified in the future and further molecular and physiological studies are needed to include/exclude some controversial elements in the list of essential plant micronutrients.

Although the essentiality of most micronutrients in plant metabolism is well established, there is evidence that other elements such as Si, Ni, Se, and Al (Tripathi et al. 2015; Gómez-Merino and Trejo-Téllez 2018) can be beneficial at certain concentration for plant growth. Previous literature surveys described the role of some beneficial elements *in planta* (Pilon-Smits et al. 2009; Broadley et al. 2012; Vatansever et al. 2017; Gómez-Merino and Trejo-Téllez 2018), while this chapter aims at describing the targeted ameliorative effect of 10 beneficial elements, namely, Ni, Mo, Cd, Al, Ti, La, Ce, Si, Se, and I, for the photosynthesis of plants growing in optimal or stressful conditions.

---

* Authors contributed equally to the book chapter.

Beneficial Chemical Elements of Plants: Recent Developments and Future Prospects, First Edition. Edited by Sangeeta Pandey, Durgesh Kumar Tripathi, Vijay Pratap Singh, Shivesh Sharma, and Devendra Kumar Chauhan.

The 10 beneficial elements have been divided in two main categories: (i) metals (including transition, basic metals, and lanthanides – Table 5.1) and (ii) non-metals (including semi-metals and non-metals *sensu stricto* – Table 5.2). For all these elements, the chemical form and concentration used in each experiment are reported as well as the species under investigation, given that a species-specific effect could be ascribable to both the ion forms as well as the plant species. In addition, the beneficial effect of these elements was reported in the case of plants grown under both optimal and stressful conditions.

The aim of this chapter was to summarize existing literature on the matter and figure out the possible ameliorative role of these elements for the photosynthetic process in healthy and/or stressed plants – this in order to fully understand the key role of some elements that are actually not considered as essential for plant development. In addition, it is conceivable that the basic concept of 'plant nutrient' should be re-examined, and the plant mineral nutrient list not only could include those elements necessary for completing the plant life cycle but also should include those elements that promote biomass development, plant yield, and/or lead to the reduction of the requirement of an essential element.

Effect of Metal Beneficial Elements

Some metals like Cu, Fe, and Mn are essential for photosynthesis as they are constituents or cofactors of many proteins and enzymes and consequently are essential for maintaining an active plant photosynthetic machinery (Aggarwal et al. 2012). Although the essentiality of these metals, as depicted in Table 5.1, it has been observed that other metals (transition metals such as Ni, Mo, and Cd; basic metals such as Al and Ti; and lanthanides such as Ce and La) improve the plant photosynthetic process. Unfortunately, for most of these beneficial metals, the intimal molecular/biochemical mechanism(s) through which they confer benefits to the plant photosynthetic apparatus has not been established yet, and only their physiological effects on photosynthetic parameters and performance have been described.

Overall, in plants grown under optimal conditions, the supply of all the analyzed beneficial elements (Cd excluded) improve the value of net photosynthesis (P_n), though in some cases it was related to the improvement of stomatal conductance (g_s) and therefore a higher CO_2 uptake (e.g. spray of Ti in *Rapahnus stivus* plants, Tighe-Neira et al. 2020; La-sprayed leaves of *Amorphophallus sinensis,* Li et al. 2020; Ce-treated plants of *Spinacea oleracea*, Liu et al. 2007), while in other cases to a higher chloroplast electron transport rate (ETR) and photosystem II (PSII) efficiency (Φ_{PSII}) (e.g. Ni-treated plants of *Pisum sativum*, Srivastava et al. 2012) or improvement of both stomatal and biochemical patterns (e.g. La-treated *Zea mays* plants, Cui et al. 2019).

Other authors observed that the effect of beneficial elements was attributable to their ability to promote the accumulation of photosynthetic pigments, thereby increasing the light harvesting by chloroplast. For example, Siqueira Freitas et al. (2018) measured a higher chlorophyll index in *Glycine max* plants treated with Ni incorporated into the substrate. Raliya et al. (2015) reported a similar stimulation of chlorophylls in Ti-sprayed leaves of *Vigna radiate*, and Li et al. (2020) found higher levels of both chlorophylls and carotenoids in La-treated plants of *A. sinensis*. Interestingly, the increase of light-harvesting

Table 5.1 Effect of beneficial transition metals (nickel, molybdenum, cadmium), basic metals (aluminium, titanium), and lanthanides (cerium, lanthanum) on photosynthetic-related traits.

Beneficial element	Treatment type	Chemical form and concentration	Species	Experimental conditions	Observation	Source
Nickel	Incorporated into the substrate	0.5 mg kg^{1} Ni	*Glycine max* (L.) Merr.	No stress	Increase of ETR and Chl index (SPAD)	Siqueira Freitas et al. (2018)
	Added to the nutrient solution	0.01 mM $NiCl_2{\cdot}6H_2O$	*Pisum sativum* L.	No stress	Increase in Chl, P_n, PSII, and PSI activity	Srivastava et al. (2012)
Molybdenum	Incorporated into the soil	0.3 mg K_2MoO_4 pot^{-1}	*Hordeum vulgare* L.	Salt stress (NaCl to obtain 5, 10, and 15 dS m^{-1})	Higher F_v/F_m	Bagheri and Jafari (2012)
	Incorporated into the soil	2 mg Mo kg^{-1}	*Glycine max* L. Merrill	Sulphur stress (0.05 µL SO_2 L^{-1})	Higher P_n	Gupta et al. (1991)
	Incorporated into the soil	50 or 100 mg NH_4MoO_3 kg^{-1}	*Brassica napus* L. cv. Zaqin No. 1	Cd stress (0.5 or 6 mg $CdCl_2$ kg^{-1} soil)	Higher F_v/F_o, F_v/F_m, Φ_{PSII}, ETR, and qP	Han et al. (2020)
	Incorporated into the soil	0.15–0.30 mg Mo kg^{-1}	*Brassica napus* L. Zhongyouza No. 2	No stress and P fertilizer	Higher P_n	Liu et al. (2010)
	Incorporated into the soil	0.15 mg $(NH_4)_6Mo_7O_{24}{\cdot}4H_2O$ kg^{-1}	*Brassica napus* L.	No stress	Higher P_n, g_s, E	Qin et al. (2017)
Cadmium	Incorporated into the nutrient solution	25, 50, 100 µM $CdCl_2$	*Gossypium hirsutum* L.	No stress	Higher C_i and Chl *a/b*	Liu et al. (2014)
Aluminium	Incorporated into the substrate	$Al_2(SO_4)_3{\cdot}18H_2O$	*Camellia sinensis* (L.) O. Kuntze	No stress	Increase of expression of genes associated with photosynthesis	Xu et al. (2017)

(*Continued*)

Table 5.1 (Continued)

Beneficial element	Treatment type	Chemical form and concentration	Species	Experimental conditions	Observation	Source
Titanium	Sprayed on leaves	2.5 mg $Plant^{-1}$ ionic Ti	(*Glycine max* (L.) Merr.)	Low light	Higher Rubisco activase activity	Hussain et al. (2019)
	Sprayed on leaves	10-1500 mg L^{-1} TiO_2 NPs	*Raphanus sativus* L.	No stress	Transient increase of P_n and g_s	Tighe-Neira et al. (2020)
	Sprayed on leaves	50–100 mg kg^{-1} TiO_2	*Fragaria* × *ananassa* Duch.	No stress	Increase in Chl content	Choi et al. (2015)
	Sprayed on leaves	Solution with 0.25% TiO_2 NPs	*Arabidopsis thaliana* (L.) Heynh.	No stress	Increase of (LHCII) b gene expression and LHCII content	Ze et al. (2011)
	Sprayed on leaves	10 mg L^{-1} TiO_2 NPs	*Vigna radiata* L.	No stress	Increase in Chl content	Raliya et al. (2015)
	Seeds soaked into solution Sprayed on leaves	Solution with 0.03% TiO_2 NPs	*Spinacia oleracea* L.	No stress	Higher P_n, Rubisco carboxylase, and activase activity	Gao et al. (2008)
Lanthanum	Sprayed on leaves	100 mg L^{-1} La	*Pseudostellaria heterophylla* (Miquel) Pax	No stress	Higher P_n, g_s, F_v/F_m, and Φ_{PSII}	Ma et al. (2017)
	Seeds soaked into solution	400–800 μmol L^{-1} $LaCl_3$	*Zea mays* L.	No stress	Higher Chl, Car, P_n, g_s, F_v/F_m, and Φ_{PSII}	Cui et al. (2019)
	Sprayed on leaves	160 mg L^{-1} $LaCl_3$	*Amorphophallus sinensis* Belval	No stress	Higher Chl, Car, P_n, and g_s	Li et al. (2020)
	Seeds soaked into solution Sprayed on leaves	500 μmol L^{-1} $CeCl_3$	*Spinacia oleracea* L.	No stress	Promote the energy transfer from LHCII to PSII and accelerate water photolysis and O_2 evolution	Liu et al. (2007)

Cerium	Infiltrated into leaves	CeO_2 NPs with low Ce^{3+}/Ce^{4+} ratio	*Arabidopsis thaliana* (L.) Heynh.	Heat stress Excess light Dark chilling	Higher P_n, V_{cmax}, F_v/F_m, and Φ_{PSII}	Wu et al. (2017)
	Sprayed on leaves	10 mg L^{-1} CeO_2 NPs	*Sorghum bicolor* (L.) Moench	C	Higher Chl index (SPAD), P_n, g_s, and F_v/F_m	Djanaguiraman et al. (2018)
	Seeds soaked into solution Sprayed on leaves	20 µmol L^{-1} $CeCl_3$	*Arabidopsis thaliana* (L.) Heynh.	No stress	Higher Chl, increase of (*LHCII*)-*b* gene expression, and LHCII content	Xiaoqing et al. (2009)
	Seeds soaked into solution Sprayed on leaves	20 µmol L^{-1} $CeCl_3$	*Spinacia oleracea* L.	No stress	Promote the energy transfer from LHCII to PSII and accelerate water photolysis and O_2 evolution	Liu et al. (2007)
	Added to the nutrient solution	0.1 µmol L^{-1} $Ce(NO_3)_3.6H_2O$	*Phaseolus vulgaris* L.	Drought	Higher Chl, P_n, and g_s	Salgado et al. (2020)

Abbreviations: ETR, electron transport rate; F_v/F_m, maximal quantum yield of PSII photochemistry; F_v/F_o, potential activities of PSII; F_m, maximal fluorescence yield in the dark-adapted state; Φ_{PSII}, effective quantum yield of PSII photochemistry; qP, photochemical quenching; Chl, chlorophylls; Car, carotenoids; LHCII, light-harvesting complex II; P_n, net photosynthetic rate; g_s, stomatal conductance; E, transpiration rate; C_i, intercellular CO_2 concentration; V_{cmax}, maximum carboxylation rate.

Table 5.2 Effect of a semi-metal (silicon) and non-metals (selenium, iodine) on photosynthetic-related traits.

Beneficial element	Treatment type	Chemical form and concentration	Species	Experimental conditions	Observation	Source
Silicon	Sprayed on leaves	200 mg kg^{-1} Na_2SiO_3	*Glycine max* (L.) Merr.	Low light	Higher P_n and g_s	Hussain et al. (2021)
	Applied to both substrate and leaves	Soil: 0.08 kg m^{-2} of fertilizer with 25% of SiO_2 Leaves: 2 mM Na_2SiO_3	*Glycine max* (L.) Merr.	No stress	Higher P_n and g_s	Tripathi et al. (2021)
	Sprayed on leaves	6 mM Na_2SiO_7	*Triticum aestivum* L.	Drought	Higher P_n and g_s	Maghsoudi et al. (2016)
	Incorporated into the substrate	2 mM Si added as SiO_2	*Sorghum* cv. BRS3 32	Drought	Higher P_n, g_s, and P_n/C_i	Avila et al. (2020)
	Added to the nutrient solution	1.5 mM added as K_2SiO_3	*Oryza sativa* L.	Zn-stress	Increase of expression of genes associated with photosynthesis	Song et al. (2014)
	Added to the nutrient solution	2 mM Si	*Oryza sativa* L.	No stress	Higher P_n and g_m	Detmann et al. (2012)
	Incorporated into the soil	100–500 mg L^{-1} CaO. SiO_2	*Saccharum officinarum* L.	Drought	Higher P_n and g_s	Verma et al. (2020)
	Added to the nutrient solution	2 mM Si	*Oryza sativa* L.	Biotic stress	Higher P_n and g_m	Pereira et al. (2020)
	Incorporated into the substrate	1 g kg^{-1}(soil) CaO.SiO_2	*Piper nigrum* L.	Biotic stress	Higher P_n, g_s, and WUE	D'Addazio et al. (2020)

Selenium	Incorporated into the nutrient solution	1 µM Na_2SeO_3 or Na_2SeO_4	*Solanum lycopersicum* L. cultivar Micro-Tom	Cd stress (0.5 mM $CdCl_2$)	Higher E, g_s, and P_n	Alves et al. (2020)
	Foliar application	1 g Na_2SeO_4 m^{-3}	*Fagopyrum esculentum* (Moench) cv. Darja and *Fagopyrum tataricum* (L.) Gaertn.	UV-B radiation	Higher Φ_{PSII}	Breznik et al. (2005)
	Incorporated into the soil	0.5, 1, and 2 mg Se kg^{-1}	*Oryza sativa* L. cv. IAC 202	Drought	Higher P_n and E	Andrade et al. (2018)
	Incorporated into the nutrient solution	2, 4, 8 mg $Na_2SeO_3{\cdot}5H_2O$ L^{-1}	*Raphanus sativus* (L.) cv. Cherry Belle	Cd stress (5–10 mg L^{-1} $3CdSO_4{\cdot}8H_2O$)	Higher Φ_{PSII} and F_v/F_m	Auobi Amirabad et al. (2020)
	Foliar application	75 mg Na_2SeO_4 L^{-1}	*Sorghum bicolor* (L.) Moench	High temperature stress (40/30 °C day/night)	Higher P_n, g_s, and E	Djanaguiraman et al. (2010)
	Foliar application	0.017 g Se L^{-1}	*Pyrus bretschneideri*, *Vitis vinifera*, and *Prunus persica*	No stress	Higher P_n, g_s, E, and C_i	Feng et al. (2015)
	Incorporated into the nutrient solution	2 µM Na_2SeO_4	*Brassica napus* var. *oleifera*	Cd stress (400 µM $CdCl_2$)	Higher F_v/F_m, Φ_{PSII}, and qP	Filek et al. (2010)
	Foliar application	40 mg Na_2SeO_3 L^{-1}	*Oryza sativa* L. cv. WYHZ	No stress	Higher g_s and E	Gao et al. (2018)
	Incorporated into the nutrient solution	0.1–0.3 mg Na_2SeO_3 or Na_2SeO_4 L^{-1}	*Helianthus annuus* L. cv. Arena PR	No stress	Higher P_n, E, F_v/F_m, and F_v/F_0	Garousi et al. (2016)

(Continued)

Table 5.2 (Continued)

Beneficial element	Treatment type	Chemical form and concentration	Species	Experimental conditions	Observation	Source
	Foliar application	10 mg Na_2SeO_4 L^{-1}	*Solanum tuberosum* L. cv. Desiree	Drought	Higher Φ_{PSII}	Germ et al. (2007)
	Plant application	10, 50, and 100 µg Na_2SeO_4 ml^{-1}	*Brassica oleracea var.* *botrytis L.*	No stress	Higher F_m	Ghasemi et al. (2016)
	Aerial part application	30 g Na_2SeO_4 ha^{-1}	*Hordeum vulgare L.* cv. Rihane-03	Drought	Higher F_v/F_m and g_s	Habibi (2013)
	Incorporated to the soil	5 mg Na_2SeO_4 kg^{-1}	*Helianthus annuus* L. cv. Hysun33	Salt stress (NaCl to obtain 15 dS m^{-1})	Higher F_v/F_o	Habibi (2017a)
	Incorporated into the nutrient solution	1 mg Na_2SeO_4 L^{-1}	*Petroselinum crispum* L.	Salt stress (80 mM NaCl)	Higher F_v/F_m	Habibi (2017b)
	Incorporated into the nutrient solution	4, 6, 8 mg $SeCl_2$ L^{-1}	*Caspicum annum* L. Dutch Simins 301	Heat stress (35 ± 2 °C, for 4 h a day)	Higher F_v/F_m, E, g_s, and C_i	Haghighi et al. (2019)
	Incorporated into the soil	3–6 mg Na_2SeO_3 kg^{-1}	*Nicotiana tabacum* L. cv. Yunyan 87	No stress	Higher P_n, g_s, and C_i	Jiang et al. (2015)
	Incorporated into the nutrient solution	1 µM Na_2SeO_3	*Zea mays* L. cv. Nongda 108	Salt stress (100 mM NaCl)	Higher P_n, g_s, and E	Jiang et al. (2017)
	Foliar application	10, 20, 30, 40, and 50 µM Na_2SeO_4	*Oryza sativa* L.	No stress	Higher P_n and g_s	Luo et al. (2019)

Foliar application	40 mg Na_2SeO_4 L^{-1}	*Triticum aestivum* L. cv. Pasban-90	Drought	Higher E, P_n, and g_s	Nawaz et al. (2015)
Foliar application	50–150 mg Na_2SeO_4 L^{-1}	*Olea europaea* L. cv. Maurino	Drought	Higher E, P_n, and g_s	Proietti et al. (2013)
Incorporate into the soil	20, 40, 80 μM	*Brassica juncea* cv. Krishna Kranti	10^{-8} M Epibrassinolide solution through the foliar application	Higher P_n, g_s, C_i, WUE	Naz et al. (2015)
Foliar application	1 g Na_2SeO_4 m^{-3}	*Fagopyrum esculentum* Moench.	Drought	Higher g_s and Φ_{PSII}	Tadina et al. (2007)
Incorporated into the soil	10, 20, 50 g Na_2SeO_3 ha^{-1}	*Oryza sativa* L.	No stress	Higher P_n, C_i, E, Fv, F_m, F_v/F_m, and F_v/F_o	Zhang et al. (2014)
Incorporated into the nutrient solution	5 ppm Na_2SeO_4	*Allium sativum* L.	Cd stress (10^{-44} M $CdCl_2$)	Higher P_n, E, and g_s	Sepehri and Gharehbaghli (2019)
Incorporated into the nutrient solution	10 μM $Na_2SeO_3 \cdot 5H_2O$	*Solanum lycopersicum* L.	Cd stress (150 mg L^{-1} $CdSO_4 \cdot 5H_2O$)	Higher Chl, Car, P_n, g_s, E, F_v/F_m, Φ_{PSII}, and qP	Alyemeni et al. (2018)
Incorporated into the soil	2, 4, 6 μM Kg^{-1} Na_2SeO_4	*Brassica juncea* L.	Cr stress (300 μM Kg^{-1})	Higher Chl, Car, P_n, g_s, C_i, E	Handa et al. (2019)
Foliar application	20, 40 mg L^{-1} Na_2SeO_4	*Zea mays* L. cv. Agaiti-2002	Salt stress (EC = 12 dS m^{-1})	Higher Chl	Ashraf et al. (2018)
Incorporated into the nutrient solution	1.5, 6 μM Na_2SeO_3	*Vicia faba* L. cv. cv. Nadwislanski	Pb stress (50 μM)	Higher Chl	Mroczek-Zdyrska et al. (2017)

(*Continued*)

Table 5.2 (Continued)

Beneficial element	Treatment type	Chemical form and concentration	Species	Experimental conditions	Observation	Source
Iodine	Incorporated into the nutrient solution	20, 40 μmol KI L^{-1} and 20, 40, 80 μmol KIO_3 L^{-1}	*Lactuca sativa* L. var. longifolia	No stress	Higher P_n, g_s, E	Blasco et al. (2011)
	Incorporated into the nutrient solution	0.20 or 10 μM KIO_3	*Arabidopsis thaliana,* ecotype Columbia 0	No stress	Presence of iodinated proteins involved in photosynthesis (PsbA, PsbB, PsbC, PsbD, PsbO, PsbP, PsbQ, PsbR, CAB3, LHCB2.1, LHCB1B1, LHCB3, LHCB5, Psb27, Psb29, Psb31, Psb33, MPH1, PsaB, PsaE, PsaF, PsaH, PetA, PetC, PETE2, FNR1, and PGR2-like A)	Kiferle et al. (2020)

Abbreviations: F_v/F_m, maximal quantum yield of PSII photochemistry; F_v/F_o, potential activities of PSII; F_m, maximal fluorescence yield in the dark-adapted state; Φ_{PSII}, effective quantum yield of PSII photochemistry; qP, photochemical quenching; Chl, chlorophylls; Car, carotenoids; P_n, net photosynthetic rate; g_s, stomatal conductance; E, transpiration rate; C_i, intercellular CO_2 concentration; g_m, mesophyll conductance; WUE, water use efficiency.

complex of *PSII (LHCII)-b* gene expression and LHCII content was observed by Ze et al. (2011) in *Arabidopsis* plants treated with a solution of 0.25% Ti nanoparticles (NPs) over the leaf lamina. Both the increase of *LHCII-b* gene expression and higher level of chlorophyll content were also reported by Xiaoqing et al. (2009) in *Arabidopsis* Ce-treated plants.

In relation to the improvement of light harvesting, another effect described for Ce and La was their capability to promote the energy transfer from LHCII to PSII reaction centres and accelerate the activity of the oxygen-evolving complex (OEC), thereby increasing the capability to act as an electron donor for chlorophylls (Liu et al. 2007). This is related to the findings of Hong et al. (2002) who observed that La or Ce can replace Mg in chlorophyll formation of spinach and support PSII more significantly than PSI. Further study indicated that La or Ce entering the chloroplasts not only increased the Mg-chlorophyll content but also formed Ce-chlorophylls or La-chlorophylls because La or Ce coordinated with nitrogen in the porphyrin rings. Due to the presence of 4f electrons and the alternation valence, Ce treatment induced the highest improvement in light absorption (as compared to La), energy transfer from LHCII to PSII, excitation energy distribution from PSII to PSI, and fluorescence quantum yield around 680 nm (Hong et al. 2002, 2005).

Besides the promotion of CO_2 stomatal uptake, a higher ETR, and photochemical efficiency of PSII, another positive effect has been revealed for some beneficial metals (e.g. Ti) that consists in the capability of Ti-treated plants of *S. oleracea* to increase the rate of ribulose-1,5-*bis*phosphate carboxylase/oxygenase (Rubisco) activase and the carboxylation activity of Rubisco enzyme (Gao et al. 2008). The authors observed that TiO_2 treatment increased by 42% the amount of Rubisco activase, thereby improving Rubisco carboxylation activity and, consequently, leading to a higher rate of photosynthesis. Of note, the authors observed that bulk TiO_2 effect was not as significant as nano-anatase TiO_2, as the nano-anatase TiO_2 size (5 nm) is much smaller than that of bulk TiO_2, which entered spinach cell more easily.

Considering the positive role of some beneficial metals, as reported above, other researches have investigated the possibility that beneficial-metal-promoted photosynthetic performances could be advantageous in plants subjected to other external stressors. For example, Han et al. (2020) postulated that the mitigation effect of Mo in plants of *Brassica napus* subjected to Cd stress consisted in the capacity of Mo to support the synthesis of enzymes correlated with the photosynthesis and the photosynthetic cycle in the rape plant. Bagheri and Jafari (2012) showed that salinity induced the decline of nitrate reductase activity and nitrogen content in barley plants. This might have decreased the activity of photosynthesis-related enzymes, and by an increase in mesophilic resistance thereby reducing the maximal PSII photochemical efficiency, maximal quantum yield of PSII photochemistry (F_v/F_m) ratio, a decrease that is expectable due to the impairment of ferredoxin-related processes (like N reduction cycle). Mo application partially relieved the photoinhibition of *Hordeum vulgare* plants likely due to the capability to promote nitrogenase activity of plants inoculated with *Azospirillium brasilense*. Similarly, Gupta et al. (1991) observed that the treatment with Mo in SO_2 stressed plants (0.05 μl L^{-1}) resulted in an increase of specific root nodule nitrogenase activity and, in this way, partially restoring the P_n level of *G. max* plants.

Wu et al. (2017) observed that *Arabidopsis* plants subjected to different stressors (heat, light excess, chilling) benefit from Ce treatments that resulted effectively in promoting plant photosynthesis. Overall, plants treated with Ce and exposed to abiotic stress exhibited an increase of up to 19% in Φ_{PSII}, 67% in carbon assimilation rate, and 61% in Rubisco carboxylation rate as compared to control plants. The authors hypothesized that Ce particles are transported into chloroplasts via non-endocytic pathways, influenced by the electrochemical gradient of the plasma membrane potential, and they are able to scavenge stress-triggered reactive oxygen species (ROS), including hydrogen peroxide, superoxide anion, and hydroxyl radicals. Of note, only lower Ce^{3+}/Ce^{4+} ratio (35%) resulted effectively in counteracting ROS levels, whilst higher Ce^{3+}/Ce^{4+} ratio (61%) increased overall leaf ROS levels and did not protect photosynthesis from oxidative damage during abiotic stress. The capability of Ce to reduce the level of oxidative stress was also proposed by plants subjected to drought (Djanaguiraman et al. 2018; Roychoudhury and Tripathi 2020; Salgado et al. 2020).

Effect of Non-metal Beneficial Elements

Other non-metal elements (semi-metals including Si and non-metal *sensu stricto*, i.e. Se and I) have shown to improve the photosynthetic performances of plants, both in optimal and limiting conditions (Table 5.2). In particular, Si and Se have been deeply investigated in crop species but, differently to the aforementioned beneficial metals, researches dealing with beneficial non-metals are mainly devoted to test their possible ameliorative effect in stressful conditions.

From Table 5.2, it emerges that one of the most consistent effects induced by treatments with beneficial non-metal elements was the improvement of P_n related to the enhancement of g_s, in both stressful and optimal conditions. For example, Hussain et al. (2021) found that application of Si increased P_n, g_s, and transpiration rate (E) of soya beans; reduced intercellular carbon dioxide concentration (C_i); and increased chlorophyll content. Tripathi et al. (2021) found similar stimulatory effects to photosynthesis in soya bean as well as an increase of the nodule number and size associated with improved root morphological traits. Based on this evidence, the authors proposed that enhancement of water and mineral uptake from the soil due to Si treatment and nodulation activity was responsible for the increase of soya bean photosynthetic rate and yield. Moreover, on rice (*Oryza sativa*), a high-Si-accumulating plant, some authors found that Si had a feed-forward effect on photosynthesis, due to an increased mesophyll conductance (g_m) (Detmann et al. 2012; Pereira et al. 2020). The g_m reflects the entire diffusive CO_2 pathway in the leaf, from the substomatal air spaces to the carboxylation sites located in the chloroplast stroma, and then through biophysical barriers (e.g. intercellular air spaces, cell wall thickness) (Gago et al. 2020). Although there is the evidence that Si might affect the cell wall thickening (Yamamoto et al. 2012), how Si could influence g_m is still unclear, and future studies related to this aspect might help to understand Si-photosynthesis interactions. The understandings of the mechanistic ameliorative effect induced by Si in plants is of outmost importance given that silicon dioxide comprises about 60% of the Earth's crust; therefore, it can be present at relevant concentrations in the rooting zones of plants.

Selenium is likely the most studied beneficial element in plants, given that it is one of the most important elements used for plant biofortification due to its relevance in the human diet (Malagoli et al. 2015). Selenium is involved in antioxidant and ROS regulation, photosynthesis increase, heavy metal uptake, and transport inhibition and construction of chloroplast components and cell membrane (Pilon-Smits 2015). In plants, selenate could be transported by sulphate transporters since both (sulphate and selenate) showed chemical similarity (White et al. 2004). Selenium usually has a dual role in plants: it plays as a pro-oxidant at high levels and causes detrimental effects on plants. On the other hand, at lower concentrations, Se improves plant growth and performance and alleviates the damage caused by various abiotic stressors (Roychoudhury and Tripathi 2019; Auobi Amirabad et al. 2020).

As reported in Table 5.2, Se, either added to the soil or provided by foliar spray, increased the photosynthetic performances in plants subjected to a plethora of stresses, such as drought (e.g. Germ et al. 2007; Andrade et al. 2018), heat (Haghighi et al. 2019), salinity (Jiang et al. 2017), or the excess of other metals, i.e. Cd (Filek et al. 2010; Auobi Amirabad et al. 2020). Overall, the addition of Se correlates with a better maintenance of PSII performances (higher F_v/F_m and Φ_{PSII}) as well as a higher level of g_s, E, and P_n measured by gas exchange.

The capability of Se to positively modulate stomatal conductance, CO_2 absorption, and water use efficiency (WUE) could be particularly useful for plant experiencing osmotic stressors such as drought or salinity. Andrade et al. (2018) found that rice plants subjected to drought showed a decline of P_n, g_s, E, and WUE, while Se application attenuated these reductions. The protective Se effect consisted in minimizing damages to the chloroplast structure and promoting the electron transport system compared to the control. Similar effects exerted by Se were reported by Nawaz et al. (2015) and Proietti et al. (2013) in *Triticum aestivum* and *Olea europea* plants, respectively, once subjected to conditions of limited water availability. Under salinity conditions, Habibi (2017b) found that the PSII photochemical efficiency improvement in the Se-supplied plants of *Petroselinum crispum* was associated with lower damages to PSII due to both (i) higher dissipation of light energy absorbed by photosynthetic pigments by non-photochemical quenching and (ii) lower occurrence of oxidative stress. Indeed, it was recently observed that a low amount of Se is able to improve the activities of some antioxidant enzymes such as glutathione peroxidase and superoxide dismutase in tomato (Castillo-Godina et al. 2016); ascorbate peroxidase, monodehydroascorbate reductase, dehydroascorbate reductase, and glutathione reductase in maize; and glutathione reductase and glutathione peroxidase in wheat (Balakhnina and Nadezhkina 2017; Yildiztugay et al. 2017).

In case of heat stress, the increasing crop yield observed in Se-sprayed plants of *Sorghum bicolor* was due to an enhanced photosynthetic rate, which was achieved by lower ROS concentrations in leaf coupled with higher antioxidant enzyme activity (Djanaguiraman et al. 2010). The authors demonstrated that Se is needed to reduce the ROS level as well as the extent of lipid peroxidation and membrane damage, thereby enhancing the photosynthetic rate and antioxidant enzyme activities under heat stress.

Though I biofortification has also been deeply investigated (Izydorczyk et al. 2021), less information is available on the ameliorative effect of I on photosynthesis as compared to Se, likely due to higher toxicity for plants and the difficulty to find adequate concentrations that both improves I accumulation and, at the same time, does not cause damages to the plant. Similarly to other beneficial elements, Blasco et al. (2011) observed that administration of I as KI or KIO_3 promoted g_s and E, which correlated with higher P_n, but only when 20 and 40 μM I^- were supplied. In this case, an enhancement of sucrose pathway was also reported. Conversely, a higher level of I^- (80 μM) did not improve the aforementioned gas exchange parameters nor the sucrose pattern, but Se induced an increment of WUE in treated lettuce plants (Blasco et al. 2011).

A deep investigation by Kiferle et al. (2020) in *Arabidopsis* plants reveals the occurrence of iodinated proteins and that several iodinated proteins were involved in photosynthesis such as (i) constituents of PSII, i.e. proteins of OEC and the reaction centre; (ii) components of the LHCII; (iii) factors involved in the assembly/preservation of the photosystems; and (iv) proteins involved in PSII protection from photodamage or in the degradation of the photodamaged D1 reaction centre. Other iodinated proteins were components of PSI complex, cytochrome b6/f complex, plastocyanin electron carrier, ferredoxin-plastoquinone reductase involved in cyclic electron flow around PSI, and components of PGRL1-like of cyclic electron flow PGR5/PGRL1 complex. The authors suggested that the incorporation of I in iodinated proteins and the maintenance of their functionality are supportive for the fact that I could be inserted among plant nutrients.

Conclusion

The present chapter aims at reviewing the role of beneficial elements in plant photosynthesis. As reported herein, the mechanisms through which some of those beneficial elements improve the photosynthetic traits of investigated species are far to be fully elucidated. In most cases, only some beneficial effects on plant physiological parameters are reported, without the explanation of the mechanistic bases of these ameliorative roles. Therefore, it is essential that further molecular and physiological studies will be aimed to clarify the intimal mechanisms through which those beneficial elements improve plant photosynthesis, which is a pivotal process to ensure the proper plant development and, finally, the plant yield. The regulatory processes in which those elements are involved need to be clarified in the condition of plant stress too, in order to possibly exploit those ameliorative effects to counteract some common stressors that can affect crop species in field cultivation. Finally, it is conceivable that beneficial elements should be included in the list of mineral nutrients because, though they are not essential, they serve a functional role in plant nutrition.

References

Aggarwal, A., Sharma, I., Tripati, B.N. et al. (2012). Metal toxicity and photosynthesis. In: *Photosynthesis: Overviews on Recent Progress & Future Perspectives* (ed. A. Aggarwal, I. Sharma, B.N. Tripathi, et al.), 229–236. New Delhi: I K International Publishing House Pvt. Ltd.

Alves, L.R., Rossatto, D.R., Rossi, M.L. et al. (2020). Selenium improves photosynthesis and induces ultrastructural changes but does not alleviate cadmium-stress damages in tomato plants. *Protoplasma* 257 (2): 597–605. https://doi.org/10.1007/s00709-019-01469-w.

Alyemeni, M.N., Ahanger, M.A., Wijaya, L. et al. (2018). Selenium mitigates cadmium-induced oxidative stress in tomato (*Solanum lycopersicum* L.) plants by modulating chlorophyll fluorescence, osmolyte accumulation, and antioxidant system. *Protoplasma* 255: 459–469.

Auobi Amirabad, S., Behtash, F., and Vafaee, Y. (2020). Selenium mitigates cadmium toxicity by preventing oxidative stress and enhancing photosynthesis and micronutrient availability on radish (*Raphanus sativus* L.) cv. Cherry Belle. *Environmental Science and Pollution Research* 27 (11): 12476–12490. https://doi.org/10.1007/s11356-020-07751-2.

Andrade, F.R., da Silva, G.N., Guimarães, K.C. et al. (2018). Selenium protects rice plants from water deficit stress. *Ecotoxicology and Environmental Safety* 164: 562–570. https://doi.org/10.1016/j.ecoenv.2018.08.022.

Arnon, D.I. and Stout, P.R. (1939). The essentiality of certain elements in minute quantity for plants with special reference to copper. *Plant Physiology* 14: 371–375.

Ashraf, M.A., Akbar, A., Parveen, A. et al. (2018). Phenological application of selenium differentially improves growth, oxidative defense and ion homeostasis in maize under salinity stress. *Plant Physiology and Biochemistry* 123: 268–280.

Avila, R.G., Magalhães, P.C., da Silva, E.M. et al. (2020). Silicon supplementation improves tolerance to water deficiency in sorghum plants by increasing root system growth and improving photosynthesis. *Silicon* 12: 2545–2554. https://doi.org/10.1007/s12633-019-00349-5.

Bagheri, A. and Jafari, A. (2012). Effect of salinity and molybdenum application on photosynthesis, nitrogenase activity and yield of barley inoculated with *Azosprillium brasilense*. *Cereal Research Communications* 40 (2): 235–245. https://doi.org/10.1556/CRC.40.2012.2.8.

Balakhnina, T. and Nadezhkina, E. (2017). Effect of selenium on growth and antioxidant capacity of *Triticum aestivum* L. during development of lead-induced oxidative stress. *Russian Journal of Plant Physiology* 64: 215–223.

Blasco, B., Rios, J.J., Leyva, R. et al. (2011). Photosynthesis and metabolism of sugars from lettuce plants (*Lactuca sativa* L. var. *longifolia*) subjected to biofortification with iodine. *Plant Growth Regulation* 65 (1): 137–143. https://doi.org/10.1007/s10725-011-9583-0.

Breznik, B., Germ, M., Gaberscik, A., and Kreft, I. (2005). Combined effects of elevated UV-B radiation and the addition of selenium on common (*Fagopyrum esculentum* Moench) and tartary [*Fagopyrum tataricum* (L.) Gaertn.] buckwheat. *Photosynthetica* 43 (4): 583–589. https://doi.org/10.1007/s11099-005-0091-1.

Broadley, M., Brown, P., Cakmak, I. et al. (2012). Beneficial elements. In: *Mineral Nutrition of Higher Plants* (ed. H. Marschner), 249–269. Elsevier https://doi.org/10.1016/B978-0-12-384905-2.00008-X.

Castillo-Godina, R., Foroughbakhch-Pournavab, R., and Benavides-Mendoza, A. (2016). Effect of selenium on elemental concentration and antioxidant enzymatic activity of tomato plants. *Journal of Agricultural Science and Technology* 18: 233–244.

Choi, H.G., Moon, B.Y., Bekhzod, K. et al. (2015). Effects of foliar fertilization containing titanium dioxide on growth, yield and quality of strawberries during cultivation. *Horticulture, Environment, and Biotechnology* 56: 575–581. https://doi.org/10.1007/s13580-015-0023-3.

Cui, W., Kamran, M., Song, Q. et al. (2019). Lanthanum chloride improves maize grain yield by promoting photosynthetic characteristics, antioxidants enzymes and endogenous hormone at reproductive stages. *Journal of Rare Earths* 37: 781–790. https://doi.org/10.1016/j.jre.2018.12.006.

D'Addazio, V., Silva, J.V.G., Jardim, A.S. et al. (2020). Silicon improves the photosynthetic performance of black pepper plants inoculated with *Fusarium solani* f. sp. *piperis*. *Photosynthetica* 58: 692–701. https://doi.org/10.32615/ps.2019.182.

Detmann, K.C., Araújo, W.L., Martins, S.C.V. et al. (2012). Silicon nutrition increases grain yield, which, in turn, exerts a feed-forward stimulation of photosynthetic rates via enhanced mesophyll conductance and alters primary metabolism in rice. *New Phytologist* 196: 752–762. https://doi.org/10.1111/j.1469-8137.2012.04299.x.

Djanaguiraman, M., Prasad, P.V., and Seppanen, M. (2010). Selenium protects sorghum leaves from oxidative damage under high temperature stress by enhancing antioxidant defence system. *Plant Physiology and Biochemistry* 48 (12): 999–1007. https://doi.org/10.1016/j.plaphy.2010.09.009.

Djanaguiraman, M., Nair, R., Giraldo, J.P., and Prasad, P.V.V. (2018). Cerium oxide nanoparticles decrease drought-induced oxidative damage in *Sorghum* leading to higher photosynthesis and grain yield. *ACS Omega* 3: 14406–14416. https://doi.org/10.1021/acsomega.8b01894.

Epstein, E. (1965). Mineral metabolism. In: *Plant Biochemistry* (ed. J. Bonner and J.E. Varner), 438–466. New York, USA: Academic Press.

Feng, T., Chen, S.S., Gao, D.Q. et al. (2015). Selenium improves photosynthesis and protects photosystem II in pear (*Pyrus bretschneideri*), grape (*Vitis vinifera*), and peach (*Prunus persica*). *Photosynthetica* 53 (4): 609–612. https://doi.org/10.1007/s11099-015-0118-1.

Filek, M., Kościelniak, J., Łabanowska, M. et al. (2010). Selenium-induced protection of photosynthesis activity in rape (*Brassica napus*) seedlings subjected to cadmium stress. Fluorescence and EPR measurements. *Photosynthesis Research* 105 (1): 27–37. https://doi.org/10.1007/s11120-010-9551-y.

Gago, J., Daloso, D.M., Carriquí, M. et al. (2020). Mesophyll conductance: the leaf corridors for photosynthesis. *Biochemical Society Transactions* 48: 429–439. https://doi.org/10.1042/BST20190312.

Gao, F., Liu, C., Qu, C. et al. (2008). Was improvement of spinach growth by nano-TiO_2 treatment related to the changes of Rubisco activase? *Biometals* 21: 211–217. https://doi.org/10.1007/s10534-007-9110-y.

Gao, M., Zhou, J., Liu, H. et al. (2018). Foliar spraying with silicon and selenium reduces cadmium uptake and mitigates cadmium toxicity in rice. *Science of the Total Environment* 631: 1100–1108. https://doi.org/10.1016/j.scitotenv.2018.03.047.

Garousi, F., Kovacs, B., and Veres, S. (2016). Investigation of photosynthesis status of sunflower plants up-taking different forms of selenium. *Advances in Plants and Agriculture Research* 3 (1): 00083. https://doi.org/10.15406/apar.2016.03.00083.

Germ, M., Kreft, I., Stibilj, V., and Urbanc-Berčič, O. (2007). Combined effects of selenium and drought on photosynthesis and mitochondrial respiration in potato. *Plant Physiology and Biochemistry* 45 (2): 162–167. https://doi.org/10.1016/j.plaphy.2007.01.009.

Ghasemi, Y., Ghasemi, K., Pirdashti, H., and Asgharzadeh, R. (2016). Effect of selenium enrichment on the growth, photosynthesis and mineral nutrition of Broccoli. *Notulae Scientia Biologicae* 8 (2): 199–203. https://doi.org/10.15835/nsb.8.2.9804.

Gómez-Merino, F.C. and Trejo-Téllez, L.I. (2018). The role of beneficial elements in triggering adaptive responses to environmental stressors and improving plant performance. In: *Biotic and Abiotic Stress Tolerance in Plants* (ed. S. Vats), 137–172. Singapore: Springer https://doi.org/10.1007/978-981-10-9029-5_6.

Gupta, M. and Gupta, S. (2017). An overview of selenium uptake, metabolism, and toxicity in plants. *Frontiers in Plant Science* 1–14: https://doi.org/10.3389/fpls.2016.02074.

Gupta, G., Grund, B., and Narayanan, R. (1991). Photosynthesis and nitrogenase activity in soybean treated with sulphur dioxide and molybdenum. *Plant Science* 79 (2): 157–161. https://doi.org/10.1016/0168-9452(91)90102-E.

Habibi, G. (2013). Effect of drought stress and selenium spraying on photosynthesis and antioxidant activity of spring barley. *Acta Agriculturae Slovenica* 101 (1): 31. https://doi.org/10.2478/acas-2013-0004.

Habibi, G. (2017a). Physiological, photochemical and ionic responses of sunflower seedlings to exogenous selenium supply under salt stress. *Acta Physiologiae Plantarum* 39: 1–9. https://doi.org/10.1007/s11738-017-2517-3.

Habibi, G. (2017b). Selenium ameliorates salinity stress in *Petroselinum crispum* by modulation of photosynthesis and by reducing shoot Na accumulation. *Russian Journal of Plant Physiology* 64: 368–374. https://doi.org/10.1134/S1021443717030086.

Haghighi, M., Ramezani, M.R., and Rajaii, N. (2019). Improving oxidative damage, photosynthesis traits, growth and flower dropping of pepper under high temperature stress by selenium. *Molecular Biology Reports* 46 (1): 497–503. https://doi.org/10.1007/s11033-018-4502-3.

Han, Z., Wei, X., Wan, D. et al. (2020). Effect of molybdenum on plant physiology and cadmium uptake and translocation in rape (*Brassica napus* L.) under different levels of cadmium stress. *International Journal of Environmental Research and Public Health* 17 (7): 2355. https://doi.org/10.3390/ijerph17072355.

Handa, N., Kohli, S.K., Sharma, A. et al. (2019). Selenium modulates dynamics of antioxidative defence expression, photosynthetic attributes and secondary metabolites to mitigate chromium toxicity in *Brassica juncea* L. plants. *Environmental and Experimental Botany* 161: 180–192. https://doi.org/10.1016/j.envexpbot.2018.11.009.

Hong, F.S., Wang, L., Meng, X.X. et al. (2002). The effect of cerium on the chlorophyll formation of spinach. *Biological Trace Element Research* 89: 263–276. https://doi.org/10.1385/bter:89:3:263.

Hong, F.S., Wang, L., and Tao, Y. (2005). Mechanism of $LaCl_3$ on increasing photosystem II activity of spinach. *Chinese Journal of Chemistry* 23: 617–621. https://doi.org/10.1002/cjoc.200590617.

Hussain, S., Iqbal, N., Brestic, M. et al. (2019). Changes in morphology, chlorophyll fluorescence performance and Rubisco activity of soybean in response to foliar application of ionic titanium under normal light and shade environment. *Science of the Total Environment* 658: 626–637. https://doi.org/10.1016/j.scitotenv.2018.12.182.

Hussain, S., Shuxian, L., Mumtaz, M. et al. (2021). Foliar application of silicon improves stem strength under low light stress by regulating lignin biosynthesis genes in soybean (*Glycine max* (L.) Merr.). *Journal of Hazardous Materials* 401: 123256. https://doi.org/10.1016/j.jhazmat.2020.123256.

Izydorczyk, G., Ligas, B., Mikula, K. et al. (2021). Biofortification of edible plants with selenium and iodine. A systematic literature review. *Science of the Total Environment* 754: 141983. https://doi.org/10.1016/j.scitotenv.2020.141983.

Jiang, C., Zu, C., Shen, J. et al. (2015). Effects of selenium on the growth and photosynthetic characteristics of flue-cured tobacco (*Nicotiana tabacum* L.). *Acta Societatis Botanicorum Poloniae* 84 (1): https://doi.org/10.5586/asbp.2015.006.

Jiang, C., Zu, C., Lu, D. et al. (2017). Effect of exogenous selenium supply on photosynthesis, Na^+ accumulation and antioxidative capacity of maize (*Zea mays* L.) under salinity stress. *Scientific Reports* 7 (1): 1–14. https://doi.org/10.1038/srep42039.

Kiferle, C., Martinelli, M., Salzano, A.M. et al. (2020). Evidences for a nutritional role of iodine in plants. *bioRxiv* https://doi.org/10.1101/2020.09.16.300079.

Li, X.X., Yu, B., Dong, Y.Y. et al. (2020). Lanthanum chloride enhances the photosynthetic characteristics and increases konjac glucomannan contents in *Amorphophallus sinensis*. Belval. *Photosynthetica* 58: 165–173. https://doi.org/10.32615/ps.2019.184.

Liu, X., Su, M., Liu, C. et al. (2007). Effect of 4f electron characteristics and alternation valence of rare earths on photosynthesis: regulating distribution of energy and activities of spinach chloroplast. *Journal of Rare Earths* 25: 495–501. https://doi.org/10.1016/S1002-0721(07)60463-3.

Liu, H., Hu, C., Sun, X. et al. (2010). Interactive effects of molybdenum and phosphorus fertilizers on photosynthetic characteristics of seedlings and grain yield of *Brassica napus*. *Plant and Soil* 326 (1): 345–353. https://doi.org/10.1007/s11104-009-0014-1.

Liu, L., Sun, H., Chen, J. et al. (2014). Effects of cadmium (Cd) on seedling growth traits and photosynthesis parameters in cotton (*Gossypium hirsutum* L.). *Plant Omics* 7 (4): 284. https://doi.org/10.3316/INFORMIT.580389605356279.

Luo, H.W., He, L.X., Du, B. et al. (2019). Foliar application of selenium (Se) at heading stage induces regulation of photosynthesis, yield formation, and quality characteristics in fragrant rice. *Photosynthetica* 57 (4): 1007–1014. https://doi.org/10.32615/ps.2019.114.

Ma, Y., Zou, H., Gu, H. et al. (2017). Stimulatory effect of lanthanum nitrate on the root tuber yield of *Pseudostellaria heterophylla* via improved photosynthetic characteristics. *Journal of Rare Earths* 35: 610–620. https://doi.org/10.1016/S1002-0721(17)60954-2.

Maghsoudi, K., Emam, Y., and Pessarakli, M. (2016). Effect of silicon on photosynthetic gas exchange, photosynthetic pigments, cell membrane stability and relative water content of different wheat cultivars under drought stress conditions. *Journal of Plant Nutrition* 39: 1001–1015. https://doi.org/10.1080/01904167.2015.1109108.

Malagoli, M., Schiavon, M., dall'Acqua, S., and Pilon-Smits, E.A.H. (2015). Effects of selenium biofortification on crop nutritional quality. *Frontiers in Plant Science* 6: https://doi.org/10.3389/fpls.2015.00280.

Marschner, P. (2012). *Mineral Nutrition of Higher Plants*, 483–643. London, Waltham, UK: Academic Press.

Mroczek-Zdyrska, M., Strubińska, J., and Hanaka, A. (2017). Selenium improves physiological parameters and alleviates oxidative stress in shoots of lead-exposed *Vicia faba* L. minor plants grown under phosphorus-deficient conditions. *Journal of Plant Growth Regulation* 36: 186–199. https://doi.org/10.1007/s00344-016-9629-7.

Nawaz, F., Ahmad, R., Ashraf, M.Y. et al. (2015). Effect of selenium foliar spray on physiological and biochemical processes and chemical constituents of wheat under drought stress. *Ecotoxicology and Environmental Safety* 113: 191–200. https://doi.org/10.1016/j.ecoenv.2014.12.003.

Naz, F.S., Yusuf, M., Khan, T.A. et al. (2015). Low level of selenium increases the efficacy of 24-epibrassinolide through altered physiological and biochemical traits of *Brassica juncea* plants. *Food Chemistry* 185: 441–448. https://doi.org/10.1016/j.foodchem.2015.04.016.

Pereira, L.F., Martins, S.C.V., Aucique-Pérez, C.E. et al. (2020). Silicon alleviates mesophyll limitations of photosynthesis on rice leaves infected by *Monographella albescens*. *Theoretical and Experimental Plant Physiology* 32: 163–174. https://doi.org/10.1007/s40626-020-00178-7.

Pilon-Smits, E.A.H. (2015). Selenium in plants. In: *Progress in Botany, Progress in Botany* (ed. U. Lüttge and W. Beyschlag), 93–107. Cham: Springer International Publishing https://doi.org/10.1007/978-3-319-08807-5_4.

Pilon-Smits, E.A., Quinn, C.F., Tapken, W. et al. (2009). Physiological functions of beneficial elements. *Current Opinion in Plant Biology* 12: 267–274. https://doi.org/10.1016/j.pbi.2009.04.009.

Proietti, P., Nasini, L., Del Buono, D. et al. (2013). Selenium protects olive (*Olea europaea* L.) from drought stress. *Scientia Horticulturae* 164: 165–171. https://doi.org/10.1016/j.scienta.2013.09.034.

Qin, S., Hu, C., Tan, Q., and Sun, X. (2017). Effect of molybdenum levels on photosynthetic characteristics, yield and seed quality of two oilseed rape (*Brassica napus* L.) cultivars. *Soil Science and Plant Nutrition* 63 (2): 137–144. https://doi.org/10.1080/00380768.2017.1286232.

Raliya, R., Biswas, P., and Tarafdar, J.C. (2015). TiO_2 nanoparticle biosynthesis and its physiological effect on mung bean (*Vigna radiata* L.). *Biotechnology Reports* 5: 22–26. https://doi.org/10.1016/j.btre.2014.10.009.

Roychoudhury, A. and Tripathi, D.K. (2019). *Molecular Plant Abiotic Stress: Biology and Biotechnology*, vol. 1, 1–480. John Wiley & Sons.

Roychoudhury, A. and Tripathi, D.K. (2020). *Protective Chemical Agents in the Amelioration of Plant Abiotic Stress: Biochemical and Molecular Perspectives*, vol. 1, 1–704. John Wiley & Sons.

Salgado, O.G.G., Teodoro, J.C., Alvarenga, J.P. et al. (2020). Cerium alleviates drought-induced stress in *Phaseolus vulgaris*. *Journal of Rare Earths* 38: 324–331. https://doi.org/10.1016/j.jre.2019.07.014.

Sepehri, A. and Gharehbaghli, N. (2019). Selenium alleviate cadmium toxicity by improving nutrient uptake, antioxidative and photosynthetic responses of garlic. *Russian Journal of Plant Physiology* 66 (1): 152–159. https://doi.org/10.1134/S1021443719010151.

Siqueira Freitas, D., Wurr Rodak, B., Rodrigues dos Reis, A. et al. (2018). Hidden nickel deficiency? Nickel fertilization via soil improves nitrogen metabolism and grain yield in soybean genotypes. *Frontiers in Plant Science* 9: 614. https://doi.org/10.3389/fpls.2018.00614.

Song, A., Li, P., Fan, F. et al. (2014). The effect of silicon on photosynthesis and expression of its relevant genes in rice (*Oryza sativa* L.) under high-Zinc stress. *PLoS ONE* 9: e113782. https://doi.org/10.1371/journal.pone.0113782.

Srivastava, G., Kumar, S., Dubey, G. et al. (2012). Nickel and ultraviolet-B stresses induce differential growth and photosynthetic responses in *Pisum sativum* L. seedlings. *Biological Trace Element Research* 149: 86–96. https://doi.org/10.1007/s12011-012-9406-9.

Tadina, N., Germ, M., Kreft, I. et al. (2007). Effects of water deficit and selenium on common buckwheat (*Fagopyrum esculentum* Moench.) plants. *Photosynthetica* 45 (3): 472–476. https://doi.org/10.1007/s11099-007-0080-7.

Tighe-Neira, R., Reyes-Díaz, M., Nunes-Nesi, A. et al. (2020). Titanium dioxide nanoparticles provoke transient increase in photosynthetic performance and differential response in antioxidant system in *Raphanus sativus* L. *Scientia Horticulturae* 269: 109418. https://doi.org/10.1016/j.scienta.2020.109418.

Tripathi, D.K., Singh, S., Singh, S. et al. (2015). Micronutrients and their diverse role in agricultural crops: advances and future prospective. *Acta Physiologiae Plantarum* 37 (7): 1–4.

Tripathi, P., Na, C.I., and Kim, Y. (2021). Effect of silicon fertilizer treatment on nodule formation and yield in soybean (*Glycine max* L.). *European Journal of Agronomy* 122: 126172. https://doi.org/10.1016/j.eja.2020.126172.

Vatansever, R., Ozyigit, I.I., and Filiz, E. (2017). Essential and beneficial trace elements in plants, and their transport in roots: a review. *Applied Biochemistry and Biotechnology* 181: 464–482. https://doi.org/10.1007/s12010-016-2224-3.

Verma, K.K., Wu, K.C., Verma, C.L. et al. (2020). Developing mathematical model for diurnal dynamics of photosynthesis in *Saccharum officinarum* responsive to different irrigation and silicon application. *PeerJ* 8: e10154. https://doi.org/10.7717/peerj.10154.

White, P.J., Bowen, H.C., Parmaguru, P. et al. (2004). Interactions between selenium and sulphur nutrition in *Arabidopsis* thaliana. *Journal of Experimental Botany* 55 (404): 1927–1937. https://doi.org/0.1093/jxb/erh192.

Wu, H., Tito, N., and Giraldo, J.P. (2017). Anionic cerium oxide nanoparticles protect plant photosynthesis from abiotic stress by scavenging reactive oxygen species. *ACS Nano* 11: 11283–11297. https://doi.org/10.1021/acsnano.7b05723.

Xiaoqing, L., Hao, H., Chao, L. et al. (2009). Physico-chemical property of rare earths—effects on the energy regulation of Photosystem II in *Arabidopsis thaliana*. *Biological Trace Element Research* 130: 141–151. https://doi.org/10.1007/s12011-009-8321-1.

Xu, Q., Wang, Y., Ding, Z. et al. (2017). Aluminum induced physiological and proteomic responses in tea (*Camellia sinensis*) roots and leaves. *Plant Physiology and Biochemistry* 115: 141–151. https://doi.org/10.1016/j.plaphy.2017.03.017.

Yamamoto, T., Nakamura, A., Iwai, H. et al. (2012). Effect of silicon deficiency on secondary cell wall synthesis in rice leaf. *Journal of Plant Research* 125: 771–779. https://doi.org/10.1007/s10265-012-0489-3.

Yildiztugay, E., Ozfidan-Konakci, C., Kucukoduk, M., and Tekis, S.A. (2017). The impact of selenium application on enzymatic and non-enzymatic antioxidant systems in *Zea mays* roots treated with combined osmotic and heat stress. *Archives of Agronomy and Soil Science* 63: 261–275. https://doi.org/10.1080/03650340.2016.1201810.

Ze, Y., Liu, C., Wang, L. et al. (2011). The regulation of TiO_2 nanoparticles on the expression of Light-Harvesting Complex II and photosynthesis of chloroplasts of *Arabidopsis thaliana*. *Biological Trace Element Research* 143: 1131–1141. https://doi.org/10.1007/s12011-010-8901-0.

Zhang, M., Tang, S., Huang, X. et al. (2014). Selenium uptake, dynamic changes in selenium content and its influence on photosynthesis and chlorophyll fluorescence in rice (*Oryza sativa* L.). *Environmental and Experimental Botany* 107: 39–45. https://doi.org/10.1016/j.envexpbot.2014.05.005.

6

Aluminium Stress in Plants: Consequences and Mitigation Mechanisms

Akbar Hossain[1], Sagar Maitra[2], Sukamal Sarker[3], Abdullah Al Mahmud[4], Zahoor Ahmad[5], Reza Mohammad Emon[6], Hindu Vemuri[7], Md Abdul Malek[6], M. Ashraful Alam[8], Md Atikur Rahman[9], Md Jahangir Alam[4], Nasrin Jahan[10], Preetha Bhadra[11], Debojyoti Moulick[12], Saikat Saha[13], Milan Skalicky[14], and Marian Brestic[14,15]

[1] *Department of Agronomy, Bangladesh Wheat and Maize Research Institute, Dinajpur, Bangladesh*
[2] *Department of Agronomy, Centurion University of Technology and Management, Odisha, India*
[3] *School of Agriculture and Rural Development, Faculty Centre for IRDM, Ramakrishna Mission Vivekananda Educational and Research Institute, Ramakrishna Mission Ashrama, Narendrapur, Kolkata, India*
[4] *On-Farm Research Division, Bangladesh Agricultural Research Institute, Gaibandha, Bangladesh*
[5] *Department of Botany, University of Central Punjab, Punjab Group of Colleges, Bahawalpur, Pakistan*
[6] *Plant Breeding Division, Bangladesh Institute of Nuclear Agriculture, Mymensingh, Bangladesh*
[7] *International Maize and Wheat Improvement Center, Hyderabad, India*
[8] *Plant Breeding Division, Spices Research Centre, Bangladesh Agricultural Research Institute, Bogura, Bangladesh*
[9] *Spices Research Center, Bangladesh Agricultural Research Institute (BARI), Bogra, Bangladesh*
[10] *Plant Genetic Resources Centre, Bangladesh Agricultural Research Institute, Gazipur, Bangladesh*
[11] *Department of Biotechnology, Centurion University of Technology and Management, Paralakhemundi, Odisha, India*
[12] *Plant Stress Biology and Metabolomics Laboratory, Assam University, Silchar, Assam, India*
[13] *Nadia Krishi Vigyan Kendra, Bidhan Chandra Krishi Viswavidyalaya, Gayeshpur, West Bengal, India*
[14] *Department of Botany and Plant Physiology, Faculty of Agrobiology, Food and Natural Resources, Czech University of Life Sciences Prague, Prague, Czechia*
[15] *Department of Plant Physiology, Slovak University of Agriculture, Nitra, Slovak Republic*

Introduction

Aluminium (Al) is the primary reason for toxicity in acid soils due to the phytotoxicity of these soils along with high mineral content (Rahman et al. 2018). There is reduction in nutrient and water uptake in the root tip of plant caused by Al toxicity because of cell division and elongation (Yang et al. 2013). Many crops and/or crop species showed tolerance to Al for its natural variation (Singh et al. 2017). Varietal improvement of specific traits has limiting factor drawbacks of taking more exercise. Introduced of gene(s) are characterized may not specific to Al express in maximum cases, and it may also be initiated by another stress situations (Samac and Tesfaye 2003). All over the world crop production is hampered due to acid soils. It is estimated that around 30–40% of cultivable lands in the world have a below 5.5 pH, and it is increasing day by day leading to enhancement of soil acidity

Beneficial Chemical Elements of Plants: Recent Developments and Future Prospects, First Edition. Edited by Sangeeta Pandey, Durgesh Kumar Tripathi, Vijay Pratap Singh, Shivesh Sharma, and Devendra Kumar Chauhan.

(Von Uexküll and Mutert 1995). Basic cations (Ca^{2+}, Mg^{2+}, K^{+}, Na^{+}) are naturally low due to these elements leaching off from soil. It also reduces pH and the buffering retention of soil. When there are decreases in the soil pH, Al is dissolved and the ratio of phytotoxic Al ions increases in the soil (Kidd and Proctor 2001). Acid soils cause plants to uptake phosphorus, nitrogen, calcium, magnesium, and potassium. Due to acid sensitivity of symbiotic *rhizobia*, legume crops face more dissenting (Hartel and Bouton 1989).

To manage acid soils various types of strategies are applied. In North America and Europe, lime (calcium carbonate) is applied for increasing soil pH and the transformation of Al to minimize toxic form and apply phosphorus (P) fertilizer. But this type of soil operation is not applicable for highly erodible slopes, nor are they economical where large areas require amendment or where transportation costs are prohibitive (Samac and Tesfaye 2003). Organic soil management performed an interim ameliorative effect on Al toxicity. Organic residues may cause increasing soil pH for a spell, and for that reason complexing of protons with organic acids and destruction of it with decarboxylation of organic acids (Haynes and Mokolobate 2001).

With the increasing demand of acid soil production in many areas, it is obligatory to develop crop species with Al tolerance. Arable land under acidic soil is necessary to enhance crop production. Overutilization of lands for cultivation by inorganic fertilizers, not maintaining crop rotation, excessive use of underground water have resulted in declining soil fertility. Recovery of arable land for acidity, salinity, alkalinity, and nutrients are necessary. Increasing demand for food should increase production in the selected cultivable area (Ahmed et al. 2001). Recovery trends in high yielding varieties and declining natural resources. Furthermore, Al stress-tolerant crop varieties need to adapt in acidic conditions everywhere. On the other hand, attempts to introgression of stress-tolerant genes from tolerant lines into cultivable popular high yielding crop varieties would be an option to develop crop varieties that will increase crop production despite increasing Al-induced acidic soils. The current chapter focuses on the adverse effects of Al-induced oxidative stress in plants and its mitigation strategies through physiological, biochemical, and cellular mechanisms.

An Overview of Al Toxicity in Plants

Effect on Root Growth

During the past era, root growth inhibition and its upshot concern to aluminium (Al) in many crop varieties have been reported (Kikui et al. 2005). As a result, Al toxicity has been assessed by root growth inhibition extensively (Figure 6.1). The combination of cell division and elongation leads to root growth. Researchers have begun to look at Al-induced (de)regulation at the cell cycle in the past decade with some research concentrating distorts in the cell division process (mitosis phase) and very few on other phases (Silva et al. 2000).

The reduction of mitotic action has been reported in wheat (Li et al. 2008), in maize (Doncheva et al. 2005), in barley (Budikova and Durcekova 2004), and in bean (Marienfeld et al. 2000), because Al has been exposed in root tips. The suppression of cell elongation has to be the primary mechanism leading to root growth inhibition, which was defined by Ciamprova (2002). Awasthi et al. (2017) conducted experiments on Al toxicity in rice and found that Al toxicity is a severe issue in rice production because higher concentration

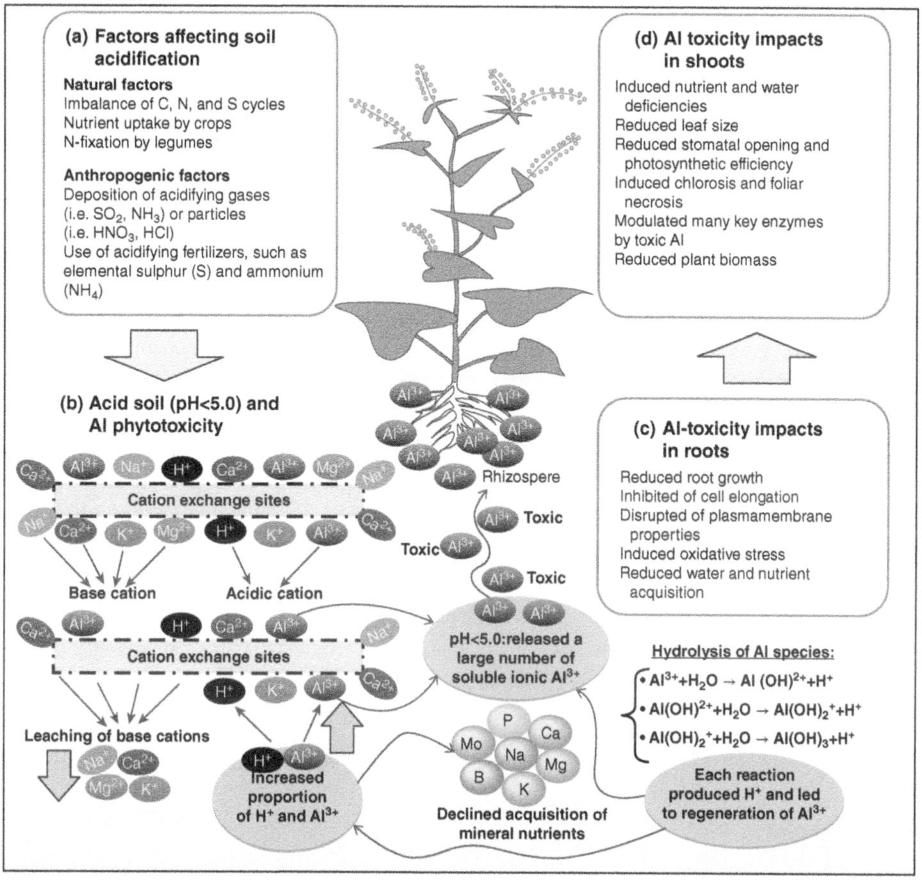

Figure 6.1 Al toxicity in plants: (a) causes of soil acidification; (b) increasing exchangeable Al^{3+} leads to the inability of plant nutrients in soil solution; (c) Al injury in roots of plants; (d) Al injury in shoots. *Source:* Rahman et al. (2018)/MDPI/CC BY-4.0.

Al reduced the rice root green and dry biomass as well as relative water content greatly in sensitive rice varieties/cultivars. Llugany et al. (1995) found root growth inhibition in Al-sensitive maize in a short period of exposure (30 minutes), which was caused by the slowdown of the cell division course. It was noted by Doncheva et al. (2005) that in the proximal meristem the suppression of cell division was found after five minutes, Al exposure, and again the suppression of the root cell division in the apical meristem was found within 10 or 30 minutes of Al exposure. Moreover, in the meristematic region of the root tip, Al can accumulate in the cells' nuclei within 30 minutes (Silva et al. 2000). Thus, it is still unclear whether the root growth inhibition is caused by the inhibition of cell elongation or cell division.

Moreover in recent times, there have been reports by Yi et al. (2010) that Al exposure has resulted in irregular make advance in cell division (mitosis) and caused micronuclei are formed in *Vicia faba* roots, which is in accord along with Al-induced chromosomal abnormalities in wheat roots (Bulanova et al. 2001). It is clear that Al directs towards cell cycle disruptions; however, numerous queries fall short to explain. So, for instance, it is still unclear how and where Al affects all over the cell cycle; however, these changes are species and region dependent, and the majority of the research was conducted in root tips, how to make the presumed changes are applied over time, and/or whether they can be reversible after Al elimination.

Root growth suppression and root diameter increase have been seen in roots subjected to Al (Zobel et al. 2007) pointed out that the plant cytoskeleton might be a cellular mark of Al toxicity (Rahman et al. 2018; Figure 6.1). Al-induced effects on microtubules and actin microfilaments were studied by Horst et al. (1999) and demonstrated that exposure to Al, microtubules, and microfilaments are changed, in their steadiness, association, and polymerization. Frantzios et al. (2005) observed Al treatment leading to the ineffectiveness of actin filaments and the creation of actin deposits in *Triticum turgidum* (Frantzios, et al. 2005). Zhang et al. (2007) showed that Al inhibited actin and profilin genes.

Profilin acts as a binding protein that provides cells with the capability to renovate the cytoskeleton (Krishnan and Moens 2009). An elongation defect was observed in *Arabidopsis thaliana* by the reduction in profilin expression (Ramachandran et al. 2000). The utmost susceptible root zone to Al toxicity is in considerable interest. It was believed that a key role in the process of Al toxicity or protection was played by the root cap was described (Bennet and Breen 1991). Nevertheless, the Al-induced inhibition of root growth in maize was not affected by the removal of the root cap was revealed by Ryan et al. (1993). They also proposed that the meristem is the main site of aluminium toxicity. Later on, Sivaguru and Horst (1998) applied the Al to root sections in an Al-sensitive maize cultivar and stated that accumulated Al in the distal transition zone resulted in the quick suppression of the root lengthening and pointed out that this root zone is targeted by aluminium.

Oxidative Stress

Al-stimulated oxidative stress is caused after any situation interferes with cellular redox homeostasis (Yamamoto et al. 2001). The reactive oxygen species (ROS) has the potential to oxidize the various cellular components that leads to cellular death. Metals are commonly known to function as catalysts in ROS production as well as to cause oxidative damage in plants. Aluminium is not in itself a transition metal and will not be able to catalyze redox reactions; though, aluminium exposure is leading to oxidative stress (Liu et al. 2008). The reason being, aluminium particles produce electrostatic bonds preferably by oxygen donor ligands, pectin in the cell wall, and the plasma membrane (PM) in the outer surface appears to be the main target of aluminium (Yamamoto et al. 2001). Again, Jones and Kochian (1997) revealed that Al binding with bio-membranes leads to rigidification that appears to accelerate the radical chain reactions by Fe ions and improve the lipid peroxidation (Yamamoto et al. 2003). Similar research findings have been indicated in various crops by Peixoto et al. (1999) in sorghum, Kuo and Kao (2003) in rice, Guo et al. (2004) in barley, and Liu et al. (2008) in triticale. In *Pisum*, for Al-treated seedlings in a simple Ca solution, the comparable allocation was found on the root tip surface area and was complemented by suppression in root growth was reported by Yamamoto et al. (2001). Though, the damage of membrane virginity was only spotted at the boundary of the ruptures over the surface of the root tip.

Moreover, Yamamoto et al. (2003) resolved that aluminium-induced lipid peroxidation is the main signal of Al deposition and gives the impression to induce partially callose production but does not induce root growth suppression. Later, however, Boscolo et al. (2003) reported that in maize lipid peroxidation was not induced by Al treatment, which indicated that lipids do not belong to the key cellular target of oxidative stress. Thus, it seems that the responses differ in plant species of cellular target of oxidative stress. Some researchers stated the production of ROS and changes in the antioxidants because of aluminium exposure. In pea seedlings, after two hours of Al exposure, ROS is detected in the root tip and intensified with time (Yamamoto et al. 2003). Again, Jones et al. (2006) stated that in all epidermal cells of maize roots, ROS production rate accelerates, only within 10 minutes exposure to aluminium.

Nutrient Imbalances

Aluminium-induced nutritional imbalances in many plant species were reported. Aluminium accrual-dependent various nutritional imbalances were reported in the eleven families of pteridophytes accrual (Olivares et al. 2009; Rahman et al. 2018; Figure 6.1). Mariano and Keltjens (2005) found that in maize uptake of macronutrients (Ca, Mg) and micronutrients (Mn, Zn) was affected by aluminium. Furthermore, the higher concentration of Ca, Mg (Mariano and Keltjens 2005), and K (Giannakoula et al. 2008) were found to be accumulated in Al-tolerant maize genotypes than in susceptible genotypes. Once again, in wheat roots both susceptible and tolerant genotypes displayed a reduction in K and Mg contents, while Ca, Al, and Si contents were enhanced by Al exposure (Silva et al. 2011).

However, in both roots and shoots the greater nutritional imbalances and Al accumulation were observed in the sensitive wheat genotypes than the tolerant one (Silva et al. 2011). They also observed an increase of Ca accumulation in rye-sensitive genotype, contrary to the tolerant rye genotype resulting from aluminium exposure. Though other findings stated dissimilar effects in Al-induced nutritional imbalances in maize, amplification in nutrient elements (except K, Mn, and Zn) in Al-treated roots and the rest in shoots bit differences of Ca and Mg were revealed by Lidon et al. (2000). In different parts of the tomato plant (roots, shoots, and leaves) aluminium exposure resulted from the decrease in the content of macro (Ca, K, and Mg) and micro elements (Mn, Fe, and Zn) was recounted by (Simon et al. 1994). Again, Zobel et al. (2007) related the adjustments in fine root diameter with fluctuation in the intensity of a few nutrients (as N, P, and Al). It appears that the difference between the tolerances to aluminium could be due to their differences in the uptake and the ability to maintain the appropriate concentrations and to make use of the nutrients more effectively. The variations in nutrient uptake, accumulation, and translocation are obvious among plant species and within each species. Furthermore, every author used various Al dilutions, different nutritive solutions, and time exposures. So it is hard to make a common and precise example of nutritional imbalances caused by aluminium.

Mechanisms for Al Stress Tolerance in Plants

Next to oxygen and silicon, Al is the third amplest element in the lithosphere as well as the most plenty metallic element in soil (Sahoo et al. 2019). Along with arsenic and cadmium, and abiotic stresses like drought and salinity, Al toxicity is also a challenging issue that attracts researchers from around the world (Chowardhara et al. 2019a, b;

Moulick et al. 2019, 2020). Depending upon the soil's acidity or alkalinity, the behaviour of Al is also changed. In acidic soil where pH is less than 5.5, generation of trivalent Al, i.e. Al^{3+} from various Al bearing ores, gets smoothen especially in agricultural land of tropics and subtropics. Whereas in alkaline or near-neutral soil Al is non-toxic. Upon exposure to Al in acidic soil, there is reduction in root elongation, decreasing crop yield mainly by modulating water and bioavailable nutrient accumulation (Saha et al. 2019).

It is observed that root cells are the primary contact point of Al and from there Al shifted to internal organelles like cytoskeleton, nucleus, and mitochondria (Dai et al. 2013). The movement and deposition of Al actually triggers a series of complex interactions that leads in higher ROS generation, deforms signalling pathway, and thus hampers the healthy growth of plants (Ma 2007; Guo et al. 2017). Plants have evolved to cope up with the adverse consequences of Al exposure by employing various means.

Speaking of Al tolerance or detoxification mechanisms, three distinct classes or approaches are well recognized. Among these mechanisms are (a) detoxified and confiscations of Al into vacuoles or other organelles located in roots and shoots and (b) Al segregation from the root employing chelating agents (oxalate, malate, and citrate) in a wide range of plants (Liu et al. 2014; Ma et al. 2014). It has been noticed that exposure of Al in acidic soil condition leads to a significant amount of Al accumulation taking place (within a hundred mg kg^{-1} on dry weight basis). Whereas Al accumulator and hyperaccumulators belonging to different families grown in the tropical and sub-tropical area usually accumulate 1000–10 000 mg kg^{-1} of Al on a dry weight basis in the leaf (Souza et al. 2018).

Phenotyping for Al-toxicity Tolerance in Plants

Roots are considered as an essential part of water and nutrient uptake from soil and trigger plant productivity under a wide range of growing environments. A new task for plant breeding has evolved to incorporate desirable root characteristics due to the underground location of the rhizosphere (Paez-Garcia et al. 2015). Despite several difficulties, different traits have been incorporated in plant roots by using phenotyping (Zhao et al. 2019). The rhizosphere is considered to be the crucial space for the discovery of innovative characteristics traits that are pertinent and appreciable to fields and farming systems. Essential criterion suggested for a rhizosphere phenotype is root-influenced soil treatment through chemical, biological, and structural changes in the rhizosphere. For instances, malate exudation for aluminium tolerance is a rhizosphere phenotype (Larsen et al. 2005).

Physiological Mechanisms of Al Tolerance in Plants

Cell wall modification is also meant to withstand Al stress in various plants. As the cell wall holds the majority of Al content amongst the entire roots (Yang et al. 2008; Rangel et al. 2009), it is judicious that plants manage with Al stress by cell wall modifications. Cellulose, pectins, and hemicelluloses are the main constituents of plant cell wall (Sasidharan et al. 2011). Though cellulose alone is considered not capable of trapping Al, both pectins (with negatively charged carboxyl or COOH- groups) and hemicelluloses with highly branched structures can trap Al^{3+} ions (Sasidharan et al. 2011; Yang et al. 2011).

The negatively charged COOH- of pectins can crosslink with cations like Ca^{2+} ions. However, in the case of trivalent Al^{3+} ions, the bonding is much more durable to pectin as compared to the Ca^{2+} ion. In acidic soil environment, the exchange of Al^{3+} with Ca^{2+} from pectin-Ca^{2+} complex makes the cell wall thicker and thus interrupts the root elongation process (Tabuchi and Matsumoto 2001). Pectin methylesterases (PMEs), an enzyme responsible for methylation of pectin, enhance the Al^{3+} entrapment capacity of pectin. The Al-tolerant species differed in methylated pectin content compared with the Al-sensitive species like rice and buckwheat (Yang et al. 2008, 2011). Hemicelluloses were also known to be a key site for Al accumulation, significantly executing its role in Al adsorption and root elongation in *Arabidopsis* (Yang et al. 2011). Sasidharan et al. (2011) reported that XEH (xyloglucan endohydrolase) and XET (xyloglucan endotransglucosylase) enzyme (xyloglucan endotransglucosylase-hydrolase or XTH genes) mediated hemicellulose (xyloglucan) – cellulose network also contributes to a great extent in Al tolerance. Zhu et al. (2012) and Pitaksaringkarn et al. (2014) reported that plant hormones gibberellin and auxin stimulate the expression of XTH gene, whereas Al inhibits the expression.

Besides oxalic acid (OA) and other biochemical components (cell wall), some organic compounds are also capable of chelate formation with Al^{3+}, such as phenolics and benzoxazinoids. Catechol, catechin, and quercetin (phenolic compound by nature) can also form a chelate with Al^{3+} (Kidd et al. 2001). Hydroxide radicals (OH^-) in these compounds provide possible targets for Al^{3+}chelation. Benzoxazinoids (HA) in the form of salicylic-acid-activated root efflux reported to form chelate and enhance protection against Al in maize (Zhao et al. 2019).

Morpho-physiological Mechanisms

Citric, malic, and oxalic acids are three prevalent organic acids present in plants involved in external as well as internal Al detoxification. These OAs or organic acids are having 5–6 bonds along with hydroxyl and carboxyl functional groups that help them to trap Al within a ring-like structure. Unique variation in the recreation of OAs has been noticed among plant species, and this variation ranges between 15 and 30 minutes from exposure of Al (malate and oxalate in beet and buckwheat) delayed up to several hours in plants like *Citrus junos* (citrate releases after 4 hours) and 10 hours in rye (Deng et al. 2009; Silva et al. 2013). The differential secretion pattern of OA in response to Al stress may be attributed to the stimulation of previously operative transport mechanism (applicable to malate) as well as the contribution of anion channels involved in OA transport or yet to be defined pathway(s). Researchers like Liao et al. (2006)and Dong et al. (2008) believed that Al stress in response to the secreted OA share a significant positive correlation. De La Fuente et al. (1997) have the opinion that upon exposure to Al, certain stress enzymes (citrate synthase and malate dehydrogenase) are involved in OA synthesis. Rout et al. (2001) suggests that carboxylates of OAs form a complex with the Al and thus prevent transportation within the root cells. Whereas Gupta et al. (2013) reported that Al present in the media could trigger the release of carboxylates from plants. Kochian et al. (2004) reported that compared to the Al stress-prone plants, *Phaseolus vulgaris* (snap bean), an Al-resistant species, could secrete citrate from the roots as much as eight times more.

Biochemical Mechanisms

Al is mostly remained in the form of aluminosilicates or oxides at mild acidic or in neutral pH values (Ligaba et al. 2006). However, phytotoxic elements of Al are released when soils become more acidic. Phytotoxic elements from Al mainly inhibit root elongation in plants initially. Thus, Al toxicity is considered a significant problem in acid soils (Gruber et al. 2010). Soluble Al exists in various ionic forms in aqueous solutions. However, recent researches indicate that Al initiates in the form of symplasmic solution above pH 7, comprising a blend of inorganic and organic ions along with macromolecules (Yokosho et al. 2011). $Al(OH)^{2+}$ ion of Al is considered as the primary toxic species for dicotyledons, followed by Al^{3+} seems to be considerably less toxic. Whereas in monocot species, Al^{3+} is the primary toxic ion than others mentioned in the case of dicots (Liu et al. 2009). It is observed in different studies that the application of cation through growth regulators helps in ameliorating Al^{3+} toxicity in plants. Correctly, Ca^{2+} and Mg^{2+} play an active role in ameliorating the toxic activity of Al^{3+} in solutions (Zhao et al. 2003). Apart from this, the application of citrate or malate in some cases alleviates inhibition of root growth due to Al toxicity has been suggested in different researches.

Cellular Mechanisms

Plants have different types of mechanisms such as detoxification or tolerance for minimizing Al stress at the cellular level. These may involve avoiding the build-up of toxic concentration at the cell wall's sensitive site that will help to prevent the damaging effect of Al rather than to develop protein to resist the effect. In case of tolerance, PM is involved through reducing the heavy metal's uptake or stimulating the efflux pumping of Al that has entered the cytosol. Within the protoplast different types of mechanism exist such as heat shock protein for repairing stress-damaged cell and organic acid, amino acids, or peptide for chelation of metals or transport into the vacuole through compartmentation of metabolic process. To minimize the toxic effect of Al, plants have developed two types of mechanism. One is mechanism of exclusion or resistance to Al, and the second one is internal tolerance mechanism.

Mechanisms of Exclusion Adjustment in rhizosphere pH. In plant solubility, toxicity and content of Al decrease with a small increment in the pH of the rhizosphere through dismissal of the metal in the root apoplast (Yang et al. 2011). In squash (*Cucurbita pepo*) and wheat, H+ATPase of the PM regulates the pH transformation in rhizosphere and resistance to Al stress (Yang et al. 2011). In the presence of other metals as well as Al, rhizosphere pH is increased or reduced by OA (Zhang et al. 2008b).

Transition in composition of cell wall and properties of plasma membrane. In soil solution the cell wall of root is directly in contact with the Al so the limited uptake or absorption of Al to cell wall has a positive effect at the surface of PM on Al activity (Horst et al. 2010; Yang et al. 2011). The root cell wall is the target of Al toxicity because it is the main binding site of Al and exclusion in plants (Kopittke et al. 2015). The resistance to Al alternation of cell wall composition contributes an important role (Chen and Liao 2016). Polysaccharide content of the cell wall is induced by Al that ultimately reduces the uptake of water and nutrient content, as well as elasticity of cellular wall (Nguyen et al. 2005). In the root apex, a positive correlation between content of polysaccharide (pectin and hemicellulose) contributed to tolerance of Al in rice, indicating the importance of the cell wall composition for exclusion of Al (Yang et al. 2008). In cell wall the differences in resistance to Al is regulated by the content of pectin and methylation of pectin (Yang et al. 2011).

In tolerance of the plants to Al, the PM's physicochemical as well as physiological properties also play a significant role. The lipid composition of the PM affects the Al tolerance (Wagatsuma et al. 2015). In PM the binding of Al is enhanced due to a negative charge, which is created by the phospholipid on the surface of PM and an increase in the sensitivity to Al. In rice and timber species, Al tolerance had a negative correlation with the proportion of phospholipids in PM's root cell (Maejima et al. 2014). Similarly, the content of sterol plays a vital role for tolerance of Al stress. In different plant species, low contents of phospholipids with high sterol content is a common strategy for tolerance of Al (Wagatsuma et al. 2015).

Root exudates secretion. Root exudates play an important role for tolerance of Al. In relation with organic acids, root secretion detoxifies the Al that has a great importance to Al tolerance. In case of buckwheat when Al stress occurs oxalate acid is secreted by the root and non-toxic Al^{-} oxalate accumulates in leaves (Ma et al. 1997). In maize and wheat through PM anion channel that is activated by Al root exudate secretion occurs (Ma et al. 2001). Besides OA there are some other organic compounds that are exudates from root for Al^{-} chelation (Kochian et al. 2015).

Elimination of chelating molecules and mucilage secretion. In case of Al exclusion mechanism, mainly in cereals OA release in the root is the most common phenomenon in plants (Kochian et al. 2015). From the root the organic anions, oxalate, malate, and citrate are secreted and non-toxic Al-OA complex are formed through chelating Al that protects the root apex and induces root growth. In all plant cells, malate and citrate are omnipresent and in mitochondria malate and citrate become important components of the tricarboxylic acid (TCA) cycle, while oxalate engages in the regulation of Ca and metal detoxification (Brunner and Sperisen 2013). Besides that many studies find that OA efflux confers Al resistance (Yang et al. 2013; Brunner and Sperisen 2013) in cultivable plants and timber trees. Secretion of low molecular weight ligand from the root that binds to Al in *Eucalyptus camaldulensis* is confirmed by the scientist (Tahara et al. 2008). The increased secretion of phenolic compound (caffeine) is observed when tea plants are exposed to Al stress (Morita et al. 2011). Root mucilage is a gelatinous material exudate from root apex external layers and composed of high molecular weight polysaccharides that contributes a major role in resistance against metals (Morel et al. 1986).

Protection by border cells and CAP type structures in the root apex. When plants face stress from microbes and soil, the border cells and root cap act as a protector of the root apex's mother cells and also play role as a receiver and transmitter of environmental signals that will eventually determine the growth of root (Hawes et al. 2016). The root cap and border cells in many plant species mucilage are produced and exuded, which have more polysaccharides and can help to bind with metallic cations (Cai et al. 2013; Kumpf and Nowack 2015).

In bean (*P. vulgaris*), soya bean, barley, and castor (*Ricinus communis*), mucilage secretion by the border cell and root cap plays a vital role for Al resistance mechanism through immobilization and detoxification of Al (Cai et al. 2013; Silva et al. 2014). As a mechanism related to the exclusion of Al, the role of border cells is observed by removing the border cells, which has a potential impact on increased sensitivity and absorption of Al in pea (*P. sativum*) plants (Yang et al. 2016).

Mitigation of aluminium-induced toxicity. Recently, many studies have revealed that exogenous addition and availability of certain elements (NH_3, Si, Mg, B, P, etc.) prevent toxicity of Al in plants. In rice, when urea is applied, ammonium (NH^{4+}) by changing the pH reduces Al accumulation in root that was induced by the uptake of NH^{4+} as well as Al, and NH^{4+} shows direct competition for the binding sites of cell wall (Wang et al. 2015). The Al content in the roots, stem, and leaves is reduced with the application of Si in pea and rice (Shen et al. 2014). In corn, elongation of root and deposition of callous are inhibited through hydroxy aluminosilicates formation in root apex by the exogenous addition of Si (Wang et al. 2004). Mg that is chlorophyll's central component and cofactor of many enzymes prevents phytotoxicity of Al. Al toxicity is reduced by high concentrations of Mg (mM) through uptake competition and interaction in the cell wall's binding sites and the PM (Rengel et al. 2015). In large number of plants, with the supplementation of B the Al toxicity has been removed. Boron plays a role in collaboration with Ca and that will hinder binding of Al in cell wall (Yu et al. 2009b; Horst et al. 2010). The metal toxicity increases with high P that increases when Al-P precipitates on the surface of root, while Al toxicity reduces due to the P deficiency by changing PM and cell wall's properties (Shao et al. 2015). In the tolerance to stress by Al, hormones and polyamines have a major function. In wheat, the Al accumulation in root apex reduces by the addition of auxins exogenously (indole acetic acid [IAA]) (Wang et al. 2013a,b). With the decreased ethylene production, root inhibition due to Al could be alleviated (Yu et al. 2016). In case of Al tolerance in plants, putrescine is found as a major signalling molecule (Wang et al. 2013a,b;Yu et al. 2016).

Mechanisms of Internal Tolerance Chelation of aluminium in the cytosol (organic acids and other organic ligands). The chelation of metal with Organic acid (OA) and subsequent integration into the vacuole are reported as the mechanism of Al internal detoxification. OA is used by many tolerant plants for the integration of Al in root cells cytosol and for remobilization or translocation of Al in shoots. Oxalate is used for internal and external chelation of Al in *Fagopyrum esculentum* (Wang et al. 2015). In case of tea, in the root cytosol oxalate is the major ligand and Al-oxalate compounds are formed (Morita et al. 2004); through formation of Al-citrate or Al-malate, Al is translocated towards the shoots (Morita et al. 2004, 2008). In *Camellia oleifera*, through phloem Al is transported. Besides OAs, other phenolic compounds can form a complex through binding to Al in the cytosol such as catechin makes a complex with Al in tea plants (Nagata et al. 1992). In the root the hydroxamates can tie up to Al (Poschenrieder et al. 2005). The Al-F compound has been found as a mechanism of F tolerance in the sap of tea leaves (Yang et al. 2016).

Aluminium transporters in PM and compartmentalization in vacuole. Transport proteins are required for transportation through biological membranes. ABC transporters act as transporters binding to ATP, AtABCI17/AtSATR1/AtALS1, AtABCI16/AtALS3, and OsALS1, and transporters Nrat1 (Nramp Family) are considered as contributors for Al detoxification in plants (Hwang et al. 2016). In epidermal cells PM of the root cortex and in phloem cells AtALS3, a partial type of ABC transporter, is located. It has been reported that inside the plant, AtASL3 distributes Al from the root apex (sensitive to Al). In fact, as an exclusion mechanism, the efflux of Al has been proposed (Arunakumara et al. 2013). In vacuoles of root cells as well as plant vascular cells, AtALS1 is a partial type ABC protein that may participate in Al intracellular transportation (Delhaize et al. 2012). OsALS1 is an ortholog of AtALS1 in rice that is located in the tonoplast of root cells and through formation of Al-OsALS1 and participates in Al internal detoxification through compartmentalization of Al in vacuoles (Huang et al. 2009, 2012).

The Nrat1 transporter is located in the PM of all cells of root apex except in epidermal cells (Xia et al. 2010). It has been reported that by OsALS1 Al-influx to the root cells and vacuole compartmentalization, Nrat1 plays a significant role for Al detoxification in the cell wall (Xia et al. 2010; Huang et al. 2012). However, it has been suggested that in rice, for internalization of Al in root apex, endocytic vesicular traffic can contribute and also OsPIN2 overexpression increases Al tolerance through the Al internalization via vesicular traffic (Wu et al. 2015).

Phytohormones-based Aluminium Stress Tolerance in Plants

Expression of plant responses to Al toxicity is dependent on phytohormones. But under Al toxicity, the role of phytohormones was given less consideration (Matsumoto et al. 2015). This chapter focused plant responses to Al toxicity, the role of phytohormones in plants, particularly against Al toxicity. Measures of toxicity including changes in root elongation rate (RER) as influenced by short-term alterations in phytohormones are important measures for this purpose. Hopefully, this information will help in determining the role of phytohormones for expressing rhizo-toxicity of Al, through further experimentation. In addition, this chapter will also highlight some information about mechanisms of Al tolerance and toxicity, for further investigations (Liu et al. 2014; Kochian et al. 2015). Klimashevskii and Chernysheva (1980) presented the earliest work relating to the effect of Al toxicity on phytohormones. Klimashevskii (1983) stated that in response to Al stress, alteration in hormonal processes was a 'non-specific' response. As proved, when 74–440 μM Al (2–12 mg L^{-1}) was applied for three days to Al-sensitive cultivars of pea (*P. sativum*), barley (*Hordeum vulgare*), and maize (*Zea mays*). In comparison to Al-tolerant cultivars, there was up to fivefold increase in abscisic acid levels in Al-intolerant cultivars (Klimashevskii and Chernysheva 1980). In another early study, it was proved that in root tissues, basipetal transport of auxin is also a function of Al; Hasenstein and Evans (1988) examined curvature of root in response to Al application in maize. Hasenstein and Evans (1988) also detected that other metals like La that bind to hard ligands (Kopittke et al. 2014) also influence root curvature. According to Kopittke et al. (2016), although Al concentration, used by Hasenstein and Evans (1988), is much higher (~100–1000 fold) than that required to reduce root elongation, the actual soluble concentration of Al is not known because it was applied in agar block. Hasenstein et al. (1988) observed that Al supplied in an agar block at a concentration of 1 mM Al showed negligible effect on root curvature. As roots are the primary site for Al toxicity, Hasenstein and Evans (1988) focused on root growth, but Pan et al. (1988) and Pan et al. (1989) determined its impact on shoot growth and cytokinins. In Al toxic soils, application of synthetic cytokinins reversed the inhibitory impacts of Al on lateral shoot development but in reducing its negative impacts on total shoot dry matter production exogenous application of cytokinin was unable (Pan et al. 1989).

Antioxidants-based Aluminium Stress Tolerance in Plants

Al toxicity usually brings oxidative stress that accompanied with redoxes of glutathione (GSH) and ascorbate decreases thiol content, lipid peroxides, and the rapid accumulation of ROS (O^{2-} and H_2O_2). Loss of growth and ATP depletion are results of Al-induced ROS production, as it causes root inhibition and damage to cellular components (Sharma and Dubey 2007). In plants for scavenging ROS, various mechanisms relating to non-enzymatic and enzymatic regulation were evolved. Al-tolerant capacity becomes higher with enhanced antioxidant ability. Superoxide dismutase (SOD) induces dismutation of O^{2-} into H_2O_2; APX uses ascorbate as an electron donor to reduce H_2O_2. The GSH-ascorbate cycle (GR, DHAR, and MDAR) regenerates reduced ascorbate from its oxidized forms; therefore, ascorbate is a radical-free scavenger and a major antioxidant.

In rice, under the stress of Al there was an increase in the activities of APX and SOD (Sharma and Dubey 2007). By upregulation of GR, two APX isoforms (Yang et al. 2013) and two SOD isoforms (Mn-SOD and Cu/Zn-SOD) (Zhou et al. 2009; Wang et al. 2013c) occur due to these increased activities. While in ASR-RNAi transgenic rice that is Al sensitive, reduction of APXs occurs (Arenhart et al. 2013). Similarly, in various plants under Al stress, upregulation of enzymes involved in the GSH-ascorbate cycle also occurs (Duressa et al. 2011; Karuppanapandian et al. 2012). In seedlings of rice, increase in Al concentration induces an increased GSH pool (Yang et al. 2007). For the detoxification of Al many metabolic processes involve GSH. For example, for the biosynthesis of phytochelatins (PCs) and high-affinity metal chelator, GSH acts as a core precursor (Cobbett and Goldsbrough 2002). PCs detoxify heavy metal by forming complexes with metals in the vacuole. In GSH synthesis one of the key steps involves γ-glutamylcysteine synthetase (γ-GCS). In Al-resistant barleys XZ16, γ-GCS is present abundantly, while in Al-sensitive barley XZ61, its abundance is much decreased (Dai et al. 2013). Thus, to enhance Al tolerance, there should be a higher level of GSH in the cells. In many detoxification reactions, there is a participation of glutathione S-transferases (GSTs) using GSH as a coenzyme or cosubstrate (Dixon et al. 2002). To exclude the toxin into the vacuole, GST catalyzes a conjugation process and forms a *S*-gluthionylated product. In Al-tolerant *Lotus corniculatus* (Navascues et al. 2012), *A. thaliana* (Karuppanapandian et al. 2012), soya bean (Duressa et al. 2011), Koshihikari (Wang et al. 2013c), and rice XN1 (Yang et al. 2007), Al detoxification capacity increases with the upregulation of GSTs under Al stress. Thus, for Al tolerance, the activity of GSTs and the content of GSH play an important role.

Potential Transgenic Approach for Aluminium Toxicity Improvement

Al toxicity is an active research around the globe where acid soil is a major problem. While cropping in acid soils, using Al^{3+}-resistant variety is a major strategy. Choice of the crop is also important as they vary in performance under acid soil. Though a number of research findings indicated a mechanism to cope with the Al toxicity, organic acid secretion in expelling Al from the root zone and/or Al binding in cell wall carries great importance (Delhaize et al. 1993a). These approaches were studied in a wide range of species like rice, wheat, maize, and bean. (Wang et al. 2004; Yang et al. 2008; Ryan et al. 2009; Rangel et al. 2010; Yokosho et al. 2011). However, in the process of developing Al^{3+}-resistant new varieties, the transgenic approach is getting more attention to confer resistance to Al stress.

Genes Responsive Under Aluminium Toxicity

Al tolerance in plants is controlled by genes. To date, several genes were identified by researchers around the world in many crops. At least fourteen genes from seven different species were identified as having functions involving tolerance and resistance to Al^{3+} stress (Ryan et al. 2011); though there are a lot have been identified as candidates. In some crops, Al tolerance is mediated by a single dominant gene, while the quantitative nature of this trait was also observed in many crops controlled by multiple genes. Al-tolerance gene was first identified in wheat named ALMT1 (Sasaki et al. 2004). Since then lot of genes have been identified in many crops that confer Al tolerance; some of them are listed in Table 6.1. The single dominant gene controlling Al tolerance was detected in wheat (Aniol and Gustafson 1984), oat (Nava et al. 2006), barley (Ma et al. 2004; Furukawa et al. 2007), and sorghum (Magalhaes et al. 2004, 2007). Whereas multiple genes controlling Al tolerance were identified in crops like maize (Pandey et al. 1994; Borrero et al. 1995;

Table 6.1 List of some genes involved in aluminium resistance or tolerance in plants.

Crops	Gene	Function	References
Arabidopsis	*AtALMT1*	Malate transport	Hoekenga et al. (2006)
Arabidopsis	*AtMATE*	Citrate transport	Liu et al. (2009)
Arabidopsis	*AtSTOP1*	Zn finger transcription factor	Martinez Lucci et al. (2008)
Arabidopsis	*AtALMT12*	Malate transport	Sasaki et al. (2010)
Arabidopsis	*AtALMT6*	Malate transport	Meyer et al. (2011)
Arabidopsis	*AtALMT9*	Malate transport	Kovermann et al. (2007)
Barley	*HvAACT1*	Citrate transport	Furukawa et al. (2007)
Brassica napus	*BnALMT1* and *BnALMT2*	Malate transport	Ligaba et al. (2006)
Maize	*ZmMATE1*	Citrate transport	Maron et al. (2010)
Rice	*OsSTAR1* and *OsSTAR2*	UDP-glucose transport	Huang et al. (2009)
Rice	*ART1*	Zn finger transcription factor	Yamaji et al. (2009)
Rye	*ScALMT*	Malate transport	Collins et al. (2008)
Rye	*ScMATE2*	Citrate transport	Yokosho et al. (2010)
Sorghum	*SbMATE1*	Citrate transport	Magalhaes et al. (2007)
Soya bean	*GmALMT1 to GmALMT34*	Malate transport	Peng et al. (2018)
Soya bean	*GmMATE*	Citrate transport	Wang et al. (2016)
Tomato	*SlALMT4*	Malate transport	Sasaki et al. (2016)
Tomato	*SlALMT5*	Malate transport	Sasaki et al. (2016)
Wheat	*TaALMT1*	Malate transport	Sasaki et al. (2004)
Wheat	*TaMATE1*	Citrate transport	Ryan et al. (2009)

Ninamango-Cárdenas et al. 2003; Krill et al. 2010) and rice (Wu et al. 2000; Ma et al. 2002; Nguyen et al. 2001, 2003). The multigenic nature of Al tolerance in wheat has recently also been identified in different studies (Garvin and Carver 2003; Cai et al. 2008; Ryan et al. 2009), but phenotypic variation mostly has been explained by a single locus that controls the Al^{3+}-activated efflux of malate from the root apices (Delhaize et al. 1993a,b; Ryan et al. 1995). The identified genes are mostly associated with organic acid production, but a recent study suggested that different functions by genes in conferring Al tolerance are identified in rice (Ma et al. 2014).

Gene Family Variation

Genes conferring Al^{3+} resistance in different species are of two different families. One is Al-activated malate transporter (ALMT) and another is a multidrug and toxic compound extrusion/Al-activated citrate transporter (MATE/AACT) (Delhaize et al. 2012). The first identified gene in wheat TaALMT1 controls the malate efflux (Sasaki et al. 2004; Yamaguchi et al. 2005). In contrast, MATE family gene was found to be involved with citrate efflux (Furukawa et al. 2007). Apart from wheat ALMT family members have also been identified in maize named ZmALMT1 and ZmALMT2 (Pineros et al. 2008; Ligaba et al. 2012), in rye as ScALMT1 (Collins et al. 2008), and BnALMT1 and BnALMT2 from rape (*Brassica napus*) (Panda et al. 2009). HvALMT1 in barley (Gruber et al. 2010) and in Yorkshire fog named HlALMT1 (Chen et al. 2013). The MATE family genes have also been identified in different crops. A family member HvAACT1 was detected in barley (Furukawa et al. 2007). The other homologs of these genes are SbMATE identified in sorghum (Magalhaes et al. 2007), ZmMATE1 in maize, ScFRDL2 in rye (Yokosho et al. 2010), OsFRDL4 in rice (Yokosho et al. 2011), and TaMATE1B in wheat (Tovkach et al. 2013). Overexpression of both family member genes leads to a better tolerance to Al toxicity (Delhaize et al. 2004, 2009; Pereira et al. 2010; Zhou et al. 2013).

Interference in the Resistance Mechanism

The sensitive apices of root are protected by chelating the Al^{3+} and turning it non-toxic. This process generally controlled by the ALMT gene family encodes the Al^{3+}-activated anion channel permeable to malate. This response was well studied in a range of species including wheat (Sasaki et al. 2004; Raman et al. 2005; Zhang et al. 2008a), *B. napus* (Ligaba et al. 2006), *Arabidopsis* (Hoekenga et al. 2006), and rye (Fontecha et al. 2007). A different gene family named as multidrug and toxic compound extrusion (MATE) also confers resistance to Al stress in plants through organic anion efflux (Furukawa et al. 2007; Magalhaes et al. 2007; Wang et al. 2007; Liu et al. 2009). Of the two different families, ALMT is generally involved in malate efflux, while MATE is associated with citrate efflux. In plants, a different expression of gene family was observed in different crops, but there are species where members of both the family contribute to the Al resistance such as wheat and *Arabidopsis* (Liu et al. 2009; Ryan et al. 2009). One similarity in both families is resistance to Al stress that is associated with the higher expression of genes regardless of the crops (Raman et al. 2005; Sasaki et al. 2006; Hoekenga et al. 2006; Magalhaes et al. 2007; Fujii et al. 2009).

Expression and Regulation of Gene Families

There is a difference in the expression of genes, point of expression, and regulation mechanism in different crops (Ma et al. 2014) (Table 6.2). In some species, the expression is upregulated by Al, whereas in some others expression was not induced by Al. The expression of TaALMT1 in wheat was not induced by Al (Sasaki et al. 2004), and it is similar for HvAACT1 in barley (Fujii et al. 2012). On the other hand, upregulation of expression was observed in other species including OsFRDL4 in rice (Yokosho et al. 2011), in SbMATE in sorghum (Magalhaes et al. 2007), ScFRDL2 and ScALMT1 in rye (Collins et al. 2008; Yokosho et al. 2010), and HlALMT1 in Yorkshire fog (Chen et al. 2013). Both the gene families vary for point of expression from root apices to the mature zone. In wheat, expression of TaALMT1 occurs within 0–5 mm of the root tips (Sasaki et al. 2004). MATE family member TaMATE1B is expressed in the basal root regions of wheat (Fernando Pereira et al. 2015). The expression of HvAACT1 within 3 mm of the root zone was concluded in different studies (Furukawa et al. 2007; Fujii et al. 2012). In root tips and mature zone, expression of OsFRDL4 was observed in rice (Yokosho et al. 2011). In a different study, SbMATE is expressed in epidermal and outer cortical layers in sorghum (Magalhaes et al. 2007; Sivaguru et al. 2013).

The higher expression of genes was generally regulated by an increase in sequence repeats, copy number, transposon insertion, and cis-acting elements (Ma et al. 2014). An increase in the resistance to Al stress by the ALMT member TaALMT1 in wheat was regulated by an increase in a tandem repeat of the sequence (Sasaki et al. 2004). Whereas an increase in the expression of genes was recorded by an increase in the insertion of transposon elements in wheat for TaMATE1B (Fernando Pereira et al. 2015), in sorghum for

Table 6.2 Al stress-responsive genes expression pattern, point of expression, and regulation mechanism in different crops.

Gene	Crops	Expression	Localization	Regulated by	References
TaALMT1	Wheat	Not induced by Al	Root tips (0–5 mm)	Tandem repeat	Sasaki et al. (2004)
TaMATE1B	Wheat	—	Basal root region	Transposon-like element insertion	Tovkach et al. (2013)
HvAACT1	Barley	Not induced by Al	Root zone (3 mm)	Transposon insertion	Furukawa et al. (2007), Fujii et al. (2012)
OsFRDL4	Rice	Upregulated by Al	Root tips and mature zone	—	Yokosho et al. (2011)
SbMATE	Sorghum	Upregulated by Al	Epidermal cell layers	Inverted repeat transposable elements	Magalhaes et al. (2007), Sivaguru et al. (2013)
ZmMATE1	Maize	—	Root zone	Increase in copy number	Maron et al. (2013)
ScFRDL2	Rye	Upregulated by Al	Root zone	—	Yokosho et al. (2010)
ScALMT1	Rye	Upregulated by Al	Root tip	Increase copy number	Collins et al. (2008)
HlALMT1	Yorkshire fog	Upregulated by Al	Root zone	Alteration in cis-acting element	Chen et al. (2013)

SbMATE (Magalhaes et al. 2007; Sivaguru et al. 2013), and in barley for HvAACT1 (Furukawa et al. 2007; Fujii et al. 2012). In maize and rye, the higher expression of gene Zm MATE1 and ScALMT1, respectively, is mediated by an increased copy number (Collins et al. 2008; Maron et al. 2013). Alteration in a cis-acting element can also play a role in Al resistance by regulating the expression of gene such as HlALMT1 in the case of Yorkshire fog (Chen et al. 2013).

Genetic Engineering

Improvement in crop performance is a prime need for growing under acid soil. Altering crop genetic architecture (i.e. transgenic crops) could be a great option in this case. It could improve the crop's response to Al stress. The reflection of Al^{3+} resistance with higher gene expression validates the exploitation of transgenic strategies to improve the performance of the crop under Al stress (Pereira et al. 2010). There are shreds of evidence of such improvements in barley (Delhaize et al. 2004) and tobacco (Sasaki et al. 2004; Zhang et al. 2008a), where TaALMT1 is expressed in suspension cells that conferred an improvement in resistance to Al^{3+} by mediating the Al^{3+}-activated efflux of malate. The transgenic plant doubled the production of barley than the untransformed one (Delhaize et al. 2004). Apart from this, other studies including genes encoding *Arabidopsis* blue copper-binding protein showed tolerance to Al stress in yeast (Ezaki et al. 2000). In rice, resistance to Al stress was controlled by two genes namely STAR1 and STAR2 (Huang et al. 2009), while under Al stress, one or two dominant genes showed Al tolerance in oat (Ezaki et al. 2000).

One of the great examples of genetic engineering food crops conferring Al^{3+} resistance was achieved by overexpressing the TaALMT1 in wheat that encodes Al^{3+}-activated malate efflux from root (Pereira et al. 2010). Stable and consistent resistance to Al^{3+} was observed in subsequent generations. An earlier study in the same crop by overexpressing the resistance gene SbMATE from sorghum encoding citrate transporter was not as robust as TaALMT1 in conferring Al^{3+} resistance (Magalhaes et al. 2007). Other reports stated increased resistance to Al stress in plants through overexpression of Al-resistant genes that were associated with the function contributing to membrane physiology and stress response (Ezaki et al. 2000; Basu et al. 2001; Stival da Silva et al. 2006; Ryan et al. 2007). An increase in plants resistance to Al stress also concluded through mutagenesis (Rounds and Larsen 2008). Apart from this, genes associated with organic anion synthesis have also been targeted and consequently improve the Al resistance in plants (de la Fuente et al. 1997; Koyama et al. 1999; Tesfaye et al. 2001; Anoop et al. 2003; Barone et al. 2008; Trejo-Tellez et al. 2010). In answering Al toxicity, transgenic approaches could be a great option in enhancing tolerance despite the limited successful transgenic events. Though it was doubtful whether the achievement of high levels of resistance could be possible by altering the expression of a single gene (Delhaize et al. 2001, 2003; Gupta and Kumar, 2013), combination of transgenic approach exploiting multiple genes could increase the tolerance to Al stress in plants (Ezaki et al. 2001).

Pyramiding of Genes

One of the most obvious goals for plant breeders is to accumulate several desirable characters into one variety. In the same way, quite a similar trend was observed for transgenic crops for complex traits. For example, in potato plants enhanced tolerance to abiotic stress was reported when overexpressing three genes that of monogenic or oligogenic transformants (Ahmad et al. 2010). For developing new genotypes, this approach of accumulating transgenes is becoming a point of interest for plant scientists. The same approach can also be exploited for Al^{3+} resistance by introgressing different Al^{3+} resistance loci into a single transgenic line (Pereira et al. 2010). One of the Al^{3+} resistance mechanism is associated with efflux of citrate (Wang et al. 2007; Ryan et al. 2009); accumulating genes encoding efflux of citrate controlled by TaMATE1 could bring additional benefits if introgressed to a transgenic plant encoding malate efflux controlled by TaALMT1 (Pereira et al. 2010). However, knowledge on Al stress resistance is expanding day by day, and new findings are enhancing our understanding regarding stress resistance mechanism in different crops. Refinement of present strategies by exploiting multigenic controls of resistance could provide consistent results. Additionally, the search for more stress-responsive genes needs to be continued and existing gene families should be exploited in a wide range of plant species.

Phytoremediation of Al Stress in Plants

Phytoremediation is a process of reclaiming soil, air, or water that has already been contaminated by heavy metals (Cd, Pb, Al, As, Sb, etc.) with the help of living hyperaccumulating plants and/or microorganisms where 'phyto' means plant and 'remediation' means to remedy. Indian rhododendron (its own roots and leaves can store over 1000 $mgkg^{-1}$ of Al without causing phytotoxicity), rooibos tea (can tolerate Al levels between 110 and 275 $\mu gAlg^{-1}$ concentration), snap bean, maize, wheat, taro, tobacco, mustard, alpine pennycress, hemp, and pigweed are well-known hyperaccumulating plants (Watanabe et al. 1997; Ezaki et al. 2001; Denton 2007; Kanu et al. 2013). Phytoremediation is now considered as one of the environmentally friendly (Arshad et al. 2008), cost-effective, aesthetically pleasing (Hong-Bo et al. 2010), and easy to use technology for cleaning up of contaminated soil particularly in highly polluted urban areas. Plants clean up the soil by phytoextraction, phytostabilization, phytodegradation, rhizodegradation, phytostimulation, phytovolatilization, rhizofiltration, biological hydraulic containment, phytoaccumulation, and phytodesalination where plants may undergo oxidative stress, integrity destruction of cell membranes, hindrance in nutrient uptake, resistance in photosynthesis, and decline in plant chlorophyll content (Ali et al. 2013). Under Al or heavy metal toxicity, plants show tolerance either externally by reducing uptake or internally by higher uptake of Al and sequestrating into the body. For exhibiting external tolerance, the plant may receive help from microorganisms available in the rhizosphere soil to immobilize toxic Al by chelating, arbuscular mycorrhizae (AM), phosphate solubilization, redox changes, or any other ways (Hong-Bo et al. 2010). For internal tolerance mechanism, plants may uptake a higher quantity of Al in Al stress condition and secrete several sorts of enzymes to sequestrate Al into cell vacuole. But transgenic plants may change the history; nitroreductase genes encoded from a bacterium showed rapid removal of TNT (2, 4, 6-trinitrotoluene) when inserted into tobacco resulting in improvement in tolerance against harmful effects of TNT (Hannink et al. 2001). However, the total mechanisms of plants to show Al tolerance under Al stress condition is yet not fully clear to the researchers (Horst et al. 2010).

Hyperaccumulation (generally low biomass and slow-growing plants) of Al occurred in plant mainly in acidic ($pH < 5$) environments affecting fauna (Jolly et al. 2014) and flora (Radić et al. 2010) adversely where speciation and concentration rely on soil pH and soil solution chemistry (Kisnierienė and Lapeikaitė 2015). On the other hand, Al stands for third position among the naturally occurred plentiful element in the world pedon in the form of Al_2SiO_5 or Al_2O_3 just after oxygen and silicon (Bojórquez-Quintal et al. 2017), but it is not essential for plants' growth and development (Zheng et al. 2014). Al stress can limit crop yield and potentially occurs in approximately 40% of arable land of the world in toxic forms (Ma et al. 2001; Zheng et al. 2014). About 63% of the land of South Asia and the Pacific is acidic, which is increasing in an alarming rate (Kochian et al. 2004; Uexktill and Mutert 1995). Al becomes available in acidic soil as $Al(OH)_2^+$, $Al(OH)_2^+$, and $Al(H_2O)_3^+$ (Samac and Tesfaye 2003), but the extent of Al toxicity is primarily due to Al^{3+} ions (Grevenstuk and Romano 2013; Mossor-Pietraszewska 2001; Singh et al. 2017). Among the forms of Al available in soil solution, Al^{3+} is extremely toxic to plant in acidic soils and can reduce crop production and even quality (Kinraide and Parker 1989; Kochian et al. 2004). At high concentration, Al mainly spells on plant root to inhibit its growth, which is occurred due to the interference in cell division of root tip (Panda and Matsumoto 2007). Thus, cellular and metabolic functions of plants have been degraded due to the disturbed root system making plant unable to uptake nutrients and water (Blancaflor et al. 1998; Barcelo and Poschenrieder 2002; Kochian et al. 2005). Long-term exposure to Al stress brings morphological alteration of roots, reduces cell division, damages to DNA and cell nuclei, and results in eventual death of the cells of the roots of the plant (Ezaki et al. 2001; Ciamprova 2002; Panda and Matsumoto 2007). Higher manufacture of reactive oxygen forms like anion of superoxide (O_2^-), perhydroxyl radical (HO_2^-), hydrogen peroxide (H_2O_2), hydroxyl radical (OH^-), and single oxygen ($^1O_2$) are accountable for causing oxidative pressure in plants that is common where Al^{3+} concentration is high (Achary et al. 2008; Pereira et al. 2010). Al is considered as one of the major contaminants of fresh water and acidic soils (10–140 $\mu mol\ L^{-1}$) caused (Jolly et al. 2014) by natural and anthropogenic activities (Haug 1984; Gensemer and Playle 1999; Ma 2000). So, it is essential to remove Al from the plant growing areas where phytoremediation is one of the most studied, sustainable, and cost-effective technology. But for doing that it is of great concern to know what is happening in the plant when it is acting as the phytoremediating agent and how it would be more efficient (Chakraborty et al. 2019). Physically or externally root cells of plants may thicken their cell wall to block Al^{3+} uptake (Kochian et al. 2004) or excrete biochemicals such as organic acid (OA) to make Al-chelate in Al toxic soil (Ma 2000). On the other hand, internally plants may produce harmless complex with organic ligands (OLs) and sequestrating Al as an organelle (vacuole) for detoxification or rapid repair of internal damage (Ma 2000; Delhaize et al. 2012). However, mechanisms of Al stress-induced OA secretion are not known (Zheng et al. 2014). It is therefore important to know the tolerance mechanism of plant for growing in Al toxicity and studying extensively on developing Al stress-resistant crop varieties.

For example, rooibos tea (*Aspalathus linearis*) in acidic soil (pH 3–5.3) can exhibit a unique tolerance against Al toxicity by its greater efficiency in sequestrating Al^{3+} into the root (1262–4078 $\mu g Al g^{-1}$) and exude OA into the rhizosphere such as malonic, citric, and malic acids to form Al-chelate and also secrete HCO_3^- and OH^- anions to increase pH of soil for immobilization of Al through complex formation. Moreover, rooibos tea is susceptible to AM that deters Al to be taken up by the plant roots (Kanu et al. 2013). Chakraborty et al. (2019) studied *Azolla microphylla* Kaulf–*Anabaena azollae* association after exposing it to 750, 500, 250, 100, and 0 (control) μM $AlCl_3$ for six days at pH 4.5 and found that mean fresh weight, root length, dry weight, filament length, heterocyst frequency, chlorophyll content, photosynthetic efficiency, starch content, sucrose content, total nitrogen, protein content, and nitrogenase activity of *A. microphylla* were decreased by 67.75, 62.82, 66.41, 62.37, 54.94, 58.71, 73.86, 59.94, 58.91, 73.30, 76.46, and 44.05% at 750 μM concentration over control, respectively. But intracellular Al content, creation of H_2O_2, electrolyte leakage, and malondialdehyde (MDA) content were observed to be increased by 242.33 $\mu gAl\ g^{-1}$ FW, 82.77, 120.78, and 125.19%, respectively, at 750 μM concentrations over control. However, SOD, CAT, and APX activity; phenolics production; and flavonoid production were observed to be increased by 94.87, 103.90, 55.95, 39.51, and 28.65% up to 500 μM concentration. Activities of enzymes (SOD, CAT, and AP), phenolics, and flavonoid production were extremely decreased at 750 μM Al. *A. microphylla – A. azollae* association creates flavonoids and phenolics and engages equipment of antioxidation to resist the Al entry into the cells and detoxifying the radicals of free oxygen gained in Al exposing environment.

Rice is amongst the most stress withstanding crops across all small grain plants (Sasaki and Yamamoto 2007; Simoes et al. 2012) due to its higher efficiency of proteins in antioxidation, detoxification, Al^{3+} sequestration in vacuole, Pi abundance, utilization of NO (nitric oxide), and signalling transduction (Yang et al. 2007). Moreover, most of the dicot plants are more sensitive to Al stress than rice such as tomato (*Solanum esculentum*) is less efficient in Al toxicity because ROS (O_2^- and H_2O_2) scavenging is decreased due to lower inducement of ROS enhancing enzymes (dehydroascorbate reductase, glutathione reductase, and catalase) and decreased abundance of germin-like protein (oxalate oxidase-OXO), which is responsible for degradation of oxalate to H_2O_2. Thus, forming deformed root tips and mortality of root cells under long-term exposure to the Al toxicity/stress is observed in tomato. Less photosynthesis due to reduced production of related enzymes (oxygen-evolving complex having 33 kDa precursor protein and ribulose bisphosphate carboxylase/oxygenase) is also observed in Al stress tomato plants (Zheng et al. 2014). Citrate secretion produced in TCA cycle of the root cell can reduce Al toxicity by means of chelating Al, but in most of the cases citrate production is lower in most of the soya bean varieties (Ma 2000) and growth may be reduced up to 54% at 10 μM Al only (Kopittke et al. 2015). Secretion of OA or oxalate from the root tip cell is identified as a key factor for showing Al^{3+} toxicity tolerance by tea plant and buckwheat (Prasad and Shivay 2017). For any crops exposed in Al stress, tolerance efficiency will be dependent on the ability of root tip cells to secrete OAs (citrate, oxalate, malate, etc.), which can prevent Al to enter into the root by chelating and immobilizing Al on root surface (Zheng et al. 2014).

The rapid development of industrialization in the twenty first century has been a major cause of environmental pollution (Gavrilescu et al. 2015) and increasing concentration of different metallic elements including Al in the soil, all of which are hazardous to health (Kelishadi et al. 2014). About 7% of the earth crust is constituted by Al (Arunakumara et al. 2013). Apart from using chemical or mechanical ways of cleaning up Al-induced toxic soils, it is conducive to use Al stress-tolerant plants and microorganisms as a phytoremediation agent (Kang, 2014; Gavrilescu et al. 2015). Biomass of hyperaccumulating plants can be a source of metals to be extracted indicating sustainable recycling though these plants may be poisonous to human health if eaten as food (Hong-Bo et al. 2010). Several research programs on studying the mechanisms of tolerance of plant when exposed to Al stress need to be executed. Varieties of Al stress-tolerant crops need to be evolved to rescue a vast quantity of arable land by phytoremediation process after adopting genetic engineering tools like cisgenesis and intragenesis (Espinoza et al. 2013).

Microorganism-mediated Aluminium Stress Tolerance in Plants

Aluminium is available in nature as Al silicates; a follow measure of component is required to do the natural procedures by living life forms, and it exists in the $^{+2}$ and $^{+3}$ states, along these lines having high fondness towards contrarily charged particles and ionic gatherings, for example, fluoride (F^-) and hydroxyl gathering (OH^-), separately (Krahl and Kemnitz 2017). Different gatherings having a liking towards Al for complex bond arrangement are carboxyl, carbonyl, and phosphate (Vitorello et al. 2005). The ionic type of Al with $^{+3}$ charge is viewed as the dissolvable one, and in this manner it is harmful to plants (Gupta et al. 2013). The ionic structure of Al is unsafe to plants and highly toxic (Bojórquez-Quintal et al. 2017; Palani et al. 2018). A few plants have advanced systems to endure Al poisonousness through the development of natural corrosive edifices with Al (Ma et al. 2001; Brunner and Sperisen 2013). In acidic soil, Al limits plant cell division (Jaskowiak et al. 2018) and the principal focus of the component for development hindrance is the base of plant (Poschenrieder et al. 2008). Being extremely responsive, Al focuses on the plasma layer, core, and cytoskeleton of plant root cells and influences cell divider, plasma film, signal transduction pathway, and calcium homeostasis (Sivaguru et al. 1999). By stabilizing superoxide radicals, Al damages DNA (Booth et al. 2015). In addition, it is to some degree hard to recognize the signs and manifestations of Al poisonousness in plants, which at times resemble lack of calcium and decrease in transport (Rout et al. 2001). Adjustment of calcium homeostasis by Al influences the sign transduction pathways through the restraint of compound phospholipase C (Kochian et al. 2004). In rice, wheat, and sorghum, Al plenitude prompts iron insufficiency. The roots lose their fine stretching capacity in view of monomeric Al poisonousness. Under an alternate acidic pH extend, different complex ionic types of Al with hydroxyl bunches are found in soil. Al harmfulness in plants additionally influences mitochondrial working due to the creation of ROS, consequently causing respiration inhibition and ATP depletion (Yamamoto et al. 2002).

In acidic soils, Al harms the growth and productivity of rice, though among different cereals of *Poaceae* family rice expresses the maximum level of Al tolerance (Ma et al. 2002). Research results indicated that plant growth-promoting bacteria (PGPB) of the sort *Bacillus, Burkholderia,* and *Stenotrophomonas* were fit for decreasing the impacts of Al harmfulness (Ren et al. 2019). Al was chelated by the arrival of organic acids in which PGPB plays an important role as they are known to produce organic acids and PGPB produce organic acids. The organic acids, namely, malic, oxalic, and citric acids by PGPB and plant roots are responsible to chelate Al and form compounds like Al-citrate ($C_6H_5AlO_7$) and Al-malate ($C_4H_4AlO_5$), which are not penetrable to plant roots (Pineros and Kochian 2009). Further, polysaccharides and plant hormones like IAA produced by PGPB increased pH of soil solution and enhanced plant growth, respectively (Panhwar et al. 2015). The arrival of natural acids as root exudates is a process in which plants defend themselves from Al harmfulness; this wonder is likely constrained by the hereditary cosmetics of rice. The decrease of Al poisonousness was additionally upgraded by the nearness of PGPB. Microscopic organisms produced natural acids, exopolysaccharides, and plant hormones. The plant hormones help in the development of root architecture that facilitates coping with the adverse environmental factors. The best natural corrosive for reducing toxicity of Al impacts was citrus, followed by oxalic and tartaric acids; however, malonic, malic, and salicylic acids also detoxify Al (Yang et al. 2013). The bacterium *Desulfovibrio desulfuricans* has the capability to remove Al from the solution when sulphate anions remain there (Martins et al. 2012). Al-tolerant strains of PGPR have the capability of detoxifying Al as observed with *Pseudomonas fluorescens*. A strain of *P. fluorescens* produces extracellular lipid to sequester and depollute Al (Piña and Cervantes 1996), and the potential of the bacterial strain can further be studied for Al stress relief to plants. An in vitro investigation suggested that high Al focus seriously influenced the development of rice seedlings during transplanting. Rice seedlings immunized with PGPB were less influenced by the nearness of high convergence of Al. *P. fluorescens* has the capability to withstand under higher concentrations of Al (Auger et al. 2013), whereas other bacteria are adversely affected. Al toxicity is frequently identified with phosphorus insufficiency in light of the fact that a soil with Al fixation will diminish the accessibility of P because of Al-Fe-phosphate connection (Ward et al. 2008). That is the reason the most perceived side effect of Al toxicity is phosphorus-related hindrance of root development. Further, it was seen that bacterial populace was marginally influenced within the sight of high Al focus. The microbial populace was marginally diminished, most presumably because of the expanded inhibitory impacts by delivering free Al (Rousk et al. 2009), which is reliable with the past researches (Pietri and Brookes 2008). Sufficient investigations have not been carried out on microorganism-mediated Al tolerance of crops; future research may be intensified in this direction targeting exploitation of the potential of microbes towards agricultural sustainability.

Agronomic Management for Mitigating Aluminium Stress in Plants

Al toxicity is an extreme constraint or limiting aspect for crop cultivation under acidic soil. The aluminosilicate and Al hydroxide minerals in soil started to dissolve when pH goes below 5.5 and produce Al species such as Al^{3+}, $Al(OH)_2{}^{+}$, $Al(OH)^{2+}$, $Al(OH)^{3+}$, and $Al(H_2O)^{3+}$ that are considered the most toxic forms and commonly affect Al toxicity (Wang et al. 2016; Pavlů et al. 2019). Al toxicity in the soil causes the prohibition of both cell division and elongation of plant root, noted to root stunting associated with reducing the uptake of plant nutrient and water by plant roots. The toxicity of Al in the soil is a crucial limiting factor for crop production. Therefore, there are some agronomic management practices to mitigate Al toxicity or to minimize the toxicity level and try to develop more congenial environment for plants. There are several approaches to reduce Al toxicity for plants grown in acidic soil. These approaches can be split into two groups. One is inorganic changes by applying calcium, magnesium, phosphorus, sulphur, boron, and silicon and also soil oxide and hydroxide, and the other one is organic modifications by adding organo-mineral fertilizers, compost, growth-promoting bacteria in plants, biochar, etc. (Guo et al. 2017; Teng et al. 2018). The following sections describe the management of some agronomic practices to mitigate Al stress in plants.

Role of Inorganic Amendments for Mitigating Al Toxicity in Plants

Calcium (Ca) as a Mitigator of Al Toxicity

Calcium (Ca) is considered as an essential macro-element for plants both under stress and non-stressed conditions. It performs a coupled activity, as a basic component of the cell membrane and cell walls performing as an intracellular nutrient carrier as well as to help elongate stem and root. It is also modulated in response to many signs such as biotic and abiotic stress, light, and mechanical disruption (Sanders et al. 2002). Ca^{2+} connects with phospholipids, balances lipid bilayer, maintains structural purity of cell membranes, and controls the portability of cell membrane by cooperation with phospholipids and protein of the cell membrane (Hepler 2005). In another, Ca^{2+} works as a vital role in extra and intracellular warning (Yocum 2008). Ca^{2+} plays a key role to mitigate Al toxicity in plants (White and Broadley 2003). It can mitigate Al toxicity in different crops growing in acid soil (pH < 5.5) and act as a regulator of plant growth and development (Brady et al. 1993; Hepler 2005). In this section, we discussed some of the Ca amendments, e.g. lime (calcite and dolomite), phosphogypsum, and gypsum application to mitigate Al toxicity of plants grown in acidic soils.

Liming Liming is a significant and common pathway around the world for correcting acidic soils (Scott et al. 2000); improving Ca availability for plants (Mora et al. 1999, 2002); alleviating Al, Fe, and Mn toxicity (Manson and Findlay 2015; Goulding 2016); boosting the P, Mo, Ca, and Mg availability; and avoiding the leaching of some elements such as Cu, Pb, and Cd in contaminated soils (Campbell et al. 2006). Liming encourages organic carbon mineralization through enhancement of soil pH, Al detoxification, and raises the microbial availability through increasing carbon use efficiency. Liming also stimulates organic N mineralization and atmospheric nitrogen fixation. It is also helpful for soil structure. Liming is broadly proficient for crops grown in dry land, e.g. wheat, maize, legumes, and oil seeds. During soil acidity, this practice can also diminish N_2O emissions than that of CO_2 emissions. The major liming materials are common liming oxides (CaO), calcium hydroxides [$Ca(OH)_2$], calcites ($CaCO_3$), magnesium carbonate ($MgCO_3$), and dolomites [$(CaMg(CO_3)_2$]. Besides these, wood ash, press mud from sugar mills, lime sludge from paper mills, and sludge from water treatment plants can also be used as liming materials. Among these, calcite ($CaCO_3$) and dolomite [$Ca\ Mg\ (CO_3)_2$] are the major limestones used in soil (von Willert and Stehouwer 2003). The direct benefits of the application of liming in soil are to increase the pH level in acidic soil especially those having pH level < 5.5 (Mora et al. 2006) and reduce Al toxicity in plants (Mora et al. 2002). On the other hand, the imperceptible advantages of liming are increase in cation exchange capacity (CEC) (de Castro et al. 1999) and increase in phosphorus (P) availability of soils by reducing iron (Fe) and Al (Tisdale et al. 1985). There have been well-documented beneficial effects of liming in acid soils on different crops. Liming increased the plant growth and suppressed weed infestation (Mora et al. 2002), reduced trace element availability for plants (Meriño-Gergichevich et al. 2010), reduced Pb (20–50%) in plants meaningfully (Han et al. 2007), expanded the new roots and reduced degraded lignin (Rosenberg et al. 2003), and reduced the frequency of fungus (*Pythium coloratum*) (El-Tarabily et al. 1997). Liming doses varied depending on the soil pH level and usually ranged from 0 to 4.5 t ha^{-1} (Rosenberg et al. 2003). The amount of liming varies on pH level, organic matter, and clay content of the soil. Power factor (PF) is essential for the efficacy of liming, as a result, larger molecule (>1.7 mm) can prevail in the soil and can effectively perform for the succeeding years. In addition, limestone is assumed to last about 3 mm and takes longer to neutralize soil acidity (Rahman et al. 2018). The mode of actions of liming starts with neutralization of H^+ in both soil solutions by OH^- or HCO_3^- produced from liming materials. The continued removal of H^+ from the soil resulted in the precipitation of free Al and iron into insoluble hydroxide forms and replacement by basic cations in soil solution. Therefore, the soil pH increases with the increase of percent base saturation. Consequently, liming should be applied at least two months ago before planting (Rahman et al. 2018) and further liming may not be required within three years.

Phosphogypsum (PG) Phosphogypsum (PG) is a by-product of the phosphate fertilizer industry (Rashad 2017). It is the composition of Ca^{2+} (116 800 mg kg^{-1} total concentration and 45 430 mg kg^{-1} exchangeable concentration) and SO_4^{2-} (83 200 mg kg^{-1}) with a less quantity of Si (silicon), P (phosphorus), and F (fluoride) (Campbell et al. 2006). The composition may vary depending on the rock phosphate used in the production process (Korcak 1988). There is also a little quantity of Ba (barium), Cr (chromium), Cu, Ni (nickel), and Zn (zinc) (Garrido et al. 2003). PG is mostly available across the world (Carvalho and van Raij 1997). PG can be suggested as a mediator to Ca^{2+} and SO_4^{2-} to reduce soil acidity and also Al toxicity in soil with F at low pH with a range of 4.1–5.3 (Garrido et al. 2003). Reports are available on the doses of PG, Mays and Morrvedt (1986) suggested the recommended dose of PG varies from 500 to 1000 kg ha^{-1}. On the contrary, PG with calcium carbonate and gypsum (2.5 t ha^{-1} reduced Al^{3+} concentration up to 5 cm of soil depth (Smith et al. 1994); and also showed the effects up to a depth of 60 cm with the increase of root density as well as reduce the Al toxicity in apple trees (Pavan et al. 1987). In another, PG doses suggested in maize and soya bean (12 t ha^{-1}) (Vicensi et al. 2016); 2 t ha^{-1} in *Medicago sativa* L. and *Glycine max* L. (Pavan et al. 1987); 2.5 t ha^{-1} *Trifolium* spp. (Smith et al. 1994) can be used as the eliminator of Al toxicity.

Gypsum (G) Gypsum ($CaSO_4 \cdot 2H_2O$), commonly known as plaster of paris (Mandal and Mandal 2002), an interesting option as a soil amendment is widely used in acidic soil because of its solubility in water. The solubility of gypsum (2.5 g L^{-1} in water) is five times higher than that of the calcite (0.5 g L^{-1} in water) lime (Korcak 1988). Gypsum (G) is a great source of Ca^{2+} and S (sulphur), which can increase the mineral content (N, P, K, Ca, Mn, S, and Zn) in plant tissues (Tuna et al. 2007). The application of gypsum has some benefits in soil by changing its chemical properties. It also increases the amount of subsoil Ca (Caires et al. 2006), reduces soil acidity (Toma et al. 1999), and alleviates Al toxicity (Toma et al. 2005; Tuna et al. 2007). Besides, the other benefits of gypsum are improvement in some of the physical properties of soil such as increased soil infiltration (Sahin et al. 2003; Chen et al. 2009), soil aggregation (Chen et al. 2009), hydraulic conductivity (Sahin et al. 2003), development of root (Takahashi et al. 2006), reduced Na adsorption and soil compaction (Gambaudo 2004), and reduction in metal toxicities (Campbell et al. 2006). However, G can reduce the Al toxicity in the subsoil. There is an important role of SO_4 in decreasing Al toxicity in acid subsoils (Alva et al. 1991). After gypsum application, OLs have been identified with the increase of $AlSO_4^+$ and decrease of Al toxicity; enhanced higher yield of corn (Shamshuddin et al. 1991), ryegrass (Mora et al. 1999), and blueberry (Lyrene and Muñoz 1997); and 50% reductio in Al toxicity (Mora et al. 1999). Interestingly, Al toxicity has been decreased by gypsum application in acidic soil, and soya bean roots were elongated due to the application of 500 mM gypsum (Brady et al. 1993).

G not only contributes to alleviate Al toxicity but also helps to supply some nutrients for plants in acidic soil. Doses of G are different in various crops. The application of G at the rate 11 t ha^{-1} in maize (*Z. mays* L.) (Farina and Channon 1988), 2 t ha^{-1} in perennial ryegrass (*Lolium perenne* L. cv. Nui) (Mora et al. 1999), 4.4 t ha^{-1} in highbush blueberry (*Vaccinium corymbosum* L.) (Reyes-Díaz et al. 2011), 1 t ha^{-1} in peanut (*Arachis hypogaea* L.) (Syed-Omar et al. 1991), and 22.2 t ha^{-1} in alfalfa (*M. sativa* L.) (Tirado-Corbalá et al. 2017) can alleviate Al toxicity. Besides, the combined use of G @ 4.0 $t{\cdot}ha^{-1}$ with 0.3 $t{\cdot}ha^{-1}$ of NPK fertilizer has resulted in increased t nutrient (N, P, K, S, Ca, and Mn) in leaves and also increases stem growth, bud survival, and blossom quantity on lowbush blueberry (*Vaccinium augustifolium*) grown in acid soil (Sanderson and Eaton 2004). On the other hand, the level of pH and Ca in soil was increased for five years after G application (0.60 $t{\cdot}ha^{-1}$), whereas Ca status was not consistent in leaves and fruits (Korcak 1992). Conversely, Hanson and Berkheimer (2004) stated that G can only increase the Ca levels in soils but not affect the leaves and fruits of lowbush blueberry. Recently, it has been reported that gypsum (22.2 t ha^{-1}) can provide both calcium and sulphur in soils that contribute to long time (12 years) soil fertility on alfalfa (Tirado-Corbalá et al. 2017).

Phosphorus (P) as a Mitigator of Al Toxicity

Phosphorus (P), another important macro-element, contributes significantly to plant growth and development in both normal and/or stressful conditions (Razaq et al. 2017, Tariq et al. 2017) and also ameliorates Al toxicity in plants (Iqbal 2014). Under soil acidic condition, inorganic phosphate (Pi) may fix Al or Fe, resulting in an intensive limitation to soil (Chen and Liao 2016). The application of P in acidic soil may alleviate Al toxicity in wheat (Iqbal 2013), buckwheat (Zheng et al. 2005), and sorghum (Tan and Keltjens 1990). At low Al concentration (0.4 $mg{\cdot}L^{-1}$), there was no effect on plant biomass production but is acutely constrained in sorghum at high Al concentration (1.6 $mg{\cdot}L^{-1}$) (Tan and Keltjens 1990). Phosphorus application can reduce the toxicity of Al in plants by increasing plant nutrition and root respiration. Reports are available on reclamation of Al toxicity by application of phosphorus. The P content (150 mg $AlCl_3$ kg^{-1} soil) in wheat seedlings was greatly reduced due to Al stress as reported by Iqbal (2013). Similarly, Chen et al. (2012) reported that P can alleviate Al toxicity when the P and Al molar proportion exceeds 5 in the plant root cells. Recently, the application of P (8 g $plant^{-1}$) was noted to promote plant growth and development in a maple tree in nursery conditions (Razaq et al. 2017).

Magnesium (Mg) as a Mitigator of Al Toxicity

Magnesium is another essential macro-element, can play an important role to ameliorate Al phytotoxicity (Bose et al. 2011). The adding of Mg^{2+} at higher doses can eliminate Al toxicity in many plants (Kinraide 2003; Kinraide et al. 2004). The increase of Mg^{2+} in soil caused a decrease in Al toxicity (Noble and Sumner 1988), decrease in Al^{3+} concentration in soil (Grauer and Horst 1992), and decrease in Al activity in plant root cell (Kinraide 2003; Kinraide et al. 2004) and progress of root growth (Silva et al. 2001; Yang et al. 2007).

Boron (B) as a Mitigator of Al Toxicity

Boron (B), the other micro-element, can alleviate Al toxicity in various plant species (Li et al. 2018). Hossain et al. (2004) determined the effect of B application (200 μM) at low pH (4.5) on Al toxicity among fifteen wheat genotypes in Bangladesh and reported that malate exudation rate was widened in the roots of all wheat varieties under Al (100 μM Al) toxicity. The most tolerant wheat cultivar emanates six times more malate than that of sensitive wheat cultivar, and 200 μM boron reduces the toxicity of Al in highly sensitive wheat cultivars. Another reported that B (25 μM B as H_3BO_3) eliminates Al toxicity by increasing antioxidant enzymes activities such as catalase, peroxidase, and ascorbate peroxidase and reduces the toxicity of Al in rapeseed (Riaz et al. 2018).

Sulphur (S) as a Mitigator of Al Toxicity

Sulphur (S) is another macro-element that can be used to alleviate Al toxicity in acidic soil. Several studies reported that S-containing material alleviates Al toxicity in different crops like wheat (Zhang et al. 2010), barley (Dawood et al. 2012), oilseed rape (Qian et al. 2014), and citrus trees (Guo et al. 2017). The process of reducing Al toxicity is done by increasing antioxidant activity, decreasing lipid peroxidation, increasing several nutrients uptakes (P, Mg, and Ca), reducing Al uptake by roots and shoots, and improving Al secretion in roots of the plant (Rahman et al. 2018). At present, Guo et al. (2017) reported that the addition of S (0.5 mM sulphur as magnesium sulphate and/or sodium sulphate) may be associated with increased minerals (Ca, Mg, and P), Al and H_2O_2 depletion, metabolism of S and antioxidant enzymes, and elimination of Al toxicity by increasing the relative water contents in citrus plant.

Silicon (Si) as a Mitigator of Al Toxicity

Silicon (Si), treated as a crucial component, has some favourable influence on numerous stress resilience in plants (Ma 2004). Several researchers exhibited that the application of Si can be decreased by the uptake of Al by plants. The addition of Si alleviated Al toxicity in rice during Al stress (Singh et al. 2011). Recently, Qian et al. (2016) reported that Si-rich biochar amendment can alleviate Al phytotoxicity in wheat. In addition, Si decreased the exchangeable Al in soil and interrupted the level of Al toxicity in various plants.

Role of Organic Amendments for Mitigating Al Toxicity in Plants

Biochar as a Mitigator of Al Toxicity

Biochar is charcoal (contained 40% carbon), made by heating any biomass without oxygen, and may be recommended for soil amelioration through carbon sequestration and improvement of soil health for medium- to long-term period. Biochar is beneficial for many soil microorganisms promoting good soil and plant health. Biochar can enhance the quality of water, reduce greenhouse gas emissions, nutrient extracting (Hale et al. 2013;

Martinsen et al. 2014), soil acidity (Yuan et al. 2011), and irrigation and fertilizer requirements. It also improves the water-holding capacity of soils (Karhu et al. 2011), increases available water for plant, and improves the aggregation of soil (Herath et al. 2013, Obia et al. 2016). Biochar may be an effective approach to alleviating Al toxicity in plants. Generally, the effective Al in soils rises with the decrease of soil pH. The inhibitory effect was mainly due to low soil pH and increase in the dissolved organic carbon in the soil (Liu et al. 2019), resulting in the improvement of the pH buffering capacity as well as suppressing the exchangeable and soluble Al through re-acidification. This higher dissolved organic carbon may increase the dissolution of Al due to greater binding with Al^{3+} (Hagvall et al. 2015; Cardiano et al. 2017) and form soluble Al organic complex (Zhang et al. 2019), which have no toxicity (Kopittke et al. 2016). The increased dissolved organic carbon decreased the Al concentration in soil and alleviated Al toxicity. The application of biochar (wood made by 20 t ha^{-1}) seemed to be highly consistent in reducing Al toxicity as reported by Shetty and Prakash (2020). Similarly, Yamato et al. (2006) also reported that the application biochar (20 t ha^{-1}) can increase the soil pH (3.9–5.1), with a reduction of exchangeable Al^{3+} concentration (2.67–0.12 $cmol_c\ kg^{-1}$), and H^+ (0.26–0.12 $cmol_c\ kg^{-1}$) with a doubled yield of maize (10 t ha^{-1}) than that of the control (5 t ha^{-1}) on tropical soils.

Compost or Organic Matter as a Mitigator of Al Toxicity

Organic matter (OM) is a component made by any plant parts or animal waste products through the decomposition procedure. OM provides numerous benefits to the soil and is present to improve the physical, chemical, and biological properties of the soil. OM or compost application also helps to alleviate Al toxicity in plants. After the decomposition of organic material, organic compounds are released that contributed to form organic complexes with active Al^{3+} that are supposed to be non-toxic to plants. Gale (2018) reported that compost (made by filter cake) was very efficient in increasing the soil pH and P availability for plants, decreasing the rate of exchangeable Al, and increasing the baby corn yield. In another, Adeleke and Akinrinde (2011) stated that application of organic manure with 50 μM $AlCl_3$ confirmed the minimization of Al toxicity in legumes in acid soil. On the other hand, the combined use of lime (4 mg g^{-1} of soil) and compost (1.25 g) has significantly reduced the exchangeable and dissolved Al and enriched of ammonium nitrogen (NH_4^+ - N) and nitrated nitrogen (NO_3^- - N) content in acid sulphate soil (Khoi et al. 2010).

Conclusion

One of the foremost disastrous penalties as a result of global industrialization is soil acidity. Besides industrialization, soil acidity also occurs due to the imbalanced use of synthetic fertilizers and also acid rain. Globally, about 50% of arable land are affected by soil acidity, particularly due to excessive Al in soil solutions. In acidic soils, Al excessiveness limits crop productivity by limiting physiological and metabolic functions in plants leading to hampering the yield of crops. Initiation of soil acidity ($pH < 5$) brings solubilization of toxic forms of phototoxic-Al^{3+} into the soil solution, which rapidly hinders the root growth, subsequently, hampering the uptake of plant-available nutrients and water from soil. While, during Al toxicity, tolerant plants could survive through following several mechanisms including morphological, physiological, and molecular mechanisms. Also, several managements can also reduce the adverse effect of Al toxicity. The current chapter provides modern updates of the consequences of Al toxicity and also highlights the alleviation mechanisms for the sustainability of crop production in the modern era of changing climate.

Conflict of Interest

The authors declare no conflict of interest

References

Achary, V.M., Jena, S., Panda, K.K., and Panda, B.B. (2008). Aluminium induced oxidative stress and DNA damage in root cells of *Allium cepa* L. *Ecotoxicol. Environ. Saf.* 70: 300–310.

Adeleke, K.A. and Akinrinde, E.A. (2011). Use of organic-based amendments to ameliorate aluminium toxicity in legume production on a typic paleudalf of south-western Nigeria. *J. Agron.* 10: 56–61.

Ahmad, R., Kim, Y.H., Kim, M.D. et al. (2010). Simultaneous expression of choline oxidase, superoxide dismutase and ascorbate peroxidase in potato plant chloroplasts provides synergistically enhanced protection against various abiotic stresses. *Physiol. Plant.* 138: 520–533.

Ahmed, I.M., Viraktamath, B.C., Ramesha, M.S., and Mishra, B. (2001). *Hybrid Rice in India: Present Status and Future Prospects*, DRR bulletin 2001-3. Rajendranagar, Hyderabad: Directorate of Rice Research.

Ali, H., Khan, E., Sajad, M. A. (2013). Phytoremediation of heavy metals-Concepts and applications. *Chemosphere* 91(7):869–881. doi: https://doi.org/10.1016/j.chemosphere.2013.01.075.

Alva, A.K., Kerven, G.L., Edwards, D.G., and Asher, C.J. (1991). Reduction in toxic aluminum to plants by sulfate complexation. *Soil Sci.* 5: 351–359.

Aniol, A. and Gustafson, J.P. (1984). Chromosome location of genes controlling aluminum tolerance in wheat, rye and triticale. *Can. J. Genet. Cytol.* 26: 701–705.

Anoop, V.M., Basu, U., McCammon, M.T. et al. (2003). Modulation of citrate metabolism alters aluminum tolerance in yeast and transgenic canola overexpressing a mitochondrial citrate synthase. *Plant Physiol.* 132: 2205–2217.

Arenhart, R.A., Lima, J.C., Pedron, M. et al. (2013). Involvement of ASR genes in aluminium tolerance mechanisms in rice. *Plant Cell Environ.* 36: 52–67.

Arshad, M., Silvestre, J., Pinelli, E. et al. (2008). A field study of lead phytoextraction by various scented Pelargonium cultivars. *Chemosphere* 71: 2187–2192.

Arunakumara, K.K.I.U., Walpola, B.C., and Yoon, M.-H. (2013). Aluminum toxicity and tolerance mechanism in cereals and legumes—a review. *J. Korean Soc. Appl. Biol. Chem.* 56: 1–9.

Auger, C., Han, S., Appanna, V.P., Thomas, S.C., Ulibarri, G., Appanna, V.D. (2013). Metabolic reengineering invoked by microbial systems to decontaminate aluminum: Implications for bioremediation technologies. *Biotechnol. Adv.* 31: 266–273. doi: https://doi.org/10.1016/j.biotechadv.2012.11.008

Awasthi, J.P., Saha, B., Regon, P. et al. (2017). Morpho-physiological analysis of tolerance to aluminum toxicity in rice varieties of North East India. *PLoS One* 12 (4): e0176357.

Barcelo, J. and Poschenrieder, C. (2002). Fast root growth responses, root exudates, and internal detoxification as clues to the mechanisms of aluminium toxicity and resistance: a review. *Environ. Exp. Bot.* 48: 75–92.

Barone, P., Rosellini, D., LaFayette, P. et al. (2008). Bacterial citrate synthase expression and soil aluminum tolerance in transgenic alfalfa. *Plant Cell Rep.* 27: 893–901.

Basu, U., Good, A.G., and Taylor, G.J. (2001). Transgenic *Brassica napus* plants overexpressing aluminium-induced mitochondrial manganese superoxide dismutase cDNA are resistant to aluminium. *Plant Cell Environ.* 24: 1269–1278.

Bennet, R.J. and Breen, C.M. (1991). The aluminium signal: new dimensions to mechanisms of aluminium tolerance. *Plant and Soil* 134 (1): 153–166.

Blancaflor, E.B., Jones, D.L., and Gilroy, S. (1998). Alterations in the cytoskeleton accompany aluminum-induced growth inhibition and morphological changes in primary roots of maize. *Plant Physiol.* 118 (1): 159–172.

Bojórquez-Quintal, E., Escalante-Magaña, C., Echevarría-Machado. I., Martínez-Estévez, M. (2017). Aluminum, a friend or foe of higher plants in acid soils. *Front. Plant Sci.* 8:1767. doi: https://doi.org/10.3389/fpls.2017.01767.

Booth, S.C., Weljie, A.M., Turner, R. J. (2015). Metabolomics reveals differences of metal toxicity in cultures of *Pseudomonas pseudoalcaligenes* KF707 grown on different carbon sources. *Front. Microbiol.*, 6: 827. doi: https://doi.org/10.3389/fmicb.2015.00827.

Borrero, J.C., Pandey, S., Ceballos, H. et al. (1995). Genetic variances for tolerance to soil acidity in a tropical maize population. *Maydica* 40: 283–288.

Boscolo, P.R.S., Menossi, M., and Jorge, R.A. (2003). Aluminum induced oxidative stress in maize. *Phytochemistry* 62 (2): 181–189.

Bose, J., Babourina, O., and Rengel, Z. (2011). Role of magnesium in alleviation of aluminium toxicity in plants. *J. Exp. Bot.* 62 (7): 2251–2264.

Brady, D.J., Edwards, D.G., Asher, C.J., and Blamey, F.P.C. (1993). Calcium amelioration of aluminium toxicity effects on root hair development in soybean [Glycine max (L.) Merr.]. *New Phytol.* 123: c531–c538.

Brunner, I. and Sperisen, C. (2013). Aluminum exclusion and aluminum tolerance in woody plants. *Front. Plant Sci.* 4: 172. https://doi.org/10.3389/fpls.2013.00172.

Budikova, S. and Durcekova, K. (2004). Aluminium accumulation in roots of Al-sensitive barley cultivar changes root cell structure and induces callose synthesis. *Biologia* 59: 215–220.

Bulanova, N.V., Synzynys, B.I., and Koz'min, G.V. (2001). Aluminum induces chromosome aberrations in cells of wheat root meristem. *Russ. J. Genet.* 37 (12): 1455–1458.

Cai, S., Bai, G.H., and Zhang, D. (2008). Quantitative trait loci for aluminum resistance in Chinese wheat landrace FSW. *Theor. Appl. Genet.* 117: 49–56.

Cai, M., Wang, N., Xing, C., Wang, F., Wu, K., and Du, X. (2013). Immobilization of aluminum with mucilage secreted by root cap and root border cells is related to aluminum resistance in *Glycine max* L. *Environ. Sci. Pollut. Res.* 20, 8924–8933. doi: https://doi.org/10.1007/s11356-013-1815-6

Caires, E.F., Churka, S., Garbuio, F.J. et al. (2006). Soybean yield and quality as a function of lime and gypsum applications. *Sci. Agrie. 63*: 370–379.

Campbell, C.G., Garrido, F., Illera, V., and García-González, M.T. (2006). Transport of Cd, Cu and Pb in an acid soil amended with phosphogypsum, sugar foam and phosphoric rock. *Appl. Geochem.* 21: 1030–1043.

Cardiano, P., Cigala, R.M., Crea, F. et al. (2017). Sequestration of Aluminium (III) by different natural and synthetic organic and inorganic ligands in aqueous solution. *Chemosphere* 186: 535–545.

Carvalho, M.C.S. and van Raij, B. (1997). Calcium sulphate, phosphogypsum and calcium carbonate in the amelioration of acid subsoils for root growth. *Plant and Soil* 192: 37–48.

de Castro, O.M., de Camargo, O.A., Vieira, S.R., and Filho, J.V. (1999). Effect of two types of lime on some soil physical attributes of an Oxisol from Brazil. *Commun. Soil Sci. Plan. 30*: 2183–2195.

Chakraborty, S., Mishra A, Verma E, Tiwari B, Mishra AK, Singh SS. (2019). Physiological mechanisms of aluminum (Al) toxicity tolerance in nitrogen-fixing aquatic macrophyte *Azolla microphylla* Kaulf: phytoremediation, metabolic rearrangements, and antioxidative enzyme responses. *Environ. Sci. Pollut. Res.* 26:9041–9054. doi: https://doi.org/10.1007/s11356-019-04408-7

Chen, Z.C. and Liao, H. (2016). Organic acid anions: an effective defensive weapon for plants against aluminum toxicity and phosphorus deficiency in acidic soils. *J. Genet. Genomics* 43: 631–638.

Chen, L., Ramsier, C., Bighamc, J. et al. (2009). Oxidation of FGD-CaSC>3 and effect on soil chemical properties when applied to the soil surface. *Fuel* 88: 1167–1172.

Chen, R.F., Zhang, F.L., Zhang, Q.M. et al. (2012). Aluminium-phosphorus interactions in plants growing on acid soils: does phosphorus always alleviate aluminium toxicity? *J. Sci. Food Agric. 92*: 995–1000.

Chen, Z.C., Yokosho, K., Kashino, M. et al. (2013). Adaptation to acidic soil is achieved by increased cis-acting element numbers regulating ALMT1 expression in *Holcus lanatus*. *Plant J.* 76: 10–23.

Chowardhara, B., Borgohain, P., Saha, B. et al. (2019a). Phytotoxicity of Cd and Zn on three popular Indian mustard varieties during germination and early seedling growth. *Biocatal. Agric. Biotechnol.* 21: 101349.

Chowardhara, B., Borgohain, P., Saha, B. et al. (2019b). Assessment of phytotoxicity of zinc on Indian mustard (Brassica juncea) varieties during germination and early seedling growth. *Ann. Plant Soil Res.* 21 (3): 239–244.

Ciamprova, M. (2002). Morphological and structural responses ofplant roots to aluminium at organ, tissue, and cellular levels. *Biol. Plant* 45: 161–171.

Cobbett, C. and Goldsbrough, P. (2002). Phytochelatins and metallothioneins: roles in heavy metal detoxification and homeostasis. *Annu. Rev. Plant Biol.* 53: 159–182.

Collins, N.C., Shirley, N.J., Saeed, M. et al. (2008). An ALMT1gene cluster controlling aluminum tolerance at the Alt4 locus of rye (*Secale cereale* L.). *Genetics* 179: 669–692.

Dai, H., Cao, F., Chen, X. et al. (2013). Comparative proteomic analysis of aluminum tolerance in Tibetan wild and cultivated barleys. *PLoS One* 8 (5): e63428.

Dawood, M., Cao, F., Jahangir, M.M. et al. (2012). Alleviation of aluminum toxicity by hydrogen sulfide is related to elevated ATPase and suppressed aluminum uptake and oxidative stress in barley. *J. Hazard. Mater. 209–210*: 121–128.

De La Fuente, M., Ramirez-Rodriguez, V., Cabrera-Ponce, J.L., and Herrera-Estrella, L. (1997). Aluminum tolerance in transgenic plants by alteration of citrate synthesis. *Science* 276: 1566–1568.

Delhaize, E., Craig, S., Beaton, C.D. et al. (1993a). Aluminum tolerance in wheat (*Triticum aestivum* L.). 1. Uptake and distribution of aluminum in root apices. *Plant Physiol.* 103: 685–693.

Delhaize, E., Ryan, P.R., and Randall, P.J. (1993b). Aluminum tolerance in wheat (*Triticum aestivum* L.). 2. Aluminum-stimulated excretion of malic acid from root apices. *Plant Physiol.* 103: 695–702.

Delhaize, E., Hebb, D.M., and Ryan, P.R. (2001). Expression of a pseudomonas aeruginosacitrate synthase gene in tobacco is not associated with either enhanced citrate accumulation or efflux. *Plant Physiol.* 125: 2059–2067.

Delhaize, E., Ryan, P.R., Hocking, P.J., and Richardson, A.E. (2003). Effects of altered citrate synthase and isocitrate dehydrogenase expression on internal citrate concentrations of tobacco (*Nicotiana tabacum* L.). *Plant and Soil* 248: 137–144.

Delhaize, E., Ryan, P.R., Hebb, D.M. et al. (2004). Engineering high-level aluminum tolerance in barley with the ALMT1gene. *Proc. Natl. Acad. Sci. U. S. A.* 101: 5249–15254.

Delhaize, E., Taylor, P., Hocking, P.J. et al. (2009). Transgenic barley (*Hordeum vulgare* L.) expressing the wheat aluminium resistance gene (TaALMT1) shows enhanced phosphorus nutrition and grain production when grown on an acid soil. *Plant Biotechnol. J.* 17: 391–400.

Delhaize, E., Ma, J.F., and Ryan, P.R. (2012). Transcriptional regulation of aluminium tolerance genes. *Trends Plant Sci.* 17: 341–348.

Deng, W., Luo, K., Li, Z. et al. (2009). Overexpression of Citrus junos mitochondrial citrate synthase gene in Nicotiana benthamiana confers aluminum tolerance. *Planta* 230: 355–365.

Denton, B. (2007). Advances in phytoremediation of heavy metals using plant growth promoting bacteria and fungi. MMG 445 Basic. *Biotechniques* 3: 1–5.

Dixon, D.P., Lapthorn, A., and Edwards, R. (2002). Plant glutathione transferases. *Genome Biol.* 3: 1–10.

Doncheva, S., Amenos, M., Poschenrieder, C., and Barcelo, J. (2005). Root cell patterning: a primary target for aluminium toxicity in maize. *J. Exp. Bot.* 56 (414): 1213–1220.

Dong, X.Y., Shen, R.F., Chen, R.F. et al. (2008). Secretion of malate and citrate from roots is related to high Al resistance in Lespedeza bicolor. *Plant and Soil* 306: 139–147.

Duressa, D., Soliman, K., Taylor, R., Senwo, Z. (2011). Proteomic analysis of soybean roots under aluminum stress. *Int. J. Plant Genomics*, 2011 pp.1–12. doi:https://doi.org/10.1155/2011/282531

El-Tarabily, K.A., Hardy, G.E., Sivasithamparam, K., and McKay, A.G. (1997). Amendment of soil with lime or gypsum and its effect on cavity spot disease of carrots *(Daucus carota* L.) caused by *Pythium coloratum*. *Aust. J. Exp. Agric.* 37: 265–270.

Espinoza, C., Schlechter, R., Herrera, D. et al. (2013). Cisgenesis and intragenesis: new tools for improving crops. *Biol. Res.* 46 (4): 323–331.

Ezaki, B., Gardner, R.C., Ezaki, Y., and Matsumoto, H. (2000). Expression of aluminum-induced genes in transgenic Arabidopsis plants can ameliorate aluminum stress and/or oxidative stress. *Plant Physiol.* 122: 657–665.

Ezaki B, Katsuhara M, Kawamura M and Matsumoto H. (2001). Different mechanisms of four aluminum (Al)-resistant transgenes for Al toxicity in Arabidopsis. *Plant Physiol.*, 127:918–927. doi:https://doi.org/10.1104/pp.010399

Farina, M. and Channon, P. (1988). Acid-subsoil amelioration: II. Gypsum effects on growth and subsoil chemical properties. *Soil Sci. Soc. Am. J. 52*: 175–180.

Jorge Fernando Pereira, J. Fernando Pereira, Diliane Barichello, D. Barichello, Jéssica Rosset Ferreira, J. Rosset Ferreira, Jorge González Aguilera, J. González Aguilera, Luciano Consoli, L. Consoli, José Pereira da Silva Júnior, J. Pereira da Silva Júnior, Sandro Bonow, S. Bonow, & Adeliano Cargnin, A. Cargnin. (2015). TaALMT1 and TaMATE1B allelic variability in a collection of Brazilian wheat and its association with root growth on acidic soil. *Molecular Breeding*, 35, 169. doi: https://doi.org/10.1007/s11032-015-0363-9

Fontecha, G., Silva-Navas, J., Benito, C. et al. (2007). Candidate gene identification of an aluminum-activated organic acid transporter gene at the Alt4 locus for aluminum tolerance in rye (*Secale cereale* L.). *Theor. Appl. Genet.* 114: 249–260.

Frantzios, G., Galatis, B., and Apostolakos, P. (2005). Aluminium causes variable responses in actin filament cytoskeleton of the root tip cells of Triticum turgidum. *Protoplasma* 225 (3–4): 129–140.

Fujii, M., Yamaji, N., Sato, K., and Ma, J.F. (2009). Mechanism regulating HvAACT1 expression in barley. In: *Plant – Soil Interactions at Low pH: A Nutriomic Approach – Proceedings of the 7th International Symposium of Plant – Soil Interactions at Low pH* (ed. H. Liao, X. Yan and L.V. Kochian), 165–166. Guangzhou: South China University of Technology Press.

Fujii M, Yokosho K, Yamaji N, Saisho D, Yamane M, Takahashi H, Sato K, Nakazono M, Ma JF. (2012). Acquisition of aluminium tolerance by modification of a single gene in barley. *Nat. Commun.*, 3:713. doi: https://doi.org/10.1038/ncomms1726.

Furukawa, J., Yamaji, N., Wang, H. et al. (2007). An aluminum-activated citrate transporter in barley. *Plant Cell Physiol.* 48: 1081–1091.

Gale, D. (2018). The potential for compost to ameliorate aluminium toxicity and decreased phosphorus availability in agricultural production systems on acid sulfate soils in Vietnam. Masters Thesis. Agriculture and wine science, Charles Stuart University, Australia. https://researchoutput.csu.edu.au/en/publications/the-potential-for-compost-to-ameliorate-aluminium-toxicity-and-de.

Gambaudo, S. (2004). Calidad del yeso natural para uso agrícola. Inta - Estación experimental agropecuaria Rafaela. Información técnica cultivos de verano. Campaña 2006. *Publicación miscelánea* N ° 2004, 106.

Garrido, F., Illera, V., Vizcayno, C., and García-González, M.T. (2003). Evaluation of industrial by-products as soil acidity amendments: chemical and mineralogical implications. *Eur. J. Soil Sci. 54*: 411–422.

Garvin, D.F. and Carver, B.F. (2003). Role of genotype in tolerance to acidity and aluminum toxicity. In: *Handbook of Soil Acidity* (ed. Z. Rengel), 387–406. New York: Marcel Dekker Inc.

Gavrilescu, M., Demnerova, K., Aamand, J. et al. (2015). Emerging pollutants in the environment: present and future challenges in biomonitoring, ecological risks and bioremediation. *N. Biotechnol.* 32: 147–156.

Gensemer, R.W. and Playle, R.C. (1999). The bioavailability and toxicity of aluminum in aquatic environments. *Crit. Rev. Environ. Sci. Technol.* 29: 315–450.

Giannakoula, A., Moustakas, M., Mylona, P. et al. (2008). Aluminum tolerance in maize is correlated with increased levels of mineral nutrients, carbohydrates and proline, and decreased levels of lipid peroxidation and Al accumulation. *J. Plant Physiol.* 165 (4): 385–396.

Goulding, K.W.T. (2016). Soil acidification and the importance of liming agricultural soils with particular reference to the United Kingdom. *Soil Use Manage. 32*: 390–399.

Grauer, U.E. and Horst, W.J. (1992). Modeling cation amelioration of aluminum phytotoxicity. *Soil Sci. Soc. Am. J. 56*: 166–172.

Grevenstuk, T. and Romano, A. (2013). Aluminium speciation and internal detoxification mechanisms in plants: where do we stand? *Metallomics* 5 (12): 1584–1594.

Gruber, B.D., Ryan, P.R., Richardson, A.E. et al. (2010). HvALMT1 from barley is involved in the transport of organic anions. *J. Exp. Bot.* 61: 1455–1467.

Guo, T., Zhang, G., Zhou, M. et al. (2004). Effects of aluminum and cadmium toxicity on growth and antioxidant enzyme activities of two barley genotypes with different Al resistance. *Plant and Soil* 258 (1–2): 241–248.

Guo, P., Li, Q., Qi, Y. P., Yang, L. T., Ye, X., Chen, H. H., Chen, L. S. (2017). Sulfur-mediated-alleviation of aluminum-toxicity in Citrus grandis seedlings. *Int. J. Mol. Sci. 18*, 2570. doi: https://doi.org/10.3390/ijms18122570

Gupta, N. and Kumar, A. (2013). Aluminium toxicity and stress resistance: transgenic approach. *Int. J. Eng. Res. Technol. (IJERT)* 2 (12): 1178–1188.

Gupta, N., Gaurav, S.S., Kumar, A. (2013). Molecular basis of aluminium toxicity in plants: a review. *Am J. Plant Sci.* 2013, 4, 21–37. doi: https://doi.org/10.4236/ajps.2013.4.12A3004.

Hagvall, K., Persson, P., and Karlsson, T. (2015). Speciation of aluminum in soils and stream waters: the importance of organic matter. *Chem. Geol. 417*: 32–43.

Hale, S.E., Alling, V., Martinsen, V. et al. (2013). The sorption and desorption of phosphate-P, ammonium-N and nitrate-N in cacao shell and corn cob biochars. *Chemosphere* 91: 1612–1619.

Han, W.Y., Shi, Y.Z., Ma, L.R. et al. (2007). Effect of liming and seasonal variation on lead concentration of tea plant *(Camellia sinensis* (L.) O. Kuntze). *Chemosphere 66*: 84–90.

Hannink N, Rosser SJ, French CE, Basran A, Murray JA, Nicklin S, Bruce NC (2001) Phytodetoxification of TNT by transgenic plants expressing a bacterial nitroreductase. *Nat. Biotechnol.* 19(12): 1168–1172, doi:https://doi.org/10.1038/nbt1201-1168

Hanson, E.J. and Berkheimer, S.R. (2004). Effect of soil calcium applications on blueberry yield and quality. *Small Fruits Rev. 3*: 133–139.

Hartel, P.G. and Bouton, J.H. (1989). *Rhizobium meliloti* inoculation ofalfalfa selected for tolerance to acid, aluminum-rich soils. *Plant and Soil* 116: 283–285.

Hasenstein, K.H. and Evans, M.L. (1988). Effects of cations on hormone transport in primary roots of Zea mays. *Plant Physiol.* 86: 890–894.

Hasenstein, K.H., Evans, M.L., Stinemetz, C.L. et al. (1988). Comparative effectivness of metal ions in inducing curvature of primary roots of Zea mays. *Plant Physiol.* 86: 885–889.

Haug A. (1984). Molecular aspects of aluminum toxicity. *Crit. Rev. Plant Sci.* 1:345–373. doi:https://doi.org/10.1080/07352688409382184

Hawes, M., Allen, C., Turgeon, B.G. et al. (2016). Root border cells and their role in plant defense. *Annu. Rev. Phytopathol.* 54: 143–161.

Haynes, R.J. and Mokolobate, M.S. (2001). Amelioration of Al toxicityand P deficiency in acid soils by additions of organic residues: a critical review of the phenomenon and the mechanisms involved. *Nutr. Cycl. Agroecosyst.* 59: 47–63.

Hepler, P.K. (2005). Calcium: a central regulator of plant growth and development. *Plant Cell 17*: 2142–2155.

Herath, H., Camps-Arbestain, M., and Hedley, M. (2013). Effect of biochar on soil physical properties in two contrasting soils: an Alfisol and an Andisol. *Geoderma 209*: 188–197.

Hoekenga, O.A., Maron, L.G., Piñeros, M.A. et al. (2006). AtALMT1, which encodes a malate transporter, is identified as one of several genes critical for aluminum tolerance in Arabidopsis. *Proc. Natl. Acad. Sci. U. S. A.* 103: 9738–9743.

Hong-Bo S, Li-Ye C, Cheng-Jiang R, Hua L, Dong-Gang G, Wei-Xiang L (2010) Understanding molecular mechanisms for improving phytoremediation of heavy metal-contaminated soils. *Crit. Rev. Biotechnol.* 30(1):23–30. doi: https://doi.org/10.3109/07388550903208057

Horst, W.J., Schmohl, N., Kollmeier, M. et al. (1999). Does aluminium affect root growth of maize through interaction with the cell wall—plasma membrane— cytoskeleton continuum? *Plant and Soil* 215 (2): 163–174.

Horst, W.J., Wang, Y., and Eticha, D. (2010). The role of the root apoplast in aluminum-induced inhibition of root elongation and aluminum resistance of plants: a review. *Ann. Bot.* 106: 185–197.

Hossain, A.K.M.Z., Hossain, M.A., Koyama, H., and Hara, T. (2004). Effects of aluminum and boron supply on growth of seedlings among 15 cultivars of wheat (*Triticum aestivum* L.) grown in Bangladesh. *Soil Sci. Plant Nutr. 50*: 189–195.

Huang, C.F., Yamaji, N., Mitani, N. et al. (2009). A bacterial-type ABC transporter is involved in aluminum tolerance in rice. *Plant Cell* 21: 655–667.

Huang, C.F., Yamaji, N., Chen, Z., and Ma, J.F. (2012). A tonoplast-localized half-size ABC transporter is required for internal detoxification of aluminum in rice. *Plant J.* 69: 857–867.

Hwang, J.-U., Song, W.-Y., Hong, D., Ko, D., Yamaoka, Y., Jang, S., et al. (2016). Plant ABC transporters enable many unique aspects of a terrestrial plant's lifestyle. *Mol. Plant* 9:338–355. doi: https://doi.org/10.1016/j.molp.2016.02.003

Iqbal, M.T. (2013). Phosphorus enhances aluminium tolerance in both aluminium-tolerant and aluminium-sensitive wheat seedlings. *S. Afr. J. Plant Soil 30*: 13–21.

Iqbal, M.T. (2014). Phosphorus alleviates aluminum toxicity in Al-sensitive wheat seedlings. *Commun. Soil Sci. Plant Anal. 45* (4): 437–450.

Jaskowiak, J., Tkaczyk, O., Slota, M. et al. (2018). Analysis of aluminum toxicity in *Hordeum vulgare* roots with an emphasis on DNA integrity and cell cycle. *PLoS One 13* (2): e0193156. https://doi.org/10.1371/journal.pone.0193156.

Jolly, S., Jaffal, A., Delahaut, L. et al. (2014). Effects of aluminium and bacterial lipopolysaccharide on oxidative stress and immune parameters in roach, *Rutilus rutilus* L. *Environ. Sci. Pollut. Res.* 21: 13103–13117.

Jones, D.L. and Kochian, L.V. (1997). Aluminum interaction with plasma membrane lipids and enzyme metal binding sites and its potential role in Al cytotoxicity. *FEBS Lett.* 400 (1): 51–57.

Jones D. L., Blancaflor E. B., Kochian L. V., Gilroy S. (2006). Spatial coordination of aluminium uptake, production of reactive oxygen species, callose production and wall rigidification in maize roots. *Plant Cell Environ.* 29:1309–1318. doi: https://doi.org/10.1111/j.1365-3040.2006.01509.x.

Kang, J.W. (2014). Removing environmental organic pollutants with bioremediation and phytoremediation. *Biotechnol. Lett.* 36 (6): 1129–1139.

Kanu SA, Okonkwo JO, Dakora FD. (2013). *Aspalathus linearis* (Rooibos tea) as potential phytoremediation agent: a review on tolerance mechanisms for aluminum uptake. *Environ. Rev.* 21(2):85–92. doi: https://doi.org/10.1139/er-2012-0055

Karhu, K., Mattila, T., Bergström, I., and Regina, K. (2011). Biochar addition to agricultural soil increased CH_4 uptake and water holding capacity – results from a short-term pilot field study. *Agric. Ecosyst. Environ.* 140: 309–313.

Karuppanapandian, T., Rhee, S.J., Kim, E.J. et al. (2012). Proteomic analysis of differentially expressed proteins in the roots of Columbia-0 and Landsberg erecta ecotypes of Arabidopsis thaliana in response to aluminum-toxicity. *Can. J. Plant Sci.* 92: 1267–1282.

Kelishadi, R., Moeini, R., Poursafa, P. et al. (2014). Independent association between air pollutants and vitamin D deficiency in young children in Isfahan. *Iran. Paediatr Int Child Health* 34: 50–55.

Khoi, C.M., Guong, V.T., Trung, P.N.M., and Nilsson, S.I. (2010). Effects of compost and lime amendment on soil acidity and N availability in acid sulfate soil. *19th World Congress of Soil Science, Soil Solutions for a Changing World* (1–6 August 2010), Brisbane, Australia.

Kidd, P. and Proctor, J. (2001). The growth response of ecotypes of Holcus lanatus L. from different soil types in northwestern Europe to aluminium and silicon. *Fresen. Environ. Bull.* 10 (6): 590–594.

Kidd, P.S., Llugany, M., Poschenrieder, C. et al. (2001). The role of root exudates in aluminium resistance and silicon-induced amelioration of aluminium toxicity in three varieties of maize (Zea mays L.). *J. Exp. Bot.* 52: 1339–1352.

Kikui, S., Sasaki, T., Maekawa, M. et al. (2005). Physiological and genetic analyses of aluminium tolerance in rice, focusing on root growth during germination. *J. Inorg. Biochem.* 99 (9): 1837–1844.

Kinraide, T.B. (2003). Toxicity factors in acidic forest soils: attempts to evaluate separately the toxic effects of excessive Al^{3+} and H^+ and insufficient Ca^{2+} and Mg^{2+} upon root elongation. *Eur. J. Soil Sci.* 54: 323–333.

Kinraide, T.B. and Parker, D.R. (1989). Assessing the phytotoxicity of mononuclear hydroxy-aluminum. *Plant Cell Environ.* 12: 479–487.

Kinraide, T.B., Pedler, J.F., and Parker, D.R. (2004). Relative effectiveness of calcium and magnesium in the alleviation of rhizotoxicity in wheat induced by copper, zinc, aluminum, sodium, and low pH. *Plant and Soil 259*: 201–208.

Kisnierienė, V. and Lapeikaitė, I. (2015). When chemistry meets biology: the case of aluminium- a review. *Chem. Aust.* 26: 148–158.

Klimashevskii, E.L. (1983). Identification of plant resistance to soil acidity. *Sov. Agric. Sci.* 10: 1–5.

Klimashevskii, E.L. and Chernysheva, N.F. (1980). Content of organic acids and physiologically active substances in plantswhich differ in sensitivity toAl^{3+} toxicity. *Sov. Agric. Sci.* 2: 5–7.

Kochian, L.V., Hoekenga, O.A., and Pineros, M.A. (2004). How do crop plants tolerate acid soils? Mechanisms of Al tolerance and phosphorous efficiency. *Annu. Rev. Plant Biol. 55*: 459–493.

Kochian, L., Pineros, M., and Hoekenga, O. (2005). The physiology, genetics and molecular biology of plant aluminum resistance and toxicity. *Plant and Soil* 274: 175–195.

Kochian, L.V., Piñeros, M.A., Liu, J., and Magalhaes, J.V. (2015). Plant adaptation to acid soils: the molecular basis for crop aluminum resistance. *Annu. Rev. Plant Biol.* 66: 571–598.

Kopittke, P.M., Menzies, N.W., Wang, P. et al. (2014). The rhizotoxicity of metal cations is related to their strength of binding to hard ligands. *Environ. Toxicol. Chem.* 33: 268–277.

Kopittke, P.M., Moore, K.L., Lombi, E. et al. (2015). Identification of the primary lesion of toxic aluminum in plant roots. *Plant Physiol.* 167 (4): 1402–1411.

Kopittke, P.M., Menzies, N.W., Wang, P., and Blamey, F.P.C. (2016). Kinetics and nature of aluminium rhizotoxic effects: a review. *J. Exp. Bot. 67*: 4451–4467.

Korcak, R.F. (1988). Fluidized bed material applied at disposal levels: effects on an apple orchard. *J. Am. Soc. Hortic Sci. 113*: 189–193.

Korcak, R.F. (1992). Short-term response of blueberry to elevated soil calcium. *J. Small Fruit Viticulture 1*: 9–20.

Kovermann, P., Meyer, S., Hörtensteiner, S. et al. (2007). The Arabidopsis vacuolar malate channel is a member of the ALMT family. *Plant J.* 52: 1169–1180.

Koyama, H., Takita, E., Kawamura, A. et al. (1999). Over expression of mitochondrial citrate synthase gene improves the growth of carrot cells in Al-phosphate medium. *Plant Cell Physiol.* 40: 482–488.

Krahl, T. and Kemnitz, E. (2017). Aluminium fluoride – the strongest solid Lewis acid: structure and reactivity. *Cat. Sci. Technol. 7*: 773–796.

Krill, A.M., Kirst, M., Kochian, L.V. et al. (2010). Association and linkage analysis of aluminum tolerance genes in maize. *PLoS One* 5: e9958.

Krishnan, K. and Moens, P.D.J. (2009). Structure and functions of profilins. *Biophys. Rev.* 1: 71–81.

Kumpf, R. P., and Nowack, M. K. (2015). The root cap: a short story of life and death. *J. Exp. Bot.* 66: 5651–5662. doi: https://doi.org/10.1093/jxb/erv295.

Kuo, M.C. and Kao, C.H. (2003). Aluminum effects on lipid peroxidation and antioxidative enzyme activities in rice leaves. *Biol. Plant.* 46 (1): 149–152.

Larsen, P.B., Geisler, M.J.B., Jones, C.A., and Williams, K.M. (2005). ALS3 encodes a phloem-localized ABC transporter- like protein that is required for aluminum tolerance in Arabidopsis. *Plant J.* 41: 353–363.

Li, Y., Yang, G.X., Luo, L.T. et al. (2008). Aluminium sensitivity and tolerance in model and elite wheat varieties. *Cereal Res. Commun.* 36 (2): 257–267.

Li, X., Li, Y., Mai, J. et al. (2018). Boron alleviates aluminum toxicity by promoting root alkalization in transition zone via polar auxin transport. *Plant Physiol. 177*: 1254–1266.

Liao, H., Wan, H., Shaff, J. et al. (2006). Phosphorus and aluminum interactions in soybean in relation to aluminum tolerance. Exudation of specific organic acids from different regions of the intact root system. *Plant Physiol.* 141: 674–684.

Lidon, F.C., Azinheira, H.G., and Barreiro, M.G. (2000). Aluminum toxicity in maize: biomass production and nutrient uptake and translocation. *J. Plant Nutr.* 23 (2): 151–160.

Ligaba, A., Katsuhara, M., Ryan, P.R. et al. (2006). The BnALMT1and BnALMT2 genes from rape encode aluminum-activated malate transporters that enhance the aluminum resistance of plant cells. *Plant Physiol.* 142: 1294–1303.

Ligaba, A., Maron, L., Shaff, J. et al. (2012). Maize ZmALMT2 is a root anion transporter that mediates constitutive root malate efflux. *Plant Cell Environ.* 35: 1185–1200.

Liu, Q., Yang, J.L., He, L.S. et al. (2008). Effect of aluminum on cell wall, plasma membrane, antioxidants and root elongation in triticale. *Biol. Plant.* 52: 87–92.

Liu, J.P., Magalhaes, J.V., Shaff, J., and Kochian, L.V. (2009). Aluminum-activated citrate and malate transporters from the MATE and ALMT families function independently to confer Arabidopsis aluminum tolerance. *Plant J.* 57: 389–399.

Liu, J., Piñeros, M.A., and Kochian, L.V. (2014). The role of aluminum sensing and signaling in plant aluminum resistance. *J. Integr. Plant Biol.* 56: 221–230.

Liu, C., Wang, H., Li, P. et al. (2019). Biochar's impact on dissolved organic matter (DOM) export from a cropland soil during natural rainfalls. *Sci. Total Environ. 650*: 1988–1995.

Llugany, M., Poschenrieder, C., and Barcelo, J. (1995). Monitoring of aluminium-induced inhibition of root elongation in four maize cultivars differing in tolerance to aluminium and proton toxicity. *Physiol. Plant.* 93 (2): 265–271.

Lyrene, P.M. and Muñoz, C. (1997). Blueberry production in Chile. *J. Small Fruit Viticulture* 1997 (5): 1–20.

Ma JF (2000) Role of organic acids in detoxification of aluminum in higher plants. *Plant Cell Physiol.* 41(4):383–390. doi:https://doi.org/10.1093/pcp/41.4.383

Ma, J.F. (2004). Role of silicon in enhancing the resistance of plants to biotic and abiotic stresses. *Soil Sci. Plant Nutr.* 2004 (50): 11–18.

Ma, J.F. (2007). Syndrome of aluminum toxicity and diversity of aluminum resistance in higher plants. *Int. Rev. Cytol.* 264: 225–252.

Ma, J.F., Hiradate, S., Nomoto, K. et al. (1997). Internal detoxification mechanism of Al in hydrangea—identification of Al form in eaves. *Plant Physiol.* 113: 1033–1039.

Ma, J.F., Ryan, P.R., and Delhaize, E. (2001). Aluminium tolerance in plants and the complexing role of organic acids. *Trends Plant Sci. 6* (6): 273–278.

Ma, J.F., Shen, R., Zhao, Z. et al. (2002). Response of rice to Al stress and identification of quantitative trait loci f or Al tolerance. *Plant Cell Physiol.* 43: 652–659.

Ma, J.F., Nagao, S., Sato, K. et al. (2004). Molecular mapping of a gene responsible for Al-activated secretion of citrate in barley. *J. Exp. Bot.* 55: 1335–1341.

Ma JF, Chen ZC, Shen RF. (2014). Molecular mechanisms of Al tolerance in gramineous plants. *Plant and Soil* 381:1–12. Doi: https://doi.org/10.1007/s11104-014-2073-1.

Maejima, E., Watanabe, T., Osaki, M., and Wagatsuma, T. (2014). Phosphorus deficiency enhances aluminum tolerance of rice (*Oryza sativa* L.) by changing the physicochemical characteristics of root plasma membranes and cell walls. *J. Plant Physiol.* 171 (2): 9–15.

Magalhaes, J.V., Garvin, D.F., Wang, Y.H. et al. (2004). Comparative mapping of a major aluminum tolerance gene in sorghum and other species in the Poaceae. *Genetics* 167: 1905–1914.

Magalhaes, J.V., Liu, J., Guimarães, C.T. et al. (2007). A gene in the multidrug and toxic compound extrusion (MATE) family confers aluminum tolerance in sorghum. *Nat. Genet.* 39: 1156–1161.

Mandal, P.K. and Mandal, T.K. (2002). Anion water in gypsum (CaSC>4 $2H_20$) and hemihydrate ($CaS0_4$ $1/2H_20$). *Cem. Concr. Res. 32*: 313–316.

Manson, A. and Findlay, N. (2015). *Agricultural Uses of Lime and Gypsum*, 1–3. Cedara, South Africa: KZN Department of Agriculture & Rural Development, Soil Fertility Research, Analytical Services.

Mariano, E.D. and Keltjens, W.G. (2005). Long-term effects of aluminum exposure on nutrient uptake by maize genotypes differing in aluminum resistance. *J. Plant Nutr.* 28 (2): 323–333.

Marienfeld, S., Schmohl, N., Klein, M. et al. (2000). Localisation of aluminium in root tips of *Zea mays* and *Vicia faba*. *J. Plant Physiol.* 156 (5-6): 666–671.

Maron, L.G., Piñeros, M.A., Guimarães, C.T., Magalhaes, J.V., Pleiman, J.K., Mao, C., Shaff, J., Belicuas, S.N. and Kochian, L.V., 2010. Two functionally distinct members of the MATE (multi-drug and toxic compound extrusion) family of transporters potentially underlie two major aluminum tolerance QTLs in maize. *Plant J.*, 61(5), pp.728–740.

Maron, L.G., Guimarães, C.T., Kirst, M. et al. (2013). Aluminum tolerance in maize is associated with higher MATE1 gene copy number. *Proc. Natl. Acad. Sci. U. S. A.* 110: 5241–5246.

Martinez Lucci, J., Amano, R.S., Rohatgi, P. and Schultz, B., 2008, January. Self-healing in an aluminum alloy reinforces with microtubes. In Energy Nanotechnology International Conference (Vol. 43239, pp. 79-88).

Martins, M., Taborda, R., Silva, G. et al. (2012). Aluminum and removal by a highly Al-resistant dissimilatory sulphate-reducing bacteria community. *Biodegradation 23*: 693–703.

Martinsen, V., Mulder, J., Shitumbanuma, V. et al. (2014). Farmer-led maize biochar trials: effect on crop yield and soil nutrients under conservation farming. *J. Plant Nutr. Soil Sci. 177*: 681–695.

Matsumoto, H., Riechers, D.E., Lygin, A.V. et al. (2015). Aluminum signaling and potential links with safener-induced detoxification in plants. In: *Aluminum Stress Adaptation in Plants* (ed. S.K. Panda and F. Baluska), 1–35. New York: Springer.

Mays, D.A. and Morrvedt, J.J. (1986). Crop response to soil applications of phosphogypsum. *J. Environ. Qual. 15*: 78–81.

Meriño-Gergichevich, C., Alberdi, M., Ivanov, A.G., and Reyes-Diaz, M. (2010). Al^{3+}- Ca^{2+} Interaction in plants growing in acid soils: Al-phytotoxicity response to calcareous amendments. *J. Soil Sci. Plant Nutr. 10*: 217–243.

Meyer S, Scholz-Starke J, De Angeli A, Kovermann P, Burla B, Gambale F, Martinoia E. (2011). Malate transport by the vacuolar AtALMT6 channel in guard cells is subject to multiple regulation. *Plant J.* 67(2):247–57. doi: https://doi.org/10.1111/j.1365-313X.2011.04587.x.

Mora, M.L., Schnettler, B., and Demanet, R. (1999). Effect of liming and gypsum on soil chemistry, yield, and mineral composition of ryegrass grown in an acidic Andisol. *Commun. Soil Sci. Plant Anal.* 30: 1251–1266.

Mora, M.L., Cartes, P., Demanet, R., and Cornforth, I.S. (2002). Effects of lime and gypsum on pasture growth and composition on an acid Andisol in Chile, South America. *Commun. Soil Sci. Plant Anal. 33*: 2069–2081.

Mora, M.L., Alfaro, M.A., Jarvis, S.C. et al. (2006). Soil aluminum availability in Andisols of Southern Chile and its effect on forage production and animal metabolism. *Soil Use Manage.* 22: 95–101.

Morel, J.L., Mench, M., and Guckert, A. (1986). Measurement of Pb^{2+}, Cu^{2+} and Cd^{2+} binding with mucilage exudates from maize (*Zea mays* L.) roots. *Biol. Fertil. Soils* 2 (1): 29–34.

Morita, A., Horie, H., Fujii, Y., Takatsu, S., Watanabe, N., Yagi, A., et al. (2004). Chemical forms of aluminum in xylem sap of tea plants (*Camellia sinensis* L.). *Phytochemistry* 65:2775–2780. doi: https://doi.org/10.1016/j.phytochem.2004.08.043

Morita, A., Yanagisawa, O., Takatsu, S., Maeda, S., and Hiradate, S. (2008). Mechanism for the detoxification of aluminum in roots of tea plant (*Camellia sinensis* (L.) Kuntze). *Phytochemistry* 69, 147–153. doi: https://doi.org/10.1016/j.phytochem.2007.06.007

Morita, A., Yanagisawa, O., Maeda, S. et al. (2011). Tea plant (*Camellia sinensis* L.) roots secrete oxalic acid and caffeine into medium containing aluminum. *Soil Sci. Plant Nutr.* 57 (6): 796–802.

Mossor-Pietraszewska, T. (2001). Effect of aluminium on plant growth and metabolism. *Acta Biochim. Pol.* 48 (3): 673–686.

Moulick, D., Chowardhara, B., and Panda, S.K. (2019). Agroecotoxicological aspect of Arsenic (As) and Cadmium (Cd) on field crops and its mitigation: current status and future prospect. In: *Plant-Metal Interactions* (ed. S. Srivastava, A. Srivastava and P. Suprasanna), 217–246. Cham: Springer https://doi.org/10.1007/978-3-030-20732-8_11.

Moulick, D., Samanta, S., Saha, B. et al. (2020). Salinity stress responses in three popular field crops belonging to fabaceae family. In: *Current Status and Future Prospect. The Plant Family Fabaceae* (ed. M. Hasanuzzaman, S. Araújo and S. Gill), 519–541. Singapore: Springer https://doi.org/10.1007/978-981-15-4752-2_20.

Nagata, T., Hayatsu, M., and Kosuge, N. (1992). Identification of aluminium forms in tea leaves by 27Al NMR. *Phytochemistry* 31:1215–1218. doi: https://doi.org/10.1016/0031-9422(92)80263-E

Nava, L.C., Delatorre, C.A., de Lima Duarte, I.T. et al. (2006). Inheritance of aluminum tolerance and its effects on grain yield and grain quality in oats (*Avena sativa* L.). *Euphytica* 148: 353–358.

Navascues, J., Perez-Rontome, C., Sanchez, D.H. et al. (2012). Oxidative stress is a consequence, not a cause, of aluminum toxicity in the forage legume Lotus corniculatus. *New Phytol.* 193: 625–636.

Nguyen, V.T., Burow, M.D., Nguyen, H.T. et al. (2001). Molecular mapping of genes conferring aluminum tolerance in rice (*Oryza sativa* L.). *Theor. Appl. Genet.* 102: 1002–1010.

Nguyen, B.D., Brar, D.S., Bui, B.C. et al. (2003). Identification and mapping of the QTL for aluminum tolerance introgressed from the new source, *Oryza rufipogon* Griff, into indica rice (*Oryza sativa* L). *Theor. Appl. Genet.* 106: 583–593.

Nguyen, N.T., Hiep, N.D., and Fujita, K. (2005). Iron enhances aluminum-induced leaf necrosis and plant growth inhibition in Eucalyptus camaldulensis. *Plant and Soil* 277 (1): 139–152.

Ninamango-Cárdenas, F.E., Guimaraes, C.T., Martins, P.R. et al. (2003). Mapping QTLs for aluminum tolerance in maize. *Euphytica* 130: 223–232.

Noble, A.D. and Sumner, M.E. (1988). Calcium and Al interactions and soybean growth in nutrient solutions. *Commun. Soil Sci. Plant Anal. 19*: 1119–1131.

Obia, A., Mulder, J., Martinsen, V. et al. (2016). In situ effects of biochar on aggregation, water retention and porosity in light-textured tropical soils. *Soil Tillage Res. 155*: 35–44.

Olivares, E., Pena, E., Marcano, E. et al. (2009). Aluminum accumulation and its relationship with mineral plant nutrients in 12 pteridophytes from Venezuela. *Environ. Exp. Bot.* 65 (1): 132–141.

Paez-Garcia, A., Motes, C.M., Scheible, W.R. et al. (2015). Root traits and phenotyping strategies for plant improvement. *Plan. Theory* 4 (2): 334–355.

Palani, K., Balasubramanian, K., Kalaivani, R.A. (2018). Study on aluminium contamination in mettur soil and its subsequent uptake by medicinal plants. *Orient. J. Chem. 34*(6): doi: https://doi.org/10.13005/ojc/340659.

Pan, W.L., Hopkins, A.G., and Jackson, W.A. (1988). Aluminum-inhibited shoot development in soybean: a possible consequence of impaired cytokinin supply. *Commun. Soil Sci. Plant Anal.* 19: 1143–1153.

Pan, W.L., Hopkins, A.G., and Jackson, W.A. (1989). Aluminum inhibition of shoot lateral branches of Glycine max and reversal by exogenous cytokinin. *Plant and Soil* 120: 1–9.

Panda, S.K. and Matsumoto, H. (2007). Molecular physiology of aluminum toxicity and tolerance in plants. *Bot. Rev.* 73: 326–347.

Panda, S.K., Baluska, F., and Matsumoto, H. (2009). Aluminum stress signaling in plants. *Plant Signal. Behav.* 4: 592–597. https://doi.org/10.4161/psb.4.7.8903.

Pandey, S., Ceballos, H., Magnavaca, R. et al. (1994). Genetics of tolerance to soil acidity in tropical maize. *Crop. Sci.* 34: 1511–1514.

Panhwar, Q. A., Naher, U. A., Radziah, O., Shamshuddin, J., Razi, I. M. (2015) Eliminating aluminum toxicity in an acid sulfate soil for rice cultivation using plant growth promoting bacteria. *Molecules 20*, 3628–3646. doi:https://doi.org/10.3390/molecules20033628

Pavan, M.A., Bingham, F.T., and Peryea, F.J. (1987). Influence of calcium and magnesium salts on acid soil chemistry and calcium nutrition of Apple. *Soil Sci. Soc. Am. J.* *51*: 1526–1530.

Pavlů, L., Borůvka, L., Drábek, O., and Nikodem, A. (2019). Effect of natural and anthropogenic acidification on aluminium distribution in forest soils of two regions in the Czech Republic. *J. Forestry Res.* 32: 363–370. https://doi.org/10.1007/s11676-019-01061-1.

Peixoto, P.H.P., Cambraia, J., Sant'Anna, R. et al. (1999). Aluminum effects on lipid peroxiadation and on the activities of enzymes of oxidative metabolism in sorghum. *Rev. Bras. Fisiol. Veg.* 11: 137–143.

Peng, W., Wu, W., Peng, J. et al. (2018). Characterization of the soybean GmALMT family genes and the function of GmALMT5 in response to phosphate starvation. *J. Integr. Plant Biol.* 60 (3): 216–231.

Pereira, L.B., Mazzanti, C.M.A., Gonçalves, J.F. et al. (2010). Aluminum-induced oxidative stress in cucumber. *Plant Physiol. Biochem.* 48: 683–689.

Pietri, J.A. and Brookes, P.C. (2008). Relationships between soil pH and microbial properties in a UK arable soil. *Soil Biol. Biochem.* 40 (7): 1856–1861.

Piña, R.G. and Cervantes, C. (1996). Microbial interactions with aluminium. *Biometals* *9*: 311–316.

Pineros, M.A. and Kochian, L.V. (2009). Overview of the structure-function relations underlying functionality of ALMT and MATE-type transporters involved in the organic acid release Al tolerance response. *Proceedings of the 7th International Symposium on Plant-Soil Interactions at Low pH* (17–21 May 2009), Guangzhou, China.

Pineros, M.A., Cancado, G.M.A., and Kochian, L.V. (2008). Novel properties of the wheat aluminum tolerance organic acid transporter (TaALMT1) revealed by electrophysiological characterization in Xenopus oocytes: functional and structural implications. *Plant Physiol.* 147: 2131–2146.

Pitaksaringkarn, W., Matsuoka, K., Asahina, M. et al. (2014). XTH 20 and XTH 19 regulated by ANAC 071 under auxin flow are involved in cell proliferation in incised Arabidopsis inflorescence stems. *Plant J.* 80: 604–614.

Poschenrieder, C., Tolra, R. P., and Barcelo, J. (2005). A role for cyclic hydroxamates in aluminium resistance in maize? *J. Inorg. Biochem.* 99: 1830–1836. doi: https://doi.org/10.1016/j.jinorgbio.2005.05.017

Poschenrieder, C., Gunse, B., Corrales, I., and Barcelo, J. (2008). A glance into Al toxicity and resistance in plants. *Sci. Total Environ.* *400* (1–3): 356–368.

Prasad R, Shivay YS (2017) Oxalic acid/oxalates in plants: from self-defence to phytoremediation. *Curr. Sci.*, 112(8):1665–1667. doi: https://doi.org/10.18520/cs/v112/i08/1665-1667

Qian, P., Sun, R., Ali, B. et al. (2014). Effects of hydrogen sulfide on growth, antioxidative capacity, and ultrastructural changes in oilseed rape seedlings under aluminum toxicity. *J. Plant Growth Regul. 33*: 526–538.

Qian, L., Chen, B., and Chen, M. (2016). Novel alleviation mechanisms of aluminum phytotoxicity via released biosilicon from rice straw-derived biochars. *Sci. Rep. 6*: 29346.

Radić, S., Babić, M., Skobic, D. et al. (2010). Ecotoxicological effects of aluminum and zinc on growth and antioxidants in Lemna minor L. *Ecotoxicol. Environ. Saf.* 73: 336–342.

Rahman, M.A., Lee, S.H., Ji, H.C. et al. (2018). Importance of mineral nutrition for mitigating aluminum toxicity in plants on acidic soils: current status and opportunities. *Int. J. Mol. Sci.* 2018 (19): 1–28. https://doi.org/10.3390/ijms19103073.

Ramachandran, S., Christensen, H.E.M., Ishimaru, Y. et al. (2000). Profilin plays a role in cell elongation, cell shape maintenance, and flowering in Arabidopsis. *Plant Physiol.* 124 (4): 1637–1647.

Raman, H., Zhang, K., Cakir, M. et al. (2005). Molecular characterization and mapping of ALMT1, the aluminium-tolerance gene of bread wheat (*Triticum aestivum* L.). *Genome* 48: 781–791.

Rangel, A.F., Rao, I.M., and Horst, W.J. (2009). Intracellular distribution and binding state of aluminum in root apices of two common bean (Phaseolus vulgaris) genotypes in relation to Al toxicity. *Physiol. Plant.* 135 (2): 162–173. https://doi.org/10.1111/j.1399-3054.2008.01183.x.

Rangel, A.F., Rao, I.M., Braun, H.P., and Horst, W.J. (2010). Aluminum resistance in common bean (Phaseolus vulgaris) involves induction and maintenance of citrate exudation from root apices. *Physiol. Plant.* 138 (2): 176–190.

Rashad, A.M. (2017). Phosphogypsum as a construction material. *J. Clean. Prod. 166*: 732–743.

Razaq, M., Zhang, P., Shen, H.L., and Salahuddin. (2017). Influence of nitrogen and phosphorous on the growth and root morphology of Acer mono. *PLoS One* 12: e0171321.

Ren, X.M., Guo, S.J., Tian, W. et al. (2019). Effects of Plant Growth-Promoting Bacteria (PGPB) inoculation on the growth, antioxidant activity, Cu uptake, and bacterial community structure of rape (*Brassica napus* L.) grown in Cu-contaminated agricultural soil. *Front. Microbiol. 10*: 1455. https://doi.org/10.3389/fmicb.2019.01455.

Rengel, Z., Bose, J., Chen, Q., and Tripathi, B. N. (2015). Magnesium alleviates plant toxicity of aluminium and heavy metals. *Crop Pasture Sci.* 66:1298–1307. doi: https://doi.org/10.1071/CP15284.

Reyes-Díaz, M., Meriño-Gergichevich, C., Alarcón, E. et al. (2011). Calcium sulfate ameliorates the effect of aluminum toxicity differentially in genotypes of highbush blueberry (*Vaccinium corymbosum* L.). *J. Soil Sci. Plant Nutr. 11*: 59–78.

Riaz, M., Wu, X., Yan, L. et al. (2018). Boron supply alleviates Al-induced inhibition of root elongation and physiological characteristics in rapeseed (*Brassica napus* L.). *J. Plant Interact. 13*: 270–276.

Rosenberg, W., Nierop, K.G.J., Knicker, H. et al. (2003). Liming effects on the chemical composition of the organic surface layer of a mature Norway spruce stand *(Picea abies* [L.] Karst.). *Soil Biol. Biochem. 35*: 155–165.

Rounds, M.A. and Larsen, P.B. (2008). Aluminum-dependent root-growth inhibition in Arabidopsis results from AtATR-regulated cell-cycle arrest. *Curr. Biol.* 18 (19): 1495–1500.

Rousk, J., Brookes, P.C., and Bååth, E. (2009). Contrasting soil pH effects on fungal and bacterial growth suggest functional redundancy in carbon mineralization. *Appl. Environ. Microbiol.* 75: 1589–1596.

Rout, G.R., Samantaray, S., and Das, P. (2001). Aluminum toxicity in plants: a review. *Agronomie* 21: 2–21.

Ryan, P.R., Ditomaso, J.M., and Kochian, L.V. (1993). Aluminium toxicity in roots: an investigation of spatial sensitivity and the role of the root cap. *J. Exp. Bot.* 44 (2): 437–446.

Ryan, P.R., Delhaize, E., and Randall, P.J. (1995). Characterization of Al-stimulated efflux of malate from the apices of Al-tolerant wheat roots. *Planta* 196: 103–110.

Ryan, P.R., Liu, Q., Sperling, P. et al. (2007). A higher plant Δ8 sphingolipid desaturase with a preference for (Z)-isomer formation confers aluminum tolerance to yeast and plants. *Plant Physiol.* 144 (4): 1968–1977.

Ryan, P.R., Raman, H., Gupta, S. et al. (2009). A second mechanism for aluminum resistance in wheat relies on the constitutive efflux of citrate from roots. *Plant Physiol.* 149: 340–351.

Ryan, P.R., Tyerman, S.D., Sasaki, T. et al. (2011). The identification of aluminium-resistance genes provides opportunities for enhancing crop production on acid soils. *J. Exp. Bot.* 62: 9–20.

Saha, B., Chowardhara, B., Kar, S. et al. (2019). Advances in heavy metal-induced stress alleviation with respect to exogenous amendments in crop plants. In: *Priming and Pretreatment of Seeds and Seedlings* (ed. M. Hasanuzzaman and V. Fotopoulos), 313–332. Singapore: Springer.

Sahin, U., Oztas, T., and Anapali, O. (2003). Effects of consecutive applications of gypsum in equal, increasing, and decreasing quantities on soil hydraulic conductivity of a saline-sodic soil. *J. Plant Nutr. Soil Sci. 166*: 621–624.

Sahoo, S., Borgohain, P., Saha, B. et al. (2019). *Seed Priming and Seedling Pre-Treatment Induced Tolerance to Drought and Salt Stress: Recent Advances. Priming and Pretreatment of Seeds and Seedlings*, 253–263. Singapore: Springer.

Samac, D.A. and Tesfaye, M. (2003). Plant improvement for tolerance to aluminium in acid soils—a review. *Plant Cell Tiss. Org. Cult.* 75: 189–207.

Sanders, D., Pelloux, J., Brownlee, C., and Harper, J.F. (2002). Calcium at the crossroad of signaling. *Plant Cell 14*: 401–407.

Sanderson, K.R. and Eaton, L.J. (2004). Gypsum: an alternative to chemical fertilizers in lowbush blueberry production. *Small Fruits Rev. 3*: 57–71.

Sasaki, T. and Yamamoto, Y. (2007). Molecular mechanisms of aluminum tolerance in plants: the principal control by organic anion transporter. *Tanpakushitsu Kakusan Koso* 52: 619–624.

Sasaki, T., Yamamoto, Y., Ezaki, B. et al. (2004). A wheat gene encoding an aluminum-activated malate transporter. *Plant J.* 37: 645–653.

Sasaki, T., Ryan, P.R., Delhaize, E. et al. (2006). Sequence upstream of the wheat (*Triticum aestivum* L.) ALMT1gene and its relationship to aluminum resistance. *Plant Cell Physiol.* 47: 1343–1354.

Sasaki, T., Mori, I.C., Furuichi, T. et al. (2010). Closing plant stomata requires a homolog of an aluminum-activated malate transporter. *Plant Cell Physiol.* 51: 354–365.

Sasaki, T., Tsuchiya, Y., Ariyoshi, M. et al. (2016). Two members of the aluminum-activated malate transporter family, SlALMT4 and SlALMT5, are expressed during fruit development, and the overexpression of SlALMT5 alters organic acid contents in seeds in tomato (*Solanum lycopersicum*). *Plant Cell Physiol.* 57 (11): 2367–2379. https://doi.org/10.1093/pcp/pcw157. Epub 2016 Sep 11. PMID: 27615796.

Sasidharan, R., Voesenek, L.A., and Pierik, R. (2011). Cell wall modifying proteins mediate plant acclimatization to biotic and abiotic stresses. *Crit. Rev. Plant Sci.* 30: 548–562.

Scott, B.J., Ridley, A.M., and Conyers, M.K. (2000). Management of soil acidity in long-term pastures of south-eastern Australia: a review. *Aust. J. Exp. Agric. 40*: 1173–1198.

Shamshuddin, J., Fauziah, I.C., and Sharifuddin, H.A.H. (1991). Effects of limestone and gypsum application to a Malaysian ultisol on soil solution composition and yields of maize and groundnut. *Plant and Soil 134*: 45–52.

Shao, J. F., Che, J., Chen, R. F., Ma, J. F., and Shen, R. F. (2015). Effect of in planta phosphorus on aluminum-induced inhibition of root elongation in wheat. *Plant and Soil* 395:307–315. doi: https://doi.org/10.1007/s11104-015-2566-6

Sharma, P. and Dubey, R.S. (2007). Involvement of oxidative stress and role of antioxidative defense system in growing rice seedlings exposed to toxic concentrations of aluminum. *Plant Cell Rep.* 26: 2027–2038.

Shen, X., Xiao, X., Dong, Z., and Chen, Y. (2014). Silicon effects on antioxidative enzymes and lipid peroxidation in leaves and roots of peanut under aluminum stress. *Acta Physiol. Plant.* 36:3063–3069. doi: https://doi.org/10.1007/s11738-014-1676-8

Shetty, R. and Prakash, N.B. (2020). Effect of different biochars on acid soil and growth parameters of rice plants under aluminium toxicity. *Sci. Rep. 10*: 12249.

Silva IR, Smyth TJ, Moxley DF, Carter TE, Allen NS, Rufty T.W. (2000). Aluminum accumulation at nuclei of cells in the root tip. Fluorescence detection using lumogallion and confocal laserscanning microscopy. *Plant Physiol.*, 123(2):543–552. doi: https://doi.org/10.1104/pp.123.2.543.

Silva, I.R., Smyth, T.J., Israel, D.W. et al. (2001). Magnesium ameliorates aluminum rhizotoxicity in soybean by increasing citric acid production and exudation by roots. *Plant Cell Physiol. 42*: 546–554.

Silva, S., Santos, C., Matos, M., and Pinto-Carnide, O. (2011). Al toxicity mechanisms in tlerant and sensitive rye genotypes. *Environ. Exp. Bot.* 75: 89–97.

Silva, S., Pinto, G., Correia, B. et al. (2013). Rye oxidative stress under long term Al exposure. *J. Plant Physiol.* 170: 879–889.

Silva, G. E. A., Ramos, F. T., de Faria, A. P., and França, M. G. C. (2014). Seeds' physicochemical traits and mucilage protection against aluminum effect during germination and root elongation as important factors in a biofuel seed crop (*Ricinus communis*). *Environ. Sci. Pollut. Res.* 21:11572–11579. doi: https://doi.org/10.1007/s11356-014-3147-6

Simoes, C.C., Melo, J.O., Magalhaes, J.V., and Guimaraes, C.T. (2012). Genetic and molecular mechanisms of aluminum tolerance in plants. *Genet. Mol. Res.* 11: 1949–1957.

Simon, L., Smalley, T.J., Jones, J.B., and Lasseigne, F.T. (1994). Aluminum toxicity in tomato Part 1. Growth and mineral-nutrition. *J. Plant Nutr.* 17 (2-3): 293–306.

Singh, V.P., Tripathi, D.K., Kumar, D., and Chauhan, D.K. (2011). Influence of exogenous silicon addition on aluminium tolerance in rice seedlings. *Biol. Trace Elem. Res. 144*: 12601274.

Singh, S., Tripathi, D.K., Singh, S. et al. (2017). Toxicity of aluminium on various levels of plant cells and organism: a review. *Environ. Exp. Bot.* 137: 177–193.

Sivaguru, M. and Horst, W.J. (1998). The distal part of the transition zone is the most aluminum-sensitive apical root zone of maize. *Plant Physiol.* 116 (1): 155–163.

Sivaguru, M., Baluska, F., Volkmann, D. et al. (1999). Impacts of aluminum on the cytoskeleton of the maize root apex, short-term effects on the distal part of the transition zone. *Plant Physiol. 119* (3): 1073–1082. https://doi.org/10.1104/pp.119.3.1073.

Sivaguru, M., Liu, J., and Kochian, L.V. (2013). Targeted expression of SbMATE in the root distal transition zone is responsible for sorghum aluminum resistance. *Plant J.* 76: 297–307.

Smith, C.J., Peoples, M.B., Keerthisinghe, G. et al. (1994). Effect of surface applications of lime, gypsum and phosphogypsum on the alleviating of surface and subsurface acidity in a soil under pasture. *Aust. J. Soil. Res. 32*: 995–1008.

Souza, M.C., Correa Scalon, M., Poschenrieder, C. et al. (2018). Mechanisms of storage and detoxification of Al in two mistletoes. *Environ. Exp. Bot.* 150: 37–45.

Stival da Silva, A.L., Sperling, P., Horst, W. et al. (2006). A possible role of sphingolipids in the aluminium resistance of yeast and maize. *J. Plant Physiol. 163*: 26–38.

Syed-Omar, S.R., Shamsuddin, Z.H., Zuraidah, J.Y. et al. (1991). Use of lime, gypsum and their combinations to improve nodulation and yield of groundnut in an acidic soil. *Plant-Soil Interactions at Low pH, Proceedings of the Second International Symposium on Plant-Soil Interactions at Low pH* (24–29 June 1990), Beckley, VA, USA; Wright, R. J., Baligar, V. C., Murrmann, R. P., Eds.; Springer: Dordrecht, The Netherlands, pp. 275–280.

Tabuchi, A. and Matsumoto, H. (2001). Changes in cell-wall properties of wheat (Triticum aestivum) roots during aluminum-induced growth inhibition. *Physiol. Plant.* 112: 353–358.

Tahara, K., Yamanoshita, T., Norisada, M. et al. (2008). Aluminum distribution and reactive oxygen species accumulation in root tips of two melaleuca trees differing in aluminum resistance. *Plant and Soil* 307 (1–2): 167–178. https://doi.org/10.1007/s11104-008-9593-5.

Takahashi, T., Ikeda, Y., Nakamura, H., and Nanzyo, M. (2006). Efficiency of gypsum application to acid Andosols estimated using aluminum release rates and plant root growth. *J. Soil Sci. Plant Nutr. 52*: 584–592.

Tan, K. and Keltjens, W.G. (1990). Interaction between aluminium and phosphorus in sorghum plants. *Plant and Soil 124*: 25–32.

Tariq, A., Pan, K., Olatunji, O.A. et al. (2017). Phosphorous application improves drought tolerance of Phoebe Zhennan. *Front. Plant Sci. 8*: 1561. https://doi.org/10.3389/fpls.2017.01561.

Teng, W., Kang, Y., Hou, W. et al. (2018). Phosphorus application reduces aluminum toxicity in two Eucalyptus clones by increasing its accumulation in roots and decreasing its content in leaves. *PLoS One* 13: e0190900.

Tesfaye, M., Temple, S.J., Allan, D.L. et al. (2001). Overexpression of malate dehydrogenase in transgenic alfalfa enhances organic acid synthesis and confers tolerance to aluminum. *Plant Physiol.* 127: 1836–1844.

Tirado-Corbalá, R., Slater, B.K., Dick, W.A., and Barker, D. (2017). Alfalfa responses to gypsum application measured using undisturbed soil columns. *Plan. Theory 6*: 29. https://doi.org/10.3390/plants6030029.

Tisdale, S.L., Nelson, W.L., and Beaton, J.D. (1985). *Soil Fertility and Fertilizers*, 4e, 329–339. New York: Mac Millan Publishing Company.

Toma, M., Sumner, M.E., Weeks, G., and Saigusa, M. (1999). Long-term effects of gypsum on crop yield and subsoil chemical properties. *Soil Sci. Soc. Am. J. 39*: 891–895.

Toma, M., Saigusa, M., Gafoku, N., and Sumner, M.E. (2005). Effects of gypsum on amelioration of subsoil acidity in Andisols. *JIFS 2*: 69–71.

Tovkach, A., Ryan, P.R., Richardson, A.E. et al. (2013). Transposon-mediated alteration of TaMATE1B expression in wheat confers constitutive citrate efflux from root apices. *Plant Physiol.* 161: 880–892.

Trejo-Tellez, L.I., Stenzel, R., Gomez-Merino, F.C., and Schmitt, J.M. (2010). Transgenic tobacco plants overexpressing pyruvate phosphate dikinase increase exudation of organic acids and decrease accumulation of aluminum in the roots. *Plant and Soil* 326: 187–198.

Tuna, A.L., Kaya, C., Ashraf, M. et al. (2007). The effects of calcium sulphate on growth, membrane stability and nutrient uptake of tomato plants grown under salt stress. *Environ. Exp. Bot. 59*: 173–178.

Uexktill, H.R.V. and Mutert, E. (1995). Global extent, development and economic impact of acid soils. *Plant and Soil* 171: 1–15.

Vicensi, M., Müller, M.M., Kawakami, J. et al. (2016). Do rates and splitting of phosphogypsum applications influence the soil and annual crops in a no-tillage system? *Rev. Bras. Ciênc. Solo* 40: https://doi.org/10.1590/18069657rbcs20150155.

Vitorello, V.A., Capaldi, F.R., and Stefanuto, V.A. (2005). Recent advances in Al toxicity and resistance in higher plants. *Braz. J. Plant Physiol. 17* (1): 129–143. https://doi.org/10.1590/S1677-04202005000100011.

Wagatsuma, T., Maejima, E., Watanabe, T. et al. (2015). Significant role of the plasma membrane lipid bilayers in aluminum tolerance of plants. In: *Aluminum Stress Adaptation in Plants* (ed. S. Panda and F. Baluška), 99–124. Cham: Springer https://doi.org/10.1007/978-3-319-19968-9_6.

Wang, Y., Stass, A., and Horts, W. J. (2004). Apoplastic binding of aluminum is involved in silicon-induced amelioration of aluminum toxicity in maize. *Plant Physiol.* 136:3762–3770. doi: https://doi.org/10.1104/pp.104.045005

Wang, J.P., Raman, H., Zhou, M.X. et al. (2007). High-resolution mapping of the Alp locus and identification of a candidate gene HvMATE controlling aluminium tolerance in barley (*Hordeum vulgare* L.). *Theor. Appl. Genet.* 115: 265–276.

Wang, Q., Nian, F., Zhao, L., Li, F., Yang, H., and Yang, Y. (2013a). Exogenous indole-3-acetic acid could reduce the accumulation of aluminum in root apex of wheat (*Triticum aestivum* L.) under Al stress. *J. Soil Plant Nutr.* 13:534–543. doi: https://doi.org/10.4067/S0718-95162013005000042

Wang, H., Liang, W., and Huang, J. (2013b). Putrescine mediates aluminum tolerance in red kidney bean by modulating aluminum-induced oxidative stress. *Crop. Sci.* 53:120–2128. doi: https://doi.org/10.2135/cropsci2013.04.0215

Wang, C.Y., Shen, R.F., Wang, C., and Wang, W. (2013c). Root protein profile changes induced by Al exposure in two rice cultivars differing in Al tolerance. *J. Proteomics 78*: 281–293.

Wang, W., Zhao, X. Q., Chen, R. F., Dong, X. Y., Lan, P., Ma, J. F., et al. (2015). Altered cell wall properties are responsible for ammonium-reduced aluminium accumulation in rice roots. *Plant Cell Environ.* 38:1382–1390. doi: https://doi.org/10.1111/pce.12490

Wang P, Yu W, Zhang J, Zhang J, Rengel Z, Xu J, Han Q, Chen L, Li K, Yu Y, Chen Q. (2016). Auxin enhances aluminium-induced citrate exudation through upregulation of GmMATE and activation of the plasma membrane H+-ATPase in soybean roots. *Ann. Bot.* 118(5):933–940. doi:https://doi.org/10.1093/aob/mcw133.

Ward, J.T., Lahner, B., Yakubova, E. et al. (2008). The effect of iron on the primary root elongation of *Arabidopsis* during phosphate deficiency. *Plant Physiol. 147*: 1181–1191.

Watanabe T, Osaki M, Tadano T (1997) Aluminum-induced growth stimulation in relation to calcium, magnesium, and silicate nutrition in *Melastoma malabathricum* L. *Soil Sci. Pl. Nutr.* 43(4):827–837. doi:https://doi.org/10.1080/00380768. 1997.10414649

White, P. and Broadley, M.R. (2003). Calcium in plants. *Ann. Bot-London* 92: 487–511.

von Willert, F.J. and Stehouwer, R.C. (2003). Compost, limestone, and gypsum effects on calcium and aluminum transport in acidic minespoil. *Soil Sci. Soc. Am. J. 67*: 778–786.

Wu, P., Liao, C.Y., Hu, B. et al. (2000). QTLs and epistasis for aluminum tolerance in rice (*Oryza sativa* L.) at different seedling stages. *Theor. Appl. Genet.* 100: 1295–1303.

Wu, D., Shen, H., Yokawa, K., and Baluška, F. (2015). Overexpressing *OsPIN2* enhances aluminium internalization by elevating vesicular trafficking in rice root apex. *J. Exp. Bot.* 66: 6791–6801. doi: https://doi.org/10.1093/jxb/erv385.

Xia, J., Yamaji, N., Kasai, T., Ma, J. F. (2010). Plasma membrane-localized transporter for aluminum in rice. *Proc. Natl. Acad. Sci. U. S. A.* 107, 18381–18385. doi: https://doi.org/10.1073/pnas.1004949107

Yamaguchi, M., Sasaki, T., Sivaguru, M. et al. (2005). Evidence for the plasma membrane localization of Al-activated malate transporter (ALMT1). *Plant Cell Physiol.* 46: 812–816.

Yamaji, N., Huang, C.F., Nagao, S. et al. (2009). A zinc finger transcription factor ART1 regulates multiple genes implicated in aluminum tolerance in rice. *Plant Cell* 21 (10): 3339–3349.

Yamamoto, Y., Kobayashi, Y., and Matsumoto, H. (2001). Lipid peroxidation is an early symptom triggered by aluminum, but not the primary cause of elongation inhibition in Pea roots. *Plant Physiol.* 125 (1): 199–208.

Yamamoto, Y., Kobayashi, Y., Devi, S.R. et al. (2002). Aluminum toxicity is associated with mitochondrial dysfunction and the production of reactive oxygen species in plant cells. *Plant Physiol.* 128: 63–72.

Yamamoto, Y., Kobayashi, Y., Devi, S.R. et al. (2003). Oxidative stress triggered by aluminum in plant roots. *Plant and Soil* 255 (1): 239–243.

Yamato, M., Okimori, Y., Wibowo, I.F. et al. (2006). Effects of the application of charred bark of *Acacia mangium* on the yield of maize, cowpea and peanut, and soil chemical properties in South Sumatra, Indonesia. *Soil Sci. Plant Nutr. 52*: 489–495.

Yang, Q.S., Wang, Y.Q., Zhang, J.J. et al. (2007). Identification of aluminum-responsive proteins in rice roots by a proteomic approach: cysteine synthase as a key player in Al response. *Proteomics* 7: 737–749.

Yang, J.L., Li, Y.Y., Zhang, Y.J. et al. (2008). Cell wall polysaccharides are specifically involved in the exclusion of aluminum from the rice root apex. *Plant Physiol.* 146: 602–611.

Yang, J.L., Zhu, X.F., Peng, Y.X. et al. (2011). Cell wall hemicellulose contributes significantly to aluminum adsorption and root growth in Arabidopsis. *Plant Physiol.* 2011 (155): 1885–1892.

Yang, Z., Rao, I.M., and Horst, W.J. (2013). Interaction of aluminium and drought stress on root growth and crop yield on acid soils. *Plant and Soil* 372: 3–25. https://doi.org/10.1007/s11104-012-1580-1.

Yang, Y., Liu, Y., Huang, C.-F., de Silva, J., and Zhao, F.-J. (2016). Aluminium alleviates fluoride toxicity in tea (*Camellia sinensis*). *Plant and Soil* 402:179–190. doi: https://doi.org/10.1007/s11104-015-2787-8

Yi, M., Yi, H., Li, H., and Wu, L. (2010). Aluminum induces chromosome aberrations, micronuclei, and cell cycle dysfunction in root cells of *Vicia faba*. *Environ. Toxicol.* 25 (2): 124–129.

Yocum, C.F. (2008). The calcium and chloride requirements of the O2 evolving complex. *Coord. Chem. Rev. 252*: 296–305.

Yokosho, K., Yamaji, N., and Ma, J.F. (2010). Isolation and characterization of two MATE genes in rye. *Funct. Plant Biol.* 37: 296–303.

Yokosho, K., Yamaji, N., and Ma, J.F. (2011). An Al-inducible MATE gene is involved in external detoxification of Al in rice. *Plant J.* 68: 1061–1069.

Yu, M., Shen, R., Xiao, H., Xu, M., Wang, H., Wang, H., et al. (2009b). Boron alleviates aluminum toxicity in pea (*Pisum sativum*). *Plant and Soil* 314: 87–98. doi: https://doi.org/10.1007/s11104-008-9708-z

Yu, W., Kan, Q., Zhang, J., Zeng, B., and Chen, Q. (2016). Role of the plasma membrane H+-ATPase in the regulation of organic acid exudation under aluminum toxicity and phosphorus deficiency. *Plant Signal. Behav.* 11:e1106660. doi: https://doi.org/10.1080/15592324.2015.1106660

Yuan, J.H., Xu, R.K., Wang, N., and Li, J.Y. (2011). Amendment of acid soils with crop residues and biochars. *Pedosphere 21*: 302–308.

Zhang, J., He, Z., Tian, H. et al. (2007). Identification of aluminium-responsive genes in rice cultivars with different aluminium sensitivities. *J. Exp. Bot.* 58 (8): 2269–2278.

Zhang, W.H., Ryan, P.R., Sasaki, T. et al. (2008a). Characterization of the TaALMT1 protein as an Al^{3+}-activated anion channel in transformed tobacco (*Nicotiana tabacum* L.) cells. *Plant Cell Physiol.* 49: 1316–1330.

Zhang, P., Wu, Z., Zhang, G. et al. (2008b). Coagulation characteristics of polyaluminum chlorides PAC-Al30 on humic acid removal from water. *Sep. Purif. Technol.* 63 (3): 642–647.

Zhang, H., Tan, Z.Q., Hu, L.Y. et al. (2010). Hydrogen sulfide alleviates aluminum toxicity in germinating wheat seedlings. *J. Integr. Plant Biol. 52*: 556–567.

Zhang, J., Hu, A., Wang, B., and Ran, W. (2019). Al (III)-binding properties at the molecular level of soil DOM subjected to long-term manuring. *J. Soil. Sediment. 19*: 1099–1108.

Zhao, Z., Ma, J.F., Sato, K., and Takeda, K. (2003). Differential Al resistance and citrate secretion in barley (Hordeum vulgare L.). *Planta* 217: 794–800.

Zhao, Z., Gao, X., Ke, Y. et al. (2019). A unique aluminum resistance mechanism conferred by aluminum and salicylic-acid-activated root efflux of benzoxazinoids in maize. *Plant and Soil* 1–17.

Zheng, S.J., Yang, J.L., He, Y.F. et al. (2005). Immobilization of aluminum with phosphorus in roots is associated with high aluminum resistance in buckwheat. *Plant Physiol. 138*: 297–303.

Zheng L, Lan P, Shen RF, Li WF (2014). Proteomics of aluminum tolerance in plants. *Proteomics* 14: 566–578. doi:https://doi.org/10.1002/pmic.201300252

Zhou, S., Sauve, R., and Thannhauser, T.W. (2009). Aluminum induced proteome changes in tomato cotyledons. *Plant Signal. Behav.* 4: 769–772.

Zhou, G.F., Delhaize, E., Zhou, M., and Ryan, P.R. (2013). The barley MATE gene, HvAACT1, increases citrate efflux and Al tolerance when expressed in wheat and barley. *Ann. Bot.* 112: 603–612.

Zhu, X.F., Shi, Y.Z., Lei, G.J. et al. (2012). XTH31, encoding an in vitro XEH/XET-active enzyme, regulates aluminum sensitivity by modulating in vivo XET action, cell wall xyloglucan content, and aluminum binding capacity in Arabidopsis. *Plant Cell* 24: 4731–4747.

Zobel, R.W., Kinraide, T.B., and Baligar, V.C. (2007). Fine root diameters can change in response to changes in nutrient concentrations. *Plant and Soil* 297 (1–2): 243–254.

7

Mechanisms of Cobalt Uptake, Transport, and Beneficial Aspects in Plants

Zaid Ulhassan[1], Aamir Mehmood Shah[2], Ali Raza Khan[1], Wardah Azhar[1], Yasir Hamid[3], and Weijun Zhou[1]

[1] *Institute of Crop Science, Ministry of Agriculture and Rural Affairs Key Laboratory of Spectroscopy Sensing, Zhejiang University, Hangzhou, China*
[2] *State Key Joint Laboratory of Environment Simulation and Pollution Control, School of Environment, Beijing Normal University, Beijing, China*
[3] *Ministry of Education (MOE) Key Lab of Environ. Remediation and Ecol. Health, College of Environmental and Resources Science, Zhejiang University, Hangzhou, China*

Introduction

The presence of heavy metals or metalloids in soils can be taken up by plant parts along with other essential nutrients. The excessive doses/accumulation of these metal ions by plants can cause phytotoxic effects and ultimately impairs growth and yield attributes (Ahmad et al. 2017; Ali et al. 2018a; Azhar et al. 2020; Khan et al. 2019; Ulhassan et al. 2018, 2019a, b, c). Cobalt (Co) is one of the beneficial micronutrient that promotes plant growth and development, but its essentiality is still debatable in literature. Though, Co is a vital constituent of coenzyme cobalamin (vitamin B12 and related derivates) in biological nitrogen fixation (Gad 2012; Komeda et al. 1997). The chemical properties of Co are similar to nickel and utilize same carriers to enter in plasma membrane. In plants, the concentration of Co is normally 0.1–10 mg kg^{-1} (DW), which indicated the beneficial roles of Co as trace element. Globally, it ranges 8–20 mg kg^{-1} in normal soils (Chatterjee and Chatterjee 2000). In contrast, Co hyperaccumulators such as *Fabaceae*, *Asteraceae*, and *Lamiaceae* can contain > 1000 mg kg^{-1} Co per plant (Brooks 1977). Cereals, forage, and rice plants contain 2.2, 0.6–3.5, and 0.02–0.150 mg kg^{-1} Co, respectively (Palit et al. 1994). Still the mechanisms involved in Co accumulation are poorly understood due to its sharing (in accumulation) with other metal ions such as Cu (Lange et al. 2017).

Mainly, the transport of Co into root epidermal is governed by IRT1 transporter (Korshunova et al. 1999). *Iron regulated* or ferroportins metal efflux proteins (FeREG/FPN) mediate the transport of Co within vacuoles of root epidermal and cortical cells, then sequestration in roots. In addition, higher expression of *HMA3* genes in roots suggested their participation in the transport of Co within vacuoles (Morel et al. 2009). The transport

Beneficial Chemical Elements of Plants: Recent Developments and Future Prospects, First Edition. Edited by Sangeeta Pandey, Durgesh Kumar Tripathi, Vijay Pratap Singh, Shivesh Sharma, and Devendra Kumar Chauhan.

of Co from roots to aerial parts can be carried out (via xylem) by FPN1 (Morrissey et al. 2009). The sequestration of Co ions in vegetative plant parts may help in exploring its transport mechanism in higher plants.

One of the major physiological responses by Co in higher plants is the nitrogen fixation by legume crops. Co is essential for cobalamin and also involves in the activation of enzymes required in nitrogen fixation by rhizobia bacteria and cyanobacteria that reside in the roots of leguminous crops. Co can restrict the leaf senescence by inhibiting the biosynthesis of ethylene and mediates in nodule production as well as and nitrogen fixation by legumes (Palit et al. 1994). In pea plants, the supplementation of 8 mg Co in soil enhanced the plant growth, nutrients uptake, and number and weight of nodules formation along with the grain quality (Gad 2006). In ground nut (*Arachis hypogaea* L.), about 8 mg Co in soil boosted the nitrogenase activity, thereby enhancing the growth, yield, quality of pods, and oil production (Gad 2012). The application of Co (up to a certain level) can delay the ageing effects in common vetch (*Vicia* spp.), marigold (*Tagetes patula* L.), and rose (*Rosa* spp.) (Mitra 2015). However, higher Co accumulation in plants can cause notable toxic effects (Ali et al. 2018b, c; Kamran et al. 2020; Lwalaba et al. 2020; Mahey et al. 2020; Nazir et al. 2020; Wang et al. 2020; Salam et al. 2021). Co (at particular dose) can restrict the accumulation and toxicity exposed by different metal ions such as Cd and Cr (Palit et al. 1994). However, its interaction with other abiotic stresses required thorough understandings at molecular level.

In recent past decades, the transport mechanisms as well as essentiality of Co ions in plant growth and metabolic functions are less documented in literature. Therefore, we intricately overview the mechanistic approaches taken part in the uptake, transport, and beneficial effects of Co in different plants.

Mechanisms of Cobalt Uptake and Transport in Plants

The understandings of uptake and transport mechanisms as well as movements inside plant tissues are crucial for the elucidation of Co-induced beneficial roles. The potential mechanisms involved in the uptake and transport of Co within plants are summarized in Figure 7.1. Usually, plants accumulate small amounts of Co ions. Co may accumulate in plant parts such as seeds, fruits, and grains. In this way, it is essential for human and animal nutrition (McDowell 1989). In plants, metal homeostasis is crucial for cellular growth and metal transporters play crucial roles in defence against cellular perturbations linked with metal homeostasis. The uptake and transport mechanisms of Co are interrelated to plant nutrition along with yield quality. Previous studies reported that Co competes with iron for access availability to transporters in different plants (Barras and Fontecave 2011). Inside cellular organelles, IRON REGULATED1 (IRT1), an *Arabidopsis thaliana* transporter, or ferroportins metal efflux proteins (FeREG1/FPN1) or FeREG2/FPN2 helps in the detoxification of Co. It is well understood that (heavy metal associated3) AtHMA3 (a P1B-ATPase family) is abundantly expressed in *Arabidopsis* roots that may participate in the transport of Co into vacuolar membrane and mediates Co detoxification (Morel et al. 2009). FeREG2/FPN2 mediated the transport of Co within vacuoles of root epidermal as well as cortical cells (via active transport and passive diffusion), then sequester Co in roots' outermost layer (Korshunova et al. 1999). A plasma membrane protein, AtMGT1, may also

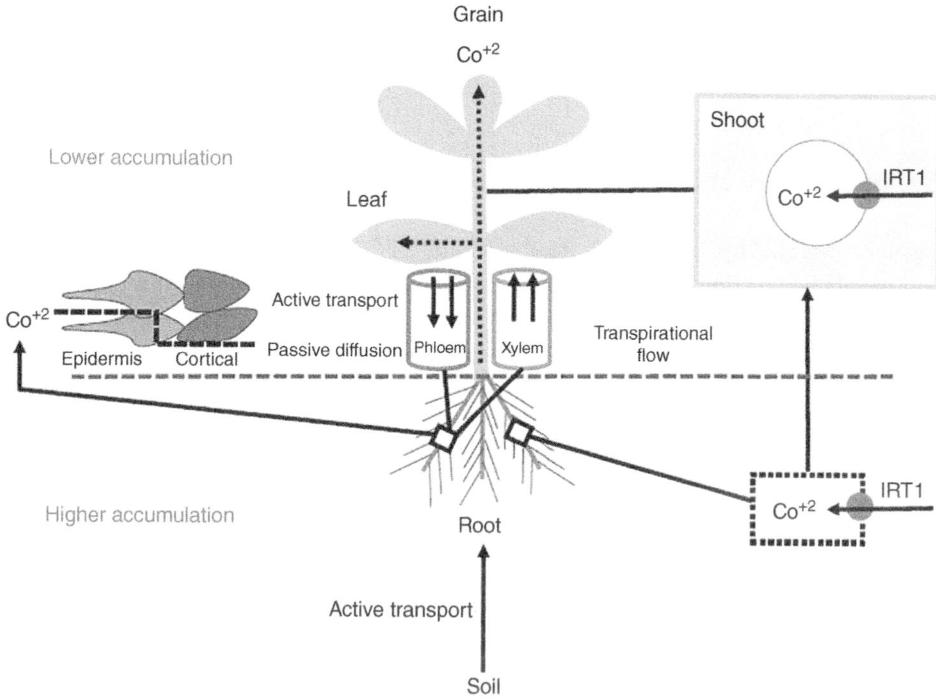

Figure 7.1 The entry of cobalt (Co) from soil to plant root system is governed by active transport. The uptake of Co ions in plant roots in carried out by IRT1, an iron regulated transporter1. Root epidermis and cortical cell mediated the transport of mobile Co ions through active transport, passive diffusion, and then taken up to aerial plant parts by transpirational flow. Lower mobility of Co ions restricts its aerial transport, which resulted in higher Co accumulation in roots.

involve in the transport of Co (Li et al. 2001). Recent investigations validate that chloroplast-localized ARG1 (ABC member gene) facilitates the Co transport (from chloroplast) and thus influences the photosynthesis functions (Li et al. 2020). Another recent study reported that CrNRAMP1 transporter is involved in the transport of Co within *Chlamydomonas reinhardtii*, a unicellular plant (Chang et al. 2020). The mobile Co can be carried to xylem tissues by FPN1 and translocated into aerial plant parts (by transpirational flow) through cortical cells via active transport and passive diffusion (Jarosick et al. 1988; Morrissey et al. 2009; Figure 7.1). Lower mobility of Co may restrict its transport to leaves (from stems) and other aerial parts. Studies demonstrated that a class of protein such as CPx type ATPases was also involved in the coordination of Co ions during transportation (Arguello et al. 2007). The knockout of ATPases encoding genes leads to the enhanced levels of cellular Co that resulted in the reduction of Co tolerance in *Synechocystis sp.* (Rutherford et al. 1999). It is not clear either ATPases are the specific contributors in Co transport. Still no specific Co transporter has been identified in plant species. But, the translocation of Co ions in whole plant system is most probably carried out by IRT1. The process of uptake, accumulation, and transportation of Co ions in plants may be affected by the physiochemical characteristics of soil along other environmental conditions, plants demand, and their physiological responses. Further investigations are required regarding the precise uptake and transport/protein channel mechanisms of Co from soil to root as well as plant organs. The pathways involved in the distribution of Co ions within plant parts should be explored rather than focusing on concentration and penetration of Co ions.

Beneficial Aspects of Cobalt in Plants

Co is not essential but required in small amounts for normal plant growth, development, and metabolic functions. It is categorized as beneficial element due to its importance in symbiotic nitrogen fixation in legume plants and require vitamin B12 (cobalamin) for stimulating the enzyme activities taking part in nitrogen fixation (Lange et al. 2017). A brief overview of Co-induced beneficial effects on growth, yield, mineral nutrient status, physio-biochemical processes, antioxidant defence system, and potential to mitigate abiotic stressors are documented in different plants (mainly legumes) (Figure 7.2).

Growth and Yield

To validate the essentiality of Co for legume plants, a recent study reported the potential of different levels of Co on the growth and yield of mung bean (*Vigna radiata* L.) plants. They verified that the exposure of Co (10, 20, and 30 ppm) in combination with slurry inoculations enhanced the plant growth traits (leaf area, plant height, number of leaves, and branches), germination percentage, seedlings emergence, nodule formation (size and weight), nodule number, and grain yields over control treatments (Vaseer et al. 2020). In *Nigella sativa*, different rates of Co (10, 20, 30, and 40 mg L^{-1}) enhanced the growth characteristics such as plant height, leaf number, capsule number, branch number, dry weight, and seed yield (Khalid and Ahmed 2016). A field study was performed to analyze the effects of Co (5, 10, 15, and 20 ppm) on the growth and yield of faba bean. They revealed that Co at 20 ppm displayed maximum values of aforementioned parameters (Kandil 2007).

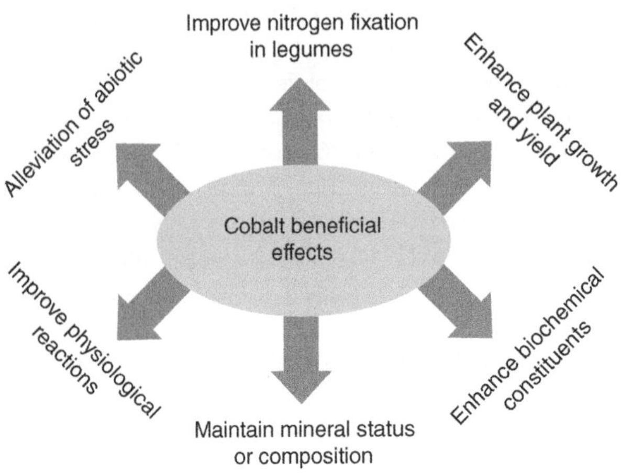

Figure 7.2 Potential beneficial effects of cobalt (Co) on morph-physiogy, biochemical traits and abiotic stress tolerance in plants.

In a pot experiment carried out by Jaleel et al. (2008), they reported that the applications of Co (50, 100, 150, 200, and 250 mg kg^{-1}) stimulated the growth attributes such as plant height, nodule number, leaf area, and biomass production. Contrary to other levels, Co especially at 50 mg kg^{-1} improves the yield traits of groundnut. In *Zea mays*, Co (50 mg kg^{-1} soil) increased the growth and yield characteristics such as number of cobs, seeds per plant, and seedling vigour (Jaleel et al. 2009). In lentil, Co at 6 kg h^{-1} significantly enhanced the plant height, number of branches, dry biomass production, pods number, and seeds per plants in comparison with control treatments (Sahay and Singh 2012). In broad bean, Co at 0.24 and 0.48 g L^{-1} (soaking and foliar spray) increased the number of seeds, seeds yield per plant, pods per plant, and straw yield per plant (Manal et al. 2016). Gad and Kandil (2008) reported that the applications of Co (5, 7.5, 10, 12.5, and 15 ppm) displayed positive effects on the growth parameters such as plant height, number of branches per plant, fresh and dry weights of plant tissues, root diameter, and root yield of sweet potato. Positive effects of Co have also been reported in soya bean plants (Gad et al. 2013; Kandil et al. 2013). They found that Co (12 mg kg^{-1}) increased the shoot and root lengths, fresh biomass, number of nodules per plant, number of leaves per plant, and number of branches per plant. These outcomes suggested that Co improves the whole plant growth characteristics at particular dose, which may vary with respect to target plant. Possibly, the increase in plant growth and yield traits is due to the improvement in photosynthesis, energy sources (sugars, carbohydrates, etc.), and minerals supply.

Nitrogen Fixation and Nodule Formation

Nitrogen fixation in plants depends upon the number of nodule formation in legumes. Therefore, the increase in nodulation would be ideal to enhance the nitrogen fixation. It has been reported that Co enhanced the rate of nitrogen fixation in legume plants (Gad 2012; Kandil 2007; Younis 2011). As compared to control treatments, Co fix 3.5% and 7% more nitrogen in two varieties of cowpea. In addition, they reported that exposure of Co at 3.5 ppm enhanced the nitrogen fixation by 13.4% (Yadav and Khanna 1983). Few studies documented that the applications of Co increased the number of nodules formation in groundnut (Gad 2012), faba bean (Kandil 2007), and *Lablab purpureus* (Younis 2011). Another pot experiment was conducted to explore the effects of Co (50, 100, 150, 200, and 200 mg kg^{-1}) on cowpea. Lower levels of Co (50 mg kg^{-1}) increased the number of nodules (63%) as compared to control treatments (Vijayarengan 2012). In pea plants, the nodule numbers were increased (about 50%) with the increase in Co levels (20 kg h^{-1}) (Akbar et al. 2013). Jayakumar (2019) reported that the application of Co (50 mg kg^{-1}) is beneficial for overall plant growth parameters as it increased the leg haemoglobin contents and root nodule formation in cluster bean. Their findings suggested that the added Co has greater potential to fix nitrogen, especially in the root nodules of legume plants.

Alterations in Nutrient Status

The stimulation in the micro and macro mineral composition (Mn, Zn, Cu, N, P, and K) was reported by the applications of Co (mainly at 7.5 ppm) in okra (*Hibiscus esculantus*) seedlings (Gad et al. 2015a). In *Zea mays*, Jaleel et al. (2009) reported that Co at the exposure of

50 mg kg^{-1} increased the macronutrients such as nitrogen (N), phosphorus (P), and potassium (K) contents as well as micronutrients such as iron (Fe), copper (Cu), zinc (Zn), and manganese (Mn). In broad bean, the (foliar and soaking) treatments of Co (0.24 and 0.48 g L^{-1}) displayed promotive effects on the uptake and accumulation of mineral nutrients (Cu, Zn, Mn), except Fe and N, P, and K (Manal et al. 2016). Another study revealed that Co (10, 20, 30, and 40 mg L^{-1}) improved the uptake and accumulation of macronutrients such as N, P, and K (especially at 30 mg L^{-1}) in *N. sativa* L. (Khalid and Ahmed 2016). Kandil et al. (2013) evaluated the potential effects of Co (4, 8, 12, 16, and 20 mg kg^{-1}) on the nutrient's status of soya bean (*Glycine max* L.). They found that Co (mainly at 12 mg kg^{-1}) increased the uptake of nutrients (N, P, K, Mn, Zn, and Cu) in the shoots and seeds. In sweet potato (*Ipomea batatas* L.), treatments of Co (5, 7.5, 10, 12.5, and 15 ppm) increased the nutrient levels (N, P, K, Cu, Mn, and Zn), except Fe. Co caused promotive effects that were noticed below 10 ppm, while concentrations higher than 10 ppm imposed negative effects on minerals status (Gad and Kandil 2008). In another study, the combination of Co (0.21 kg h^{-1}) with rhizobium and phosphobacterium inoculation significantly increased the nutrients uptake in groundnut as compared to untreated plants (Basu et al. 2006). The preceding outcomes revealed that lower levels of Co can stimulate the micronutrient and macronutrient uptake or supply in plant parts that may sustain metabolism and thus increase the plant growth.

Alterations in Physiological and Biochemical Constituents

The identification of changes in physiochemical process and chemical constituents is crucial in order to understand the plant tolerance strategies against Co applications. Gericó et al. (2020) reported that the application of Co (4 mL kg^{-1}) on peanut (*A. hypogaea* L.) seeds stimulated the photosynthetic pigments, carotenoids, sugars, sucrose, total amino acids correlated to nitrogen compounds, and biological nitrogen fixation by increasing the nitrogen forms such as nitrate and ammonia. Different levels of Co (10, 20, 30, and 40 mg L^{-1}) improved the proteins, total soluble sugars, total carbohydrates, and essential and fixed oil contents (mainly at 30 mg L^{-1}) in *N. sativa* L. (Khalid and Ahmed 2016). In *Zea mays*, concentration of Co (particularly at 50 mg kg^{-1} soil) increased the growth and yield characteristics (number of cobs, seeds per plant, seedling vigour), photosynthetic pigments (chlorophyll a, chlorophyll b, total chlorophyll contents), and biochemical traits such as starch, amino acids, total sugars, and protein contents (Jaleel et al. 2009). The okra (*H. esculantus*) seedlings were irrigated with different levels of Co (2.5, 5, 7.5, 10, and 12.5 ppm) to find out its potential impacts on the chemical constituents. Their findings showed that Co (particularly at 7.5 ppm) enhanced the total soluble solids, total soluble sugars, total carbohydrates, and vitamin 'C' and 'A' (Gad et al. 2015a). In addition, the application of Co (5, 7.5, 10, 12.5, and 15 ppm) in soil caused the collegial effects on the yield quality indicators such as carotenoids, starch, protein, total soluble sugars, total soluble solids, and L-ascorbic acids in sweet potato (*I. batatas* L.). Importantly, Co concentrations up to 10 ppm caused positive effects (Gad and Kandil 2008). The promotive effects of Co (4, 8, 12, and 16 ppm) were observed on total carbohydrates, proteins, leaf water potential, water use efficiency, and vitamin 'C' of green bean pods (Gad et al. 2019). A field experiment was conducted with the intention to explore the possible effects of foliar applications of Co (in combination with ethylene) at pre- and post-flowering stages on photosynthetic machinery and pedicel anatomy of pigeonpea (*Cajanus cajan* L.). Their results indicated that Co especially at 10 µg mol^{-1} increased the contents of photosynthetic parameters such as photosynthetic rate, intercellular CO_2, stomatal conductance, photosynthetic efficiency (F_v/F_m), and total chlorophyll (mainly at flowering stage) (Pahwa et al. 2017). These observations suggested that particular levels of Co ions can boost the production of biochemical constitutes along with other energy reservoirs/sources, metabolism of oil production, photosynthesis attributes, and thus facilitated the overall plant growth performance and higher yield.

Antioxidant Enzyme Activities and Synthesis of Hormones

To combat the oxidative damages induced by the overproduction of reactive oxygen species, plants boost their antioxidant defence system in the form of enzymatic antioxidant such as superoxide dismutase (SOD), catalase (CAT), and peroxidase (POX) and non-enzymatic antioxidants such as ascorbic acid, glutathione, carotenoids, and flavonoids. Usually, these defence systems are stimulated under environmental stresses. In tomato plants, the exposure of Co (20 mM) activated the activities of superoxide dismutase (SOD) and catalase (CAT), while the activities of ascorbate peroxidase (APOX) and glutathione reductase (GR) declined, and the overall growth attributes improved (Zhu et al. 2000). A field experiment tested that low levels of Co (especially 50 mg kg^{-1}) increased the antioxidant enzyme activities such as catalase (CAT), peroxidase (POX), and polyphenol oxidase (PPO) in cluster bean (*Cyamopsis tetragonaloba* L.) (Jayakumar 2019). The increasing concentrations of Co ions in growth media can enhance the activities of antioxidant enzymes in mung bean (*Phaseolus aureus*) (Tewari et al. 2002) and brassica napus plants (Ali et al. 2018b, c). These observations suggested that Co (at lower or higher levels) can boost the antioxidant defence systems and enhance plant tolerance potential.

Previous studies have reported the synergetic relationships between Co and phytohormones. Recent experiments verified that the irrigation of Co (4, 8, 12, and 16 ppm) significantly enhanced the endogenous production of phytohormones such as auxins, abscisic acid, and gibberellins in the tissues of mung bean plants as compared with control treatments (Gad et al. 2019). In maize plants, the exogenous applications of Co were observed to positively regulate the endogenous production of phytohormones (Gad and El-Metwally 2015b). Results from the field experiment evaluated the possible effects of Co (5, 7.5, 10, 12.5, and 15 ppm) on the synthesis of phytohormones of rosemary (*Rosmarinus officinals* L.) (Gad et al. 2014). These investigations reported that Co (up to 10 ppm) enhanced the production of endogenous levels of phytohormones such as gibberellins, cytokines, abscisic acid, and auxins in plant tissues. In addition, the promotive effects of Co (7.5 ppm) on the synthesis of gibberellins and auxins were reported in tomato plants (Gad 2005). These observations indicated that lower levels of Co improved the production of phytohormones. This suggested that Co could be utilized as beneficial element in higher plants. However, mechanistic approaches regarding the applications of Co as fertilizer are still lacking.

Protective Roles of Cobalt Against Abiotic Stresses

The efficacy of particular essential micronutrients in mitigating abiotic stresses have been reported (Gad et al. 2019; Tripathi et al. 2018; ul Hassan et al. 2017a, b). Earlier studies tested the potential of Co in curbing the salinity stress in wheat plants (Gad and Kandil 2011a).

They revealed that Co improved the wheat tolerance against salinity stress. In addition, the exposure of Co at 15 ppm significantly enhanced the plant growth, yield attributes, and quality indicators as compared with control treatments. In another study, the efficacy of Co levels (15, 17.5, 20, and 22.5 ppm) was experimented on maize plants under salinity stress (Gad and El-Metwally 2015b). They further reported that Co (up to 20 ppm) enhanced the growth, hormones synthesis, nutrients level, and enhanced maize tolerance by alleviating the salt stress. Gad et al. (2011b) also reported the efficiency of Co (5, 7.5, 10, 12.5, 15, 17.5, and 20 ppm) in alleviating the salt stress in barley (*Hordeum vulgare* L.). Their results indicated that Co (up to 10 ppm) enhanced the plant growth, nutritional status, and yield quality by minimizing the salt stress. Recent investigations also revealed that particular levels of Co improved the cucumber tolerance against salt stress (Gad et al. 2018). It has been documented that Cd stress impaired the physio-biochemical and metabolic processes in plants (Mwamba et al. 2020) and highly mobile in contaminated soils (Hamid et al. 2020). The roles of Co (inhibitor of ethylene synthesis) in mitigating the Cd toxicities of soyabean plants were evaluated by Chmielowska-Bak et al. (2014). They subjected soya bean seedlings to Cd along with Co ($CoCl_2$) treatments. Results indicated that Co regulates the transcript levels of those certain enzymes that are responsible for the synthesis of ethylene. After few hours of $CoCl_2$ treatments, the gene expressions were slightly restrained. These observations suggested that $CoCl_2$ can be used as ethylene inhibitor in response to Cd stress. The potential of Co to cope with osmotic stress was reported in potato seedlings (Li et al. 2004). Their findings revealed that osmotic stress induced membrane damages that were effectively minimized by the exogenous application of Co. Further evidences confirmed that Co ($25\,\mu M\, L^{-1}$) restricts the rate of ethylene synthesis and boosts polyamine contents as well as antioxidant enzyme activities that ultimately minimize the membrane injury. Recent investigations documented that Co (mainly at 12 ppm) enhanced the tolerance ability of green beans (*Phaseolus valgaris* L.) against drought stress (up to 60%) (Gad et al. 2019). These outcomes suggested that Co has efficiency to overcome the toxicities of heavy metals (Cd), salinity, osmotic, drought, or water stress in particular plants. However, the efficacy of Co against abiotic stresses such as temperature (high or low), ultraviolet, heat, and other heavy metals or metalloids in different plants are required for further investigations. Studies concerning the potential roles of Co against abiotic/biotic stressors would be ideal to boost our knowledge in order to develop tolerance strategies for plants.

Conclusions and Future Prospects

We conclude that adequate supply of Co is not only required for the normal plant growth and development but also enhances its internal defence systems. At lower doses, Co effectively enhances plant growth, biomass generation, photosynthesis machinery, nitrogen fixation, nutrient levels, physiological traits, biochemical constituents, essential or fix oil production, antioxidant (enzymatic or non-enzymatic) defence systems, endogenous production of phytohormones and improves plant tolerance under particular abiotic stresses. Through what mechanisms (at molecular level) Co antagonizes the abiotic stresses and improves plant tolerance (especially in higher plants) require complete investigations. Most of the previous studies have targeted the legume plants to explore the Co-induced beneficial effects. It is proposed to further elucidate, how lower levels of

Co behave as 'beneficial' element in plant metabolism, especially in higher plants. The distribution, uptake, and transport of Co may vary with the type of plant species and involved transporters. The mechanisms involved in the accumulation, uptake, and transport of Co in different plant parts are still not fully understood. Soil characteristics may drastically affect the Co bioavailability. Therefore, controls on the development of Co by plants are mandatory for typical plant cycle. This would help in maintaining daily requirements for human nutrition without health hazards. In the current era of modern research, mechanisms related to Co that cause beneficial effects deserve more attention and need further awareness or validation regarding the fertilization of crops as feed or food source.

Acknowledgments

This work was supported by the Science and Technology Department of Zhejiang Province (2022C02034), Collaborative Innovation Center for Modern Crop Production co-sponsored by Province and Ministry (CIC-MCP), and the Agriculture and Rural Affairs Department of Zhejiang Province (2022SNJF010). We thank Ping Yang from Agricultural Experiment Station of Zhejiang University for his assistance.

References

Ahmad, R., Ali, S., Hannan, F. et al. (2017). Promotive role of 5-aminolevulinic acid on chromium-induced morphological, photosynthetic, and oxidative changes in cauliflower (Brassica oleracea botrytis L.). *Environ. Sci. Pollut. Res.* 24: 8814–8824.

Akbar, F.M., Zafar, M., Hamid, A. et al. (2013). Interactive effect of cobalt and nitrogen on growth, nodulation, yield and protein content of field grown pea. *Hortic. Environ. Biotechnol.* 54: 465–474.

Ali, S., Jin, R., Gill, R.A. et al. (2018a). Beryllium stress-induced modifications in antioxidant machinery and plant ultrastructure in the seedlings of black and yellow seeded oilseed rape. *Biomed. Res. Int.* 1–14. https://doi.org/10.1155/2018/1615968.

Ali, S., Gill, R.A., Mwamba, T.M. et al. (2018b). Differential cobalt-induced effects on plant growth, ultrastructural modifications, and antioxidative response among four *Brassica napus* L. cultivars. *Int. J. Environ. Sci. Technol.* 15: 1–16. https://doi.org/10.1007/s13762-017-1629-z.

Ali, S., Gill, R.A., Ulhassan, Z. et al. (2018c). Insights on the responses of Brassica napus cultivars against the cobalt-stress as revealed by carbon assimilation, anatomical changes and secondary metabolites. *Environ. Exp. Bot.* 156: 183–196.

Arguello, J.M., Eren, E., and Gonzalez- Guerrero, M. (2007). The structure and function of heavy metal transport P1B-ATPases. *Biometals* 20: 233–248.

Azhar, W., Khan, A.R., Muhammad, N. et al. (2020). Ethylene mediates CuO NP-induced ultrastructural changes and oxidative stress in Arabidopsis thaliana leaves. *Environ. Sci.: Nano* 7 (3): 938–953.

Barras, F. and Fontecave, M. (2011). Cobalt stress in Escherichia coli and Salmonella enterica: molecular bases for toxicity and resistance. *Metallomics* 3 (11): 1130–1134.

Basu, M., Bhadoria, P.B.S., and Mahapatra, S.C. (2006). Influence of microbial culture in combination with micronutrient in improving the groundnut productivity under alluvial soil of India. *Acta Agric. Slov.* 87: 435–444.

Brooks, R.S. (1977). Copper, magnesium and cobalt uptake by Haumanistrum species. *Plant Soil* 48: 541–544.

Chang, P., Yin, H., Imanaka, T. et al. (2020). The metal transporter CrNRAMP1 is involved in zinc and cobalt transports in *Chlamydomonas reinhardtii*. *Biochem. Biophys. Res. Commun. 523* (4): 880–886.

Chatterjee, J. and Chatterjee, C. (2000). Phytotoxicity of cobalt, chromium and copper in cauliflower. *Environ. Pollut.* 109: 69–74.

Chmielowska-Bak, J., Lefèvre, I., Lutts, S. et al. (2014). Effect of cobalt chloride on soybean seedlings subjected to cadmium stress. *Acta Soc. Bot. Pol.* 83: 201–207.

Gad, N. (2005). Interactive effect of cobalt and salinity on tomato plants II- some physiological parameters as affected by cobalt and salinity. *Res. J. Agric. Bio. Sci. Pakistan* 1: 270–276.

Gad, N. (2006). Increasing the efficiency of nitrogen fertilization through cobalt application to pea plant. *Res. J. Agric. Biol. Sci.* 2: 433–442.

Gad, N. (2012). Role and importance of cobalt nutrition on groundnut (*Arachis hypogaea*) production. *World Appl. Sci. J.* 20 (3): 359–367.

Gad, N. and El-Metwally, I.M. (2015b). Chemical and physiological response of maize to salinity using cobalt supplement. *Int. J. ChemTech Res.* 8: 45–52.

Gad, N. and Kandil, H. (2008). Response of sweet potato (*Ipomoea batatas* L.) plants to different levels of cobalt. *Aust. J. Basic Appl. Sci.* 2: 945–955.

Gad, N. and Kandil, H. (2011a). Maximizing the tolerance of wheat plants to soil salinity using cobalt. 1- Growth and mineral composition. *J. Appl. Sci. Res.* 7: 1569–1574.

Gad, N., Abd El-Moez, M.R., and Kandil, H. (2011b). Barley response to salt stress at varied levels of cobalt II. Some physiological and chemical characteristics. *J. Appl. Sci. Res. 7* (11): 1447–1453.

Gad, N., El-Moez, A., Bekbayeva, L.K. et al. (2013). Effect of cobalt supplement on the growth and productivity of soybean (Glycine max L. Merrill). *World Appl. Sci. J.* 26: 926–933.

Gad, N., El-Moez, A.M., and Kandil, H. (2014). Effect of cobalt nutrition on rosemary (rosmarinus officinals L.) B-endogenous hormons, chemical and nutritional contents. *Int. J. Basic Appl. Sci. 3* (4): 479.

Gad, N., Moez, A., and Kandil, H. (2015a). Response of Okra (Hibiscus esculentus) growth and productivity to cobalt and humic acid rates. *Int. J. Chem. Tech. Res. 8* (4): 1782–1791.

Gad, N., Abdel-Moez, M.R., Fekry Ali, M.E. et al. (2018). Increasing salt tolerance in cucumber by using cobalt. *Middle East J. Appl. Sci.* 08: 345–354.

Gad, N., Abdel-Moez, M.R., and Ali, M.F. (2019). Maximization of drought tolerance of bean plants using cobalt supplementation. *Plant Archives 19* (2): 2282–2287.

Gericó, T.G., Tavanti, R.F.R., de Lima, J.P. et al. (2020). Cobalt and molybdenum stimulate compounds of primary metabolism, nitrogen forms, and photosynthetic pigments in peanut plants (*Arachis hypogaea* L.). *J. Plant Nutr. 43* (12): 1907–1922.

Hamid, Y., Tang, L., Hussain, B. et al. (2020). Cadmium mobility in three contaminated soils amended with different additives as evaluated by dynamic flow-through experiments. *Chemosphere* 261: 127763. https://doi.org/10.1016/j.chemosphere.2020.127763.

Jaleel, C.A., Jayakumar, K., Chang-xing, Z. et al. (2008). Effect of soil applied cobalt on activities of antioxidant enzymes in Arachis hypogaea. *Global J. Mol. Sci.* 3: 42–45.

Jaleel, C.A., Changxing, Z., Jayakumar, K. et al. (2009). Low concentration of cobalt increases growth, biochemical constituents, mineral status and yield in Zea mays. *J. Sci. Res. 1* (1): 128–137.

Jarosick, J., Vara, P.Z., Koneeny, J. et al. (1988). Dynamics of cobalt 60 uptake by roots of pea plants. *Sci. Total Environ.* 71: 225–229.

Jayakumar, K. (2019). Antioxidant enzyme mechanism of cluster bean (*Cyamopsis tetragonaloba* (L.) Taub). under cobalt stress. *World News Nat. Sci. 24*: 64–70.

Kamran, M., Danish, M., Saleem, M.H. et al. (2020). Application of abscisic acid and 6-benzylaminopurine modulated morpho-physiological and antioxidative defense responses of tomato (Solanum lycopersicum L.) by minimizing cobalt uptake. *Chemosphere* 263: 128169. https://doi.org/10.1016/j.chemosphere.2020.128169.

Kandil, H. (2007). Effect of cobalt fertilizer on growth, yield and nutrients status of faba bean (Vicia faba L.) plants. *J. Appl. Sci. Res.* 3: 867–872.

Kandil, H., Farid, I.M., and El-Maghraby, A. (2013). Effect of cobalt level and nitrogen source on quantity and quality of soybean plant. *J. Basic Appl. Sci. Res. 3* (12): 185–192.

Khalid, K.A. and Ahmed, A.M. (2016). Effect of cobalt on growth, yield and chemical constituents of Nigella sativa L. *J. Mater. Environ. Sci 7*: 2201–2207.

Khan, A.R., Wakeel, A., Muhammad, N. et al. (2019). Involvement of ethylene signaling in zinc oxide nanoparticle-mediated biochemical changes in Arabidopsis thaliana leaves. *Environ. Sci.: Nano 6* (1): 341–355.

Komeda, H., Kobayashi, M., and Shimizu, S. (1997). A novel transporter involved in cobalt uptake. *Proc. Natl. Acad. Sci.* 94: 36–41.

Korshunova, Y.O., Eide, D., Clark, W.G. et al. (1999). The IRT1 protein from Arabidopsis thaliana is a metal transporter with a broad substrate range. *Plant Mol. Biol. 40* (1): 37–44.

Lange, B., van der Ent, A., Baker, A.J.M. et al. (2017). Copper and cobalt accumulation in plants: a critical assessment of the current state of knowledge. *New Phytol. 213* (2): 537–551.

Li, L., Tutone, A.F., Drummond, R.S. et al. (2001). A novel family of 766 magnesium transport genes in Arabidopsis. *Plant Cell* 13 (12): 2761–2775.

Li, C.Z., Wang, D., and Wang, G.X. (2004). The protective effects of cobalt on potato seedling leaves during osmotic stress. *Bot. Bull. Acad. Sin.* 46: 119–125.

Li, H., Liu, Y., Qin, H. et al. (2020). A rice chloroplast-localized ABC transporter ARG1 modulates cobalt and nickel homeostasis and contributes to photosynthetic capacity. *New Phytol.* 228: 163–178.

Lwalaba, J.L.W., Louis, L.T., Zvobgo, G. et al. (2020). Physiological and molecular mechanisms of cobalt and copper interaction in causing phyto-toxicity to two barley genotypes differing in Co tolerance. *Ecotoxicol. Environ. Saf. 187*: 109866.

Mahey, S., Kumar, R., Sharma, M. et al. (2020). A critical review on toxicity of cobalt and its bioremediation strategies. *SN Appl. Sci. 2* (7): 1–12.

Manal, A.A., Manal, A., Alla, A. et al. (2016). Effect of zinc and cobalt applied with different methods and rates on the yield components of Vicia faba L. *World Wide J. Multidiscip. Res. Develop.* 2: 52–58.

McDowell, L.R. (1989). *Vitamins in Animal and Human Nutrition*, 2e, 523–559. Ames: Iowa State University Press.

Mitra, G.N. (2015). Cobalt (Co), Selenium (Se), Vanadium (V), Cadmium (Cd), Lead (Pb) and Titanium (Ti). In: *Regulation of Nutrient Uptake by Plants*, 189–195. New Delhi: Springer.

Morel, M., Crouzet, J., Gravot, A. et al. (2009). AtHMA3, a P1B-ATPase allowing Cd/Zn/co/Pb vacuolar storage in Arabidopsis. *Plant Physiol. 149* (2): 894–904.

Morrissey, J., Baxter, I.R., Lee, J. et al. (2009). The ferroportin metal efflux proteins function in iron and cobalt homeostasis in *Arabidopsis. Plant Cell 21* (10): 3326–3338.

Mwamba TM, Islam F, Ali B. et al. (2020). Comparative metabolomic responses of low- and high-cadmium accumulating genotypes reveal the cadmium adaptive mechanism in Brassica napus. *Chemosphere* 250: 126308. doi: https://doi.org/10.1016/j.chemosphere.2020.126308.

Nazir, M.M., Ulhassan, Z., Zeeshan, M. et al. (2020). Toxic metals/metalloids accumulation, tolerance, and homeostasis in *Brassica* oilseed species. In: *The Plant Family Brassicaceae* (ed. M. Hasanuzzaman). Singapore: Springer, https://doi.org/10.1007/978-981-15-6345-4_13.

Pahwa, K., Ghai, N., Kaur, J. et al. (2017). Influence of ethylene and cobalt chloride on photosynthetic parameters and pedicel anatomy of pigeonpea (Cajanus cajan L.) genotypes. *J. Environ. Biol. 38* (3): 367.

Palit, S., Sharma, A., and Talukder, G. (1994). Effects of cobalt on plants. *Bot. Rev.* 60 (2): 149–181.

Rutherford, J.C., Cavet, J.S., and Robinson, N.J. (1999). Cobalt dependent transcriptional switching by a dual efector MeR like protein regulates a cobalt exporting variant CPx type ATPases. *J. Biol. Chem.* 274: 25827–25832.

Sahay, N. and Singh, S.P. (2012). Effect of cobalt application on growth, yield attributes and yield of lentil cultivars. *Ann. Plant Soil Res.* 14: 39–41.

Salam, A., Khan, A.R., Liu, L. et al. (2021). Seed priming with zinc oxide nanoparticles downplayed ultrastructural damage and improved photosynthetic apparatus in maize under cobalt stress. *Journal of Hazardous Materials* 423 part A: 127021. doi: https://doi.org/10.1016/j.jhazmat.2021.127021.

Tewari, R.K., Kumar, P., Sharma, P.N. et al. (2002). Modulation of oxidative stress responsive enzymes by excess cobalt. *Plant Science* 162: 381–388.

Tripathi, D.K., Singh, S., Gaur, S. et al. (2018). Acquisition and homeostasis of iron in higher plants and their probable role in abiotic stress tolerance. *Frontiers in Environmental Science* 5: 86.

ul Hassan, Z., Ali, S., Rizwan, M. et al. (2017a). Role of Zinc in alleviating heavy metal stress. In: *Essential Plant Nutrients*, (ed. M. Naeem). Springer International Publishing AG. doi https://doi.org/10.1007/978-3-319-58841-4_14

ul Hassan, Z., Ali, S., Rizwan, M. et al. (2017b). Role of Iron in alleviating heavy metal stress. In: *Essential Plant Nutrients*, (ed. M. Naeem). Springer International Publishing AG. doi https://doi.org/10.1007/978-3-319-58841-4_13 pp 335–350.

Ulhassan, Z., Ali, S., Gill, R.A. et al. (2018). Comparative orchestrating response of four oilseed rape (*Brassica napus*) cultivars against the selenium stress as revealed by physio-chemical, ultrastructural and molecular profiling. *Ecotoxicol. Environ. Saf.* 161: 634–647.

Ulhassan, Z., Gill, R.A., Ali, S. et al. (2019a). Dual behavior of selenium: Insights into physio-biochemical, anatomical and molecular analyses of four Brassica napus cultivars. *Chemosphere* 225: 329–341.

Ulhassan, Z., Gill, R.A., Huang, H. et al. (2019b). Selenium mitigates the chromium toxicity in Brassicca napus L. by ameliorating nutrients uptake, amino acids metabolism and antioxidant defense system. *Plant Physiol. Biochem.* 145: 142–152. https://doi.org/10.1016/j.plaphy.2019.10.035.

Ulhassan, Z., Huang, Q., Gill, R.A. et al. (2019c). Protective mechanisms of melatonin against selenium toxicity in Brassica napus: insights into physiological traits, thiol biosynthesis, and antioxidant machinery. *BMC Plant Biology* 19: 507. doi:https://doi.org/10.1186/s12870-019-2110-6

Vaseer, S.G., Rasheed, M., Ansar, M. et al. (2020). Cobalt application improves the growth and development of mung bean. *Pak. J. Agric. Res.* 33 (2): 303–310.

Vijayarengan, P. (2012). Changes in growth, biochemical constitution and antioxidant potential in cowpea (*Vigna unguiculata* L.) under cobalt stress. *Int. J. Res. Environ. Sci. Tech.* 2: 74–82.

Wang, Y.M., Yang, Q., Xu, H. et al. (2020). Physiological and transcriptomic analysis provide novel insight into cobalt stress responses in willow. *Sci. Rep. 10* (1): 1–12.

Yadav, D.V. and Khanna, S.S. (1983). Effect of cobalt and phosphorous on their availability and nitrogen fixation by legumes. *Haryana Agric. Univ. J. Res.* 13: 396–401.

Younis, M. (2011). Soil applied cobalt and copper alter the Rhizobium symbiosis and growth status of Lablab purpureus. *J. Plant Nutr.* 34: 1736–1745.

Zhu, H., Shipp, E., Sanchez, R.J. et al. (2000). Cobalt (2+) binding to human and tomato copper chaperone for superoxide dismutase: implications for the metal ion transfer mechanism. *Biochemistry* 39 (18): 5413–5421.

8

Cobalt in Plant Life: Responses and Deficiency Symptoms

Xiu Hu[1], Xiangying Wei[2], Jie Ling[3], and Jianjun Chen[4]

[1] *College of Horticulture and Landscape Architecture, Zhongkai University of Agriculture and Engineering, Guangzhou, Guangdong, China*
[2] *Fujian Key Laboratory on Conservation and Sustainable Utilization of Marine Biodiversity, Fuzhou Institute of Oceanography, College of Geography and Oceanography, Minjiang University, Fuzhou, Fujian, China*
[3] *He Xiangning College of Art and Design, Zhongkai University of Agriculture and Engineering, Guangzhou, Guangdong, China*
[4] *Department of Environmental Horticulrture and Mid-Florida Research and Education Center, Institute of Food and Agricultural Sciences, University of Florida, Apopka, FL, USA*

Introduction

Cobalt (Co) is a transition metal and ranked 19th in abundance among the trace elements (Peterson and Girling 1981). Co has 1^+ to 5^+ oxidation state and one naturally occurring isotope (^{59}Co), of which 2^+ and 3^+ are the most popular states. Co^{3+} reacts with various acids and creates salts (Phipps 1976), but Co^{2+} is less reactive (Kobayashi and Shimizu 1999). Thus, Co is mainly present as Co^{2+}, which actively participates in cation exchange reactions in soils.

Cobalt in the Earth's crust varies from 15 to 30 mg kg^{-1} (Roberts and Gunn 2014). The content of Co in soils is mainly inherited from parental minerals, such as cobaltite, smaltite, and erythrite (Bakkaus et al. 2005). Kabata-Pendias and Mukherjee (2007) documented that Co in the world's surface soils ranges from 4.5 to 12 mg kg^{-1} with the highest occurring in heavy loamy soils and the lowest in organic and light sandy soils. However, Co in reference soil samples was found to vary from 5.5 to 29.9 mg kg^{-1} in the United States and 5.5 to 97 mg kg^{-1} in Chinese soils (Govindaraju 1994). Pilon-Smits et al. (2009) reported that soil concentrations of Co generally range from 15 to 25 mg kg^{-1}.

Co is an essential element for human beings, other mammals, archaea, and bacteria. Its essentiality for animals was uncovered in 1935 during the investigation of ruminant livestock nutrition in Australia (Underwood and Filmer 1935; Marschner 2012). Co is a core element of vitamin B12 or cobalamin. The structure of cobalamin was identified by Hodgkin et al. (1956) with an empirical formula of $C_{63}H_{88}O_{14}N_{14}PCo$. Five scientists were awarded Nobel Prizes for either direct or indirect investigation of vitamin B12: George Whipple, George Minot, and William Murphy in 1934; Alexander R. Todd in 1957; and

Beneficial Chemical Elements of Plants: Recent Developments and Future Prospects, First Edition. Edited by Sangeeta Pandey, Durgesh Kumar Tripathi, Vijay Pratap Singh, Shivesh Sharma, and Devendra Kumar Chauhan.

Dorothy Hodgkin in 1964 (Carpenter 2004). As shown in Figure 8.1, cobalamin is a complex and large molecule containing a modified tetrapyrrole ring called corrin with Co in the centre (Osman et al. 2021). Cobalamin commonly occurs in two biologically active forms: methylcobalamin and adenosylcobalamin. Co in the cobalamin is not interexchangeable with other metals and cannot be released from the ring unless the ring is broken (Yamada 2013), indicating the importance of Co to cobalamin. Cobalamin is a cofactor of two enzymes: methylmalonyl-CoA mutase (MCM) and methionine synthase in human beings. MCM catalyzes the reversible isomerization of l-methylmalonyl-CoA to succinyl-CoA, which requires cobalamin as a cofactor in the form of 5′-deoxy-5′-adenosylcobalamin (AdoCbl) (Takahashi-Iñiguez et al. 2012). A deficiency of MCM causes an inherited metabolism disorder, commonly referred to as methylmalonic aciduria. Methionine synthase requires cobalamin as a cofactor to regenerate methionine from homocysteine. One consequence of reduced methionine synthase activity is megaloblastic anemia (Tjong and Mohiuddin 2020). Ruminant animals, like camels, cows, deer, goats, and sheep produce vitamin B12 if there is adequate Co in their diet. About 3–13% of the Co in the diet of a ruminant animal is incorporated by rumen bacteria into vitamin B12 (Huwait et al. 2015).

Cobalamin biosynthesis is confined to some bacteria and archaea (Guo and Chen 2018). There are more than 20 cobalamin-dependent enzymes in those prokaryotes including diol dehydratase, ethanolamine ammonia-lyase, glutamate and MCM, methionine synthase, and ribonucleotide reductase (Marsh 1999). These enzymes catalyze a series of transmethylation and rearrangement reactions (Rodionov et al. 2003). Thus, Co is essential for those archaea and bacteria.

Fungi and higher plants, however, have not been found to use cobalamin as a cofactor of any enzymes (Rodionov et al. 2003), and Co has not been specifically identified to play

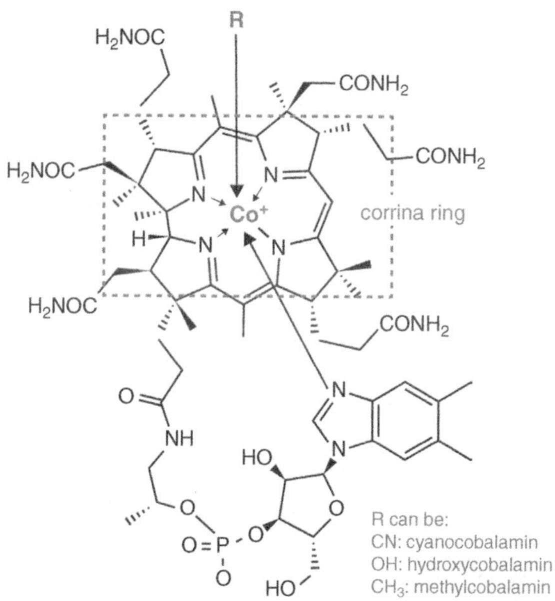

Figure 8.1 The structure of cobalamin or vitamin B12 where Co is in the centre of the corrin ring.

essential roles in plant life thus far. As early as 1952, Hewitt (1952) was unable to produce evidence that Co was essential for the growth of tomato, lettuce, or sugar beet. As a result, Co has not been considered an essential element for plant growth and development. On the other hand, Co at appropriately low concentrations have been found to positively affect plant performance. Thus, Co at present is considered a beneficial element for plant growth. However, lower plants, particularly some algae, have been reported to require Co. This chapter is intended to discuss Co in plant life, plant responses to Co, and its deficiency in and toxicity to plants.

Cobalt in Lower Plants

Lower plants are those without the vascular system, i.e. xylem and phloem; thus, they are commonly known as non-vascular plants. Non-vascular plants generally include two distinct groups: bryophytes and algae.

Bryophytes

Bryophytes include Anthocerotophyta (hornworts), Bryophyta (mosses), and Marchantiophyta (liverworts) (Davies et al. 2020). Bryophyte plants have been documented to take up metals including Co from soil, air, and water. In an early geochemical study conducted in Wisconsin and adjacent states and in Missouri and Kentucky in the United States concerning the concentrations of 33 elements in 38 samples of liverworts and mosses, Shacklette (1965) reported that the average concentration of Co among the samples was 32 mg kg^{-1}, and the concentration in the bryophytes was generally related to the amount of the element in the soil, suggesting they act as bioindicator of Co concentration in the environment (Baker 1981). In fact, mosses, such as *Bryum argenteum* and *Hypnum cupressiforme*, were proposed to be indicator plants for monitoring heavy metal contamination in the air. For example, Co concentrations in the two moss species ranged from 1.12 to 7.61 mg kg^{-1} in Podgorica (Montenegro), which were strongly correlated with Co concentrations in the polluted areas (Andic et al. 2015). Similarly, liverworts were also used as bioindicators for monitoring heavy metals in soils. Co concentrations in *Conocephalum conicum*, *Marchantia polymorpha*, and *Pellia epiphylla* were 26.0, 26.5, and 38.7 mg kg^{-1}, respectively (Samecka-Cymermann et al. 1997). There was a significant correlation between Co in soil and in *C. conicum*. Furthermore, the higher Co contents in these species did not cause any physiological damages in the plants, thus making them suitable indicators for monitoring Co contaminations. Aquatic moss was also found to be an indicator of Co concentration in stream sediments. Mosses taken from streams of the Idaho Cobalt Belt (United States) showed that Co contents in the mosses almost perfectly correlated with that in the sediments, and the maximum concentration of Co (2000 mg kg^{-1}) was observed in the ash of mosses compared to the maximum concentration of 320 mg kg^{-1} in the sediment (Erdman and Modreski 1984).

Bryophyte plants have no vascular systems, and there is little information as to how they take up Co. Pipiška et al. (2007) studied the sorption and desorption of ^{60}Co in *Polytrichum commune* and *Dicranu scoparium* through solution culture. Total amount of ^{60}Co accumulated in *P. commune* and *D. scoparium* under given conditions were 7.1 and 6.1 mg kg^{-1}, respectively. Of which ^{60}Co was found mainly in the extracellular compartment of the two species. Sequential extraction with $NiCl_2$ and HNO_3 at 20 °C and boiling HNO_3 showed that more than 95% of ^{60}Co in *D. scoparium* was localized extracellular, 70% of ^{60}Co in *P. commune* was localized extracellular, and about 20% was localized intracellular. This study showed that Co was primarily adsorbed in the extracellular sites rather than absorbed into intracellular sites, suggesting that bryophyte plants have abundant cation exchange sites for the adsorption. The predominate localization in extracellular sites may indicate that bryophyte plants develop some mechanisms allowing limited amount of Co to be transported into intracellular space. In fact, some bryophytes, primarily liverworts and hornworts can establish symbiotic relationships with cyanobacteria, commonly known as blue-green algae, such as *Nostoc* spp. Adams and Duggan (2008) reported that after infection, *Nostoc* undergoes some morphological and physiological alterations by reducing growth rate and CO_2 fixation but enhancing the fixation of atmospheric dinitrogen (N_2) as well as releasing fixed N compounds, such as ammonia to plants. Cyanobacteria, like rhizobia, require cobalamin as a cofactor for nitrogenase complex to fix N_2 (Bohme 1998). Thus, cyanobacteria-bryophyte symbioses require Co.

Algae

Algae belong to a polyphyletic group ranging from unicellular microalgae, like chlorella and diatoms to multicellular forms, such as the giant kelp, seaweeds, and charophytes (Barsanti and Gualtieri 2006). Algae have been extensively studied for their ability to take up metal ions through adsorption and absorption phases (Bilal et al. 2018). Adsorption is a process of rapid binding of ions to the outside of cell walls, and absorption is a process of slow transfer of metal ion into algal cells (Wilde and Benemann 1993). Algal cell wall carries a negative charge by different functional groups, such as amino ($-NH_2$), carboxyl (–COOH), hydroxyl (–OH), phosphoryl ($-PO_3O_2$), and sulfhydryl (–SH) (Torres 2016). Species vary in types of the functional groups, and different groups have different affinities to metal ions. Thus, metal uptake may differ by species. The absorption is usually dependent on metabolic activity of algae; thus, it is an active process (Bilal et al. 2018). Metal ions are adsorbed on cell wall, transported across cell membrane into the cytoplasm, and then stored in the vacuole (Hanikenne et al. 2005). Due to their capacity in metal sorption, microalgae have been used for remediating aquatic environments from toxic metal (Mantzorou et al. 2018).

Co is an essential element for some marine algal species, including charophyte, diatoms, and dinoflagellates (Nagpal 2004). The growth of diatoms of northeast Pacific isolates *Pseudonitzschia delicatissima* UNC1205 and *Thalassiosira* spp. UNC1203 was enhanced by addition of Co (Kellogg et al. 2020). An explanation for the enhanced growth of diatoms was due to the substitution of zinc by Co. In a study of estuarine green alga *Chlorella salina*, Garnham et al. (1992) showed that there were two phases of uptake of Co. The initial adsorption accounted for 33% of total amount of Co accumulated by *C. salina*. The greatest amount of Co was associated with cell wall. Absorption was affected by light, temperature, and metabolic inhibitors. Co concentrations in the cytosol was 0.17 but 2.89 mM in the vacuole. The authors argued that Co had its own uptake system in *C. salina*, which might be shared with Zn, Mn, and Mg. Czerpak et al. (1994) investigated responses of a freshwater green alga *Chlorella pyrenoidosa* to different concentrations of Co and found that Co in a range from 5 to 50 mM significantly stimulated *C. pyrenoidosa* growth, resulting in 150–160% and 50–60% increase in fresh and dry weights, respectively. Such increase was related to the increase of chlorophylls *a* and *b* by 45–65%, total carotenoids by 55–65%, water soluble proteins by 19–20%, and monosaccharides content by 55–60%, when compared to the culture devoid of Co. However, mechanisms behind the stimulatory effects were not elucidated.

It is well documented that some algal species, such as those in the genera *Coccomyxa* and *Elliptochloris* as well as diatoms form symbiotic relationships with cyanobacteria (Grube et al. 2017). Co is required for the growth of blue-green algae, such as *Anabaenza cylindrica* Lemm (Holm-Hansen et al. 1954) and *Prochlorococcus* (Hawco et al. 2020) as they need it for N_2 fixation through specialized cells called heterocysts using the enzyme nitrogenase. Thus, algal species symbiotic with cyanobacteria require Co in N_2-fixation. Additionally, many algal species require different combinations of cobalamin, vitamin B1, and vitamin B7 (Croft et al. 2005) because they do not have pathways to biosynthesize cobalamin or may use alternative cobalamin-independent routes bypassing the need for the vitamin (Cruz-Lopez and Maske 2016; Yao et al. 2018). As Co is a constituent of cobalamin, Co is required by those algae.

Cobalt in Higher Plants

Plants are sessile organisms, and their growth requires absorbing water and mineral elements from soils. The Earth contains 92 elements, 82 of them have been found in plants (Reimann et al. 2001). Co concentrations in plants were found to range from 0.1 mg kg^{-1} (Palit and Sharma 1994) to over 2000 mg kg^{-1} in *Alyssum murale* (Tappero et al. 2007) and *Rinorea* cf. *bengalensis* (Paul et al. 2020). Mean concentrations of Co in grasses varied from 60 to 270 μg kg^{-1} and in clover differed from 100 to 570 μg kg^{-1} across Australia, Great Britain, Finland, Germany, Japan, New Zealand, Poland, Sweden, and the United States (Kabata-Pendias and Mukherjee 2007). In general, legumes absorb more Co than grasses. However, some trees, such as persimmon and black gum, had Co in leaves as high as 2000 and 10 000 mg kg^{-1} in ash, respectively (Connor and Shacklette 1975). Plants that accumulate metals to a level 100-fold higher than those typically recorded in common plants are known as hyperaccumulators (Brooks 1998).

Root Absorption of Cobalt

Root absorption of Co depends on the physical and chemical properties of soil, Co concentrations in soils, and the ability of plants to absorb Co. In general, increasing bioavailability of Co in soils results in elevated levels of Co in plant tissues. Co content of red clover almost linearly increased as Co concentrations increased from 0.1 to 100 μg L^{-1} (Kabata-Pendias 1973). However, how plants take up Co remains inconclusive. Several Co transporters have been identified in bacteria. A novel transporter specifically for Co uptake was identified in a bacterium *Rhodococus rhodochrous* J1 (Komeda et al. 1996). An ABC-type

Co transport system was found essential for growth of *Sinorhizobium meliloti*, a Gram-negative bacterium that forms symbiotic relationships with alfalfa (*Medicago sativa*) and fixes N_2 (Cheng et al. 2011). However, Co transporters specifically for higher plants have not been reported.

Current understanding of Co transport in higher plants is largely linked to the transport of iron (Fe). In *Arabidopsis thaliana*, Co is absorbed from soil into epidermal cells of roots by a transporter called iron-regulated transporter 1(IRT1), which is commonly known for absorption of Fe (Korshunova et al. 1999). Once Co is absorbed inside cells, ferroportins, FPN1 and FPN2, are responsible for its further movement. FPN1 is situated in the plasma membrane and expressed around the vasculature. It is involved in loading Fe into the vasculature. FPN2 is localized in the vacuole and expressed in the outer root layers (Conte and Elsbeth 2011). In Fe transport, FPN2 plays a role in buffering the levels of Fe in the cells by sequestering Fe into the vacuole when it is in excess. For Co transport, FPN2 is responsible for transporting Co into vacuoles of epidermal and cortical cells of roots, resulting in the sequestration of Co in the outer most layers of roots. On the other hand, FPN1 is able to load mobile Co into the xylem and translocate it to the shoots (Morrissey et al. 2009). In addition, HMA3 (heavy metal ATPase 3) is also able to transport Co into vacuoles (Morel et al. 2009). Thus, root absorption of Co from soils is first regulated by IRT1 and subsequently controlled by FPN1 and FPN2 to determine if it should be stored in vacuole or loaded into xylem.

Cobalt Transport in Plants

Mobile Co that is being transported to shoots is maintained as Co^{2+} and chelated with ligands. Evidence suggests that Co has little affinity with phytochelatins (Chen et al. 1997; Cheng et al. 2005), thus the ligands are probably not Co-S bonds. A study conducted by Collins et al. (2010) showed that Co^{2+} complexed with carboxylic acids was transported from wheat or tomato roots to shoots. Other likely ligands are citrate or malate as well as non-proteinogenic amino acids, such as nicotianamine. Co has a rather low mobility within the leaf tissue and is predominately distributed in the vascular system of tomato and wheat leaves (Collins et al. 2010).

Experimental results showed that transport of absorbed Co from roots to shoots is closely controlled. Page and Feller (2005) investigated Co transport in wheat plants using radiolabelled ^{57}Co. They found that after four days of culture, 80% of ^{57}Co remained in the root system, and 50% of which was retained in the roots after 50 days. During that period, some ^{57}Co moved to the apical part of the main roots. These results suggest that the loading of Co to the xylem is well controlled, probably by FPN1 in wheat plants. In another study, Collins et al. (2010) found that tomato and wheat plants grown in a nutrient solution containing 2.94 mg L^{-1} Co had 4423 and 9319 μg kg^{-1} of Co in roots, respectively, but shoot concentrations of Co were 1581 and 395 μg kg^{-1}, respectively. This means that 35.7% of Co absorbed by tomato and 4.2% of Co absorbed wheat plants are transported from roots to shoots. For the 1581 μg kg^{-1} Co in tomato shoots, 846 μg kg^{-1} was in stem, 492 μg kg^{-1} in old leaves, only 243 μg kg^{-1} in young leaves, indicating that only 5.5% of absorbed Co is transported to actively growing shoots of tomato plants. These transport patterns are like those of titanium (Lyu et al. 2017), which are strictly controlled by plants. Due to its toxicity at higher concentrations, such a controlled transport and distribution would ensure that only appropriate amount of Co could be transported to active growing shoots. As to why more Co was transported to dicot tomato shoots than monocot wheat shoots is unknown. It is possible that different plants have different ligands for complexing Co, and Co complexed by ligands in tomato was more mobile than that in wheat. Another explanation is that tomato plants need more Co to play some unidentified roles in shoots. Further research is needed to verify these propositions.

For plants that hyperaccumulate Co, substantial amount of Co can be transported to shoots. The highest concentration of Co was found in old leaves of *Rinorea* cf. *bengalensis* and localized in the teeth at the margin of the leaves. Co distribution in old leaves was heterogenous since the vein system had 2000 mg kg^{-1} of Co, whereas the lamina had 1000 mg kg^{-1} Co. Co in young leaves was relatively homogeneous, ranging from 200 to 400 mg kg^{-1} (Paul et al. 2020). Another study conducted with *Glochidion* cf. *sericeum* showed that this species can hyperaccumulate Co up to 1500 mg kg^{-1}, and Co was localized in lesions on the leaf surface and complexed with carboxylic acid (Van der Ent et al. 2018a).

Cobalt Effects on Plant Growth

Cobalt at appropriately low concentrations can enhance plant growth and performance. The growth enhancement was first documented by Bolle-Jones and Mallikarjuneswara (1957) with rubber plant (*Hevea brasiliensis*) where application of 0.005 mg kg^{-1} Co to purified sand culture significantly improved plant growth. Two years later, Co was found to be required for N_2 fixation in legumes (Ahmed and Evans 1959; Ahmed and Evans 1960; Ahmed and Evans 1961). Additionally, Co has been reported to alleviate abiotic stresses and improve vase life of some ornamental plants (Palit and Sharma 1994; Liu 1998; Akeel and Jahan 2020).

Cobalt is Essential for N_2 Fixation in Nodulated Legumes

The requirement of Co for N_2 fixation in nodulated leguminous crops was established in the 1950s and 1960s (Ahmed and Evans 1960; Reisenauer 1960). Soybean seedlings inoculated with rhizobia were transplanted in 10-L pots containing a nutrient solution supplemented with 1 μg L^{-1} Co, the seedlings were healthy, and four seedlings per pot produced 25.3 g of dry weight. On the contrary, rhizobia-inoculated four plants grown in the same nutrient solution devoid of Co exhibited nitrogen deficiency and produced only 16.6 g of dry weight, a 34.4% reduction in biomass due to the absence of Co (Ahmed and Evans 1959). Reisenauer (1960) reported that alfalfa grown in nutrient solutions with a limited supply of N and without Co addition developed severe N-deficiency symptoms and accumulated 49% less dry weight than plants supplied with Co. Subsequently, cobalamin was isolated from root nodules of legumes by Kliewer and Evans (1963b), and the authors demonstrated that there was an interdependent relationship among Co supply, cobalamin concentrations in *Rhizobium*, leghaemoglobin formation, and N_2 fixation (Kliewer and Evans 1963a). Consequently, it was established that *Rhizobium* and other N_2-fixing bacteria require Co as an essential element whether or not they grow within nodules or they rely on supply of N from N_2 fixation or from mineral sources.

The role of cobalamin in N_2 fixation is schematically presented in Figure 8.2. Soybean plants grown in soil establish a symbiotic relationship with rhizobia, characterized by the abundant nodules on roots. N_2 in the air is converted into ammonia or other related compounds through nitrogenase in rhizobia. Ammonia or other related compounds are used by plants for N metabolism. Three enzymes: methyl malonyl-CoA mutase, methionine synthase, and ribonucleotide reductase are known to be cobalamin dependent and significantly affect nodulation and N_2 fixation. Methyl malonyl-CoA mutase is responsible for the synthesis of leghaemoglobin. If there is a deficient supply of Co, leghaemoglobin synthesis is directly affected, resulting in reduced N_2 fixation and ultimately a shortfall of N supply. This is because leghaemoglobin can protect nitrogenase from oxygen by limiting its supply (Hopkins 1995). Co deficiency also affects methionine synthase by reducing methionine synthesis, resulting in lower protein synthesis and smaller-sized bacteroids (bacteria in the nodules capable of N_2 fixation) (Marschner 2012). Therefore, Co is essential for the growth of rhizobia and N_2 fixation in legumes.

Cobalt has been documented to increase nodule numbers, tissue N contents, and growth of leguminous plants. Co application significantly enhanced the growth of sweet lupin (*Lupinus angustiflius*) and N concentrations in shoots (Robson et al. 1979). Application of Co increased nodule number in faba beans (*Vicia faba*) (Hala 2007), hyacinth bean (*Lablab purpureus*) (Younis 2011), and groundnut (*Arachis hypogaea*) (Gad 2012). Supplying

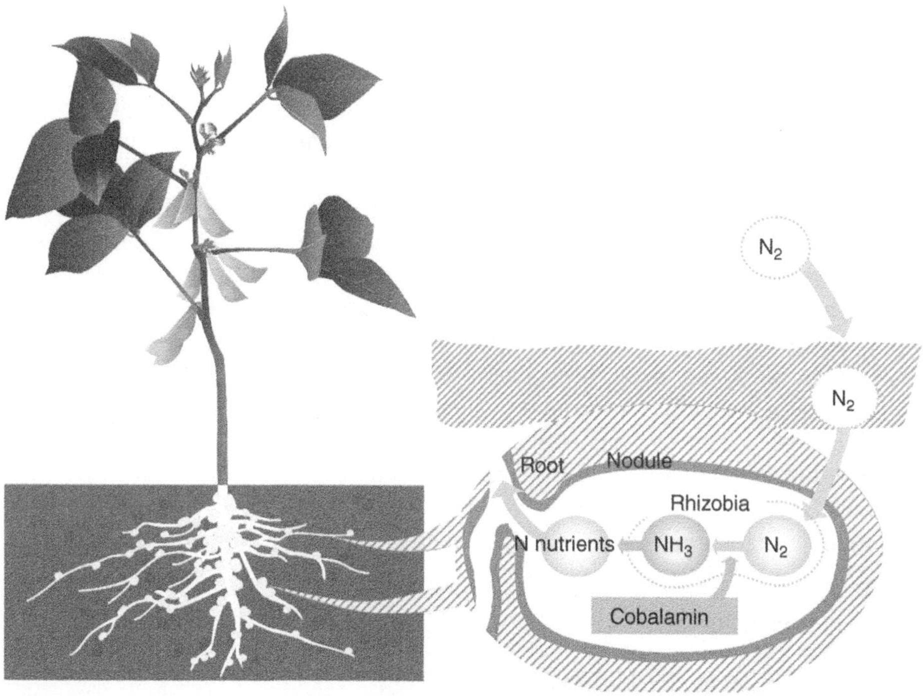

Figure 8.2 The role of cobalamin in nitrogen fixation. Molecular dinitrogen (N_2) from the air is absorbed into a nodule and converted into ammonia or other nitrogenous compounds through the action of nitrogenase. Cobalamin is a cofactor of three enzymes involved in nodulation and N_2 fixation. The converted N compounds are used by plants in N metabolism.

Vicia faba plants with 20 mg kg^{-1} Co resulted in the total protein increase by 47.82 and 44.54% in the first and the second seasons as well as grain yield over 200% compared to the control treatment without Co (Hala 2007). Similar beneficial effects on soybean were reported by Jayakumar and Jaleel (2009). Plants of chickpea (*Cicer arietinum*) were applied with varied levels of Co and sampled after 60 days of growth. Co at concentration 5.9 mg kg^{-1} significantly increased the number of nodules, their dry weights, leghaemoglobin concentrations, and the activity of nitrogenase (Ali et al. 2010). The activities of glutamine synthetase, glutamate synthase, and glutamate dehydrogenase also increased in nodules and leaves. However, Co concentration greater than 5.9 mg kg^{-1} generated oxidative stress with increased lipid peroxidation. Mung bean (*Vigna radiata*) plants treated with 2.9 mg kg^{-1} of Co in sand culture showed improved growth by increasing leaf, stem, and total dry weights as compared to the control treatments. However, applying Co with concentrations of 5.9 mg kg^{-1} and above resulted in gradual decrease in dry weight, of which roots were more significantly affected than shoots (Tewari et al. 2002). The improved plant growth was related to the increased concentration of chlorophyll *a* and *b*, while the decreased growth was associated with decreased concentrations of chlorophyll *a* and *b* as well as carotenoid contents.

Cobalt Enhances Growth of Non-Leguminous Crops

Cobalt at low concentrations has also been documented to promote growth of non-leguminous crops. Application of Co to a sandy soil at 1.0 mg kg^{-1} increased shoot and root dry weights of wheat by 33.7 and 35.8%, respectively, compared to the control (Aery and Jagetiya 2000), and the same Co rate applied to a sandy loam soil increased shoot and root dry weights of wheat by 27.9 and 39.6%, respectively, comparted to the control. Helmy and Gad (2002) reported that parsley (*Petroselinum crispum*) yield and essential oil contents increased considerably after application of Co at 25 mg kg^{-1}. Co applied at 20 and 40 mg kg^{-1} significantly increased plant height, branch numbers, and fruit numbers as well as anthocyanin and flavonoids contents of *Hibiscus sabdariffa* (Aziz et al. 2007). Corn plants (*Zea mays*) were evaluated after application of 50, 100, 150, 200, and 250 mg kg^{-1} Co (Jaleel et al. 2009), and results showed that root length, shoot height, number of cobs, and seeds per plant increased when plants were applied with 50 mg kg^{-1} Co, but these parameters decreased with 100 mg kg^{-1} Co and above. Additionally, Co applied at 10 mg kg^{-1} significantly enhanced the growth of two onion cultivars, bulb yields, and bulb quality, such as nutrients and essential oils content as well as bulb length, bulb diameter, and bulb weights were much higher than the control treatment (Attia et al. 2014), but Co concentrations higher than 10 mg kg^{-1} significantly reduced the promotive effects.

Possible Mechanisms

Mechanisms underlying the improved growth of leguminous plants include the cobalamin involvement in N_2 fixation as discussed earlier. Additionally, Co application has also been shown to increase uptake of other nutrients in legumes. Applying Co with a *Rhizobium* strain enhanced N, phosphorus (P), and potassium (K) absorption in groundnut crops (Yadav and Khanna 1988). Co applied at 12 mg kg^{-1} significantly enhanced N, P, and K contents in soybean crops compared to the control plants (Kandil et al. 2013). The enhanced uptake of other nutrients in legumes could be related to the improved overall plant growth. Co application increased nodule numbers, leghaemoglobin and chlorophyll contents, net assimilation rates, increased growth rate in chickpea (*Cicer arietnum*) (Rod et al. 2019), pigeon pea (*Cajanus cajan*), and groundnut (Raj 1987; Gericó et al. 2020). Plants with high growth vigour will demand a continuous supply of other nutrients. As a result, healthy plants acquire more nutrients from soils.

Explanations for enhanced growth of other plants include the following: (i) Appropriate application of Co can alleviate abiotic stresses of plants, such as drought, salinity, and heavy metals. Co addition has been shown to increase plant drought tolerance by reducing transpiration rate (Gad et al. 2018). Co was used to reduce harmful effects of salinity on tomato plants (Hussien 1984; Gad and Kandil 2011), and Co improved wheat tolerance to salinized soil (Gad and Kandil 2011). Attia et al. (2014) observed that salinity stress decreased onion dry weight and regeneration capacity; however, Co application increased free proline accumulation, which counteracted the salt stress. Additionally, Co has been reported to suppress plant uptake of cadmium (Chmielowska-Bąk et al. 2014). In general, abiotic stresses often cause plant imbalance between production and accumulation of reactive oxygen species (ROS), including superoxide radical (O_2^-), singlet oxygen (O_2), hydroxyl radical (OH^-), and hydrogen peroxide (H_2O_2) (Sachdev et al. 2021). ROS can activate plant's antioxidant system, thus minimizing the damages (Tewari et al. 2002; Choudhury et al. 2017). The antioxidant system includes (a) enzymatic antioxidants, such as ascorbate peroxidase, catalase, dehydroascorbate reductase, general peroxidases, glutathione reductase, monodehydroascorbate reductase, and superoxide dismutase, and (b) non-enzymatic antioxidants, mainly ascorbic acid, α-tocopherol, carotenoids, reduced glutathione, plastoquinone/ubiquinone, and flavonoids (Garcia-Caparros et al. 2020). The action of the antioxidant system could be the first line of defence against the adverse effects. Therefore, it is not surprising to notice increased activities of ascorbate peroxidase, catalase, peroxidase, and superoxide dismutase (Hasanuzzaman et al. 2020). (ii) Co may substitute other nutrient elements when such nutrients have limited availability. Baxter et al. (2008) showed that when *Arabidopsis* plants are grown under a low Fe concentration, the shoot concentration of Co increased, which was concomitant with the increased expression of Fe transporter IRT1. Recently, a gene called *ARG1* (albino-revertible green 1), a member of the ATP-binding cassette (ABC) transporter family, was identified in rice (Li et al. 2020). It is localized in chloroplast and is responsible for the transport of Co and nickel (Ni) from chloroplasts to avoid excessive Co and Ni to compete with essential metal cofactors in chlorophyll and also metal-binding proteins acting in photosynthesis. Although the exact role Co plays in this case is unclear, it does indicate the interaction of Co with Ni in maintaining cell homeostasis to prevent cells from oxidative damages.

Clearly, some progress has been made in our understanding of mechanisms behind Co enhanced plant growth (Wang et al. 2020), but we still have no clear picture as to what exact roles Co plays in plant growth except that it is a constituent of cobalamin and a cofactor of several enzymes in archaea and bacteria as discussed earlier. Co as a cofactor of a plant enzyme has not been documented. The identified *ARG1* appears not to be specific for transport of Co, although the role of ABCI subfamily members in plants has not been well studied (Li et al. 2020). As a result, the function of Co in plant biology requires further investigation.

Other Beneficial Effects on Plants

Other beneficial effects have been documented in the literature. Co has been used for improving seed germination. Application of Co enhanced seed germination of cucumber and tomato (Helmy et al. 1994), onion (Attia et al. 2014), and sunflower (Singh and Rao 1993) and prevented floral malformation in mango (Zora et al. 1993; Zora et al. 1994). However, physiological and molecular bases for Co-mediated improved seed germination and flower development were not elaborated.

Cobalt has been used to improve postharvest quality of floriculture crops and prolong vase life of cut flowers. Cut fronds of *Adiantum raddianum* placed in water wilted in four days due to vascular blockage at the basal end of the petiole (Fujino and Reid 1983). This wilting was delayed, and vase life was increased to eight days in the deionized water control or by Co addition at 58.93 mg L^{-1} compared to only three days in control treatments (Fujino and Reid 1983). Co has been reported to significantly reduce transpiration rates in the leaves of beans, tomatoes, and melons (Reddy 1988). Co addition to preservative solutions increased leaf diffusive resistance, reduced xylem blockage, sustained water flow and uptake, and prolonged vase life of cut flowers of *Rosa hybrida* 'Samantha'. Reddy (1988) suggested that partial closure of stomata by Co was responsible for reducing the water loss/water uptake ratio, and thereby maintained a higher water potential in the cut roses. Co was also reported to slow the senescence process in harvested lettuce (Tosh et al. 1979). The beneficial effects of Co on prolonged vase life of cut flowers are attributed to its inhibitory activity to 1-aminocyclopropane-1-carboxylic acid (ACC) oxidase. Ethylene is synthesized from amino acid methionine by two key enzymes, ACC synthase and ACC oxidase. Co can block the conversion of ACC to ethylene by inhibiting ACC oxidase activity in the ethylene biosynthesis pathway (Lau and Yang 1976; Serek et al. 2006). Another explanation for the increased vase life was the antibacterial activity of Co. Co can suppress bacterial growth, thus reducing the vascular blockage in cut stem or fronds (Fujino and Reid 1983). Co^{3+} has been reported to form Co-complexes, which have antiviral and antibacterial activities (Chang et al. 2010).

Cobalt Deficiency in Plants

Soils generally meet the need of plants for Co as Co in surface soils ranges from 15 to 25 mg kg^{-1} and plants only need 0.1–10 mg kg^{-1} Co based on dry weight (Pilon-Smits et al. 2009). Due to the uneven distribution of Co on the Earth, Co deficiency has been reported in soils in Sierra Leone, Malta, New Zealand, Finland, and South Australia (Sillanpaa and Jansson 1992). Additionally, plant species vary in Co uptake (Liu 1998), and the absorption is significantly affected by soil drainage status, pH, and content of humus and MnO_2 (MeKenzie 1972; Palit and Sharma 1994). As a result, Co deficiency occurs in plants, which has significant implication in human and animal health.

Cobalt deficiency symptoms resemble N deficiency in plants, including leaf chlorosis and necrosis, growth retardation, and reduced crop yield (Liu 1998). Co deficiency commonly occurs in leguminous crops. Co-deficient legumes have reduced plant size, smaller and pale-yellow leaves, and smaller pods compared to those without Co deficiency

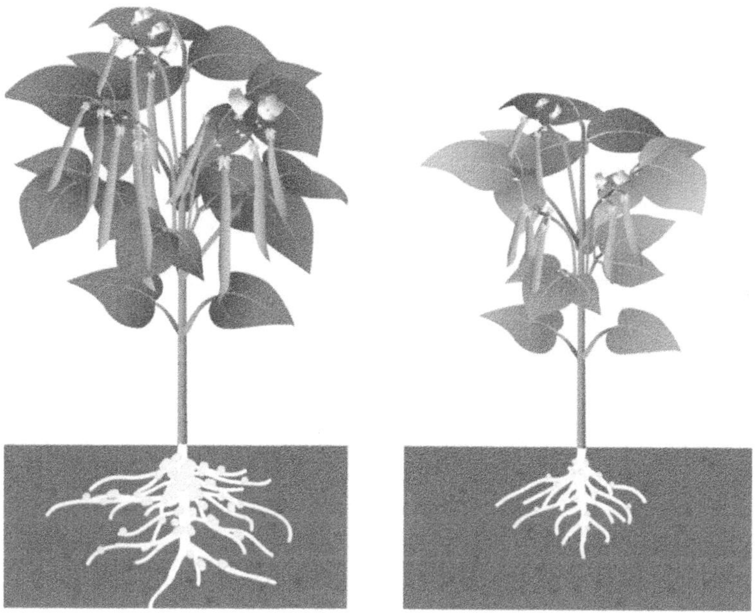

Figure 8.3 Common bean (*Phaseolus vulgaris*) plants without Co deficiency (left) and with Co deficiency (right).

(Figure 8.3). Furthermore, root growth is affected by exhibiting overall reduction of root volume and root lengths. Nodule size and numbers are less abundant than the plants without Co deficiency. Co deficiency causes reduced synthesis of methionine, thus limiting protein synthesis and contributing to the smaller-sized bacteroids (Marschner 2012). Sweet lupin is particularly sensitive to Co deficiency (Robson et al. 1979). In field grown lupins, Co deficiency reduced bacteroid number per gram of nodule (Chatel et al. 1978) and affected nodule development and function at different levels (Dilworth et al. 1979). Co deficiency in legumes can be assessed by analysis of Co contents in shoots. In general, deficient symptom appears when shoot Co falls in a range from 0.04 (Ozanne et al. 1963) to 0.02 mg kg^{-1} based on dry weight (Robson et al. 1979).

Cobalt deficiency also occurs in non-leguminous plants. Symptoms of Co deficiency in wheat and non-nodulated clover showed leaf chlorosis and reduced growth (Wilson and Nicholas 1967), which is illustrated in Figure 8.4. Low leaves may become necrotic, and root systems are reduced. Co deficiency causes growth retardation in rubber trees and tomato plants (Wilson and Nicholas 1967). Grasses with low contents of Co result in Co deficiency of sheep and cattle. For countries, like South Australia, Sierra Leone, Malta, New Zealand, and Finland, where soils have low Co contents (Sillanpaa and Jansson 1992), application of Co could improve forage grass growth and enrich tissue Co content. Thus, the feeding of ruminants with the healthy grass can reduce Co deficiency (Lee 1951; Dewey et al. 1958). Due to low Co concentrations in plants, Co deficiency in grazing animals may occur, which can be corrected by mixing Co salts with fertilizers or sand carrier to broadcast it over grazed pastures. For improving legume growth, Co application rates are very low, ranging from 1.8 to 145.6 g per hectare (Havlin et al. 2013). Other methods to enhance

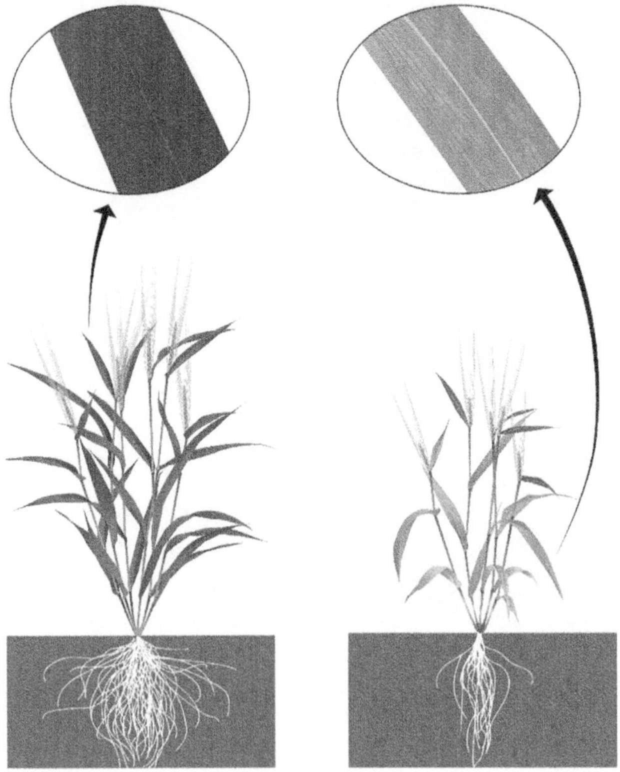

Figure 8.4 Common wheat (*Triticum aestivum*) plants without Co deficiency (left) and with Co deficiency (right).

Co concentrations in plants include seed treatment with Co compounds or foliar application of Co solution. Additionally, providing Co supplements directly to grazing animals can also mitigate Co deficiencies. Production of legumes in mixture with forage species and/or grasses is another way to supplement Co to grazing livestock.

Cobalt Toxicity in Plants

Cobalt at high concentrations causes cytotoxicity and phytotoxicity in plants. Cytotoxicity includes the inhibition of mitosis and damage of chromosomes and disruption of endoplasmic reticulum of root tip cells (Rauser 1981; Smith and Carson 1981; Akeel and Jahan 2020). Phytotoxicity symptoms vary depending on plant species and concentration of Co in plant organs. Leguminous plants generally exhibit chlorosis or pale-white color on young leaves, and tomatoes show either interveinal chlorosis or diffused chlorosis on young leaves (Akeel and Jahan 2020). Figure 8.5 shows seed germination and germinated young cucumber seedlings in response to increased concentrations of Co from 0, 5, 10, 20, 40 to 240 mg kg^{-1}, respectively. Six seeds germinated in the control soil, seedlings growth was robotic, and the first true leaf was dark green. However, the true leaf became yellowish

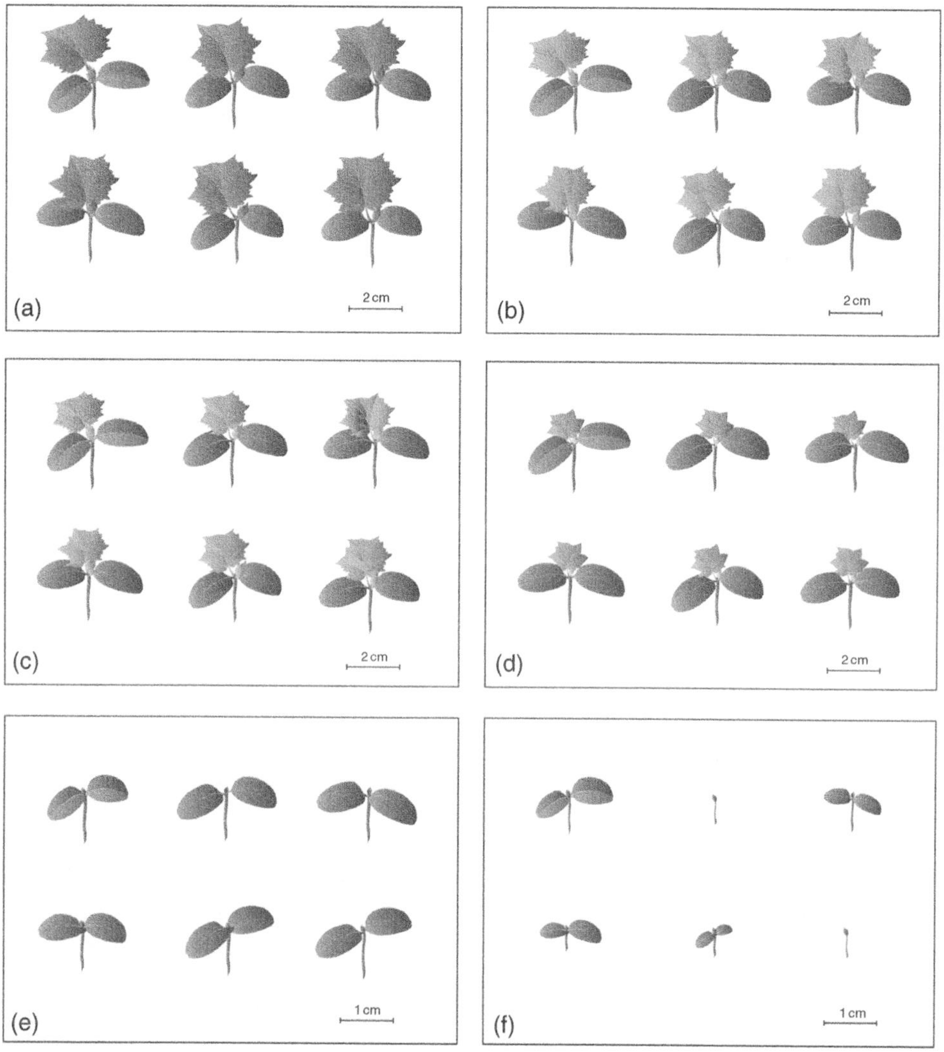

Figure 8.5 Cucumber (*Cucumis sativus*) seed germination and seedling growth in a soil amended with Co at 0.0 mg kg^{-1} (a), 5 mg kg^{-1} (b), 10 mg kg^{-1} (c), 20 mg kg^{-1} (d), 40 mg kg^{-1} (e), and 240 mg kg^{-1} (f) levels. Source: Adapted from Hu et al. (2010).

green after seeds were germinated in the soil amended with Co. With the increased concentrations of Co, the true leaf became smaller, and there was no true leaf after seeds were germinated in the soil with 40 mg kg^{-1} Co (Hu 2010). Furthermore, some seeds failed to germinate at the highest concentration of Co (240 mg kg^{-1}). Thus, high concentrations of Co can suppress the germination of cucumber seeds.

Cobalt toxicity in plants is generally uncommon in natural soils, but it occurs when plants grow in Co-contaminated soils (Ormrod et al. 1980). Soil contamination by Co is mainly from mining and smelting activities, disposal of sewage sludge, and the use of chemical fertilizers (Hamilton 1994). As discussed earlier, plants develop mechanisms to control Co absorption, transport, and distribution. However, once Co in contaminated soils becomes highly available, Co may gain a competitive advantage over Fe, resulting in more Co being absorbed than Fe through IRT1. With increasing concentrations of Co inside cells, FPN2 may not be able to effectively sequester Co into the vacuole, resulting in more Co to transport from roots to shoots. Li et al. (2009) showed that Co concentrations in shoots of barley, oil seed rape (*Brassica napus*), and tomato were linearly correlated with the soil solution Co. As a result, excessive Co in shoots may initially cause oxidative stress, resulting in increased antioxidative enzyme activities (Tewari et al. 2002). As the stress progresses, Co may compete with Fe or Mg in the chloroplast by decreasing chlorophyll content (Lwalaba et al. 2017), which causes newly growing leaves with a yellowish color as shown in Figure 8.5. As reported by Sree et al. (2015), Co is able to inhibit the activity of enzymes involved in the biosynthesis of chlorophyll intermediates, like 5-aminolevulinic acid and protoporphyrin, which will reduce net photosynthetic activities. Co also adversely affects the translocation of P, S, Cu, Mn, and Zn from roots to shoots (Chatterjee and Chatterjee 2000). All these effects act together, resulting in the phytotoxicity and significant reduction of plant growth.

Different plants show different abilities to tolerate Co. Oat (*Avena sativa*) plants were adversely affected when grown in a soil solution containing 0.14 mg L^{-1} Co (Anderson et al. 1973). Rice (*Oryza sativa*) plants would develop toxic symptoms when grown in soils with a Co concentration ranging from 25 to 50 mg kg^{-1} (Kitagishi and Yamane 1981). The contents of Co could be used for predicting the development of toxicity (Akeel and Jahan 2020). Toxic symptoms occurred in bush beans when tissue Co contents ranged from 43 to 142 mg kg^{-1} (Wallace et al. 1977) and similarly, 6 mg kg^{-1} in barley seedlings (Davis et al. 1978) and 19–32 mg kg^{-1} in Sudan grass (Gough et al. 1979). In general, tissue Co contents between 30 and 40 mg kg^{-1} are considered critical levels for potential development of Co toxicity (Macnicol and Beckett 1985). However, due to the evolutionary adaptation, Co hyperaccumulators do not develop toxic symptoms at this concentration level. Co contents in leaves of *Rinorea* cf. *bengalensis* can be 1200 mg kg^{-1} (Paul et al. 2020), and *G.* cf. *sericeum* can accumulate 1500 mg kg^{-1} Co (Van der Ent et al. 2018b). Co hyperaccumulators are not the focus of this chapter. The reader is referred to publications by Brooks et al. (Brooks 1977; Brooks et al. 1977; Brooks et al. 1980), Baker (1981, 1987), Lange et al. (2017), and Yamaguchi et al. (2019) for more information.

Conclusions and Future Perspectives

Cobalt contents in soils range from 15 to 25 mg kg^{-1}. Plant roots can absorb Co from soils and transport absorbed Co from roots to shoots. Shoot concentrations of Co vary with plant species. Co has been considered an essential element for prokaryotes, human beings, and other mammals, but its essentiality for plants has not been established. Accumulating evidence, however, shows that Co at low concentrations improves plant growth. Co is a constituent of cobalamin, and it is required for N_2 fixation in nodulated leguminous crops and free N_2-fixation bacteria. Co is also a beneficial element to other plants. At low concentrations, Co can promote plant growth, reduces plant uptake of cadmium, and plays a role in inhibition of ethylene biosynthesis. Co deficiency is characterized by pale-yellow leaves when plants, particularly nodulated legumes, grow in soils with limited available Co. Conversely, Co is toxic to plants at higher concentrations. High levels of Co result in the loss of leaves, discolored leaves along with Fe deficiency. This chapter outlines the current understanding of Co in plant biology, but further research is needed to uncover molecular mechanisms underlying the beneficial effects of Co and probably its essentiality to plant growth and development.

References

Adams, D.G. and Duggan, P.S. (2008). Cyanobacteria–bryophyte symbioses. *J Exp Bot* 59: 1047–1058. https://doi.org/10.1093/jxb/ern005.

Aery, N.C. and Jagetiya, B.L. (2000). Effect of cobalt treatments on dry matter production of wheat and dtpa extractable cobalt content in soils. *Commun Soil Sci Plant Anal* 31: 1275–1286.

Ahmed, S. and Evans, H.J. (1959). Effect of cobalt on growth of soybeans in absence of supplied nitrogen. *Biochem Biophys Research Comm* 1: 271–275.

Ahmed, S. and Evans, H.J. (1960). The essentiality of co for soybean plants grown under symbiotic conditions. *Proc Natl Acad Sci U. S. A.* 47: 24–36.

Ahmed, S. and Evans, H.J. (1961). The essentiality of cobalt for soybean plants grown under symbiotic conditions. *Proc Natl Acad Sci U. S. A.* 47: 24–36.

Akeel, A. and Jahan, A. (2020). Role of cobalt in plants: its stress and alleviation. In: *Contaminants in Agriculture* (ed. M. Naeem, A.A. Ansari and S.S. Gill), 339–352. Springer Nature Switzerland AG, https://doi.org/10.1007/978-3-030-41552-5_17.

Ali, B., Hayat, S., Hayat, Q., and Ahmad, A. (2010). Cobalt stress affects nitrogen metabolism, photosynthesis and antioxidant system in chickpea (*Cicer arietinum* L.). *J Plant Interact* 5: 223–231.

Anderson, A., Meyer, D., and Mayer, F. (1973). Heavy metal toxicities: levels of nickel, cobalt and chromium in the soil and plants associated with visual symptoms and variation in growth of an oat crop. *Aust J Agric Res* 24: 557–571.

Andic, B., Dragicevic, S., Stesevic, D. et al. (2015). Comparative analysis of trace elements in the mosses - *Bryum argenteum* Hedw. and *Hypnum cupressiforme* Hedw. in Podgorica (Montenegro). *J Mater Environ Sci* 6: 333–342.

Attia, S.A.A., Gad, N., and Abdel-Rahman, H.M. (2014). Effect of cobalt on growth, yield and production quality with mitotic and meiotic divisions in two onion cultivars. *Current Sci Int* 3: 122–131.

Aziz, E.E., Gad, N., and Nadia, M.B. (2007). Effect of cobalt and nickel on plant growth, yield and flavonoid. *Aust J basic Appl Sci* 1: 73–78.

Baker, A. (1981). Accumulators and excluders-strategies in the response of plants to heavy metals. *J Plant Nutr* 3: 643–654.

Baker, A. (1987). Metal tolerance. *New Phytol* 106: 93–111.

Bakkaus E, Gouget B, Gallien JP, Khodja H, Carrot F, Morel JL, Collins R. (2005) Concentration and distribution of cobalt in higher plants: the use of micro-pixe spectroscopy. *Nucl Instrum Methods Phys Res, Sect. B* 231: 350–356; doi: https://doi.org/10.1016/j.nimb.2005.01.082.

Barsanti, L. and Gualtieri, P. (2006). *Algae: Anatomy, Biochemistry and Biotechnology*. Boca Raton, FL: Taylor & Francis Group.

Baxter, IR, Vitek O, Lahner B, Muthukumar B, Borghi M, Morrissey J, Guerinot ML, Salt DE. (2008). The leaf ionome as multivariable system to detect a plant's physiological status. *Proc Natl Acad Sci USA* 105:12081–12086. doi:https://doi.org/10.1073/pnas.0804175105

Bilal M, Rasheed T, Sosa-Hernández JE, Raza A, Nabeel F, Iqbal HMN. (2018) Biosorption: an interplay between marine algae and potentially toxic elements - a review *Mar Drugs* 16: 65; doi: https://doi.org/10.3390/md16020065.

Bohme, H. (1998). Regulation of nitrogen fixation in heterocyst-forming cyanobacteria. *Trends Plant Sci* 3: 346–351.

Bolle-Jones, E. and Mallikarjuneswara, V. (1957). A beneficial effect of co on the growth of the rubber plant. *Nature* 179: 738–739.

Brooks, R.R. (1977). Copper and cobalt uptake by haumanistrum species. *Plant Soil* 48: 541–544.

Brooks, R.R. (1998). *Plants that Hyperaccumulate Heavy, Metals: Their Role in Phytoremediation, Microbiology, Archaeology, Mineral Exploration and Phytomining*. Oxon: UK: CAB International.

Brooks, R.R., McCleave, J.A., and Schofield, E.K. (1977). Cobalt and nickel uptake by the Nyssaceae. *Taxon* 26: 197–201.

Brooks, R.R., Reeves, R.D., Morrison, R.S., and MalaJsse, F. (1980). Hyperaccumulation of copper and cobalt: a review. *Bull Soc Roy Bot Belg* 113: 166–172.

Carpenter, K.J. (2004). The nobel prize and the discovery of vitamins. (accessed 5 December 2020).

Chang, E., Simmers, C., and Knight, D. (2010). Cobalt complexes as antiviral and antibacterial agents. *Pharmaceuticals* 3: 1711–1728.

Chatel, D.L., Robson, A.D., Gartrell, J.W., and Dilworth, M.J. (1978). The effect of inoculation on cobalt application on the growth of and nitrogen fixation of sweet lupins. *Aust J Agric Res* 29: 1191–1202.

Chatterjee, J. and Chatterjee, C. (2000). Phytotoxicity of cobalt, chromium and copper in caulifower. *Environ Pollut* 109: 69–74.

Chen J, Zhou J, Goldsbrough P. (1997) Characterization of phytochelatin synthase from tomato. *Physiol Plant* 101: 165–192; doi: https://doi.org/10.1111/j.1399-3054.1997.tb01833.x.

Cheng, Y.S., Yan, Y.B., and Liu, J.Y. (2005). Spectroscopic characterization of metal bound phytochelatin analogue (Glu-Cys)$_4$-Gly. *J Inorg Biochem* 99: 1952–1962. https://doi.org/10.1016/j.jinorgbio.2005.06.016.

Cheng, J., Poduska, B., Morton, R.A., and Finan, T.M. (2011). An ABC-type cobalt transport system is essential for growth of *Sinorhizobium meliloti* at trace metal concentrations. *J Bacteriol* 193: 4405–4416.

Chmielowska-Bąk, J., Gzyl, J., Rucińska-Sobkowiak, R. et al. (2014). The new insights into cadmium sensing. *Front Plant Sci* 5: 245.

Choudhury, F.K., Rivero, R.M., Blumwald, E., and Mittler, R. (2017). Reactive oxygen species, abiotic stress and stress combination. *Plant J* 90: 856–867.

Collins, R.N., Bakkaus, E., Carrière, M. et al. (2010). Uptake, localization, and speciation of cobalt in *Triticum aestivum* L. (wheat) and *Lycopersicon esculentum* L. (tomato). *Environ Sci Technol* 44: 2904–2910.

Connor, J.J. and Shacklette, H.T. (1975). Background geochemistry of some rocks, soils, plants, and vegetables in the conterminous united states. In: *U.S. Geol. Survey Professoinal Paper 574-F*, 168 p (ed. R.J. Ebens, J.A. Erdman, A.T. Miesch, et al.). Washington, DC: The U.S. Government Printing Office.

Conte, S. and Elsbeth, W.L. (2011). Transporters contributing to iron trafficking in plants. *Mol. Plant* 4: 464–476.

Croft, M.T., Lawrence, A.D., Raux-Deery, E. et al. (2005). Algae acquire vitamin B_{12} through a symbiotic relationship with bacteria. *Nature* 438: 90–93.

Cruz-Lopez, R. and Maske, H. (2016). The vitamin B_1 and B_{12} required by the marine dinoflagellate Lingulodinium polyedrum can be provided by its associated bacterial community in culture. *Front Microbiol* 7: 560.

Czerpak, R., Bajguz, A., Chodkowski, K., and Popow, H. (1994). Influence of nickel and cobalt on the growth and biochemical changes of *Chlorella pyrenoidosa* (Chlorophyceae). *Pol Arch Hydrobiol* 41: 161–169.

Davies, K.M., Jibran, R., Zhou, Y. et al. (2020). The evolution of flavonoid biosynthesis: a bryophyte perspective. *Front. Plant Sci* 11: 7.

Davis, R., Beckett, P., and Wollan, E. (1978). Critical levels of twenty potentially toxic elements in young spring barley. *Plant Soil* 49: 395–408.

Dewey, D.W., Lee, H.J., and Marston, H.R. (1958). Provision of cobalt to ruminants by means of heavy pellets. *Nature* 181: 1367–1371.

Dilworth, M.J., Robson, A.D., and Chatel, D.L. (1979). Cobalt and nitrogen fixation in *Lupinus angustifolius* L. II. Nodule formation and function. *New Phytol* 83: 63–79.

van der Ent A, Mak R, de Jonge MD, Harris HH. (2018a) Simultaneous hyperaccumulation of nickel and cobalt in the tree *Glochidion* cf. *Sericeum* (Phyllanthaceae): elemental distribution and chemical speciation. *Sci Reps* 8: 9683; doi: https://doi.org/10.1038/s41598-018-26891-7.

van der Ent, A., Przybyłowicz, W., De Jonge, M. et al. (2018b). X-ray elemental mapping techniques for elucidating the ecophysiology of hyperaccumulator plants. *New Phytol* 218: 432–452.

Erdman, J.A. and Modreski, P.J. (1984). Copper and cobalt in aquatic mosses and stream sediments from the Idaho cobalt belt. *J Geochem Explor* 20: 75–84.

Fujino, D.W. and Reid, M.S. (1983). Factors affecting the vase life of fronds of maidenhair fern. *Sci Hort* 21: 181–188.

Gad N. (2012) Physiological and chemical response of groundnut (*Arachis hypogaea*) to cobalt nutrition. *World Appl Sci J* 20: 327–335; doi: https://doi.org/10.5829/idosi.wasj.2012.20.02.2818.

Gad, N. and Kandil, H. (2011). Maximizing the tolerance of wheat plants to soil salinity using cobalt. A-Growth and mineral composition. *J Appl Sci Res* 7: 1569–1574.

Gad, N., Ali, M.E.F., Abbas, M.M., and Abdel-Moez, M.R. (2018). Maximization of drought tolerance of bean plants using cobalt supplementation a-growth, yield and nutritional status. *Midlle East J Agri Res* 7: 1819–1826.

García-Caparrós, P., De Filippis, L., Gul, A., et al. (2020). Oxidative stress and antioxidant metabolism under adverse environmental conditions: a review. *Bot. Rev.* 40: 353–362. doi:10.1007/s12229-020-09231-1

Garnham, G.W., Codd, G.A., and Gadd, G.M. (1992). Kinetics of uptake and intracellular location of cobalt, manganese and zinc in the estuarine green alga *Chlorella salina*. *Appl Microbiol Biotechnol* 37: 270–276. https://doi.org/10.1007/BF00178183.

Gericó, T.G., Tavanti, R.F.R., Lima, J.P. et al. (2020). Cobalt and molybdenum stimulate compounds of primary metabolism, nitrogen forms, and photosynthetic pigments in peanut plants (Arachis hypogaea L.). *J Plant Nutri* 43: 1907–1922. https://doi.org/10.1080/01904167.2020.1750646.

Gough, L., Shacklette, H., and Case, A. (1979). Element concentrations toxic to plants, animals and man. *LiSGS Bull* 1466: 80–84.

Govindaraju, K. (1994). Compilation of working values and sample description for 383 geostandards. *Geostand Newsletters* 18: 1–158.

Grube, M., Seckbach, J., and Muggia, L. (2017). *Algal and Cyanobacteria Symbioses*. London, UK: World Scientific Publishing Europe Ltd.

Guo, M. and Chen, Y. (2018). Coenzyme cobalamin: biosynthesis, overproduction and its application in dehalogenation-a review. *Rev Environ Sci Biotechnol* 17: 259–284. https://doi.org/10.1007/s11157-018-9461-6.

Hala, K. (2007). Effect of cobalt fertilizer on growth, yield and nutrients status of faba bean (*Vicia faba* L.) plants. *J App Sci Res* 3: 867–872.

Hamilton, E.I. (1994). The geobiochemistry of cobalt. *Sci Total Environ* 150: 7–39.

Hanikenne, M., Krämer, U., Demoulin, V., and Baurain, D. (2005). A comparative inventory of metal transporters in the green alga *Chlamydomonas reinhardtii* and the red alga *Cyanidioschyzon merolae*. *Plant Physiol* 137: 428–446.

Hasanuzzaman, M., Bhuyan, M.B., Zulfiqar, F. et al. (2020). Reactive oxygen species and antioxidant defense in plants under abiotic stress: revisiting the crucial role of a universal defense regulator. *Antioxidants* 9: 681. https://doi.org/10.3390/antiox9080681.

Havlin, J.L., Tisdale, S.L., Nelson, W.L., and Beaton, J.D. (2013). *Soil Fertility and Fertilizers*. Upper Saddle River, NJ: Prentice Hall, Inc.

Hawco, N.J., Mcilvin, M.R., Bundy, R.M. et al. (2020). Minimal cobalt metabolism in the marine cyanobacterium *Prochlorococcus*. *Proc Natl Acad Sci.* 117: 15740–15747.

Helmy, L.M. and Gad, N. (2002). Effect of cobalt fertilization on the yield, quality the essential oil composition of parsley leaves. *Arab Univ J Agric Sci Ain Shams Univ Cairo* 10: 779–802.

Helmy, Y.H., El-Abd, S.O., Abou-Hadid, A.F. et al. (1994). Ethylene production from tomato and cucumber plants under saline conditions. *Egypt J Hort* 21: 153–160.

Hewitt, E.J. (1952). *Sand and Water Culture Methods Used in the Study of Plant Nutrition*, Technical Communication 22, 189. East Malling, UK: Commonwealth Bureau of Horticulture Plantation Crops.

Hodgkin, D., Kamper, J., Mackay, M. et al. (1956). Structure of vitamin. *Nature* 178: 64–66.

Holm-Hansen, O., Gerloff, G.C., and Skoog, F. (1954). Cobalt as an essential element for blue-green algae. *Physiol Planta* 7: 665–675.

Hopkins, W. (1995). *Introduction to Plant Physiology*. New York: John Wiley and Sons.

Hu, XX. (2010). Study on the transfer characteristics of cobalt and nickel from ssoils to plants and their toxicity to plants. Master Thesis. Fujian Agriculture and Forestry University, Fuzhou, Fujian.

Hussien, L.A. (1984). Plant tolerance to salt and heavy metals. Ph.D. thesis. Ain Shams University, Cairo, Egypt.

Huwait, A.E., Kumosani, A.T., Moselhy, S.S. et al. (2015). Relationship between soil cobalt and vitamin B_{12} levels in the liver of livestock in saudi arabia: Role of competing elements in soils. *Afr Health Sci* 15: 993–998.

Jaleel CA, Jayakumar K, Zhao C-X, Iqbal M. (2009) Low concentration of cobalt increases growth, biochemical constituents, mineral status and yield in zea mays. *J Sci Res* 1: 128–137; doi: https://doi.org/10.3329/jsr.vlil.1226.

Jayakumar, K. and Jaleel, C.A. (2009). Uptake and accumulation of cobalt in plants: A study based on exogenous cobalt in soybean. *Bot Res Int* 2: 310–314.

Kabata-Pendias, A. (1973). Uptake of cobalt and copper by clover from minerals impregnated by these cations. *Roczn Glebozn* 24: 273–287.

Kabata-Pendias, A. and Mukherjee, A.B. (2007). *Trace Elements from Soils to Human*. Heidelberg: Springer-Verlag: Berlin; doi: https://doi.org/10.1007/978-3-540-32714-1.

Kandil, H., Ihab, M., and El-Maghraby, A. (2013). Effect of cobalt level and nitrogen source on quantity and quality of soybean plant. *J Basic Appl Sci Res* 3: 185–192.

Kellogg MM, Mcllvin MR, Vedamati J, Twining BS, Moffett JW, Marchetti A, Moran DM, Saito MA. (2020) Efficient zinccobalt inter-replacement in northeast pacific diatoms and. *Limnol Oceanogr* doi: https://doi.org/10.1002/lno.11471.

Kitagishi, K. and Yamane, I. (1981). *Heavy Metal Pollution in Soils of Japan*. Tokyo, Japan: Japan Science Society Press.

Kliewer, M. and Evans, H.J. (1963a). Cobamide coenzyme contents of soybean nodules and nitrogen fixing bacteria in relation to physiological conditions. *Plant Physiol* 38: 99–104.

Kliewer, M. and Evans, H.J. (1963b). Identification of cobamide coenzyme in nodules of symbionts and isolation of the B_{12} coenzyme from *Rhizobium meliloti*. *Plant Physiol* 38: 55–59.

Kobayashi, M. and Shimizu, S. (1999). Cobalt proteins. *Eur J Biochem* 261: 1–9.

Komeda, H., Hori, Y., Kobayashi, M., and Shimizu, S. (1996). Transcriptional regulation of the *Rhodococcus rhodochrous* J1 *nita* gene encoding a nitrilase. *Proc Natl Acad Sci USA* 93: 10572–10577. https://doi.org/10.1073/pnas.93.20.10572. PMID: 8855219.

Korshunova, Y.O., Eide, D., Clark, W.G. et al. (1999). The IRT1 protein from *Arabidopsis thaliana* is a metal transporter with a broad substrate range. *Plant Mol Biol* 40: 37–44.

Lange B, van der Ent A, Baker AJ, Echevarria G, Mahy G, Malaisse F, Meerts P, Pourret O, Verbruggen N, Faucon MP. (2017) Copper and cobalt accumulation in plants: a critical assessment of the current state of knowledge. *New Phytol* 213: 537–551; doi: https://doi.org/10.1111/nph.14175.

Lau, O.-L. and Yang, S.F. (1976). Inhibition of ethylene production by cobaltous ion. *Plant Physiol* 58: 114–117.

Lee, H.J. (1951). Cobalt and copper deficiencies affecting sheep in South Australia. *J Agri Sci of Austr* 54: 475–532.

Li, H., Gray, C., Mico, C. et al. (2009). Phytotoxicity and bioavailability of cobalt to plants in a range of soils. *Chemosphere* 75: 979–986.

Li, H., Liu, Y., Qin, H. et al. (2020). A rice chloroplast-localized ABC transporter ARG1 modulates cobalt and nickel homeostasis and contributes to photosynthetic capacity. *New Phytol* 228: 163–178.

Liu, J.C. (1998). Physiological effects and uptake mechanisms in plants. Ph.D. thesis. The University of Adelaide, Adelaide, Australia.

Lwalaba JLW, Zvogbo G, Mulembo M, Mundende M, Zhang G. (2017) The effect of cobalt stress on growth and physiological traits and its association with cobalt accumulation in barley genotypes differing in cobalt tolerance. *J Plant Nutri* 40: 2192–2199; doi: https://doi.org/10.1080/01904167.2017.1346676.

Lyu S, Wei X, Chen J, Wang C, Wang X, Pan D. (2017) Titanium as a beneficial element for crop production. *Front Plant Sci* 8 1–19; doi: https://doi.org/10.3389/fpls.2017.00597.

Macnicol, R. and Beckett, P. (1985). Critical tissue concentrations of potentially toxic elements. *Plant Soil* 85: 107–129.

Mantzorou, A., Navakoudis, E., Paschalidis, K., and Ververidis, F. (2018). Microalgae: a potential tool for remediating aquatic environments from toxic metals. *Int J Environ Sci Technol* 15: 1815–1830. https://doi.org/10.1007/s13762-018-1783-y.

Marschner, P. (2012). *Marschner's Mineral Nutrition of Higher Plants*. Waltham, MA: Academic Press: London, UK.

Marsh, E.N. (1999). Coenzyme B_{12} (cobalamin)-dependent enzymes. *Essays Biochem.* 34: 139–154.

MeKenzie, R.M. (1972). The manganese oxides in soils: A review. *Bodenk. & Pflanzenernfihr* 131: 221–242.

Morel, M., Crouzet, J., Gravot, A. et al. (2009). AtHMA3, a P1B-ATPase allowing Cd/Zn/Co/Pb vacuolar storage in Arabidopsis. *Plant Physiol* 149: 894–904. https://doi.org/10.1104/pp.108.130294.

Morrissey, J., Baxter, I., Lee, J. et al. (2009). The ferroportin metal efflux proteins function in iron and cobalt homeostasis in *Arabidopsis*. *Plant Cell* 21: 3326–3338.

Nagpal, N. (2004). *Water Quality Guidelines for Cobalt*. Victoria: Ministry of Water, Land and Air Protection, Water Protection Section, Water, Air and Climate Change Branch.

Ormrod, D.P., Proctor, J.T.A., Hofstra, G. et al. (1980). Air pollution effects on agricultural crops in Ontario: a review. *Can J Plant Sci* 60: 1023–1030.

Osman, D., Cooke, A., Young, T.R. et al. (2021). The requirement for cobalt in vitamin B_{12}: a paradigm for protein metalation. *Biochim. Biophys. Acta Mol Cell Res* 1868: 118896.

Ozanne, P.G., Greenwood, E.A.N., and Shaw, T.C. (1963). The cobalt requirement of subterranean clover in the field. *Aust J Agric Res* 14: 39–50.

Page V, Feller U. (2005) Selective transport of zinc, manganese, nickel, cobalt and cadmium in the root system and transfer to the leaves in young wheat plants. *Ann Bot* 96: 425–434; doi: https://doi.org/10.1093/aob/mci189.

Palit, S. and Sharma, A. (1994). Effects of cobalt on plants. *Bot. Rev.* 60: 149–181.

Paul ALD, Nkrumah PN, Echevarria G, Erskine PD, Chaney RL, Spiers KM, Sumail S, van der Ent A. (2020) Cobalt hyperaccumulation in *Rinorea* cf. *Bengalensis* (Violaceae) from sabah: accumulation potential and tissue and cellular-level distribution of cobalt. *Plant Soil* 455: 289–303; doi: https://doi.org/10.1007/s11104-020-04629-7.

Peterson, P.J. and Girling, C.A. (1981). Other trace metals. In: *Effect of Heavy Metal Pollution on Plants* (ed. N.W. Lepp), 213–278. London: Applied Science.

Phipps, D.A. (1976). *Metals and Metabolism*, Oxford Chemistry Series No. 26. Clarendon Press: 134 p.

Pilon-Smits EA, Quinn CF, Tapken W, Malagoli M, Schiavon M. (2009) Physiological functions of beneficial elements. *Curr Opin Plant Biol* 12: 267–274; doi: https://doi.org/10.1016/j.pbi.2009.04.009.

Pipiška, M., Vrtoch, L., Horník, M. et al. (2007). Uptake of ^{137}Cs and ^{60}Co by bryophytes and lichens. *Cereal Res Commun* 35 (2 PART II): 929–932.

Raj, A.S. (1987). Cobalt nutrition of pigeonpea and peanut in relation to growth and yield. *J Plant Nutrit* 10: 2137–2145.

Rauser, W. (1981). Entry of sucrose into minor veins of bean seedlings exposed to phytotoxic burdens of Co, Ni or Zn. *J Plant Nutr* 3: 319–328.

Reddy, T.V. (1988). Mode of action of cobalt extending the vase life of cut roses. *Sci Hort* 36: 303–313.

Reimann, C., Koller, F., Frengstad, B. et al. (2001). Comparison of the element compositin in several plant species and their substrate from a 1555000 km^2 area in Northern Europe. *Sci Total Environ* 278: 87–112.

Reisenauer, H.M. (1960). Cobalt in nitrogen fixation by a legume. *Nature* 186: 375–376.

Roberts, S. and Gunn, G. (2014). Cobalt. In: *Critical Metals Handbook*, 1e (ed. G. Gunn), 122–147. Keyworth, Nottingham, UK: John Wiley & Sons.

Robson, A.D., Dilworth, M.J., and Chatel, D.L. (1979). Cobalt and nitrogen fixation on *Lupinus angusfifohus* L. I. Growth nitrogen concentrations and cobalt distribution. *New Phytol* 83: 53–62.

Rod, N.K., Gudadhe, N.N., Karmakar, N. et al. (2019). Cobalt chloride enhances crop duration, increases production, and productivity of chickpea. *J Plant Nutri* 42: 40–57.

Rodionov, D.A., Vitreschak, A.G., Mironov, A.A., and Gelfand, M.S. (2003). Comparative genomics of the vitamin B_{12} metabolism and regulation in prokaryotes. *J Bio Chem* 278: 41148–41159.

Sachdev S, Ansari SA, Ansari MI, Fujita M, Hasanuzzaman M. (2021) Abiotic stress and reactive oxygen species: generation, signaling and defense mechanisms. *Antioxidants* 10:277. doi: https://doi.org/10.3390/antiox10020277.

Samecka-Cymermann, A., Marczonek, A., and Kempers, A. (1997). Bioindicators of heavy metals in soil by liverworts. *Arch Environ Contam Toxicol* 33: 162–171.

Serek, M., Woltering, E.J., Sisler, E.C. et al. (2006). Controlling ethylene responses in flowers at the receptor level. *Biotechnol Adv* 24: 368–381.

Shacklette, H.T. (1965). Element content of bryophytes. *U.S. Geological Survey Bulletin* 1198: 1–18.

Sillanpaa, M. and Jansson, H. (1992). *Status of Cadmium, Lead, Cobalt and Selenium in Soils and Plants of Thirty Countries*. Rome: FAO, 1995 p.

Singh, B.G. and Rao, G.R. (1993). Effect of chemical soaking of sunflower (Helianthes annuus) seed on vigour inder. *Indian J Agri Sci* 63: 232–233.

Smith, H. and Carson, B. (1981). *Trace Metals in the Environment*. Ann Arbor, MI: Ann Arbor Science Publishers.

Sree KS, Keresztes A, Mueller-Roeber B, Brandt R, Eberius M, Fischer W, Appenroth KJ. (2015) Phytotoxicity of cobalt ions on the duckweed lemna minor - morphology, ion uptake, and starch accumulation. *Chemosphere* 131: 149–156; doi: https://doi.org/10.1016/j.chemosphere.2015.03.008.

Takahashi-Iñiguez T, García-Hernandez E, Arreguín-Espinosa R, Flores M. (2012) Role of vitamin B_{12} on methylmalonyl-coa mutase activity. *J Zhejiang Univ Sci B* 13: 423–437; doi: https://doi.org/10.1631/jzus.B1100329.

Tappero R, Peltier E, Grafe M, Heidel K, Ginder-Vogel M, Livi KJ, Rivers ML, Marcus MA, Chaney RL, Sparks DL. (2007) Hyperaccumulator alyssum murale relies on a different metal storage mechanism for cobalt than for nickel. *New Phytol* 175: 641–654; doi: https://doi.org/10.1111/j.1469-8137.2007.02134.x.

Tewari, R., Kumar, P., Sharma, P., and Bisht, S. (2002). Modulation of oxidative stress responsive enzymes by excess. *Plant Sci* 162: 381–388.

Tjong, E. and Mohiuddin, S. (2020). Biochemistry, tetrahydrofolate. (accessed 20 November 2020.

Torres, E.M. (2016). Microalgae sorption of ten individual heavy metals and their effects on growth and lipid accumulation. All Graduate Theses and Dissertations. Paper 4956, Utah State University.

Tosh, S., Choudhuri, M.A., and Chatterjee, S.K. (1979). Retardation of letture (*Lactura sativa* L.) leaf senescence by cobalt ions. *Indian J Exp l Bio* 17: 1134–1136.

Underwood, E.J. and Filmer, J.F. (1935). The determination of the biologicaþ potent element (cobalt) in limonite. *Aust Vet J* 11: 84–92.

Wallace, A., Alexander, G.V., and Chaudhry, F.M. (1977). Phytotoxicity of cobalt, vanadium, silver and chromium. *Commun Soil Sci Plant Anal* 8: 751–756.

Wang, Y.M., Yang, Q., Xu, H. et al. (2020). Physiological and transcriptomic analysis provide novel insight into cobalt stress responses in willow. *Sci Rep* 10: 2308.

Wilde, E.W. and Benemann, J.R. (1993). Bioremoval of heavy metals by the use of microalgae. *Biotechnol Adv* 11: 781–812.

Wilson, S.B. and Nicholas, D.J.D. (1967). A cobalt reguirement for non-nodulated legumes and for wheat. *Phytochemistry* 6: 1057–1066.

Yadav, D. and Khanna, S. (1988). Role of cobalt in nitrogen fixation: a review. *Review* 9: 180–182.

Yamada, K. (2013). Cobalt: its role in health and disease. *Met Ions Life Sci* 13: 295–320. https://doi.org/10.1007/978-94-007-7500-8_9, PMID: 24470095.

Yamaguchi, T., Tsukada, C., Takahama, K. et al. (2019). Localization and speciation of cobalt and nickel in the leaves of thecobalthyperaccumulating tree Clethra barbinervis. *Trees* 33: 521e532.

Yao S, Lyu S, An Y, Lu J, Gjermansen C, Schramm A. (2018). Microalgae-bacteria symbiosis in microalgal growth and biofuel production: A review. *J Appl Microbiol* 126: 359–368. doi: https://doi.org/10.1111/jam.14095.

Younis, M. (2011). Soil applied cobalt and copper alter the rhizobium symbiosis and growth status of lablab purpureus. *J Plant Nutri* 34: 1736–1745.

Zora, S., Lakhvir, S., Singh, Z., and Singh, L. (1993). Effect of cobalt ions on floral malformation, yield and fruit quality of dusheri mango (*Mangifera indica* L.). *J Horti Sci* 68: 535–540.

Zora, S., Lakhvir, S., Arora, C.L. et al. (1994). Effect of cobalt, cadmium, and nickel as inhibitors of ethylene biosynthesis on floral malformation, yield and fruit quality of mango. *J Plant Nutrit* 17: 1659–1670.

9

Silicon Uptake, Transport, and Accumulation in Plants

Shivani Sharma[1], Muntazir Mushtaq[2], Sreeja Sudhakaran[1], Vandana Thakral[1,3], Gaurav Raturi[1,3], Ruchi Bansal[1,3], Virender Kumar[1], Sanskriti Vats[1], S. M. Shivaraj[1], and Rupesh Deshmukh[1,4]

[1] *Department of Agri-Biotechnology, National Agri-Food Biotechnology Institute, Sahibzada Ajit Singh Nagar, Punjab, India*
[2] *School of Biotechnology, Sher-e-Kashmir University of Agricultural Sciences and Technology of Jammu, Chatha, J&K, India*
[3] *Department of Biotechnology, Panjab University, Chandigarh, India*
[4] *Department of Biotechnology, Central University of Haryana, Mahendragarh, Haryana, India*

Introduction

Silicon (Si) is one of the most considerable element in the Earth's crust. Silicon uptake by plants offers many useful effects along with improving pest and pathogen resistance (Ishiguro 2001; Meyer and Keeping 2001), drought tolerance (Lux et al. 2002), heavy metal tolerance (Neumann and Zur Nieden 2001), and enhanced crop yield (Korndörfer and Lepsch 2001). Silicon was also found to provide indirect defence response to herbivores by altering the composition of plant's volatile compounds. Considering its role in plants, International Plant Nutrition Institute (IPNI) announced Si as a helpful element for plant development (www.ipni.net). The farmers started using Si as fertilizer to reinforce crop productivity. The proper utilization of Si was proposed to help sustainable development and climate adapted cropping (Deshmukh et al. 2017; Tripathi et al. 2021).

Not all plant species equally uptake Si from soil (Deshmukh et al. 2020; Mandlik et al. 2020). Primarily based on Si accumulation in leaves, plants constitute high, intermediate, and low Si accumulators. The genetic ability to take up Si is found to be associated with the taxonomical distribution of plant species (Deshmukh et al. 2020; Hodson et al. 2005). The studies have identified Si transporters in rice that helped in understanding Si uptake mechanism in plants and also the observed taxonomical variation in its accumulation (Deshmukh et al. 2020; Ma et al. 2006, 2007). Silicon is absorbed by means of roots in the form of an uncharged monomeric molecule, i.e. silicic acid, and brought up while the solution pH is beneath 9 (Ma and Takahashi 2002; Mandlik et al. 2020). Within the angiosperms, species from the commelinoid monocot orders Poales and Arecales accumulate notably more Si in their shoots while compared with the other species from monocot

Beneficial Chemical Elements of Plants: Recent Developments and Future Prospects, First Edition. Edited by Sangeeta Pandey, Durgesh Kumar Tripathi, Vijay Pratap Singh, Shivesh Sharma, and Devendra Kumar Chauhan.

clade (Deshmukh et al. 2020). In Poales, species from the Gramineae and Cyperaceae families accumulate high levels of Si (Ma and Takahashi 2002; Deshmukh et al. 2020). Maximum plants, mainly dicots, are unable to accumulate high levels of Si in their shoots. The main characteristic in Si accumulation among the species has been attributed to the variation in the Si uptake capacity of the roots (Ma and Takahashi 2002; Deshmukh et al. 2020). Even in high accumulators, Si accumulation differs among different tissues. Some plants have specialized tissues and cells that get heavily silicified (Dabney et al. 2016; Mandlik et al. 2020; Tripathi et al. 2020). The study of such specialized cells and tissues may be beneficial to understand Si-derived benefits in plants. In the present chapter, we have discussed the molecular mechanism involved in Si uptake, accumulation, and benefits associated with it.

Molecular Mechanism Involved in Silicon Uptake

In recent times, development has been made to assess the mechanism by which Si is absorbed into the plant system. In plants, Si transport is very complex than in other organisms like unicellular diatoms for the long-distance transport through specific tissues and compartments from the roots to stomata. Rice was the first model crop used to study Si transport as it is a high accumulator of Si. In rice, Si uptake is not affected by transpiration but is inhibited by metabolic inhibitors like 2,4-dinitrophenol (DNP), iodoacetate, and 2,4-dichlorophenoxyacetic acid, then through the respiratory inhibitor sodium cyanide (NaCN).

Seminal Studies Defining Uptake of Silicon in Different Plant Species

All plants that grow and develop in soils absorb Si. The Si content in plant tissues varies significantly among plants, ranging from the concentration of 0.1–10% (w/w) silicon on a dry weight basis (Epstein 1994, 1999; Deshmukh et al. 2020). To observe Si content, the plants are categorized into Si accumulator, intermediate type, and excluder species (Jones and Handreck 1967; Takahashi et al. 1990). The *Oryza sativa* is a characteristic Si accumulator that takes up Si actively (Takahashi et al. 1990; Ma et al. 2001a). Similarly, graminaceous plants like *Triticum aestivum* (Van der Vorm 1980; Jarvis 1987; Rafi and Epstein 1999; Casey et al. 2004), *Lolium perenne* (Jarvis 1987), and *Hordeum vulgare* (Barber and Shone 1966) also uptake Si actively, while certain other plants like *Avena sativa* uptake Si passively (Jones and Handreck 1965). Some dicots, for example, *Cucumis sativus, Cucumis melo, Fragaria ananassa*, and *Glycine max* (Takahashi et al. 1990; Ma et al. 2001b; Mitani and Ma 2005) uptake Si passively, and *Solanum lycopersicum* (Takahashi et al. 1990), *Phaseolus vulgaris* (Jones and Handreck 1967), and *Vicia faba* (Liang et al. 2005) act as Si excluders.

A research study reported that xylem loading of Si was facilitated by a transporter in *O. sativa*, but passive diffusion takes place in *C. sativus* and *S. lycopersicum*. These consequences deduced that a higher density of transporter for radial transport and the feature of a transporter for xylem loading is responsible for the high Si accumulation in *O. sativa* (Mitani and Ma 2005). Although all plants, mainly dicots, are incompetent to accumulate a large quantity of Si in the shoot from the soil. Hence, they do not take advantage from Si.

Silicon Influx Transporter

Bioavailable source of Si in soil is mainly monomeric, uncharged molecule of orthosilicic acid (H_4SiO_4) found in concentrations ranging from 0.1 to 0.6 mM taken up by plant roots from the soil solution (Epstein 1999). Silicon uptake by roots of the plant is primarily attributed to an influx transporter Lsi1, first reported by Ma et al. (2006) in rice. The breakthrough discovery of Si influx transporter by Ma's group led to a deeper understanding of Si accumulation and deposition in higher plants. Lsi1 is a predominant transporter responsible for taking up Si from soil solution into the plant root cells and works in substrate-specific manner. In order to establish Si uptake ability of rice, Lsi1 gene was cloned from a mutant rice plant that has lost Si uptake ability from soil (Ma et al. 2006). The Lsi1 gene from the mutated plant showed change in the nucleotide G (wild type) to A resulting in change in the amino acid from alanine to threonine at 132nd position. Alanine at 132nd position seemed to be a critical residue alteration that lead to a change in pore confirmation simultaneously inhibiting Si transportation via Lsi1. Lsi1 protein comprises 298 amino acid residues and is a member of Nod26-like major intrinsic protein (NIP) subfamily of aquaporins (AQPs) (Luu and Maurel 2005). The deduced structure of Lsi1 gene reveals six transmembrane domains including two NPA (asparagine-proline-alanine) motifs typical of all AQPs. Expression analysis and subcellular localization of Lsi1 was predicted in rice roots. Constitutive expression of Lsi1 was affected by Si concentration as expression of transporter decreases with the increase in Si supply. Lsi1 gene is located at the distal side of both exodermis and endodermis of rice roots, which helps in Si uptake from soil and further allowing its passage from aerenchyma to steel (inhibited by Casparian strip), leading to accumulation of Si in shoots (Ma et al. 2006; Mandlik et al. 2020).

Homologs of Lsi1 have been recently found in various Si accumulating crops including both monocots and dicots like barley, maize, pumpkin, strawberry, and soya bean (Figure 9.1). The rice Si influx transporter, OsLsi1, and the silicon influx transporter protein isolated from these crop organisms showed a high amino acid sequence similarity. Barley (*H. vulgare*) Si influx transporter HvLsi1 showed 81.1% sequence similarity with rice OsLsi1. HvLsi1 also found to have structure and motifs similar to OsLsi1. Expression of HvLsi1 was analyzed in Lsi1 mutated rice plant, which when transformed with HvLsi1 showed high Si influx activity similar to OsLsi1 confirming its role as a Si transporter. Unlike OsLsi1 the expression pattern of HvLsi1 is not affected by Si supply and shows distinct cellular specificity and localization. HvLsi1 is found to be localized only on the distal side of epidermal and cortical cells of seminal roots. The polarity of HvLsi1 on distal side of the cells showed its role as an influx Si transporter. In addition to this, genetic variations in Si uptake ability was found in different barley species. These variations may be attributed to Si uptake capability of roots of different species that varies depending on HvLsi1 expression level or may be controlled by other Si transporter like Lsi2, which is reported to export Si out of the cell (Chiba et al. 2009).

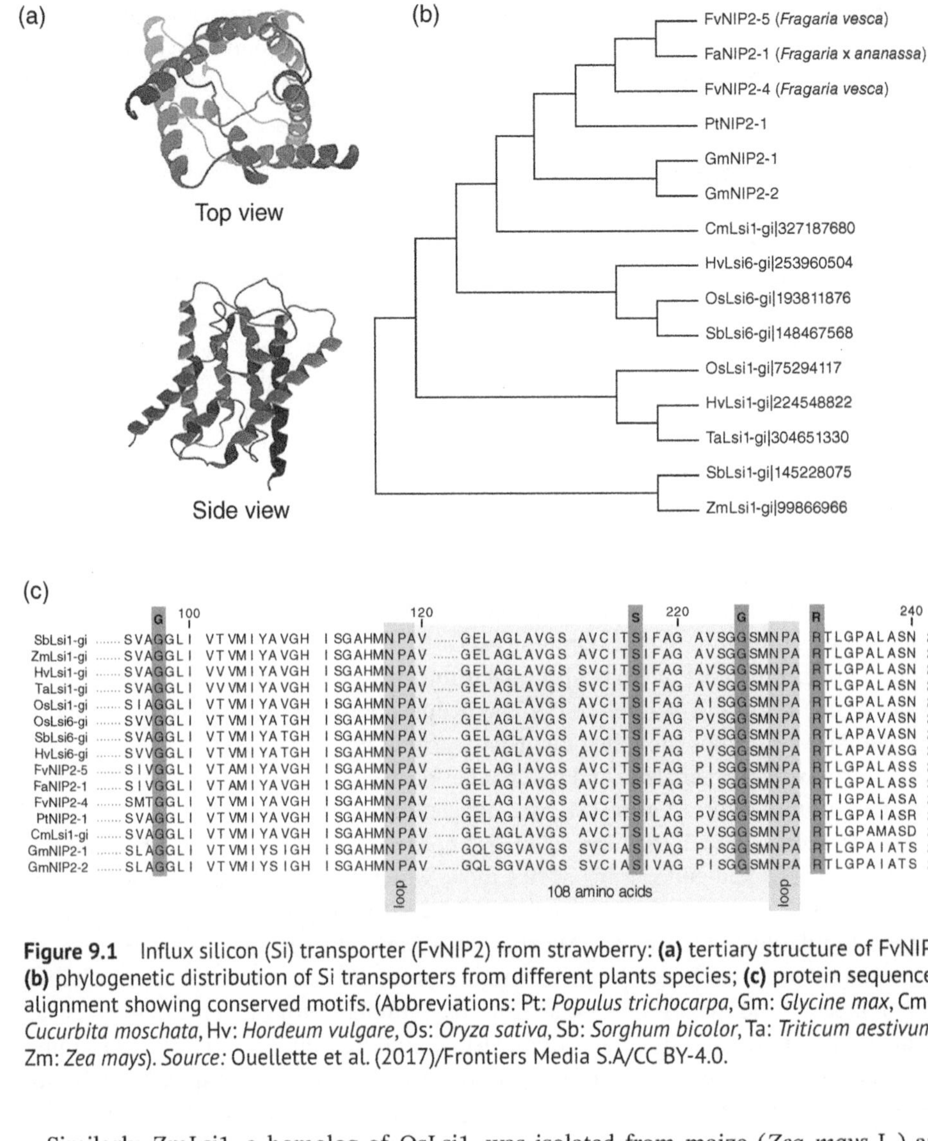

Figure 9.1 Influx silicon (Si) transporter (FvNIP2) from strawberry: **(a)** tertiary structure of FvNIP2; **(b)** phylogenetic distribution of Si transporters from different plants species; **(c)** protein sequence alignment showing conserved motifs. (Abbreviations: Pt: *Populus trichocarpa*, Gm: *Glycine max*, Cm: *Cucurbita moschata*, Hv: *Hordeum vulgare*, Os: *Oryza sativa*, Sb: *Sorghum bicolor*, Ta: *Triticum aestivum*, Zm: *Zea mays*). *Source:* Ouellette et al. (2017)/Frontiers Media S.A/CC BY-4.0.

Similarly, ZmLsi1, a homolog of OsLsi1, was isolated from maize (*Zea mays* L.) and showed 83% sequence similarity with OsLsi1. Like HvLsi1, ZmLsi1 shares structural and functional similarity with OsLsi1. Silicon influx capability of ZmLsi1 was investigated by expressing the transporter protein in *Xenopus* oocytes. ZmLsi1 expressing oocytes showed higher Si accumulation as compared to control. Similar to barley HvLsi1, the expression pattern of ZmLsi1 did not change due to Si supplementation. Cellular localization of ZmLsi1 was found primarily in epidermal and hypodermal cells of seminal roots and lateral roots of epidermal and cortical cells (Mitani et al. 2009b).

Silicon accumulation in dicots is lower and poorly understood as compared to monocots species. Very few studies have been conducted to understand the mechanism of Si accumulation in dicots. Mitani et al. (2011) did the first systematic study on Si influx transporter in dicots where they compared Si accumulation capability of two Cucurbita: pumpkin cultivars namely Shintosa and Super-unryu. Isolated CmLsi1 protein-coding sequences of both cultivars share 58% sequence similarity with OsLsi1. CmLSi1 of both cultivars contains two conserved NPA motifs with 108 amino acid residues distance and G/S/G/R selectivity filter (SF) which is important for substrate specificity of the transporter (Deshmukh et al. 2015). However, out of these two, Shintosa cultivar showed higher Si uptake capability than Super-unryu cultivar. This difference is attributed to single amino acid substitution at proline at 242 position to leucine in CmLsi1 of Super-unryu cultivar that resulted in loss of its Si uptake capability. Proline at 242 position plays a key role in protein folding, substitution of which leads to altered protein localization in the cell. The same allelic mutation was observed in seven other pumpkin cultivars out of which four cultivars showed the same mutation at position 242 and hence accumulates lower level of Si in their tissues. Pumpkin is the only species that has natural genetic variation in Si transporter protein sequence causing difference in Si accumulation. In rice, these variations are generated by mutating the gene using gamma irradiation or using a chemical mutagen like ethyl methane sulphonate (EMS)(Chiba et al. 2009; Yang et al. 2018). Unlike Si influx transporters in monocots, expression of CmLsi1 is observed in both roots and shoots, localized at distal side in endodermal and exodermal cells.

In soya bean, Deshmukh et al. (2013) predicted in silico Si influx transporter protein using comparative genomics using both Si accumulators and non-accumulators as query sequence. Subsequently, two genes, *GmNIP2-1* and *GmNIP2-2* were characterized whose protein sequence features were similar to NIPs of accumulators, inferring their ability to transport Si. In silico studies were further supplemented by cloning of coding sequences of both genes and their expression analysis in *Xenopus laevis* oocytes where they tested positive for Si permeability. Similar to OsLsi1, *GmNIP2-1* and *GmNIP2-2* expression level decreased in presence of high Si concentration confirmed using qRT-PCR of roots grown with or without Si. Further, the coding sequence of both proteins was found to be conserved and have a similar level of expression among different cultivars of soya bean. However, in presence of Si few cultivars like Hikmok sorip showed increased Si accumulation. The reason for this variation in Si uptake capability among different cultivars in soya bean is still not known.

Silicon Efflux Transporter

Silicon absorbed by plant roots and transported to different plant tissue parts via Si influx and efflux transporters. A seminal study by Ma's group identified Lsi2 (low silicon 2), a previously uncharacterized gene in *O. sativa* (rice) with Si efflux ability (Ma et al. 2007). Lsi2 is an active transporter and belongs to an uncharacterized anion transport family. Computational analysis such as BLAST and Clustal W revealed that Lsi2 has no similarity with Lsi1. Lsi2 transport activity was examined through *Xenopus oocyte* assay and found that it shows an efflux transport action and is capable of transporting Si out of the plant cell, specifying that Lsi2 is a Si efflux transporter in plants. Further, activity of Lsi2 is increased with a decrease in external pH, suggesting that Lsi2 is energy dependent. In contrast, the efflux of Si is decreased by low temperature treatments and the application of

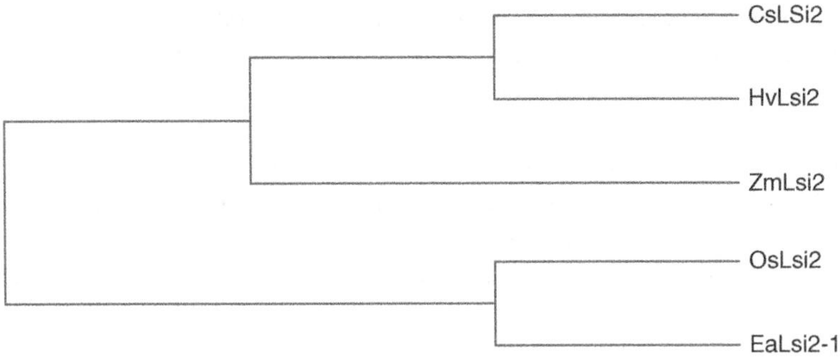

Figure 9.2 Phylogenetic distribution of Lsi2 homologs in different plant species.

proton translocators like DNP, carbonyl cyanide m-chlorophenyl-hydrazone (CCCP), and carbonyl cyanide p-trifluoro-methoxyphenyl hydrazone (FCCP) (Ma et al. 2007). Identification and characterization of Lsi2 revealed the mechanism behind the efflux of Si, but very less is known about the structural and functional features of Lsi2.

After the prime discovery of Si efflux transporter in rice, studies have been carried out to find such transporters in other plants (Figure 9.2). Later, OsLsi2-like Si efflux transporters were identified in some monocot plants like *H. vulgare* and *Z. mays*, belonging to the Graminaceae family (Mitani et al. 2009a). They show an 80% similarity with OsLsi2. Expression in *Xenopus oocyte* shows efflux transport activity in both HvLsi2 and ZmLsi2, suggesting that both share functional similarity with OsLsi2 and are Si efflux transporters. Furthermore, HvLsi2 is suggested to be an important Si efflux transporter in *H. vulgare*, and it also has an ability to improve uptake of Si in Si-efflux defective rice mutants. Both barley and maize are high accumulators of Si, and HvLsi2 and ZmLsi2 are the transport proteins responsible behind the active transport of Si (Ma and Takahashi 2002; Mitani et al. 2009a).

In dicots, OsLsi2 efflux transporter homologs were characterized from *Cucurbita moschata* (Mitani-Ueno et al. 2011) and *C. sativus* (Sun et al. 2018). Mitani-Ueno and his co-workers isolated and characterized CmLsi2-1 and CmLsi2-2 efflux transporters in *C. moschata* cultivars. These are the first efflux transporters of Si identified in eudicots. The efflux transport activity of CmLsi2-1 and CmLsi2-2 were analyzed by their heterologous expression in *Xenopus oocyte* and their role in Si efflux was confirmed (Mitani-Ueno et al. 2011). Compared with other eudicots, cucumber plants accumulate more Si, and thus it is commonly used as a dicotyledon model for Si accumulation studies. Recently, Sun and co-workers identified and characterized a Si efflux transporter gene, *CsLsi2*-1 in *C. sativus*, which helps in understanding the molecular mechanism behind Si uptake and accumulation in cucumber (Sun et al. 2018). The analysis showed that CsLsi2-1 is an energy-dependent transporter that shows efflux activity but lacks influx activity for Si. Computational analysis showed that CsLsi2-1 has a higher similarity with pumpkin Si-efflux transporters (Sun et al. 2018).

Among land plants, *Equisetum arvense* is considered as one of the main Si accumulator with highest ability to accumulate Si, and due to this, it is also used as a model for Si mineralization studies in plants (Bauer et al. 2011). Two unique Si-efflux transporters, EaLsi2-1 and EaLsi2-2, were discovered and characterized in horsetail (Vivancos et al. 2016). Heterologous expression of EaLsi2-1 and EaLsi2-2 in *Xenopus oocyte* showed efflux activity for Si. Horsetail Lsi2s share less than 50% similarity with other Lsi2s identified from other higher plants. The mechanism behind the higher accumulation of Si is the combined action efflux transporters and influx transporters present in horsetail. For understanding the benefits of Si and molecular mechanism of Si uptake in plants, we need to explore the efflux transport activity of Si in higher plants.

Cordial Activity of Silicon Influx and Efflux Transporter

In plants, Si uptake and transportation mechanism is governed by the combined and cordial activity of Si influx and efflux transporters. Although function of Si transporters is similar, their localization, expression pattern, and mode of activity vary among different plant species. The initial report of the identification of influx and efflux transporter genes for Si in plants was reported in *O. sativa* (Ma et al. 2006, 2007). In rice, OsLsi1 and OsLsi2 are mostly expressed in root cells. OsLsi1 is expressed in primary and secondary roots; however, in root hairs it is absent, suggesting that root hairs do not take part in Si uptake (Ma et al. 2002; Kaur and Greger 2019). The influx transporter, OsLsi1, is located on the distal side plasma membrane of root endodermis and exodermis cells. Whereas efflux transporter OsLsi2 is located on the proximal side of cells in endodermis and exodermis (Figure 9.3). OsLsi1 helps to get Si in its available form silicic acid from the surrounding soil into the root cells. The apoplastic transport of Si up to root stele is blocked by Casparian

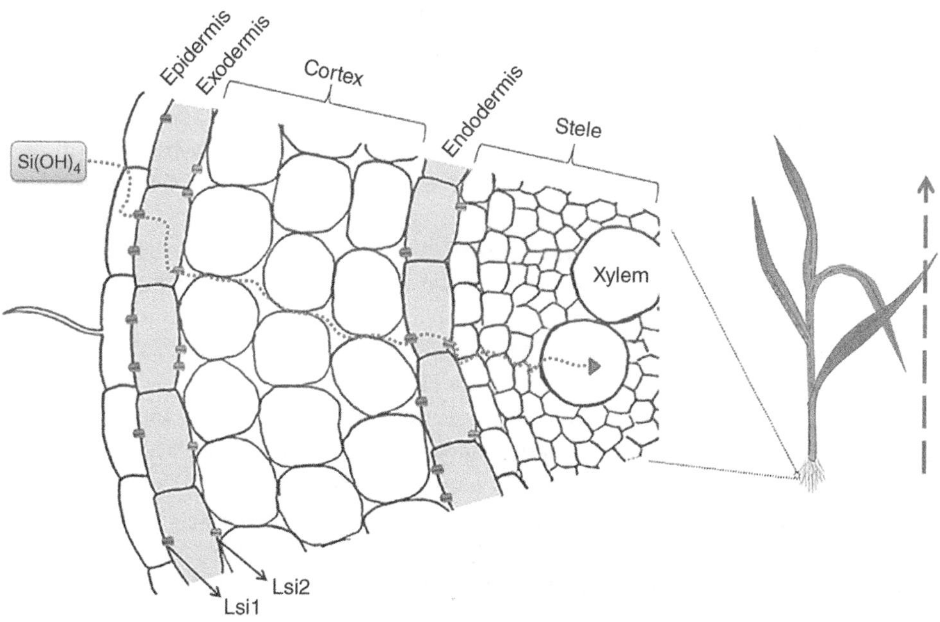

Figure 9.3 Cordial activity of silicon influx and efflux transporter involved in silicon uptake and transport in rice.

strips present on exodermis and endodermis cells, while the combined action of OsLsi1 and OsLsi2 transporters transport Si up to the root stele. OsLsi1 present at the distal side of exodermis cell membrane enables the Si influx into the exodermis, and OsLsi2 present at the proximal side releases it into the cortex. Si travels across the aerenchyma cells and reaches the endodermis through the apoplastic pathway. Then Si enters the endodermis cells through OsLsi1 and exits into the stele through OsLsi2. The efflux action of OsLsi2 is energy-dependent driven by the proton gradient. OsLsi1 and OsLsi2 localize in rice root cells help to get Si from the soil and translocate to xylem that further takes it to leaves through transpirational flow through the xylem. Another influx transporter OsLsi6 located on the adaxial side of xylem parenchyma cells in the leaf blades and sheaths is responsible for the transport of silicic acid from xylem parenchyma cells to other cells of the leaves. Mutation in OsLsi6 does not affect the activity of Si uptake through root; however, it may affect the deposition pattern of Si in leaf cells (Ma et al. 2006; Ma et al. 2007; Feng et al. 2011).

Homologs of rice Lsi1 and Lsi2 were identified in barley and maize (Chiba et al. 2009; Mitani et al. 2009a, 2009b). Rice Si transporters were localized in exodermis and endodermis cells with a specific polarity. But in barley and maize, Si influx transporters, HvLsi1 and ZmLsi1, are localized on the distal side of epidermis, hypodermis, and cortex layer cells (Chiba et al. 2009; Mitani et al. 2009b). Whereas Si efflux transporters HvLsi2 and ZmLsi2 are located only at endodermal cells without any specific polarity (Mitani et al. 2009a). From the soil, silicic acid is taken up into the root cells via HvLsi1 and ZmLsi1, and then it is exported to endodermis through symplastic pathway. Then silicic acid is released from endodermal cells by ZmLsi2 and HvLsi2 to the stele. Similar to rice, silicic acid is transported to the shoot via transpirational flow. ZmLsi6 is a Si influx transporter that is localized in the xylem parenchyma cells and is responsible for xylem unloading in maize (Mitani et al. 2009b). The structural difference in root anatomy also affects the mode of uptake system. In rice, well-developed aerenchyma cells are present between the exodermis and endodermis, and Casparian strips are present in both exodermis and endodermis. In contrast, barley and maize lack well-developed aerenchyma cells, and Casparian strips are only present in the endodermis cells of barley and maize under non-stressed conditions (Feng et al. 2011).

Pumpkin influx transporters, CmLsi1s, are localized in all root cells. Further, the subcellular localization studies indicated the localization of bloomless and bloom pumpkin transporters at the endoplasmic reticulum and plasma membrane, respectively (Mitani et al. 2011). Contrary to rice, barley, and maize Lsi2s that are expressed mainly in roots, pumpkin Lsi2s are expressed in roots as well as shoots (Mitani-Ueno et al. 2011). Further, cucumber CsLsi1 is localized on the plasma membrane of cortical cells and endodermis cells in the root tips. CsLsi1 present on the endoderm shows polarity and is localized in the distal side, whereas in cortical cells it has no polarity (Sun et al. 2017). Similar to Lsi2s present in barley and maize, CsLsi2 is localized on the plasma membrane of the root endodermal cells, with no polarity (Sun et al. 2018). The influx transporter CsLsi1 takes up silicic acid from the soil. CsLsi1 facilitates the transport of Si across cortical cells through the symplastic pathway. From the root endodermis, Si is released into the stele via CsLsi2. Thus, it is evident that the combined activity of Lsi1 and Lsi2 mediates silicic acid uptake and transport across the plant cell. Although the transporters behind xylem unloading is yet to be identified (Sun et al. 2017; Sun et al. 2018; Kaur and Greger 2019). In dicots the exact mechanism of Si uptake and accumulation is poorly understood, and the transporters for xylem unloading are still not known.

Other Homologs of Silicon Influx and Efflux Transporter

In *O. sativa*, the two major transporters, OsLsi1 and OsLsi2, are localized on roots (Ma et al. 2006; Ma et al. 2007). Subsequently, another channel protein, OsLsi6, and efflux transporter, OsLsi3, have been discovered in *O. sativa*. The OsLsi6 and OsLsi3 are known to regulate the distribution of Si to aerial plant parts (Yamaji et al. 2008; Yamaji et al. 2015). More than 90% of the Si absorbed by roots in *O. sativa* is translocated to shoots (Ma and Takahashi 2002). The characterization of transporters defined the movement of Si from soil solution to different aerial plant tissues. The OsLs2, OsLsi3, and OsLsi6 showed higher expression at node I of the shoot. The coordinated function of all three transporters is necessary to mediate the transfer of the Si from the xylem of an enlarged vascular bundle (EVB) to xylem of diffused vascular bundles (DVBs) at node I and subsequently, Si is transferred to panicle. The OsLsi6 belongs to the NIP III subgroup of aquaporins and is a homolog of OsLsi1. OsLsi6 located in the xylem transfer cells unloads the Si from EVB. Later Si moves symplastically through plasmodesmata to the adjacent layer of bundle sheath cells. In bundle sheath cells, Lsi2 that is localized at the distal side releases Si towards the apoplast, which then enters xylem of the DVB. The residual Si is transported through symplast to adjacent cell layers containing OsLsi3 that pumps it into the xylem of DVB (Yamaji et al. 2008; Yamaji et al. 2015). From DVB the Si is targeted to the panicle.

Silicon Transporters yet to be Discovered

In *O. sativa*, the Lsi1 gene located in the root endodermis takes up the Si from aerenchyma, and Lsi2 releases it into the stele (Ma and Yamaji 2015). The further loading of Si into the xylem is not understood yet. Several studies were undertaken to decipher the mechanism behind the xylem loading of Si. The concentration of Si has been measured in the xylem sap of *O. sativa* grown on 0.5 mM Si containing nutrient solution. The concentration of Si in xylem sap reached 6 mM in 30 minutes and subsequently, 18 mM in 8.5 hours (Mitani and Ma 2005). A higher concentration of Si in the rice xylem sap compared to the external solution indicated energy-dependent transport against a concentration gradient involving a transporter.

In rice, the Si from the xylem is either accumulated in the leaves or diverted to the panicle through the node I. In the node I, the function of the transporters, Lsi6, Lsi2, and Lsi3, results in the transfer of Si to panicle. In the panicle, a higher level of Si was detected in the husk as compared to grains (Yamaji et al. 2015). The mechanism underlying the distribution of Si in the panicle and restricted accumulation of Si in the husk is not yet understood completely.

Silicon Deposition in Different Tissues

Silicon Deposition in Roots

There are a number of plants mainly from grasses that accumulate the Si into root endodermis cell such as rice, wheat, bamboo, and sorghum. Among all these species the sorghum has a unique mode of solid Si accumulation. It accumulates Si inside the shape of dome aggregates having 10–15 μm diameters on inner tangential endodermal barriers. The pattern of Si deposition is basipetal, which means Si content is high at basal parts of root and decreases towards the apical parts. In case of deeply penetrated aggregates, the silicification may start at an early stage. At beginning the silica aggregates may develop with endodermal secondary wall formation and can penetrate to various degrees into secondary wall. On the other hand, silicification may start at later stages of wall thickening, and in this case mature aggregates would not penetrate deeply into completed wall (Lux et al. 2003).

In date palm, the silica aggregates/phytoliths accumulate in a special kind of small cells known as stegmata. The location of stegmata is exclusively found with the sclerified bundles of fibres in the cortex. Such bundles are mostly present in lateral and adventitious roots but rarely present in primary seminal roots. The individual bundles with 2–4 fibres are established in thinnest laterals (Figure 9.4). The SEM-EDX analysis of date palm explains the pattern of distribution in roots (Figure 9.5). Unlike the grasses wherein accumulation of Si is commonly associated with epidermal tissue, in palm the stegmata cells containing Si phytoliths are organized in a row of cells and located on the outer surface of sclerenchyma bundles. The distance between each phytolith is about 10–12 μm and has average size of 6–8 μm (Bokor et al. 2019).

Silicon Deposition in Shoot

Silicon also deposits in shoot and other plant tissues as amorphous silicates or phytolithic opal, where it can regulate the transport and imbalance of different minerals. Hu et al. (2018) found that the Si accumulation in shoot is responsible for downregulation of phosphorus transporter gene expression. The experiment was carried out using wild type and Lsi1 mutant of rice. The results clearly suggest that in wild type variety Si accumulated in shoot downregulates the expression of PT6 (phosphorus transporter). A similar kind of study carried out by Hurtado et al. (2020) suggests that the accumulated Si can alleviate the harmful effect of NaCl by decreasing Na^{+}/K^{+}, Na^{+}/Ca^{2+}, and K^{+}/Ca^{2+} ratios.

Guerriero et al. (2019) studied hemp for the silicification in shoot. The results of this study revealed the deposition of Si in the hypocotyls of hemp. In the presence of enriched Si, the deposition also occurred in xylem vessel and bast fibres. When the plant is supplemented with sodium metasilicate, the deposition of the Si in bast fibre is to be more evident in the distal side of bast cell wall and facing the stem cortex. This accumulation may be continuous on distal side and impregnation on cell wall with silica results in the formation of ring-like structure that provides strength to the whole stem. This association of Si with bast fibre in stem may have a couple of benefits in terms of plant physiology and industrial application. It can provide strength to whole stem that ultimately provides resistance against mechanical stress (Guerriero et al. 2019).

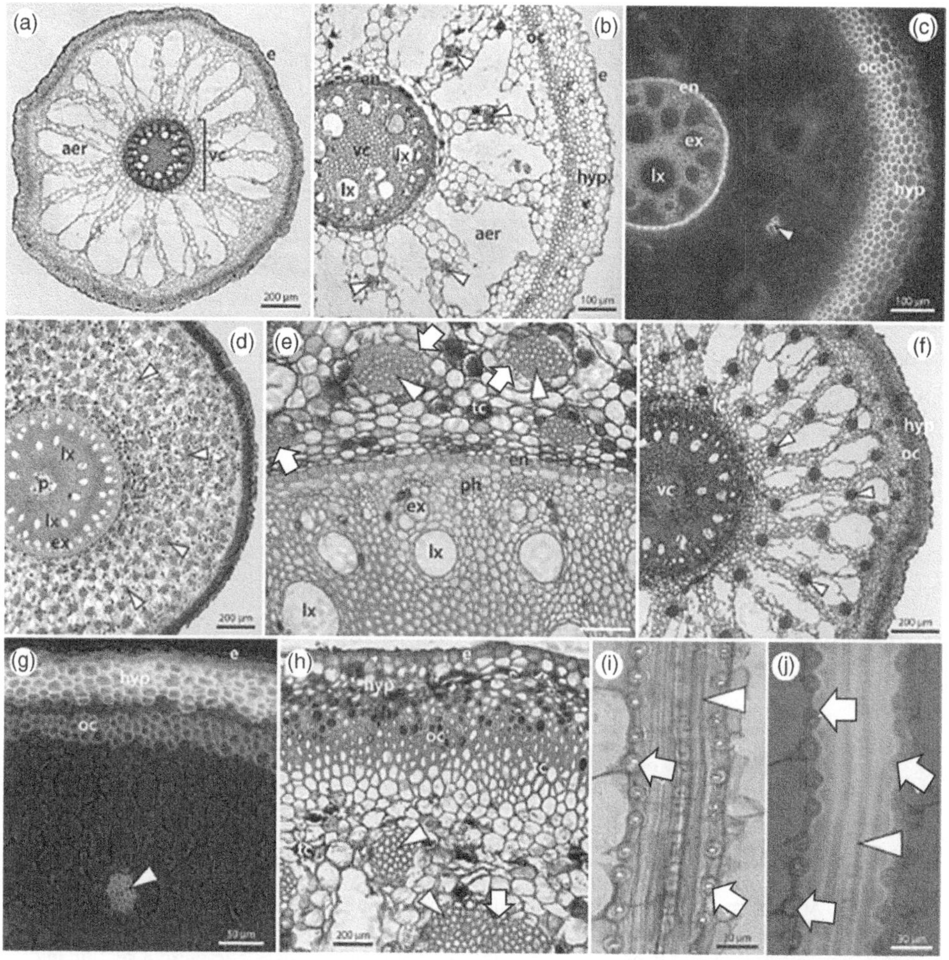

Figure 9.4 Root anatomy for representation of specialized Si-accumulating cells (Stegmata) of datepalm. **(a)** Transverse section of seminal root covered with epidermis (e) (rhizodermis) and composed hypodermis (hyp). The sclerified pith is located in the centre surrounded by polyarch vascular cylinder (vc) with 11 alternative xylem and phloem poles. Extensive aerenchyma (aer) develops by mid-cortex surrounding the VC. **(b,c)** Transverse section of lateral roots. Triangles represents fibre bands. Extensive aerenchyma (aer) is surrounded by multiple cell layers of (e) and (hyp). Endodermis (en) with thick U-shaped inner tangential walls. Broad late metaxylem vessels (lx) are shifted centripetally from the xylem poles formed by early metaxylem vessels (ex). **(d–j)** Arrowheads represent the fibre band that may count from dozens to few hundreds. Arrows represent the stegmata with silica phytoliths that are exclusively attached to fiver band **(i, j)**. Abbreviation: tannins (tc), late metaxylem (lx), alternating xylem (ex), phloem (ph), pith (p), extensive aerenchyma (aer), composed hypodermis (hyp), vascular cylinder (vc), and endodermis (en). *Source:* Reproduced with permission from Bokor et al. (2019)/Frontiers Media S.A./CC BY 4.0.

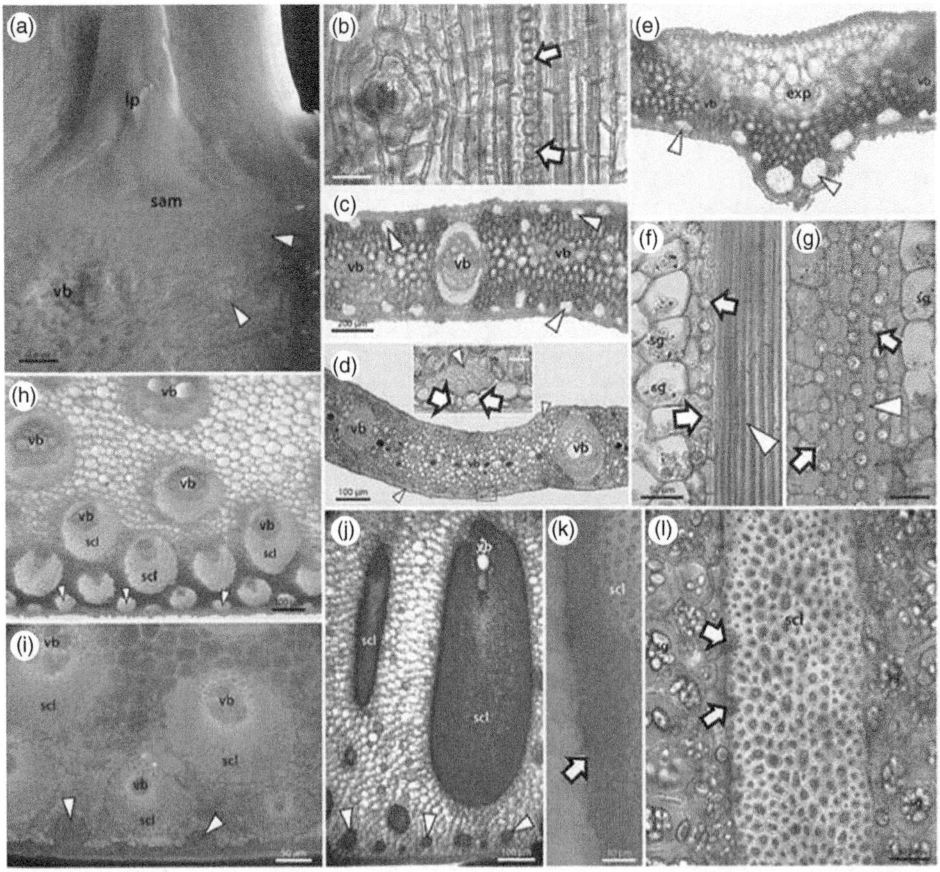

Figure 9.5 Accumulation pattern of silicon in leaf of datepalm. **(a)** Stereomicroscope image of the shoot apex representing the leaf primordia (lp), shoot apical meristem (sam), vascular bundles (vb), and the triangles are representing fibre bands. **(b)** Unstained young leaf's paradermal section with stomata (st) and the arrows representing the subepidermally located stegmata. **(c,e)** Unstained adult leaflet (from 10 year old) triangles arrowheads are representing fibre bands present in both abaxial and adaxial side **(c)**. **(d,f,g)** Adult leaflet (from 10 year old) and the triangles are representing fibre bands and white arrows stegmata. **(h,i)** leaf petiole **(h)** under white light and **(i)** under UV light. **(j–l)** 10 year old plant under white light. Abbreviation: leaf primordia (lp), shoot apical meristem (sam), vascular bundles (vb), stomata (st), expanding tissue of large parenchyma cells (exp), sclerenchyma (scl), and starch grains (sg). *Source:* Reproduced with permission from Bokor et al. (2019)/Frontiers Media S.A./CC BY 4.0.

Silicon Deposition in Leaves

Leaves are generally regarded as the most important structures for studying Si accumulation in plants. They contain different types of silicified cells such as trichomes, silica cells, and stomata. Guerriero et al. (2019) investigated the pattern of silica accumulation in hemp leaves. The results revealed a high accumulation of Si (amorphous opaline silica) at the base of hemp trichome that provided mechanical support to trichome shaft. The Si accumulated in tissues can be quantified by a method that involved wet digestion of tissue and subsequent determination of Si by colorimetric analysis. Although it is rapid and alleviates the need for costly instruments, this method sometimes reported providing highly variable and inconsistent results. Kraska and Breitenbeck (2010) identified the cause of this variability and modified the wet digestion method. The study identified that the addition of ammonium fluoride leads to release of polysilicic acid for a colorimetric determination that stabilizes within six minutes and remains stable up to five hours and provides reproducible results.

The total Si content present in plants is directly associated with rootstock (Coskun et al. 2019). The quantitative differences of Si deposition among accumulators are based on different agronomic traits such as size of the leaf, presence of different kinds of silica cells or phytoliths, growth condition, root architecture, and rooting media such as type of soil, pH, and hydroponics. The deposition of silica in soil occurs in silica cells, trichomes, and papillae. The deposited Si in leaves can provide resistance to fungal pathogen and insects and can also a play role in terms of biochemistry of the plants (Coskun et al. 2019).

Phytoliths: Biochemical Composition and Deposition Patterns

Silica precipitation and polymerization gives rise to microscopic, amorphous, intracellular, and intercellular structures called phytoliths. This process is termed as biosilicification (Nawaz et al. 2019). Biosilicification is widespread in plants and related studies have gained momentum recently (Katz 2018). At neutral pH and under ambient temperature and pressure, silica precipitates in plants, with evaporation and water metabolism playing a larger role. (He et al. 2014). Deposition of Si occurs both in the cell cytoplasm and vacuoles; however, intercellular phytoliths are deposited throughout the plant. Besides silica, non-salicaceous phytoliths are also produced by some plants that are formed of calcium oxalate (CaOX), calcium carbonate, calcium sulphate, calcium phosphate, magnesium oxalate, stronium oxalate, and stronium and barium sulphate.

Higher plants and other photosynthetic organisms contribute to overall Si cycling, and several reports revealed that biosilicification affected the overall Si cycle throughout the geological period. Terrestrial Si cycle has been mainly affected by flora, predominantly tree species, and grasses through absorption and returning back via litterfall/plant death. Thus, in terrestrial ecosystems, pedogenesis and its relationship to the plant community and variety play a key role in Si dynamics. Therefore, the global Si cycle is regulated by terrestrial biogeochemical Si cycle, which also determines deposition of silica within living plant tissues. It further spotlights the significance of considering Si merely an element existing within plants on geological and biogeochemical scale. Silicon is absorbed by the plants as silicic acid, together with water, and gets circulated to different plant parts by transpiration stream via vessels and ducts. Further, silicic acid after being concentrated in various plant tissues in the form of solid hydrated silica is precipitated as phytoliths. When there is a lack of water, these phytoliths become resolubilized and are expelled from the plant bodies, where they join the silicic acid flux into the oceans. Phytolith deposition in the ocean is a complex process involving different known and unknown means. It is crucial to understand phytolith bioavailability and the process that leads to it so that they may be employed in paleobotanic and archaeological record examinations, as well as crop improvement.

Silicon Deposition and the Phytolith Formation

Except for specialized silica cells, biosilicification is driven by cell development and tissue maturation, and it is anticipated that there is little or no Si deposition in the early developmental phases of juvenile organs. In the case of oats (*A. sativa* L.), silica polymerization occurs in stages, with silicic acid being permanently converted to amorphous silica (SiO_2*nH_2O). Silica is deposited as a non-crystalline structure called silica gel, which develops into amorphous solid formations. Long and short cell sequences emerge during the developmental process of internode intercalary meristem cells; the short cells can evolve into a pair of cork and silica cells, stomatal cells, or trichomes. For silicification, a pair of cells with identical morphologies (silica and cork cell) emerge, which then differentiate relative to size and shape. When ryegrass (*Lolium temulentum*) cells differentiate, they undergo fast changes such as vacuole enlargement and a rise in the number of rough endoplasmic reticula and mitochondria. The disappearance of the nucleus of silica cells is accompanied with the formation of Golgi bodies and other tiny vesicles. Fibrillar elements also form from damaged nuclei and other cytoplasmic materials and spread into the cell lumen, strengthening the cell wall around the cork and silica cell. It implies that metabolic activities influence silica deposition or phytolith production in general. Minerals produced in silicified cells are localized between the cell wall and the cell membrane in this situation. The silicification process has been investigated in case of sorghum as well, where transpiration is not solely responsible for silicification, since only young leaves could accumulate silica in cells; hence, some kind of biological process might be governing the local silica polymerization. Additionally, if silicification is determined by transpiration only, there should be deep silicification of stomatal guard cells and continuous transpiration from leaves accumulating silica in their cells; hence, it should occur even after the leaf has fully developed. Surprisingly, this never happens because guard cell silicification is a delayed process that advances with age, indicating that mineralization is not a uniform process across all cell types. Because tissue maturation is followed by silicification, it is critical to link it to the developing leaf tissues and cells, particularly the silica cells. There is a tissue maturation gradient in a developed leaf, which should also be interpreted as a silicification gradient. The silicic acid reaches silica cells, and this is maintained by the transpiration stream and Si transporters and is deposited intercellularly as silica. Kumar et al. (2017) and Roychoudhury and Tripathi (2019) demonstrated silica biogenesis in sorghum silica cells governed by siliplant1 protein (Slp1). This was proven in a previous study on sorghum by the same group of researchers, who revealed that silica cells are still viable in non-silicified settings and may die in such situations that firmly signify that "silica cells lose their nucleus and cytoplasm, become empty and later get silicified passively" (Kumar et al. 2017). It is clear from this study that silicification in silica cells of sorghum depends on the Slp1 protein, whereas in case of other cells, Slp1 protein does not play any role in silicification process.

Although tissue maturation and the Slp1 protein regulate silicification in silica cells, other cell types' lumens still silicify (Motomura et al. 2002). First, it has been revealed that silica deposition varies according to the cell type (Motomura et al. 2004). Second, each cell type in various tissues accumulates a variable quantity of silica, which persists in some cell types even after maturation of leaves such as bulliform cells, micro-hairs, and prickle hairs in the aging leaf tissues of bamboo grass (*Sasa veitchii*). The silica cells in this species, on the other hand, are exclusively silicified during leaf development and not after maturity

(Motomura et al. 2004). It is fascinating to learn that the basal cells of micro-hairs of grasses, such as bamboo, are silicified prior to their exposure to the outside environment, implying that processes other than transpiration are involved (Sakai and Sanford 1984). Third, the age-dependent micro-hair increase proposes transpirational dependency. Finally, it is unknown whether other types of proteins, besides silica cells in leaves, contribute in the silicification process of cells.

SIp1 is also expressed in inflorescence cells. However, it is unknown whether the protein is exclusively localized in the glume or lemma, and what type of post-transcriptional changes may be enabling silica deposition. Like the micro-hair basal cells of a bamboo leaf, the outer tangential cell wall of papillae and prickle hair tips appears to silicify at the emerging stage. Kumar et al. (2017) postulated that in the event of dead cells, the process can be completed in two steps: at the tips and the cell wall induced by wall materials, followed by spontaneous precipitation of Si inside the lumen produced by cell death and evapotranspiration. It is critical to determine if silicification occurs before or after cell death in "active cells." Further research into how protoplast disintegration affects silica precipitation in the lumen could aid in understanding the transcriptional and post-transcriptional mechanisms that lead to ongoing silicification after cell death.

Kumar et al. (2017)reviewed evidence on different forms of silicification in grasses and established that cell wall silicification can be either spontaneous or direct, while excluding paramural silicification. Organs that are mature and transpire intensively have a constant Si supply that is assisted by transpiration stream. Upon dehydration, this Si gets deposited on non-silicified cell wall passively and beyond metabolic control of the host cells. The controlled cell wall silicification has been a fascinating topic until now. The silica from the cell wall unswervingly adheres onto the carbohydrate material. Rice and horsetail both have been reported to interact with cell wall polymers and Si. Nonetheless, the structure, composition, and architecture of plant cell walls change significantly between cell types and plant species. It has been shown that there is a strong relation between Si content and cell wall polymers, and these polymers aid in Si precipitation. Callose[-(13)-D-polyglucose] causes silica deposition, and other polysaccharides are likely to play a role in cell wall silicification as well. Non-phytolith Si is thought to be deposited on cellulose microfibrils through hydrogen bonding or condensation with cellulose (Maeda et al. 2006). The molecular architectures of templates, according to Hodson (2016) characterize the silicification tendencies of specific structures that can severely silicify themselves. The silica structures are modelled, dissolved, and remodelled during this process.

Frick et al. (2019) established an analytical routine based on femtosecond laser ablation multi-collector inductively coupled plasma mass spectrometry (fsLA-MC-ICP-MS) in situ to assess the Si isotopic composition of a range of phytoliths from various plants. fsLA-MC-ICP-MS is considered as an important tool for screening and characterizing heterogeneous phytolith populations. They established that certain species such as bushgrass, common reed, and horsetail indicated large internal variations amounting up to 2% δ^{30}Si among various phytolith fragments. Apprehending such heterogeneity expands the scope to investigate Si pathway in plants developed in vivo. Furthermore, Si biogeochemical and climate cycles could be explored using phytoliths from geo-archives.

Li et al. (2014) revealed that rice crop considerably influences the phytolith composition (Si, Al, Fe, and C) in various aerial parts (grains, sheath, leaf, and stem) (Figure 9.6). Their findings suggested that rice cultivars and Si, Al, and Fe uptake might affect phytolith

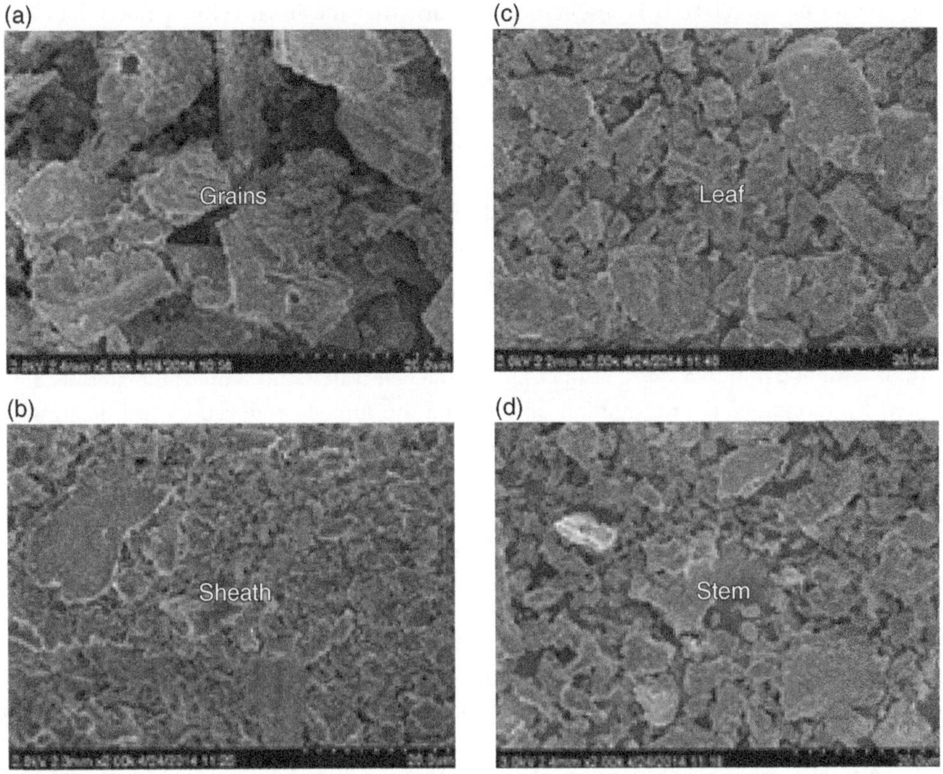

Figure 9.6 Scanning electron micrograph of phytoliths observed in different rice organs. *Source:* Reproduced with permission from Li et al. (2014)/Frontiers Media S.A./CC BY 4.0.

production and their chemical composition. The organic carbon (OC) content of phytoliths in rice cultivars appears to be considerably influenced by Si, Al, and Fe in plants. Authors demonstrated that rice organs and cultivars can largely impact Si dynamics in soil-plant systems via variable phytolith solubility. This is significant for consideration the effect of soil amendment with different plant organs and cultivars on soil OC stock and on dissolved Si delivery as authors demonstrated rice organs, viz. sheath and leaf are both characterized by a higher content of OC occluded in phytolith and higher phytolith solubility in contrast to grains and stems. This investigation reported the significance of investigating the influence of agro-management on the evolution of sinks and sources of Si and C in soils used for Si-high accumulator plants.

Role of Phytoliths in the Silicon Biogeochemical Cycle

The silicate weathering constitutes 70% of the continents and is closely associated with global carbon cycle (Berner and Kothavala 2001). Various studies reveal that weathering of silicate minerals is mainly responsible for Si dissolved in oceans (Struyf et al. 2009). During the Precambrian, oceanic Si was regulated by tectonic, weathering, and hydrothermal inputs (Robert and Chaussidon 2006). Later, in the Phanerozoic era, Si in the oceans was mainly contributed due to organic reactions by organisms secreting silica (Treguer et al. 1995). Falkowski et al. (2004) further showed that during the Paleogene, diversification of grasses and diatoms coincided probably for the reason that Si mobilized by terrestrial plants delivered an enriched nutrient supply to the oceans that were favourable to diatom radiation. On the continents, the contribution of plants to the global cycle of Si is not well known, but it is prominent that Si fixation in plants is opposing Si fixation via oceanic biogeochemical cycle (Conley 2002; Loucaides et al. 2010). Recently, Struyf and Conley (2012) calculated that the yearly export of dissolved and biogenic exports to the coastal zone is 10–40 times less than the annual fixation of dissolved Si into terrestrial vegetation.

Though, Si is not an essential element, it is rather beneficial to plants as it improves the ability to cope with both biotic and abiotic stresses (Guntzer et al. 2012; Shivaraj et al. 2021). Si is absorbed by the plant from the soil solution as silicic acid, which when translocated towards the shoot, is polymerized by evaporation and finally gets precipitated as phytoliths. Unlike many plant nutrients, Si is not subsequently redistributed within the plant, since the phytoliths remain insoluble until plant remains return to the soil and start decomposing. Naturally, dissolved Si is mainly found as $Si(OH)_4$ originating from the dissolution of primary minerals. Dissolved Si passes through a number of pathways in accordance with biogeochemical, environmental conditions: adsorption on the surface of oxides, precipitation as amorphous silica or clay minerals, release to the streams by gravity, or uptake by plants. Particulate Si reaches the soil via dust fallout or litterfall as phytoliths. Very few studies have reported the Si cycle in natural grasslands. Blecker et al. (2006) demonstrated that a large proportion of Si was recycled via litterfall, whereas output and weathering were limited. The impact of grasslands on Si recycling seems to be of a similar degree as forests since the lower biomass is compensated by larger Si concentration and turnover. It is known that grasses store more Si in their shoots in contrast to other species. The grasslands had a twofold increase in mineral dissolution over forests proposing that Si biocycling had an impact over the weathering of minerals to a larger extent in grasslands in contrast to forests(Blecker et al. 2006).

Borrelli et al. (2010) assessed the role of phytoliths on biogeochemical cycle of Si in typical Arguidolls under different vegetations of the Pampean Plain, Argentina. Results revealed that acacia and eucalyptus are not involved in phytolith production, while *Dactylis glomerata* (grass) accumulates silica. Moreover, soil phytolith concentration decreases with depth. In the case of arboreal plots, SiO_2 content in soil was higher and decreased with depth, whereas SiO_2 was lesser and increased with depth likely due to differences in the nutritional supplies and root design between vegetal species, as a result, in the different depth uptake from the soil solution. In groundwater, Si content was recorded as very high. Thus, this investigation revealed that phytoliths are very representative in typical Arguidolls and showed a greater degree of weathering and could contribute to silica content in the soil solution, affecting the silica biogeochemical cycle.

Several studies showed that phytoliths are one of the most essential soluble sources of Si in comparison to other soil minerals regardless of their low soil concentration (Fraysse et al. 2006, 2009; Fraysse et al. 2010). Certainly, out of all soil minerals, phytoliths have the peak dissolution percentage, except at $pH < 3–4$. The dissolution percentage of 0.1–1 of the phytoliths pool could easily describe the dissolved Si values present in surface waters (Fraysse et al. 2010). The higher phytolith solubility can even control the DSi (dissolved silicon) concentration in rivers as proposed for Hawaiian streams. The study was employed to determine the silica release rates of phytoliths from tropical and temperate plants (bamboos, horsetails). The authors compared results with litter degradation of horsetails and pine needles. Findings reveal that the reactivity of phytolith surface is independent of topology and geometry of local structures and disapprove the presence of preferential dissolution sites on surface. Moreover, degradation of litter suggests that the silica release rate does not depend on cellulose hydrolysis that infers the presence of phytoliths in an inorganic Q pool not complexed with organic matter (Fraysse et al. 2006).

References

Barber, D. and Shone, M. (1966). The absorption of silica from aqueous solutions by plants. *Journal of Experimental Botany* 17: 569–578.

Bauer, P., Elbaum, R., and Weiss, I.M. (2011). Calcium and silicon mineralization in land plants: transport, structure and function. *Plant Science* 180: 746–756.

Berner, R.A. and Kothavala, Z. (2001). GEOCARB III: a revised model of atmospheric CO2 over Phanerozoic time. *American Journal of Science* 301: 182–204.

Blecker, S.W., McCulley, R.L., Chadwick, O.A., and Kelly, E.F. (2006). Biologic cycling of silica across a grassland bioclimosequence. *Global Biogeochemical Cycles* 20: 1–11.

Bokor, B., Soukup, M., Vaculík, M. et al. (2019). Silicon uptake and localisation in date palm (phoenix dactylifera) – a unique association with sclerenchyma. *Frontiers in Plant Science* 10: 988.

Borrelli, N., Alvarez, M.F., Osterrieth, M.L., and Marcovecchio, J.E. (2010). Silica content in soil solution and its relation with phytolith weathering and silica biogeochemical cycle in typical argiudolls of the Pampean Plain, Argentina—a preliminary study. *Journal of Soils and Sediments* 10: 983–994.

Casey, W., Kinrade, S., Knight, C. et al. (2004). Aqueous silicate complexes in wheat, *Triticum aestivum* L. *Plant, Cell & Environment* 27: 51–54.

Chiba, Y., Mitani, N., Yamaji, N., and Ma, J.F. (2009). HvLsi1 is a silicon influx transporter in barley. *The Plant Journal* 57: 810–818.

Conley, D.J. (2002). Terrestrial ecosystems and the global biogeochemical silica cycle. *Global Biogeochemical Cycles* 16: 68-61–68-68.

Coskun, D., Deshmukh, R., Sonah, H. et al. (2019). The controversies of silicon's role in plant biology. *New Phytologist* 221: 67–85.

Dabney, C., Ostergaard, J., Watkins, E., and Chen, C. (2016). A novel method to characterize silica bodies in grasses. *Plant Methods* 12: 3.

Deshmukh, R.K., Vivancos, J., Guérin, V. et al. (2013). Identification and functional characterization of silicon transporters in soybean using comparative genomics of major intrinsic proteins in Arabidopsis and rice. *Plant Molecular Biology* 83: 303–315.

Deshmukh, R.K., Vivancos, J., Ramakrishnan, G. et al. (2015). A precise spacing between the NPA domains of aquaporins is essential for silicon permeability in plants. *The Plant Journal* 83: 489–500.

Deshmukh, R.K., Ma, J.F., and Bélanger, R.R. (2017). Role of silicon in plants. *Frontiers in Plant Science* 8: 1858.

Deshmukh, R., Sonah, H., and Belanger, R. (2020). New evidence defining the evolutionary path of aquaporins regulating silicon uptake in land plants. *Journal of Experimental Botany* 71 (21): 6775–6788.

Epstein, E. (1994). The anomaly of silicon in plant biology. *Proceedings of the National Academy of Sciences* 91: 11–17.

Epstein, E. (1999). Silicon. *Annual Review of Plant Biology* 50: 641–664.

Falkowski, P.G., Katz, M.E., Knoll, A.H. et al. (2004). The evolution of modern eukaryotic phytoplankton. *Science* 305: 354–360.

Feng, J., Yamaji, N., and Mitani-Ueno, N. (2011). Transport of silicon from roots to panicles in plants. *Proceedings of the Japan Academy, Series B* 87: 377–385.

Fraysse, F., Pokrovsky, O.S., Schott, J., and Meunier, J.-D. (2006). Surface properties, solubility and dissolution kinetics of bamboo phytoliths. *Geochimica et Cosmochimica Acta* 70: 1939–1951.

Fraysse, F., Pokrovsky, O.S., Schott, J., and Meunier, J.-D. (2009). Surface chemistry and reactivity of plant phytoliths in aqueous solutions. *Chemical Geology* 258: 197–206.

Fraysse, F., Pokrovsky, O., and Meunier, J.-D. (2010). Experimental study of terrestrial plant litter interaction with aqueous solutions. *Geochimica et Cosmochimica Acta* 74: 70–84.

Frick, D.A., Schuessler, J.A., Sommer, M., and von Blanckenburg, F. (2019). Laser ablation in situ silicon stable isotope analysis of phytoliths. *Geostandards and Geoanalytical Research* 43: 77–91.

Guerriero, G., Deshmukh, R., Sonah, H. et al. (2019). Identification of the aquaporin gene family in *Cannabis sativa* and evidence for the accumulation of silicon in its tissues. *Plant Science* 287: 110167.

Guntzer, F., Keller, C., Poulton, P.R. et al. (2012). Long-term removal of wheat straw decreases soil amorphous silica at Broadbalk, Rothamsted. *Plant and Soil* 352: 173–184.

He, C., Li, Q., Liou, K.N. et al. (2014). Black carbon radiative forcing over the Tibetan Plateau. *Geophysical Research Letters* 41: 7806–7813.

Hodson, M.J. (2016). The development of phytoliths in plants and its influence on their chemistry and isotopic composition. Implications for palaeoecology and archaeology. *Journal of Archaeological Science* 68: 62–69.

Hodson, M., White, P., Mead, A., and Broadley, M. (2005). Phylogenetic variation in the silicon composition of plants. *Annals of Botany* 96: 1027–1046.

Hu, A.Y., Che, J., Shao, J.F. et al. (2018). Silicon accumulated in the shoots results in down-regulation of phosphorus transporter gene expression and decrease of phosphorus uptake in rice. *Plant and Soil* 423: 317–325.

Hurtado, A.C., Chiconato, D.A., de Mello, P.R. et al. (2020). Silicon alleviates sodium toxicity in sorghum and sunflower plants by enhancing ionic homeostasis in roots and shoots and increasing dry matter accumulation. *Silicon* 13: 475–486.

Ishiguro, K. (2001). Review of researmicrch in Japan on the roles of silicon in conferring resistance against rice blast. *Studies in Plant Science* 8: 277–291.

Jarvis, S. (1987). The uptake and transport of silicon by perennial ryegrass and wheat. *Plant and Soil* 97: 429–437.

Jones, L. and Handreck, K. (1965). Studies of silica in the oat plant: III. Uptake of silica from soils by the plant. *Plant and Soil* 29: 79–96.

Jones, L. and Handreck, K. (1967). Silica in soils, plants, and animals. *Advances in agronomy* 19: 107–149.

Katz, O. (2018). Plant silicon and phytolith research and the earth-life superdiscipline. *Frontiers in Plant Science* 9: 1281.

Kaur, H. and Greger, M. (2019). A review on si uptake and transport system. *Plants* 8: 81.

Korndörfer, G. and Lepsch, I. (2001). Effect of silicon on plant growth and crop yield. *Studies in Plant Science* 8: 133–147.

Kraska, J. and Breitenbeck, G. (2010). Simple, robust method for quantifying silicon in plant tissue. *Communications in Soil Science and Plant Analysis* 41: 2075–2085.

Kumar, S., Milstein, Y., Brami, Y. et al. (2017). Mechanism of silica deposition in sorghum silica cells. *New Phytologist* 213: 791–798.

Li, Z., Song, Z., and Cornelis, J.-T. (2014). Impact of rice cultivar and organ on elemental composition of phytoliths and the release of bio-available silicon. *Front. Plant Sci.* 5: 529. doi: 10.3389/fpls.2014.00529

Liang, Y., Si, J., and Römheld, V. (2005). Silicon uptake and transport is an active process in Cucumis sativus. *New Phytologist* 167: 797–804.

Loucaides, S., Behrends, T., and Van Cappellen, P. (2010). Reactivity of biogenic silica: surface versus bulk charge density. *Geochimica et Cosmochimica Acta* 74: 517–530.

Luu, D.T. and Maurel, C. (2005). Aquaporins in a challenging environment: molecular gears for adjusting plant water status. *Plant, Cell & Environment* 28: 85–96.

Lux, A., Luxová, M., Hattori, T. et al. (2002). Silicification in sorghum (Sorghum bicolor) cultivars with different drought tolerance. *Physiologia Plantarum* 115: 87–92.

Lux, A., Luxová, M., Abe, J. et al. (2003). The dynamics of silicon deposition in the sorghum root endodermis. *New Phytologist* 158: 437–441.

Ma, J.F. and Takahashi, E. (2002). *Soil, Fertilizer, and Plant Silicon Research in Japan.* Amsterdam, The Netherlands: Elsevier.

Ma, J.F. and Yamaji, N. (2015). A cooperative system of silicon transport in plants. *Trends in Plant Science* 20: 435–442.

Ma, J., Miyake, Y., Takahashi, E. et al. (2001a). *Silicon in Agriculture. Silicon as a Beneficial Element for Crop Plants*, 17–34. Amsterdam: Elsevier.

Ma, J.F., Goto, S., Tamai, K., and Ichii, M. (2001b). Role of root hairs and lateral roots in silicon uptake by rice. *Plant Physiology* 127: 1773–1780.

Ma, J.F., Tamai, K., Ichii, M., and Wu, G.F. (2002). A rice mutant defective in Si uptake. *Plant Physiology* 130: 2111–2117.

Ma, J.F., Tamai, K., Yamaji, N. et al. (2006). A silicon transporter in rice. *Nature* 440: 688.

Ma, J.F., Yamaji, N., Mitani, N. et al. (2007). An efflux transporter of silicon in rice. *Nature* 448: 209.

Maeda, H., Nakajima, M., Hagiwara, T. et al. (2006). Bacterial cellulose/silica hybrid fabricated by mimicking biocomposites. *Journal of Materials Science* 41: 5646–5656.

Mandlik, R., Thakral, V., Raturi, G. et al. (2020). Significance of silicon uptake, transport, and deposition in plants. *Journal of Experimental Botany* 71: 6703–6718.

Meyer, J. and Keeping, M. (2001). Past, present and future research of the role of silicon for sugarcane in southern Africa. *Silicon in Agriculture* 8: 257–275.

Mitani, N. and Ma, J.F. (2005). Uptake system of silicon in different plant species. *Journal of Experimental Botany* 56: 1255–1261.

Mitani, N., Chiba, Y., Yamaji, N., and Ma, J.F. (2009a). Identification and characterization of maize and barley Lsi2-like silicon efflux transporters reveals a distinct silicon uptake system from that in rice. *The Plant Cell* 21: 2133–2142.

Mitani, N., Yamaji, N., and Ma, J.F. (2009b). Identification of maize silicon influx transporters. *Plant and Cell Physiology* 50: 5–12.

Mitani, N., Yamaji, N., Ago, Y. et al. (2011). Isolation and functional characterization of an influx silicon transporter in two pumpkin cultivars contrasting in silicon accumulation. *The Plant Journal* 66: 231–240.

Mitani-Ueno, N., Yamaji, N., and Ma, J.F. (2011). Silicon efflux transporters isolated from two pumpkin cultivars contrasting in Si uptake. *Plant Signaling & Behavior* 6: 991–994.

Motomura, H., Mita, N., and Suzuki, M. (2002). Silica accumulation in long-lived leaves of *Sasa veitchii* (Carrière) Rehder (Poaceae–Bambusoideae). *Annals of Botany* 90: 149–152.

Motomura, H., Fujii, T., and Suzuki, M. (2004). Silica deposition in relation to ageing of leaf tissues in Sasa veitchii (Carriere) Rehder (Poaceae: Bambusoideae). *Annals of Botany* 93: 235–248.

Nawaz, M.A., Zakharenko, A.M., Zemchenko, I.V. et al. (2019). Phytolith formation in plants: from soil to cell. *Plants* 8: 249.

Neumann, D. and Zur Nieden, U. (2001). Silicon and heavy metal tolerance of higher plants. *Phytochemistry* 56: 685–692.

Ouellette, S., Goyette, M.H., Labbé, C. et al. (2017). Silicon transporters and effects of silicon amendments in strawberry under high tunnel and field conditions. *Frontiers in Plant Science* 8: 949.

Rafi, M.M. and Epstein, E. (1999). Silicon absorption by wheat (Triticum aestivum L.). *Plant and Soil* 211: 223–230.

Robert, F. and Chaussidon, M. (2006). A palaeotemperature curve for the Precambrian oceans based on silicon isotopes in cherts. *Nature* 443: 969–972.

Roychoudhury, A. and Tripathi, D.K. (2019). *Molecular Plant Abiotic Stress: Biology and Biotechnology*, vol. 1, 1–480. John Wiley & Sons.

Sakai, W.S. and Sanford, W. (1984). A developmental study of silicification in the abaxial epidermal cells of sugarcane leaf blades using scanning electron microscopy and energy dispersive X-ray analysis. *American Journal of Botany* 71: 1315–1322.

Shivaraj SM, Mandlik R, Bhat JA, Raturi G, Elbaum R, Alexander L, Tripathi DK, Deshmukh R, Sonah H.2021. Outstanding questions on the beneficial role of silicon in crop plants. *Plant and Cell Physiology* doi: https://doi.org/10.1093/pcp/pcab145

Struyf, E. and Conley, D.J. (2012). Emerging understanding of the ecosystem silica filter. *Biogeochemistry* 107: 9–18.

Struyf, E., Smis, A., Van Damme, S. et al. (2009). The global biogeochemical silicon cycle. *Silicon* 1: 207–213.

Sun, H., Guo, J., Duan, Y. et al. (2017). Isolation and functional characterization of CsLsi1, a silicon transporter gene in Cucumis sativus. *Physiologia Plantarum* 159: 201–214.

Sun, H., Duan, Y., Qi, X. et al. (2018). Isolation and functional characterization of CsLsi2, a cucumber silicon efflux transporter gene. *Annals of Botany* 122: 641–648.

Takahashi, E., Ma, J., and Miyake, Y. (1990). The possibility of silicon as an essential element for higher plants. *Comments on Agricultural and Food Chemistry* 2: 99–102.

Treguer, P., Nelson, D.M., Van Bennekom, A.J. et al. (1995). The silica balance in the world ocean: a reestimate. *Science* 268: 375–379.

Tripathi, D.K., Singh, V.P., Lux, A., and Vaculik, M. (2020). Silicon in plant biology: from past to present, and future challenges. *Journal of Experimental Botany* 71 (21): 6699–6702.

Tripathi, D.K., Singh, V.P., Ahmad, P. et al. (2021). Metalloids in plant biology: new avenues in their research. *Journal of Hazardous Materials* 422: 126738.

Van der Vorm, P. (1980). Uptake of Si by five plant species, as influenced by variations in Si-supply. *Plant and Soil* 56: 153–156.

Vivancos, J., Deshmukh, R., Grégoire, C. et al. (2016). Identification and characterization of silicon efflux transporters in horsetail (Equisetum arvense). *Journal of Plant Physiology* 200: 82–89.

Yamaji, N., Mitatni, N., and Ma, J.F. (2008). A transporter regulating silicon distribution in rice shoots. *The Plant Cell* 20: 1381–1389.

Yamaji, N., Sakurai, G., Mitani-Ueno, N., and Ma, J.F. (2015). Orchestration of three transporters and distinct vascular structures in node for intervascular transfer of silicon in rice. *Proceedings of the National Academy of Sciences* 112: 11401–11406.

Yang, C., Zeng, D., Qin, R. et al. (2018). Rice gene, BBH/Lsi1, regulates the color of rice hull by reducing the absorption and deposition of silicon and accumulating excess flavonoid. *Plant Growth Regulation* 85: 133–142.

10

Silicon in Soil, Plants, and Environment

Mujahid Ali[1], Muhammad Zia Ur Rehman[1], Asad Jamil[1], Muhammad Ashar Ayub[2], and Muhammad Tahir Shehzad[1]

[1] *Institute of Soil and Environmental Sciences, Faculty of Agriculture, University of Agriculture, Faisalabad, Pakistan*
[2] *Institute of Agro-Industry and Environment, Faculty of Agriculture and Environment, The Islamia University of Bahawalpur, Bahawalpur, Pakistan*

Introduction

Silicon is a crystalline, brittle substance with a major use in plant science and having crystalline structure (Atta et al. 2019). Silicon has a concentration of 27.6% in earth crust making it second most abundant element after oxygen (Durga 2020), as a substantial part of the soil is formed by silicon or silicate aluminium minerals (Zhu and Gong 2014). Si is found in primary, and secondary silicate minerals, alumino-silicates, and in compound like SiO_2 and mostly not available in free state, but its soil concentration can range from 14 to 20 mg/L in soil solution (Durga 2020). Some soil types like, histosols, are extremely leached/ poor base saturation containing soils (ultisols and oxisols) and have significant Si concentration in soil solution. Due to its immense importance and necessity in agro-ecosystem, Si is regarded as "quasi-essential" or almost essential element (Khan et al. 2019). Plant's physiological and biochemical processes are regulated by the application of Si, especially under abiotic and biotic stresses (Rizwan et al. 2015; Verma et al. 2020). Role of Si in plant's body is vast ranging from crop growth promotion to better light utilization and defence modulation against abiotic and biotic stresses (Chanchal Malhotra et al. 2016; Zhang et al. 2017a, b; Sakr 2018).

Silicon has widespread application in various aspects of life and agro-ecosystem (Vasanthi et al. 2012a, b). Though its amount in soil is very significant, its bioavailable concentration which is available for plants to uptake is very limiting (Ma et al. 2011; Datnoff and Rodrigues 2015; Tubana et al. 2016; Zargar et al. 2019) The translocation of Si from soil to plant roots and from roots to shoots is very important and can account for up to 0.1–10% of the biomass of plants (Epstein 2009), as plants have developed special channels for Si translocation (Broadley et al. 2012; Gaur et al. 2020). Majority of this contents is utilized in structural modification and development, while some amount is stored under cuticle for further use (Ma et al. 2006, 2007). Major role of inter-tissue Si translocation is played by numerous influx and efflux Si-transporters present in both monocots and dicots

Beneficial Chemical Elements of Plants: Recent Developments and Future Prospects, First Edition. Edited by Sangeeta Pandey, Durgesh Kumar Tripathi, Vijay Pratap Singh, Shivesh Sharma, and Devendra Kumar Chauhan.

(Zargar et al. 2019) like LSi1, LSi2, LSi3, and LSi6 helping absorption and translocation of Si in plants (Gaur et al. 2020). The uptake of Si from the soil takes place in the form of monosilicic acid [$Si(OH)_4$] ions that are further transferred to the root and shoot via active or passive transport (Ma et al. 2011). For active transport, special proteins are present in plants body (Broadley et al. 2012), while passive transport is mainly being done in dicot plants via transpiration pull (Guntzer et al. 2012a, b).

Abiotic and biotic stresses are major agricultural problems and application of Si can be effective in tackling these problems (Etesami 2018; Shivaraj et al. 2021), as Si enhances photo-chemical, physiological, and biochemical defence processes of plants (Atkinson and Urwin 2012). Other major ways via which Si helps in mitigation of stress in plants are, nutrients provision, heavy metal (HM) complexation, decreasing HMs uptake, and modulation of antioxidant response (Adrees et al. 2015a; Rizwan et al. 2015). Silicon deposition in upper layer of plants (as silica) helps in strengthening of cell wall, thus decreasing damage by insect/pests (Massey and Hartley 2009). Furthermore, polymerization of Si in cell wall of plants also acts as a potential barrier for pathogen/pest infestation in plant's cell (Debona et al. 2017). Various innovative sources of Si are being applied in agriculture field to tackle abiotic and biotic sources like Si fertilizer and Si nanoparticles (Adrees et al. 2015a; Rizwan et al. 2015). This chapter summarizes sources, fate, and uses of Si in agro-ecosystem with special focus on its efficacy in stress mitigation in plants.

Sources of Silicon in Soil, Plants and Environment

Natural Sources

Si is the second most abundant element on the Earth's crust (Luyckx et al. 2017) making about 27.6% of the Earth's crust (Haynes 2017; Durga 2020). Silicon is found in compound form as insoluble crystalline aluminosilicates, soil fixed pool, and in mineral forms in rocks (Monger and Kelly 2002; Zhu and Gong 2014; Liang et al. 2015a, b;). Physico-chemical processes in soil are responsible for inter conversion (of mobile and immobile Si) and Si cycling in soil (Vandevenne et al. 2015). Though present in excess in soil, bioavailable Si concentration in soil is very limiting and it is divided into three phases: the liquid, solid, and adsorbed phases (Sauer et al. 2006). Only bioavailable fraction of Si in soil is needed and taken up by plants and is involved in stress mitigation and growth promotion (Zargar et al. 2019). Various weathering processes of soil make transformation of SiO_2 as monosilicic acid, which makes major part of available Si pool in soil (Tubana et al. 2016; Zargar et al. 2019). Si concentration varies from 0.1 to 0.6 mM in soil solution (Datnoff et al. 2001). Though after soil, major source of Si is plant matter (dead material) (Frantz et al. 2005; Hodson et al. 2005), but small amount of Si is also being provided by bacteria, protozoa, diatoms, and fungi (Puppe et al. 2015). This makes Si's biogenic pool play a vital role in Si cycling especially in forest and grassland ecosystem (Sommer et al. 2013; Frings et al. 2014).

Artificial/Synthetic Sources

Silicon has a potential role in smelting process thus making slag a potential source of Si (Ma and Takahashi 2002; Sousa and Korndorfer 2010; Haynes et al. 2013). Slag is a by-product of metal smelting process and being a Si source can be used as a potential Si fertilizer in agriculture, but Si contents in slag vary from place to place and mainly range between 15 and 25% (Haynes et al. 2013; Liang et al. 2015a, b). Besides slag, silicate of potassium, magnesium, calcium, calcium metal, and sodium and silica gel are some other potential sources of Si (Gascho 2001), and these Si sources can be added in soil as Si fertilizer after quality assurance (Rizwan et al. 2012; Guntzer et al. 2012a, b; Kleiber 2018).

Uses of Silicon

Industrial Use

Due to widespread availability of natural and synthetic Si sources, its applications in industry are also diverse (Rebelo et al. 2019). Materials comprising of Si have rich chemistry to be used in optical fiber industry in optical waveguide and electronic devices. Production of optical wave guide is primarily being done by the use of complementary metal oxide semiconductors, Si products are a major player in it, and Si-based optical racetrack resonator wave guide has better efficacy than others (Ariannejad et al. 2020). Besides optical and sensor industry, Si-based sensors are also being deployed and used in health care and biomedical industry (Khan et al. 2018; Benedikovic et al. 2019; Xu et al. 2019). A naturally occurring Si mineral wollastonite has being proven to be used in food processing, metallics, paints, ceramic, and plastic industries as well as replacement of asbestos (Virta 2004) due to its distinct Si (24.3%), calcium (34.3), and other metal and mineral contents (Maxim and McConnell 2005; Maxim et al. 2008). Pure Si is a major component of circuit boards and computer chips, which is obtained by reduction of SiO_2 using carbon. The Si products are also being used as a smelting agent in Al-Si alloy production industry (Vasanthi et al. 2012a).

Application in Agro-ecosystems

Silicon is characterized as a beneficial component for plants and can lead to a major increase in plant productivity and food safety insurance (Parisi et al. 2015). Application of silicon in crops can help them withstand diverse types of abiotic and biotic stress (Chen et al. 2011a, b; Gall et al. 2015; Yavaş and Aydın 2017; Kleiber, 2018) via improvement of CO_2 fixation (Chen et al. 2011b), water use efficiency (Li et al. 2018; Tibbitts 2018), and enhanced antioxidant defence (Tibbitts 2018). Upon accumulation in plant body, Si is deposited in cell wall resulting in cell wall strengthening and ultimately protecting plants against disease infestation (Etesami 2018) and inhibit electrolyte leakage under salinity stress (Khan et al. 2019).

Among various applied sources of Si, nano-Si has a unique character and can help reverse HMs stress (Tripathi et al. 2012, 2015a, b), UV-B (Tripathi et al. 2017), salinity (Abdel-Haliem et al. 2017; Khan et al. 2019), and water stress (Jullok et al. 2016). Furthermore, Si-NPs-based formulations can also act as nano-fertilizer, nano-pesticide, and nano-herbicide as nanotechnology application in food crops has gained much pace in past decade (Gogos et al. 2012; Magda and Hussein 2016; Ziaee and Ganji 2016). Action of Si-NPs can be as nano-pesticide and as nano-carrier for enhancement in efficacy of other commercial pesticides.

Toxicity of nano-silica for plant pathogens is main reason attributed to its use as nano-pesticide as observed for *Callosobruchus maculatus* (Rouhani et al. 2012)

Role of Silicon in Plant Nutrition-Growth Responses

Nutrient Acquisition

For natural growth and production, plants need ample supplies of essential nutrients. Proper nourishment of plants is an important means of optimum yield and fruit quality and quantitative and qualitative improvement of agricultural products like fruits (Brunetto et al. 2015). Nutritional imbalances hamper the development, growth, and yield of plants (Paul and Lade 2014). It is an established fact that Si supplementation can boost nutrient acquisition, retention, and utilization efficiency of crop plants (Guntzer et al. 2012a, b; Etesami and Jeong 2018). Under nutrient deficiency conditions, silicon can play a vital role in uplifting nutrient acquisition thus helping plant thrive and grow properly (Etesami and Jeong 2018; Tripathi et al. 2020). Main reason behind role of Si in boosting nutrient acquisition is its ability for root strengthening and elongation (Isa et al. 2010) and increased concentration of nutrient transported molecules, thus helping in alleviating micronutrient deficiency (Hernandez-Apaolaza 2014). It is a proven fact from literature that Si nutrition plays a vital role in managing poor nutrition and stress in crop plants and is recommended for application by literature (Etesami and Jeong 2018).

Plant Growth Promotion

Under biotic and abiotic stresses, plants suffer severely and decreased plant growth is observed. Silicon application can enhance the plant growth under conditions of stress, via various mechanisms like nutrition provision and defence strengthening (Ma et al. 2001, 2006). Silicon helps plants improve growth via nutrient homeostasis (Alsaeedi et al. 2019), alleviating sodium toxicity under salt stress (Hurtado et al. 2019) and plant physiological strengthening (Liu et al. 2015; Rizwan et al. 2015).

Gas Exchange Attributes Modulation

Plant facing any stress observes severe disturbance in its gas exchange parameters (Gupta and Huang 2014) like in salt and water stress (Liu et al. 2015; Rizwan et al. 2015). Application of Si exogenously can enhance gas exchange and CO_2 assimilation efficiency, net photosynthesis, and stomatal conductance under salt (Rizwan et al. 2015; Xie et al. 2015) and drought stress (Rizwan et al. 2015). Silicon application improves the stomatal conductance under drought stress and also increase the ability of plants to absorb and utilize more water, in plant suffering from drought stress (Gao et al. 2006; Chen et al. 2011a).

Plant Water Balance

Water and salinity stress are leading causes of water imbalance in higher plants, and Si supplementation has a significant role in tackling both stresses (Coskun et al. 2016). Silicon fertilizer application helps plant withstand water stress (as it is involved in cell wall strengthening) (Ahmed and Khurshid 2011; Ahmed et al. 2011), thus improving plant growth (Gong et al. 2003) and reverse negative effects of water stress (Pei et al. 2010). Major role of Si in water utilization management under stress is hidden in its role in transpiration management (Hattori et al. 2005), increased leaf firmness (Gong et al. 2003), improved cellular water storage (Tuna et al. 2008), and decreased electrolyte leakage (Keller et al. 2015).

Antioxidant Enzymes Activities

Under stress (especially abiotic conditions), plants observe different gene expression that leads to excessive production of reactive oxygen species (ROS) (Lucas et al. 2011; Qin et al. 2011; Roychoudhury and Tripathi 2019). To counteract these ROS, plants have evolved a defence mechanism in which antioxidant enzymes are generated that help to quench these ROS and reduce oxidative damage (Ahmed and Khurshid 2011; Ahmed et al. 2011; Rizwan et al. 2015; Roychoudhury and Tripathi 2020). Exogenous Si application can help plant, uplifting production of these antioxidants thus helping in stress mitigation. Antioxidants like CAT, APX, SOD, and others have found to be upregulated by Si application under stress conditions (Abbas et al. 2015; Khan et al. 2016; Li et al. 2016; Xu et al. 2017). Besides enzymatic antioxidants, Si application is also attributed to non-enzymatic antioxidants like glutathione like reported under HM stress (Hasanuzzaman et al. 2017).

Uptake and Translocation Mechanisms of Silicon

There are two types of Si compounds in soil, i.e.solid and liquid. The solid form consists mainly of amorphous silica (ASi), crystalline, and low crystalline; however, liquid form is present in soil solution (Sauer et al. 2006). The liquid form usually consists of mono- and polysilicic acids, which are present in a soil solution (Cornelis et al. 2011). Though a major portion of soil and Earth crust's is made up of Si, a very small portion of Si is available for plants to uptake. Only bioavailable fraction of Si is taken up by plants (Ma et al. 2001) and is involved in management of abiotic and biotic stresses (Ma et al. 2001, 2006; Zhu and Gong 2014; Rizwan et al. 2015; Dhiman et al. 2021). Major bioavailable form of Si in soil is mono silicic or ortho-silicic acid, the concentration of which in soil varies as controlled by various soil conditions (Carey and Fulweiler 2014). At soil pH = 9, the bioavailable Si pool ranges between 0.1 and 0.6 mM (Sommer et al. 2006). Slag-based Si sources are dynamic in this regard and can help in improvement of soil Si and carbon contents (Song et al. 2015). Based on accumulation of Si, plants can be categorized as accumulators, intermediates, and excluders with Si/Ca ratio of >1, 0.5–1, and <0.5, respectively (Carey and Fulweiler 2014). Plants grown on Si-rich soils can uptake a significant amount of Si, and it can range between 0.1 and 10% by weight (Ma and Yamaji 2006).

Silicon bioavailable pool is readily taken up and assimilated by plants (Ma and Yamaji 2006; Meharg and Meharg 2015) making 0.1–10% of plant's biomass (Epstein 2009). With respect to Si uptake and accumulation, plants are classified into various types, i.e. high accumulator, intermediate accumulator, and excluders (Coskun et al. 2016; Landi et al. 2017). Silicon uptake in plants is done via both active (Rains et al. 2006) and passive pathways resulting in huge variation among various species of plants. Once Si is taken up by root hairs, Si is transported to xylem (via Si transporters in endodermis), from there its translocation is governed by transpiration pull (Ma et al. 2006, 2007). Major Si transport involved in Si-inter tissue translocation are influx and efflux Si transporters (Mitani et al. 2009). Majority of plants translocate up to 90% of absorbed Si into their aerial parts (Ma and Takahashi 2002) via Si transporters (Yamaji et al. 2012). Once Si reaches shoot cells, its again converted to $SiO_2{\cdot}nH2O$ and deposited as amorphous silica at cellular levels mainly in cell wall, shoot epidermis, root endodermis, and at transpiration sites (Ma et al. 2011; Keller et al. 2015), but it can vary from plant to plant (Li et al. 2015). Figure 10.1 shows a schematic explanation of Si chemistry and effects on plants.

Role of Silicon in Agriculture

Role of Silicon in Abiotic Stress Management

Heavy Metals

HMs are potential pollutants (if they exceed certain levels), and their uptake in plant tissues is very much evident and reported (Lee et al. 2010; Adrees et al. 2015a, b). Being excessively bioavailable in soil, these HMs (like, Cd, Cu, Pb, Cr, As, and Mn) can be taken up by plants and possess severe physiological, biochemical, and oxidative stress in plants (Nascimento and Xing 2006; Nagajyoti et al. 2010; Afshan et al. 2015) affecting plant growth (Ali et al. 2015a, b).

Application of Si fertilizer in plants can help plants withstand HMs stress via reduction of HM-active concentrations, uplifting of anti-oxidant enzymes, and modulating soil pH (Liang et al. 2007; Li et al. 2012; Rizwan et al. 2012; Adrees et al. 2015a, b). Besides this, Si application also enhances plant growth (Kleiber et al. 2015b) and effectively stimulates metal translocation (Liang et al. 2007). Silicon application in soil also helps decreasing soil bioavailable metal contents via chelation (Rizwan et al. 2012; Ding et al. 2013; Shim et al. 2014; Qian et al. 2016).

Co-precipitation of various HMs with exogenously applied Si is also an effective and evident role of Si in HM stress management (Gu et al. 2011, 2012). In another role, Si application decreases cellular oxidative (Liang et al. 2007; Adrees et al. 2015a, b; Rizwan et al. 2015) and hydroxyl stress (Dragišić Maksimović et al. 2012). Boosting plant physiology, carbon assimilation, and enhanced photoassimilation are various other ways in which Si helps in remediating HMs toxicity in various plants (Liang et al. 2007; Adrees et al. 2015a, b; Afshan et al. 2015; Sanglard et al. 2016).

Salinity

Salinity or salt stress is another very important abiotic stress affecting plants growth and yield worldwide. Moderate to high salt stress leads to severe ionic imbalance, oxidative stress, and osmotic imbalance in plants (Ayub et al. 2020), which can be reversed by exogenous Si application (Liu et al. 2019). Ionic imbalance under salinity stress leads to excessive sodium intake (Na^+) and chloride (Cl^-) intake in plants creating ROS leading to oxidative stress, which can be hindered by Si application due to its distinctive cross-talk with plants' homeostasis and physiology (Murad et al. 2020). The Si mediated mitigation of abiotic tress can be attributed to its role in antioxidants' regulation (Liu et al. 2019), mineral homeostasis (Ahmed and Khurshid 2011; Ahmed et al. 2011), cell wall strengthening (Liang et al. 2015a, b; Guerriero et al. 2016),

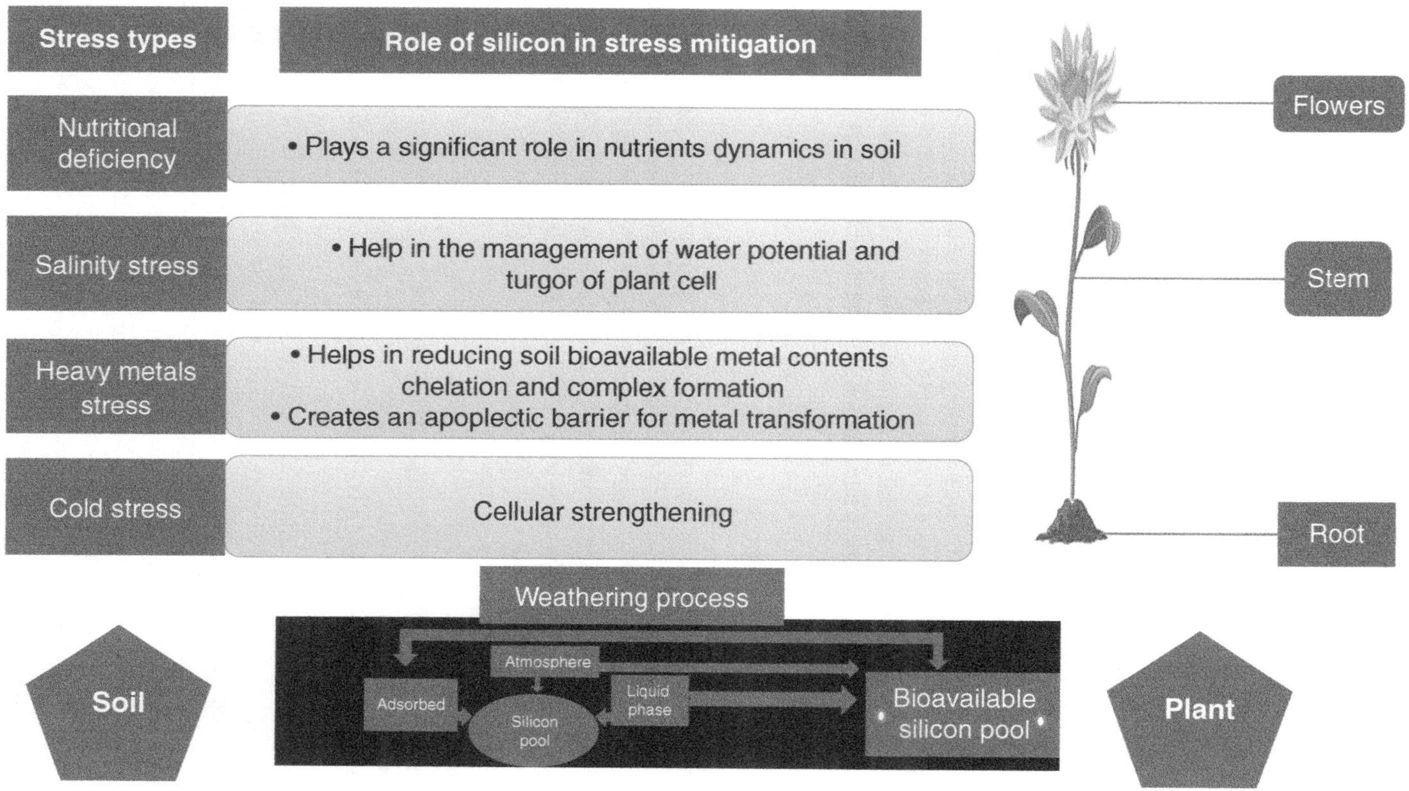

Figure 10.1 Chemistry of Si in soil–plant continuum under stress conditions.

lignification (Fleck et al. 2011, 2015), and lignin suberin-related gene expression enhancement (Fleck et al. 2011; Suzuki et al. 2012). All these processes help to create an effective apoplastic barrier for the translocation of Na in plants resulting in better efficiency of water use in many plants (Zhu et al. 2015; Shi et al. 2016).

Silicon also facilitates polyamine (PA) synthesis and accumulation (Alcázar et al. 2010; Yin et al. 2016) and sucrose/fructose biosynthesis (Yin et al. 2013) under saline stress thus helping with plant's cellular biochemical homeostasis.

Water Stress

One of the most important environmental factors affecting crop production and plant growth is the availability of water in the soil. Water shortage can be due to soil water scarcity (dry) or the issue of water absorption induced by physiological drought (as in saline conditions) (Pei et al. 2010; Shen et al. 2010). Decreasing water absorption can result in a significant reduction in turgor pressure and water content in plant cells, which can be devastating. Application of Si has shown promising results in management of water stress (Verma et al. 2019, 2020) and physiological drought (Liang et al. 2015a, b; Guerriero et al. 2016). Major phenomenon behind this action is enhanced cell wall integrity (Liang et al. 2015a, b; Guerriero et al. 2016), lignification (Fleck et al. 2011, 2015), and Ca enrichment (Kaya et al. 2006; Rizwan et al. 2015). Water scarcity severely influences plants gaseous exchange and water balance, which can be regulated via Si application (Chen et al. 2011a) as it involves chlorophyll content enhancement, proline production, and decrease in electrolyte leakage (Kaya et al. 2006). Plants do have water management system in their cells (Horie et al. 2011), and osmotic adjustments help plants escape water scarcity (Osakabe et al. 2014), and Si application can further enhance this phenomenon (Wang et al. 2015; Zhu et al. 2015). Silicon supplemented plants have shown a better stomatal conductivity, leaf water content, transpiration rate, hydraulic conductivity of the root (Yin et al. 2013; Liu et al. 2015; Wang et al. 2015; Zhu et al. 2015), reduced oxidative stress (Shi et al. 2016), and plant growth enhancement (Sonobe et al. 2010).

Temperature Stress

Stress to cold (low temperature) is also an active abiotic stress for plants, and application of Si has shown promising results (Liang et al. 2008; Jiao-Jing et al. 2009). Lower temperature is attributed to membrane damage and oxidative stress that can be managed by the application of silicon (Jiao-Jing et al. 2009). Application for silicon was attributed to a reduced accumulation of H_2O_2 and malondialdehyde (MDA) and sugar production under freezing ($-5\,°C$) temperature stress in wheat cultivars (Liang et al. 2008) and linked to wilting prevention due to cell wall strengthening (Liang et al. 2006). Post-harvest Si application also helps in fruit preservation against chilling stress (Mditshwa et al. 2013). Exogenous Si application also helps plants to accumulate osmo-protectants (Zhu et al. 2006; Liang et al. 2008), enhance chlorophyll content (Zhu et al. 2006), and enhance chances of fertilization via decreased cracking incidents (Li et al. 2005). Similarly, application of Si has been found equally effective in decreasing high temperature stress in *Fragaria × ananassa* (Muneer et al. 2017), *Lablab purpureus* (Rai et al. 2018), *Pyropia yezoensis* (Le et al. 2019), and *Triticum aestivum* (Sattar et al. 2017) via membrane strengthening and oxidative stress reduction. Table 10.1 shows reviewed literature regarding role of Si in abiotic stress management.

Table 10.1 Role of Si in abiotic stress management in plants.

Stresses	Si source	Crop plants	Effects on decreasing stresses	References
Heavy metal toxicity	Potassium silicate, $K_2SiO_3{\cdot}nH_2O$	*Oryza sativa*	Growth promotion of rice plants via split application of Si was evident. Main reason for growth promotion was decreased Cd uptake by rice plants.	Rehman et al. (2019)
	Silicon dioxide nanoparticle (SNPs)	*Zea mays*	Application of Si-NPs significantly reduced Al toxicity in maize plants.	Sousa et al. (2019)
	Salicylic acid, 24-epi-brassinolide, silicon	*Triticum aestivum*	Stress had been significantly reduced in plants treated with salicylic acid (SA) and silicon (Si) and 24-epi-brassinolide (BR).	Maghsoudi et al. (2019)
	SiO_4		Adding Si in Cr-stressed crop plants significantly uplifted plant's growth and reduced Cr toxicity compared to control.	Sarkar et al. (2020)
Salinity stress	Potassium silicate, $K_2SiO_3{\cdot}nH_2O$		Application of Si uplifted salinity-mediated oxidative stress in plants via modulation antioxidant pathways.	Alzahrani et al. (2018)
	Sodium silicate, Na_2SiO_3	*Vigna radiata*	Silicon application lifted salinity/sodicity stress via production of osmoprotectants and antioxidants to quench ROS.	Ahmad et al. (2019)
	Silicic acid	*Glycine max*	Applied Si-trigged plant-hormonal stress signalling pathways and reduced salinity stress.	Chung et al. (2019)
	Silicon dioxide, SiO_2	*Brassica napus*	Silicon application decreased oxidative stress and increased plant growth via production of osmolytes and antioxidants.	Hasanuzzaman et al. (2018)
	Na_2SiO_3 at three levels (0, 0.5, and 1 mM)	*Cicer arietinum*	When applying Si that helps the plant to elevate salinity stress, the osmoprotective contents in plants have been significantly increased.	Zamani et al. (2017)
Temperature stress	K_2SiO_3 source	*Fragaria × ananassa*	Silicon-mediated modulation of antioxidants helped plants withstand high temperature stress.	Muneer et al. (2017)

(Continued)

Table 10.1 (Continued)

Stresses	Si source	Crop plants	Effects on decreasing stresses	References
Temperature stress	Exogenous salicylic	*Lablab purpureus* L.	Silicon-mediated modulation of antioxidants helped plants withstand high temperature stress.	Rai et al. (2018)
Temperature stress	Silicic acid (50 μM)	*Pyropia yezoensis*	Silicon increased temperature stress tolerance in plants.	Le et al. (2019)
Temperature stress	Calcium silicate	*Triticum aestivum*	Modulation of antioxidants and biochemical pathways was observed under Si application helping plants withstand temperature stress.	Sattar et al. (2017)
Water stress	Sodium metasilicate (Si)	*Ocimum basilicum*	Enhancement in ROS quenching helped plants withstand water stress.	Farouk and Omar (2020)
	Si+Biochar	*Zea mays* L.	Increased plant growth and enhanced plant's physiology were key factors behind Si-mediated water stress tolerance.	Sattar et al. (2020)
	Sodium silicate	*Lycopersicon esculentum*	Enhanced contents of antioxidants under Si treatment helped plants.	Malhotra and Kapoor (2019)
	Na_2SiO_3 as the source of Si	*Poa pratensis*	Decreased lipid peroxidation and enhanced antioxidants production were key contributors.	Bae et al. (2017)
	Sodium metasilicate (Na_2SiO_3)	*Lens culinaris* Medik.	Improved hydrolytic enzyme activity and better defence/ROS quenching response were key contributors.	Biju et al. (2017)
	Calcium magnesium silicate	*Saccharum officinarum*	Cell wall strengthening by Si application helped plants withstand water stress (via decreased electrolyte leakage).	de Camargo et al. (2019)
	Sodium silicate	*Triticum aestivum*	Silicon helped plants to improve physiology and CO_2 fixation, thus helping withstand water stress.	Maghsoudi et al. (2015)
	($SiO_2{\cdot}H_2O$) orthosilicic acid (H_4SiO_4) and disilicic acid ($H_2Si_2O_5$)	*Solanum tuberosum*	Silicon application decreased oxidative stress in plants that helped in withstanding water stress.	Pilon et al. (2014)
	SiO_2 powder Biogenic silica	*Theobroma cacao* L.	Increased light harvesting and membrane stability helped plants under water-deficit conditions.	Zanetti et al. (2016)

Role of Silicon in Biotic Stress Management

Pest Attack

Unlike other animals, insects are among more than a million species of abundant living organisms and make up more than half of all known living species (Chapman 2012). Insects destroy around one-quarter of global food production in agriculture (Sallam 2008) by boring the roots, leaves, stems, and fruits; chewing leaves;, consuming vegetable juice; and spreading diverse plant diseases (Aetiba and Peter 2015). Insect damage to plant tissue can be managed via Si application in plants as it acts as cell wall strengthening agent and pesticide (Van Bockhaven et al. 2013). Silicon application have been attributed towards insect pest tolerance against green leafhopper, brown planthopper, stem borer, white-backed planthopper, as well as against spider mites that are non-insect pests (Ma and Takahashi 2002; Gomes et al. 2005; Liang et al. 2015a, b; Debona et al. 2017; Yavaş and Aydın 2017). Si deposition in leaves as phytoliths increases the stiffness of tissue and weakness of herbivores and decreases digestibility of plant leaves (Santos et al. 2015). Many of the mechanisms underlying Si-afforded plant pest resistance include Si's proposed deposition under the cuticle that also appears to work against a variety of insect pests as a physical barrier to plant disease (Sidhu et al. 2013; Reynolds et al. 2016). Moreover, Si also acts as a plant defence response inducer (Van Bockhaven et al. 2013; Yang et al. 2017) under pest attack thus decreasing plants death rate (Correa et al. 2005). In addition to biotrophic interactions (plant–herbivorous), Si also affected tri-trophic interactions (plant–herbivorous – natural enemies), which could play an additional role in controlling pests (Kvedaras et al. 2010; Reynolds et al. 2016). The use of silicon will reduce the feeding rate of insects by increasing its resistance to its natural enemies (Gatarayiha et al. 2010) or by regulating production of herbivore-induced plant volatiles (HIPVs) (Kvedaras et al. 2010).

Role of Silicon in Disease Management

Silicon (Si) has been recognized in many crops as a beneficial element and has been shown to improve disease resistance (Mburu et al. 2016). In several plant pathogen systems, silicon is known to effectively minimize severity of the disease. Main role of Si-mediated disease tolerance provision is via trigging cellular and molecular responses of plants (Shahrtash 2018). The potential impact of silicate salts on enzyme activity is also reflective of their role in plant protection against such a disease (Abd-El-Kareem et al. 2019a, b). Silicon changes the microbial activity, soil enzyme activity, and physiology of plants (photosynthetic pigments, compatible solutes, and enzyme activity), which ultimately reduces disease attacks on plants (El-Sheery 2017; Deshmukh et al. 2020). The fungal growth and spread of disease have been inhibited by silicon salts, and the consequent effect on plants damaging to strawberries. The best way to reduce the incidence of the disease and increase crop yield was to use calcium silicate and potassium silicate, respectively, as foliar spray soil treatment (Abd-El-Kareem et al. 2019a, b). According to Sarma et al. (2019), stem borer infestation (white ear head and dead heart) decreased substantially after silicon fertilizers were applied in all rice genotypes. Rice blast disease in all rice genotypes was significantly reduced compared to control after treatment with silicon solubilizes in soil. Different findings were also observed in the case of false smut disease, soil application of silicon fertilizers decreased false smut infestation and extreme false smut in all rice genotypes compared to control. According to the study reported by Nasla et al. (2019) it was revealed that the percentage of plant infection and grain discoloration was reduced by applying silicon supplements in foliar form. Among the Si-based formulations tested, Gainexa UPL has provided the best results in controlling rice grain discoloration. El-Sheery (2017) concluded that the application of potassium silicate was an effective approach in reducing the stress of *Sclerotium cepivorum* in onion. Also, the Si-treated plants displayed higher peroxidase enzyme activity compared to untreated control plants. Application of Si increases resistance to banana Xanthomonas wilt (BXW) and can provide an effective treatment strategy for diseases (Mburu et al. 2016). Improved disease tolerance in rice husk treated plants tends to be positively correlated with improved silicon and Si-enhanced phenolic content accumulation, and increased superoxide POD and PPO enzymes production in leaves (Ratnayake et al. 2018). Table 10.2 shows literature regarding role of Si in biotic stress management in plants.

Silicon-Mediated Endogenous Modifications in Plants

Silicon deposition in plant's cell makes its cell wall strong, thus facilitating decreased pest attack directly (Janicka-Russak et al. 2012; Liang et al. 2015a, b). Besides structural modification, Si application can also control plant defence signaling (Van Bockhaven et al. 2013; Yang et al. 2017), antioxidants production (Lang et al. 2017), apoplastic barrier (Lukačová et al. 2013; Tripathi et al. 2015a, b), improved lignification and silicification (Coskun et al. 2016), and phytohormone production (Kim et al. 2014). All these approaches help plant withstand biotic stresses (Cai et al. 2009; Lang et al. 2017).

C. Mechanism of Silicon-Mediated Abiotic Stress Management

Silicon created a physical barrier in plants (apoplastic pathway), thus managing metal accumulation and ionic balance in plants under abiotic stress (Gu et al. 2011; Adrees et al. 2015a, b). Si reduces HM stress through various mechanisms for managing stress from HMs, such as controlling the expressions of genes that are involved in the metal chelation and transport of metals mechanisms, participation in metal coprecipitation, modification of plant structure, and stimulation of antioxidant enzyme production (Liang et al. 2007; Guntzer et al. 2012a, b; Adrees et al. 2015a, b; Rizwan et al. 2012, 2015; Pontigo et al. 2017). For salinity management, Si helps in plant nutrition and oxidative stress management, compatible solutes production, ionic homeostasis, osmoprotectants, biosynthesis, ion absorption and transport regulation, antioxidants and polyamine synthesis, antioxidant enzyme activation, gene and channel regulation, hormone alteration, and phytohormones production (Gill and Tuteja 2010; Coskun et al. 2016; Shi et al. 2016; Farooq et al. 2019; Khan et al. 2019; Verma et al. 2019, 2020).

Under salt stress, the application of Si in rice plants reduced the toxic effect of s odium (Na) and boron (B), which resultantly reduce the oxidative stress (Farooq et al. 2019). In drought stress management application of Si helps in cell wall and cuticle strengthening (Chen et al. 2011a; Fleck et al. 2011, 2015; Liang et al. 2015a, b; Guerriero et al. 2016) and

Table 10.2 Role of Si in biotic stress management by plants.

Disease and pest stress	Si source	Crop plants	Disease/pest management	References
Fungal pathogen	Rice husk containing silicon	*Momordica charantia*	High accumulation of Si in plants under treatment of Si-rich rice husk was found to be a reason behind the decrease in fugal attack. Another factor was increased phenolic contents and enhanced antioxidants to quench ROS.	Ratnayake et al. (2018)
Xanthomonas wilt	Silicon dioxide nanoparticles (SNPs)	*Musa acuminata*	The use of Si has been shown to be an efficient approach for wilt management.	Mburu et al. (2016)
Downy mildew disease	Sodium metasilicate	*Vitis vinifera*	Silicon application decreased intensity and extent of mildew disease in vines.	Farouk et al. (2017)
Heavy metal stress	Silicon solubilizers	*Oryza sativa*	Soil application of silicon fertilizers reduced infestation of false smut and severe false smut in all rice genotypes as compared to control.	Sarma et al. (2019)
Grain discoloration disease	Orthosilicic acid	*Oryza sativa*	Results revealed that incidence of plant infection and grain discoloration percentage were reduced by foliar application of silicon supplements.	Nasla et al. (2019)
Onion white rot	Potassium silicate	*Allium cepa*	Potassium silicate was the best cure for decreasing white onion rot and growing fresh plant weight under greenhouse conditions.	El-Sheery (2017)
Fusarium wilt	Sodium silicate	*Cucumis sativus*	Cucumber seedling's resistance against wilt was found to be increased under Si treatment.	Zhou et al. (2018)
Wheat blast	Silicon dioxide (SiO_2)	*Triticum aestivum*	Silicon supplemented set of experiment showed less disease spread.	Pagani et al. (2014)

(*Continued*)

Table 10.2 (Continued)

Disease and pest stress	Si source	Crop plants	Disease/pest management	References
Strawberry leaf blight	Potassium, sodium, and calcium silicates	*Fragaria ananassa*	Both sources of Si were found to be effective in decreasing spread of disease.	Abd-El-Kareem et al. 2019a, b
Yellow stem borer larvae, *Scripophaga incertulus*	Sodium metasilicate	*Oryza sativa*	Silicon usages improved plant–pest tolerance by rising the absorption of silicon in rice plants.	Liang et al. (2006)
Stem borer (SSB), *Chilo suppressalis* Temperature Stress	Sodium silicate	*Oryza sativa*	Improves plant resistance to herbivores larvae and deteriorates insect detoxification capability. Interaction between plant and insect combats with pesticide management.	Wang et al. (2020)
Bird cherry-oat aphid (*Rhopalosiphum padi*)	Ground silica gel, SiO_2	*Triticum aestivum*	Si supplements act as refiner and induced wheat defensces against pest attack.	de Oliveira et al. (2020)
Sugarcane mite, *Oligonychus sacchari*	Foliar amendment of silicic acid	*Saccharum officinarum*	The results of application in mites and leaf dryness are significant. The silica acid appears to be a promising product for mite damage management and can be included in other integrated pest management/insect pest management (IPM) sugarcane management strategies.	Nikpay and Laane (2020)
Spider mite, *Tetranychus urticae*	Potassium silicate	*Carica papaya*	Potassium silicate application stimulated plant protection against pest and improved plant growth.	Catalani and Camargos (2017)
Bemisia tabaci	Potassium silicate, K_2SiO_3	*Gossypium hirsutum*	Si reduced the pest growth by prolonging immaturation stages of pest and defending plants against adverse impacts of pests.	Abbasi et al. (2020)
Diamondback moth (DBM), *Plutella xylostella*	Foliar fertilization with agrosilicon	*Brassica oleracea* var. capitata	Under application of Si treatments, less pest attack was observed.	Telles et al. (2019)

physiological and biochemical process enhancement (Yin et al. 2013; Liu et al. 2015; Wang et al. 2015; Zhu et al. 2015). In cold stress management as mentioned earlier, Si is involved in oxidative stress management, growth promotion, and cellular stability (Li et al. 2005; Zhu et al. 2006; Liang et al. 2008).

D. Mechanism of Silicon-Mediated Biotic Stress Management

Several studies have shown that the use of Si can generally reduce insect infestation from crops (Reynolds et al. 2009, 2016). Silicon helps in insect pest management via direct (physical barrier production) and indirect (via pest-predator influx) ways (Liang et al. 2007, 2015a; Reynolds et al. 2009, 2016). Increased silicon plant content can decrease the growth of insect pests and their effectiveness by affecting them in several ways such as: (i) digestive effectiveness by reducing ingested food transformation into the body mass, (ii) efficiency of the consumption of food by reducing adequate amounts of nutrients and water, (iii)altering and preventing insect pest feeding, (iv) the consumption, and (v) by counteracting the effect of high nitrogen levels on the effectiveness of insect pests (Reynolds et al. 2009, 2016; Liang et al. 2015a). Furthermore, Si-mediated defence signalling against abiotic stress like pest attack is also prominent in this regard (Correa et al. 2005; Fleck et al. 2011; Song et al. 2016; Wang et al. 2017).

Source of Silicon for Agricultural Application

There are many natural (biogenic) and chemical sources of Si available for application in agriculture system. Among biogenic sources, Si-rich organic matter (amendments) is important like rice husk derived products (ash, biochar, etc.) (Sohail et al. 2020a, b). Rice husk is a natural source of silicon (Si), which is further broken down into ground rice husk ash (RHA) and husk (GRH). These silicon sources are used as fertilizers in agriculture to treat downy mildew caused by *Pseudoperonospora* sp. in bitter gourd leaves (*Momordica charantia* L.) (Ratnayake et al. 2018). Biochar made from agricultural residues, such as rice and wheat straws, sugar cane residues, and their husks can contain large amounts of Si available and may be used as a natural source of silicon fertilizer (Abbas et al. 2017; Wang et al. 2018). Biochar is gaining more and more attention because of its potential environmental benefits. Besides carbon content is present in biochar, a major elementary component with abundant precursor sources and remarkable properties is silicon (Si). Role and application of these types of Si-rich biochar is well reported (Wang et al. 2019). The RH-derived biochar used in contaminated soil helps in soil management and better crop growth (Shen 2017; Azhar et al. 2019; Sohail et al. 2020b).

Among synthetic sources of Si, water soluble silicon salts are used as silicon fertilizers (Jeer et al. 2020). Silicic acid is one such chemical sources that can be applied via soil or foliar application, and its application is related to significant provision of Si to plants; thus, it acts as a bio-stimulant (Laane 2017). Besides silicic acid, zeolite minerals are also being used and considered as supplemental source of Fe and Si (Yao et al. 2017; Eroglu et al. 2017). Among other sources, slag and silicate minerals are important (Datnoff et al. 2001; Haynes 2014; Zhou et al. 2018; Zhu et al. 2019).

Recommendations for Exogenous Silicon Applications

As the quantity of silicon in crops is normally greater than that of silicon that is being released through slowly decaying silicate clay minerals. that is why it is the need of hour to apply Si sources exogenously (Song et al. 2015; Vandevenne et al. 2015). Exogenous application of Si can help plants with stand stresses (Epstein 1994; Ma et al. 2001), and its ample supply is not even toxic to plants thus can be applied in surplus (Zhu and Gong 2014; Etesami and Jeong 2018). Application of Si-rich amendments and sources is very much needed and in terms of crop economics, very beneficial for agriculture especially under contaminated soil conditions (Azhar et al. 2019; Rehman et al. 2019).

Conclusion and Future Perspectives

Whilst we understand how this overlooked dimension contributes to the countering of pathogenic invasion, pest attack, and different types of abiotic stresses, it is still important to analyze in more depth Si modulating plant physiology at the genomic and proteomic levels through the capacity of host defence. Addressing Si's impact on microbial expression and biochemistry is needed in additional studies. The mechanisms for biotic or abiotic stress, now absent from comparative studies of the interactions between Si and plants, should shed light upon this factor. The most important aspect of further Si studies should be Si's role in abiotic stress tolerance, and thus its functions in the environmental remediation should be completely exploited in plant biology. Applied research is needed to understand optimal Si concentration and acceptable methods for silicon application on plants, in particular in abiotic and biotic stress conditions.

References

Abbas, T., Balal, R.M., and Shahid, M.A. (2015). Silicon-induced alleviation of NaCl toxicity in okra (*Abelmoschus esculentus*) is associated with enhanced photosynthesis, osmoprotectants and antioxidant metabolism. *Acta Physiol. Plant.* 37 (2): 1–15.

Abbas, T., Rizwan, M., Ali, S. et al. (2017). Effect of biochar on cadmium bioavailability and uptake in wheat (*Triticum aestivum* L.) grown in a soil with aged contamination. *Ecotoxicol. Environ. Saf.* 140: 37–47.

Abbasi, A., Sufyan, M., Arif, M.J., and Sahi, S.T. (2020). Effect of silicon on oviposition preference and biology of *Bemisia tabaci* (Gennadius) (Homoptera: Aleyrodidae) feeding on *Gossypium hirsutum* (Linnaeus). *Int. J. Pest. Manag. 28* (124–134): 1–11.

Abdel-Haliem, M.E., Hegazy, H.S., Hassan, N.S., and Naguib, D.M. (2017). Effect of silica ions and nano silica on rice plants under salinity stress. *Ecol. Eng.* 99: 282–289.

Abd-El-Kareem, F., Elshahawy, I.E., and Abd-Elgawad, M.M. (2019a). Effectiveness of silicon and silicate salts for controlling black root rot and induced pathogenesis-related protein of strawberry plants. *Bull. Nat. Res. Cent.* 43: 1–8.

Abd-El-Kareem, F., Elshahawy, I.E., and Abd-Elgawad, M.M. (2019b). Management of strawberry leaf blight disease caused by *Phomopsis obscurans* using silicate salts under field conditions. *Bull. Nat. Res. Cent.* 43 (1): 1–6.

Adrees, M., Ali, S., Rizwan, M. et al. (2015a). Mechanisms of silicon-mediated alleviation of heavy metal toxicity in plants: a review. *Ecotoxicol. Environ. Saf.* 119: 186–197.

Adrees, M., Ali, S., Rizwan, M. et al. (2015b). The effect of excess copper on growth and physiology of important food crops: a review. *Environ. Sci. Pollut. Res.* 22: 8148–8162.

Aetiba, N. and Peter, J. (2015). Field evaluation of levo 2.4 sl (botanical) for the management of iInsect pests of eggplant (*Solanum melongena* L.) and Okra (*Abelmoschus esculentus* L. Moench). Doctoral dissertation.

Afshan, S., Ali, S., Bharwana, S.A. et al. (2015). Citric acid enhances the phytoextraction of chromium, plant growth, and photosynthesis by alleviating the oxidative damages in *Brassica napus* L. *Environ. Sci. Pollut. Res.* 22: 11679–11689.

Ahmad, P., Ahanger, M.A., Alam, P. et al. (2019). Silicon (Si) supplementation alleviates NaCl toxicity in mung bean [*Vigna radiata* (L.) Wilczek] through the modifications of physio-biochemical attributes and key antioxidant enzymes. *J. Plant Growth Regul. 38*: 70–82.

Ahmed, M. and Khurshid, Y. (2011). Does silicon and irrigation have impact on drought tolerance mechanism of sorghum? *Agric. Water Manag.* 98: 1808–1812.

Ahmed, M., Qadeer, U., and Aslam, M.A. (2011). Silicon application and drought tolerance mechanism of sorghum. *Afr. J. Agric. Res.* 6: 594–607.

Alcázar, R., Altabellan, T., Marco, F. et al. (2010). Polyamines: molecules with regulatory functions in plant abiotic stress tolerance. *Planta* 231: 1237–1249.

Ali, S., Bharwana, S.A., Rizwan, M. et al. (2015a). Fulvic acid mediates chromium (Cr) tolerance in wheat (*Triticum aestivum* L.) through lowering of Cr uptake and improved antioxidant defense system. *Environ. Sci. Pollut. Res.* 22: 10601–10609.

Ali, S., Chaudhary, A., Rizwan, M. et al. (2015b). Alleviation of chromium toxicity by glycinebetaine is related to elevated antioxidant enzymes and suppressed chromium uptake and oxidative stress in wheat (*Triticum aestivum* L.). *Environ. Sci. Pollut. Res.* 22: 10669–10678.

Alsaeedi, H., El-Ramady, T., Alshaal, M. et al. (2019). Silica nanoparticles boost growth and productivity of cucumber under water deficit and salinity stresses by balancing nutrients uptake. *Plant Physiol. Biochem.* 139: 1–10.

Alzahrani, Y., Kuşvuran, A., Alharby, H.F. et al. (2018). The defensive role of silicon in wheat against stress conditions induced by drought, salinity or cadmium. *Ecotoxicol. Environ. Saf. 154*: 187–196.

Ariannejad, M.M., Akbari, E., Niazi, M., and Hanafi, E. (2020). Urea sensor by racetrack silicon resonator. *Optik* 208: 164042.

Atkinson, N.J. and Urwin, P.E. (2012). The interaction of plant biotic and abiotic stresses: from genes to the field. *J. Exp. Bot.* 63: 3523–3543.

Atta, B., Rizwan, M., Sabir, A.M. et al. (2019). Silicon mediated induced resistance in plants for the management of agricultural insect pests: a review. *World J. Biol. Biotechnol.* 4: 19–27.

Ayub, M.A., Ahmad, H.R., Ali, M. et al. (2020). Salinity and its tolerance strategies in plants. In: *Plant Life Under Changing Environment*, 47–76. Academic Press.

Azhar, M., Ur Rehman, M.Z., Ali, S. et al. (2019). Comparative effectiveness of different biochars and conventional organic materials on growth, photosynthesis and cadmium accumulation in cereals. *Chemosphere 227*: 72–81.

Bae, E.J., Hong, A.R., Choi, S.M. et al. (2017). Silicon pretreatment alleviates drought stress and increases antioxidative activity in Kentucky bluegrass. *Int. Turfgrass Societ. Res. J. 13*: 591–600.

Benedikovic, D., Virot, L., Aubin, G. et al. (2019). 25 Gbps low-voltage hetero-structured silicon–germanium waveguide pin photodetectors for monolithic on-chip nanophotonic architectures. *Photonics Res.* 7: 437–444.

Biju, S., Fuentes, S., and Gupta, D. (2017). Silicon improves seed germination and alleviates drought stress in lentil crops by regulating osmolytes, hydrolytic enzymes and antioxidant defense system. *Plant Physiol. Biochem. 119*: 250–264.

Broadley, M., Brown, P., Cakmak, I. et al. (2012). Beneficial elements. In: *Marschner's Mineral Nutrition of Higher Plants*, 3e (ed. P. Marschner), 249–269. London: Academic.

Brunetto, G., GWBD, M., Toselli, M. et al. (2015). The role of mineral nutrition on yields and fruit quality in grapevine, pear and apple. *Rev. Bras. Frutic.* 37: 1089–1104.

Cai, K., Gao, D., Chen, J., and Luo, S. (2009). Probing the mechanisms of silicon-mediated pathogen resistance. *Plant Signal. Behav.* 4: 1–3.

Carey, J.C. and Fulweiler, R.W. (2014). Silica uptake by Spartina – evidence of multiple modes of accumulation from salt marshes around the world. *Front. Plant Sci.* 5: 186.

Catalani, V.M.R. and Camargos, L.S.D. (2017). Effects of potassium silicate application on papaya plants in the twospotted spider mite population. *Rev. Bras. Frutic.* 39 (2): 1–4.

Chanchal Malhotra, C., Kapoor, R., and Ganjewala, D. (2016). Alleviation of abiotic and biotic stresses in plants by silicon supplementation. *Scientia* 13 (2): 59–73.

Chapman, A. (2012). *Numbers of Living Species in Australia and the World*. Canberra: Australian Biological Resources Study 80 p.

Chen, J., Wang, W., Xu, Y., and Zhang, X. (2011a). Slow-release formulation of a new biological pesticide, pyoluteorin, with mesoporous silica. *J. Agric. Food Chem.* 59 (1): 307–311.

Chen, W., Yao, X., Cai, K., and Chen, J. (2011b). Silicon alleviates drought stress of rice plants by improving plant water status, photosynthesis and mineral nutrient absorption. *Biol. Trace Elem. Res.* 142 (1): 67–76.

Chung, Y.S., Kim, K.S., Hamayun, M., and Kim, Y. (2019). Silicon confers soybean resistance to salinity stress through regulation of reactive oxygen and reactive nitrogen species. *Front. Plant Sci. 10*: 1275.

Cornelis, J.T., Titeux, H., Ranger, J., and Delvaux, B. (2011). Identification and distribution of the readily soluble silicon pool in a temperate forest soil below three distinct tree species. *Plant Soil* 342 (1–2): 369–378.

Correa, R.S., Moraes, J.C., Auad, A.M., and Carvalho, G.A. (2005). Silicon and acibenzolar-*S*-methyl as resistance inducers in cucumber, against the whitefly *Bemisia tabaci* (Gennadius) (Hemiptera: Aleyrodidae) biotype B. *Neotrop. Entomol.* 34 (3): 429–433.

Coskun, D., Britto, D.T., Huynh, W.Q., and Kronzucker, H.J. (2016). The role of silicon in higher plants under salinity and drought stress. *Front. Plant Sci.* 7: 1072.

Datnoff, L.E. and Rodrigues, F.A. (2015). History of silicon and plant disease. In: *Silicon and Plant Diseases* (ed. F.A. Rodrigues and L.E. Datnoff), 1–5. Cham: Springer.

Datnoff, L.E., Snyder, G.H., and Korndörfer, G.H. (2001). *Silicon in Agriculture* (ed. L.E. Datnoff, G.H. Snyder and G.H. Korndörfer). Elsevier.

De Camargo, M.S., Bezerra, B.K.L., Holanda, L.A. et al. (2019). Silicon fertilization improves physiological responses in sugarcane cultivars grown under water deficit. *J. Soil Sci. Plant Nutr. 19* (1): 81–91.

Debona, D., Rodrigues, F.A., and Datnoff, L.E. (2017). Silicon's role in abiotic and biotic plant stresses. *Annu. Rev. Phytopathol.* 55: 85–107.

Deshmukh, R., Tripathi, D.K., and Guerriero, G. (2020). *Metalloids in Plants: Advances and Future Prospects*, vol. 1, 1–520. Wiley.

Dhiman, P., Rajora, N., Bhardwaj, S. et al. (2021). Fascinating role of silicon to combat salinity stress in plants: an updated overview. *Plant Physiol. Biochem.* 162: 110–123.

Ding, X., Zhang, S., Li, S. et al. (2013). Silicon mediated the detoxification of Cr on Pakchoi (*Brassica Chinensis* L.) in Cr-contaminated soil. *Procedia Environ. Sci.* 18: 58–67.

Dragišić Maksimović, J., Mojović, M., Maksimović, V. et al. (2012). Silicon ameliorates manganese toxicity in cucumber by decreasing hydroxyl radical accumulation in the leaf apoplast. *J. Exp. Bot.* 63 (7): 2411–2420.

Durga (2020). Silicon nutrition in rice. *Biotica Res. Today* 2 (7): 565–566.

El-Sheery, I. (2017). Effectiveness of potassium silicate in suppression white rot disease and enhancement physiological resistance of onion plants, and its role on the soil microbial community. *Middle-East J. Sci. Res.* 6 (2): 376–394.

Epstein, E. (1994). The anomaly of silicon in plant biology. *Proc. Natl. Acad. Sci.* 91 (1): 11–17.

Epstein, E. (2009). Silicon: its manifold roles in plants. *Ann. Appl. Biol.* 155 (2): 155–160.

Eroglu, N., Emekci, M., and Athanassiou, C.G. (2017). Applications of natural zeolites on agriculture and food production. *J. Sci. Food Agric.* 97 (11): 3487–3499.

Etesami, H. (2018). Can interaction between silicon and plant growth promoting rhizobacteria benefit in alleviating abiotic and biotic stresses in crop plants? *Agric. Ecosyst. Environ.* 253: 98–112.

Etesami, H. and Jeong, B.R. (2018). Silicon (Si): review and future prospects on the action mechanisms in alleviating biotic and abiotic stresses in plants. *Ecotoxicol. Environ. Saf.* 147: 881–896.

Farooq, M.A., Saqib, Z.A., Akhtar, J. et al. (2019). Protective role of silicon (Si) against combined stress of salinity and boron (B) toxicity by improving antioxidant enzymes activity in rice. *Silicon* 11 (4): 2193–2197.

Farouk, S. and Omar, M.M. (2020). Sweet basil growth, physiological and ultrastructural modification, and oxidative defense system under water deficit and silicon forms treatment. *J. Plant Growth Regul.* 39 (3): 1–25.

Farouk, S., Belal, B.E.A., and El-Sharkawy, H.H.A. (2017). The role of some elicitors on the management of Roumy Ahmar grapevines downy mildew disease and it's related to inducing growth and yield characters. *Sci. Hortic. 225*: 646–658.

Fleck, A.T., Nye, T., Repenning, C. et al. (2011). Silicon enhances suberization and lignification in roots of rice (*Oryza sativa*). *J. Exp. Bot.* 62 (6): 2001–2011.

Fleck, A.T., Schulze, S., Hinrichs, M. et al. (2015). Silicon promotes exodermal Casparian band formation in Si-accumulating and Si-excluding species by forming phenol complexes. *PLoS One* 10 (9): 1–18.

Frantz, J.M., Pitchay, D.D.S., Locke, J.C. et al. (2005). Silicon is deposited in leaves of New Guinea impatiens. *Plant Health Prog.* 6 (1): 15.

Frings, P.J., Clymans, W., Jeppesen, E. et al. (2014). Lack of steady-state in the global biogeochemical Si cycle: emerging evidence from lake Si sequestration. *Biogeochemistry* 117 (2–3): 255–277.

Gall, H., Philippe, F., Domon, J.M. et al. (2015). Cell wall metabolism in response to abiotic stress. *Plants* 4 (1): 112–166.

Gao, X., Zou, C., Wang, L., and Zhang, F. (2006). Silicon decreases transpiration rate and conductance from stomata of maize plants. *J. Plant Nutr.* 29 (9): 1637–1647.

Gascho, G.J. (2001). Silicon sources for agriculture. In: *Studies in Plant Science*, vol. 8 (ed. L.E. Datnoff, G.H. Snyder and G.H. KorndOrfer), 197–207. Elsevier.

Gatarayiha, M.C., Laing, M.D., and Miller, R.M. (2010). Combining applications of potassium silicate and *Beauveria bassiana* to four crops to control two spotted spider mite, *Tetranychus urticae* Koch. *Int. J. Pest. Manag.* 56 (4): 291–297.

Gaur, S., Kumar, J., Kumar, D. et al. (2020). Fascinating impact of silicon and silicon transporters in plants: a review. *Ecotoxicol. Environ. Saf.* 202: 110885.

Gill, S.S. and Tuteja, N. (2010). Polyamines and abiotic stress tolerance in plants. *Plant Signal. Behav.* 5: 26–33.

Gogos, A., Knauer, K., and Bucheli, T.D. (2012). Nanomaterials in plant protection and fertilization: current state, foreseen applications, and research priorities. *J. Agric. Food Chem.* 60 (39): 9781–9792.

Gomes, F.B., Moraes, J.C.D., Santos, C.D.D., and Goussain, M.M. (2005). Resistance induction in wheat plants by silicon and aphids. *Sci. Agric.* 62 (6): 547–551.

Gong, H.J., Chen, K.M., Chen, G.C. et al. (2003). Effects of silicon on growth of wheat under drought. *J. Plant Nutr.* 26 (5): 1055–1063.

Gu, H.H., Qiu, H., Tian, T. et al. (2011). Mitigation effects of silicon rich amendments on heavy metal accumulation in rice (*Oryza sativa* L.) planted on multi-metal contaminated acidic soil. *Chemosphere* 83 (9): 1234–1240.

Gu, H.H., Zhan, S.S., Wang, S.Z. et al. (2012). Silicon-mediated amelioration of zinc toxicity in rice (*Oryza sativa* L.) seedlings. *Plant Soil* 350 (1–2): 193–204.

Guerriero, G., Hausman, J.F., and Legay, S. (2016). Silicon and the plant extracellular matrix. *Front. Plant Sci.* 7: 463.

Guntzer, F., Keller, C., and Meunier, J.D. (2012a). Benefits of plant silicon for crops: a review. *Agron. Sustain. Dev.* 32 (1): 201–213.

Guntzer, F., Keller, C., Poulton, P.R. et al. (2012b). Long-term removal of wheat straw decreases soil amorphous silica at Broadbalk, Rothamsted. *Plant Soil* 352 (1–2): 173–184.

Gupta, B. and Huang, B. (2014). Mechanism of salinity tolerance in plants: physiological, biochemical, and molecular characterization. *Int. J. Genomics* 2014: 701596.

Hasanuzzaman, M., Nahar, K., Anee, T.I., and Fujita, M. (2017). Exogenous silicon attenuates cadmium-induced oxidative stress in *Brassica napus* L. by modulating AsA-GSH pathway and glyoxalase system. *Front. Plant Sci.* 8: 1061.

Hasanuzzaman, M., Nahar, K., Rohman, M.M. et al. (2018). Exogenous silicon protects *Brassica napus* plants from salinity-induced oxidative stress through the modulation of AsA-GSH pathway, thiol-dependent antioxidant enzymes and glyoxalase systems. *Gesunde Pflanzen 70* (4): 185–194.

Hattori, T., Inanaga, S., Araki, H. et al. (2005). Application of silicon enhanced drought tolerance in *Sorghum bicolor*. *Physiol. Plant.* 123 (4): 459–466.

Haynes, R.J. (2014). A contemporary overview of silicon availability in agricultural soils. *J. Plant Nutr. Soil Sci.* 177 (6): 831–844.

Haynes, R.J. (2017). The nature of biogenic Si and its potential role in Si supply in agricultural soils. *Agric. Ecosyst. Environ.* 245: 100–111.

Haynes, R.J., Belyaeva, O.N., and Kingston, G. (2013). Evaluation of industrial wastes as sources of fertilizer silicon using chemical extractions and plant uptake. *J. Plant Nutr. Soil Sci.* 176: 238–248.

Hernandez-Apaolaza, L. (2014). Can silicon partially alleviate micronutrient deficiency in plants? A review. *Planta* 240 (3): 447–458.

Hodson, M.J., White, P.J., Mead, A., and Broadley, M.R. (2005). Phylogenetic variation in the silicon composition of plants. *Ann. Bot.* 96: 1027–1046.

Horie, T., Kaneko, T., Sugimoto, G. et al. (2011). Mechanisms of water transport mediated by PIP aquaporins and their regulation via phosphorylation events under salinity stress in barley roots. *Plant Cell Physiol.* 52 (4): 663–675.

Hurtado, A.C., Chiconato, D.A., de Mello, P.R. et al. (2019). Silicon attenuates sodium toxicity by improving nutritional efficiency in sorghum and sunflower plants. *Plant Physiol. Biochem.* 142: 224–233.

Isa, M., Bai, S., Yokoyama, T. et al. (2010). Silicon enhances growth independent of silica deposition in a low-silica rice mutant, lsi1. *Plant Soil* 331 (1–2): 361–375.

Janicka-Russak, M., Kabała, K., and Burzyński, M. (2012). Different effect of cadmium and copper on H+-ATPase activity in plasma membrane vesicles from *Cucumis sativus* roots. *J. Exp. Bot.* 63 (11): 4133–4142.

Jeer, M., Yele, Y., Sharma, K.C., and Prakash, N.B. (2020). Exogenous application of different silicon sources and potassium reduces pink stem borer damage and improves photosynthesis, yield and related parameters in wheat. *Silicon* 13: 901–910.

Jiao-Jing, L., Shao-Hang, L., Pei-Lei, X. et al. (2009). Effects of exogenous silicon on the activities of antioxidant enzymes and lipid peroxidation in chilling-stressed cucumber leaves. *Agric. Sci. China* 8: 1075–1086.

Jullok, N., Van Hooghten, R., Luis, P. et al. (2016). Effect of silica nanoparticles in mixed matrix membranes for pervaporation dehydration of acetic acid aqueous solution: plant-inspired dewatering systems. *J. Clean. Prod.* 112: 4879–4889.

Kaya, C., Tuna, L., and Higgs, D. (2006). Effect of silicon on plant growth and mineral nutrition of maize grown under water-stress conditions. *J. Plant Nutr.* 29 (8): 1469–1480.

Keller, C., Rizwan, M., Davidian, J.C. et al. (2015). Effect of silicon on wheat seedlings (*Triticum turgidum* L.) grown in hydroponics and exposed to 0 to 30 μM Cu. *Planta* 241 (4): 847–860.

Khan, W.U.D., Aziz, T., Hussain, I. et al. (2016). Silicon: beneficial nutrient for maize crop to enhance photochemical efficiency of photosystem II under salt stress. *Arch. Agron. Soil Sci.* 63 (5): 599–611.

Khan, S.M., Gumus, A., Nassar, J.M., and Hussain, M.M. (2018). CMOS enabled microfluidic systems for healthcare based applications. *Adv. Mater.* 30 (16): 1705759.

Khan, A., Khan, A.L., Muneer, S. et al. (2019). Silicon and salinity: cross-talk in crop mediated stress tolerance mechanisms. *Front. Plant Sci.* 10: 1429.

Kim, Y.H., Khan, A.L., Waqas, M. et al. (2014). Silicon application to rice root zone influenced the phytohormonal and antioxidant responses under salinity stress. *J. Plant Growth Regul.* 33 (2): 137–149.

Kleiber, T. (2018). The role of silicon in plant tolerance to abiotic stress. In: *Plant Nutrients and Abiotic Stress Tolerance* (ed. Hasanuzzaman, M., Fujita, M., Oku, H. et al.), 253–267. Singapore: Springer.

Kleiber, T., Calomme, M., and Borowiak, K. (2015b). The effect of choline-stabilized orthosilicic acid on microelements and silicon concentration, photosynthesis activity and yield of tomato grown under Mn stress. *Plant Physiol. Biochem. 96*: 180–188.

Kvedaras, O.L., An, M., Choi, Y.S., and Gurr, G.M. (2010). Silicon enhances natural enemy attraction and biological control through induced plant defences. *Bull. Entomol. Res.* 100 (3): 367.

Laane, H.M. (2017). The effects of the application of foliar sprays with stabilized silicic acid: an overview of the results from 2003–2014. *Silicon* 9 (6): 803–807.

Landi, S., Hausman, J.F., Guerriero, G., and Esposito, S. (2017). Poaceae vs. abiotic stress: focus on drought and salt stress, recent insights and perspectives. *Front. Plant Sci.* 8: 1214.

Lang, Y., Han, Y.Q., Li, P. et al. (2017). Silicon amendment is involved in the induction of plant defense responses to a phloem feeder. *Sci. Rep.* 7: 4232.

Le, B., Nadeem, M., Yang, S.H. et al. (2019). Effect of silicon in *Pyropia yezoensis* under temperature and irradiance stresses through antioxidant gene expression. *J. Appl. Phycol. 31* (2): 1297–1302.

Lee, S.K., Sohn, E.Y., Hamayun, M. et al. (2010). Effect of silicon on growth and salinity stress of soybean plant grown under hydroponic system. *Agrofor. Syst.* 80: 333–340.

Li, W.B., Wang, H., and Zhang, F.S. (2005). Effects of silicon on anther dehiscence and pollen shedding in rice under high temperature stress. *Acta Agron. Sin.* 31: 134–136.

Li, L., Zheng, C., Fu, Y. et al. (2012). Silicate-mediated alleviation of Pb toxicity in banana grown in Pb-contaminated soil. *Biol. Trace Elem. Res.* 145 (1): 101–108.

Li, H., Zhu, Y., Hu, Y. et al. (2015). Beneficial effects of silicon in alleviating salinity stress of tomato seedlings grown under sand culture. *Acta Physiol. Plant.* 37 (4): 71.

Li, Y.T., Zhang, W.J., and Cui, J.J. (2016). *et al*, Silicon nutrition alleviates the lipid peroxidation and ion imbalance of *Glycyrrhiza uralensis* seedlings under salt stress. *Acta Physiol. Plant.* 38: 1–9.

Li, R., Petit, L., Franke, D.P. et al. (2018). A crossbar network for silicon quantum dot qubits. *Sci. Adv.* 4 (7): 3960.

Liang, Y., Zhang, W., Chen, Q. et al. (2006). Effect of exogenous silicon (Si) on H+-ATPase activity, phospholipids and fluidity of plasma membrane in leaves of salt-stressed barley (*Hordeum vulgare* L.). *Environ. Exp. Bot.* 57 (3): 212–219.

Liang, Y.C., Sun, W.C., Zhu, Y.G., and Christie, P. (2007). Mechanisms of silicon-mediated alleviation of abiotic stresses in higher plants: a review. *Environ. Pollut.* 147: 422–428.

Liang, Y., Zhu, J., Li, Z. et al. (2008). Role of silicon in enhancing resistance to freezing stress in two contrasting winter wheat cultivars. *Environ. Exp. Bot.* 64 (3): 286–294.

Liang, Y., Nikolic, M., Bélanger, R. et al. (2015a). Effect of silicon on crop growth, yield and quality. In: *Silicon in Agriculture* (ed. D.G.H. Snyder and G.H. KorndOrfer), 209–223. Dordrecht: Springer.

Liang, Y., Nikolic, M., Bélanger, R. et al. (2015b). Silicon-mediated tolerance to salt stress. In: *Silicone in Agriculture*, 123–142. Springer Science.

Liu, P., Yin, L., Wang, S. et al. (2015). Enhanced root hydraulic conductance by aquaporin regulation accounts for silicon alleviated salt-induced osmotic stress in *Sorghum bicolor* L. *Environ. Exp. Bot.* 111: 42–51.

Liu, B., Soundararajan, P., and Manivannan, A. (2019). Mechanisms of silicon-mediated amelioration of salt stress in plants. *Plants* 8 (9): 307.

Lucas, S., Durmaz, E., Akpınar, B.A., and Budak, H. (2011). The drought response displayed by a DRE-binding protein from *Triticum dicoccoides*. *Plant Physiol. Biochem.* 49 (3): 346–351.

Lukačová, Z., Švubová, R., Kohanová, J., and Lux, A. (2013). Silicon mitigates the Cd toxicity in maize in relation to cadmium translocation, cell distribution, antioxidant enzymes stimulation and enhanced endodermal apoplasmic barrier development. *Plant Growth Regul.* 70 (1): 89–103.

Luyckx, M., Hausman, J.F., Lutts, S., and Guerriero, G. (2017). Silicon and plants: current knowledge and technological perspectives. *Front. Plant Sci.* 8: 411.

Ma, J.F. and Takahashi, E. (2002). *Soil, Fertilizer, and Plant Silicon Research in Japan* (ed. J.F. Ma and E. Takahashi). Elsevier.

Ma, J.F. and Yamaji, N. (2006). Silicon uptake and accumulation in higher plants. *Trends Plant Sci.* 11 (8): 392–397.

Ma, J.F., Miyake, Y., and Takahashi, E. (2001). Silicon as a beneficial element for crop plants. In: *Studies in Plant Science*, vol. 8 (ed. L.E. Datnoff, G.H. Snyder and G.H. KorndOrfer), 17–39. Elsevier.

Ma, J.F., Tamai, K., Yamaji, N. et al. (2006). A silicon transporter in rice. *Nature* 440: 688–691.

Ma, J.F., Yamaji, N., Mitani, N. et al. (2007). An efflux transporter of silicon in rice. *Nature* 448: 209–212.

Ma, J.F., Yamaji, N., and Mitani-Ueno, N. (2011). Transport of silicon from roots to panicles in plants. *Proc. Jpn. Acad. Ser. B* 87: 377–385.

Magda, S. and Hussein, M.M. (2016). Determinations of the effect of using silica gel and nano-silica gel against Tutaabsoluta (*Lepidoptera: Gelechiidae*) in tomato fields. *Pharm. Chem. J. Res.* 8 (4): 506–512.

Maghsoudi, K., Emam, Y., and Ashraf, M. (2015). Influence of foliar application of silicon on chlorophyll fluorescence, photosynthetic pigments, and growth in water-stressed wheat cultivars differing in drought tolerance. *Turk. J. Bot. 39* (4): 625–634.

Maghsoudi, K., Arvin, M.J., and Ashraf, M. (2019). Mitigation of arsenic toxicity in wheat by the exogenously applied salicylic acid, 24-epi-brassinolide and silicon. *J. Soil Sci. Plant Nutr.* 20: 577–588.

Malhotra, C. and Kapoor, R.T. (2019). Assessment of efficacy of sodium silicate on morphological parameters of *Lycopersicon esculentum* Mill. Under water deficit conditions. *Bulg. J. Agric. Sci. 25* (6): 1224–1232.

Massey, F.P. and Hartley, S.E. (2009). Physical defences wear you down: progressive and irreversible impacts of silica on insect herbivores. *J. Anim. Ecol.* 78 (1): 281–291.

Maxim, L.D. and McConnell, E.E. (2005). A review of the toxicology and epidemiology of wollastonite. *Inhal. Toxicol.* 17 (9): 451–466.

Maxim, L.D.R., Niebo, S., LaRosa, B. et al. (2008). Product stewardship in wollastonite production. *Inhal. Toxicol.* 20: 1199–1214.

Mburu, K., Oduor, R., Mgutu, A., and Tripathi, L. (2016). Silicon application enhances resistance to xanthomonas wilt disease in banana. *Plant Pathol.* 65 (5): 807–818.

Mditshwa, A., Bower, J.P., Bertling, I. et al. (2013). The potential of postharvest silicon dips to regulate phenolics in citrus peel as a method to mitigate chilling injury in lemons. *Afr. J. Biotechnol.* 12 (13): 1482–1489.

Meharg, C. and Meharg, A.A. (2015). Silicon, the silver bullet for mitigating biotic and abiotic stress, and improving grain quality, in rice? *Environ. Exp. Bot.* 120: 8–17.

Mitani, N., Yamaji, N., and Ma, J.F. (2009). Identification of maize silicon influx transporters. *Plant Cell Physiol.* 50: 5–12.

Muneer, S., Park, Y.G., Kim, S., and Jeong, B.R. (2017). Foliar or subirrigation silicon supply mitigates high temperature stress in strawberry by maintaining photosynthetic and stress-responsive proteins. *J. Plant Growth Regul.* *36* (4): 836–845.

Murad, M., Khan, A.L., and Muneer, S. (2020). Silicon in horticultural crops: cross-talk, signaling, and tolerance mechanism under salinity stress. *Plants* 9 (4): 460.

Nagajyoti, P.C., Lee, K.D., and Sreekanth, T.V.M. (2010). Heavy metals, occurrence and toxicity for plants: a review. *Environ. Chem. Lett.* 8 (3): 199–216.

Nascimento, C.W.A.D. and Xing, B. (2006). Phytoextraction: a review on enhanced metal availability and plant accumulation. *Sci. Agric.* 63 (3): 299–311.

Nasla, M.F.F., Prasannath, K., and Gunapala, K.R.D. (2019). Exploring the efficacy of silicon supplementation on control of rice grain discoloration disease. *AGRIEAST J. Agric. Sci.* 13: 1–11.

Nikpay, A. and Laane, H.M. (2020). Foliar amendment of silicic acid on population of yellow mite, *Oligonychus sacchari* (Acari: Tetranychidae) and its predatory beetle, *Stethorus gilvifrons* (Col.: Coccinellidae) on two sugarcane commercial varieties. *Pers. J. Acarol. 9* (1): 57–66.

Oliveira, R.S., Peñaflor, M.F.G., and Gonçalves, F.G. (2020). Silicon-induced changes in plant volatiles reduce attractiveness of wheat to the bird cherry-oat aphid *Rhopalosiphum padi* and attract the parasitoid *Lysiphlebus testaceipes*. *PLoS One 15* (4): e0231005.

Osakabe, Y., Osakabe, K., Shinozaki, K., and Tran, L.S.P. (2014). Response of plants to water stress. *Front. Plant Sci.* 5: 86.

Pagani, A.P.S., Dianese, A.C., and Café-Filho, A.C. (2014). Management of wheat blast with synthetic fungicides, partial resistance and silicate and phosphite minerals. *Phytoparasitica* 42 (5): 609–617.

Parisi, C., Vigani, M., and Rodríguez-Cerezo, E. (2015). Agricultural nanotechnologies: what are the current possibilities? *Nano Today* 10 (2): 124–127.

Paul, D. and Lade, H. (2014). Plant-growth-promoting rhizobacteria to improve crop growth in saline soils: a review. *Agron. Sustain. Dev.* 34: 737–752.

Pei, Z.F., Ming, D.F., Liu, D. et al. (2010). Silicon improves the tolerance to water-deficit stress induced by polyethylene glycol in wheat (*Triticum aestivum* L.) seedlings. *J. Plant Growth Regul.* 29 (1): 106–115.

Pilon, C., Soratto, R.P., Broetto, F., and Fernandes, A.M. (2014). Foliar or soil applications of silicon alleviate water-deficit stress of potato plants. *Agron. J. 106* (6): 2325–2334.

Pontigo, S.K., Godoy, H., Jiménez, A. et al. (2017). Silicon-mediated alleviation of aluminum toxicity by modulation of al/si uptake and antioxidant performance in ryegrass plants. *Front. Plant Sci.* 8: 642.

Puppe, D., Ehrmann, O., Kaczorek, D. et al. (2015). The protozoic Si pool in temperate forest ecosystems quantification, abiotic controls and interactions with earthworms. *Geoderma* 243: 196–204.

Qian, L., Chen, B., and Chen, M. (2016). Novel alleviation mechanisms of aluminum phytotoxicity via released biosilicon from rice straw-derived biochars. *Sci. Rep.* 6: 29346.

Qin, F., Shinozaki, K., and Yamaguchi-Shinozaki, K. (2011). Achievements and challenges in understanding plant abiotic stress responses and tolerance. *Plant Cell Physiol.* 52 (9): 1569–1582.

Rai, K.K., Rai, N., and Rai, S.P. (2018). Response of *Lablab purpureus* L. to high temperature stress and role of exogenous protectants in mitigating high temperature induced oxidative damages. *Mol. Biol. Rep. 45* (5): 1375–1395.

Rains, D.W., Epstein, E., Zasoski, R.J., and Aslam, M. (2006). Active silicon uptake by wheat. *Plant Soil* 280 (1–2): 223–228.

Ratnayake, R.M.R.N.K., Ganehenege, M., Ariyarathne, H., and Daundasekera, W. (2018). Soil application of rice husk as a natural silicon source to enhance some chemical defense responses against foliar fungal pathogens and growth performance of Bitter Gourd (*Momordica charantia* L.). *Ceylon J. Sci. 47* (1): 49–55.

Rebelo, Q.H.F.C.S., Ferreira, P.L., Santos, J.A. et al. (2019). Synthesis and characterization of a nanocomposite NiO/SiO_2 from a sustainable source of SiO_2. *Particul. Sci. Tech.* 37 (8): 911–915.

Rehman, M.Z., Rizwan, M., Rauf, A. et al. (2019). Split application of silicon in cadmium (Cd) spiked alkaline soil plays a vital role in decreasing Cd accumulation in rice (*Oryza sativa* L.) grains. *Chemosphere 226*: 454–462.

Reynolds, O.L., Keeping, M.G., and Meyer, J.H. (2009). Silicon-augmented resistance of plants to herbivorous insects: a review. *Ann. Appl. Biol.* 155 (2): 171–186.

Reynolds, O.L., Padula, M.P., Zeng, R., and Gurr, G.M. (2016). Silicon: potential to promote direct and indirect effects on plant s against arthropod pests in agriculture. *Front. Plant Sci.* 7: 744.

Rizwan, M., Meunier, J.D., Miche, H., and Keller, C. (2012). Effect of silicon on reducing cadmium toxicity in durum wheat (*Triticum turgidum* L. cv. Claudio W.) grown in a soil with aged contamination. *J. Hazard. Mater.* 209: 326–334.

Rizwan, M., Ali, S., Ibrahim, M. et al. (2015). Mechanisms of silicon-mediated alleviation of drought and salt stress in plants: a review. *Environ. Sci. Pollut. Res.* 22: 15416–15431.

Rouhani, M., Samih, M.A., and Kalantari, S. (2012). Insecticidal effect of silica and silver nanoparticles on the cowpea seed beetle, *Callosobruchus maculatus* F. (Col.: Bruchidae). *J. Entomol. Res.* 4: 297–305.

Roychoudhury, A. and Tripathi, D.K. (2019). *Molecular Plant Abiotic Stress: Biology and Biotechnology*, vol. 1, 1–480. Wiley.

Roychoudhury, A. and Tripathi, D.K. (2020). *Protective Chemical Agents in the Amelioration of Plant Abiotic Stress: Biochemical and Molecular Perspectives*, vol. 1 (ed. R. Aryadeep and D.K. Tripathi), 1–704. Wiley Publishers.

Sakr, N. (2018). Silicon-enhanced resistance of plants to biotic stresses review article. *Acta Phytopathol. Entomol. Hung.* 53 (2): 125–141.

Sallam, M.N. (2008). *Insect Damage: Damage on Post-harvest.* AGSI/FAO: INPhO Available via.

Sanglard, L.M., Detmann, K.C., Martins, S.C. et al. (2016). The role of silicon in metabolic acclimation of rice plants challenged with arsenic. *Environ. Exp. Bot.* 123: 22–36.

Santos, M.C., Junqueira, M.R., de Sá, V.M. et al. (2015). Effect of silicon on the morphology of the midgut and mandible of tomato leafminer *Tuta absoluta* (Lepidoptera: Gelechiidae) larvae. *Invertebr. Surviv. J.* 12 (1): 158–165.

Sarkar, U., Tahura, S., Das, U. et al. (2020). Mitigation of chromium toxicity in wheat (*Triticum aestivum* L.) through silicon. *Gesunde Pflanzen* 72: 237–244.

Sarma, R.S., Shankhdhar, D., and Shankhdhar, S.C. (2019). Beneficial effects of silicon fertilizers on disease and insect-pest management in rice genotypes (*Oryza sativa* L.). *J. Pharmacognosy Phytochem.* 8 (3): 358–362.

Sattar, A., Cheema, M.A., Abbas, T. et al. (2017). Physiological response of late sown wheat to exogenous application of silicon. *Cereal Res. Commun. 45* (2): 202–213.

Sattar, A., Sher, A., Ijaz, M. et al. (2020). Interactive effect of biochar and silicon on improving morpho-physiological and biochemical attributes of maize by reducing drought hazards. *J. Soil Sci. Plant Nutr.* 20: 1819–1826.

Sauer, D., Saccone, L., Conley, D.J. et al. (2006). Review of methodologies for extracting plant-available and amorphous Si from soils and aquatic sediments. *Biogeochemistry* 80 (1): 89–108.

Shahrtash, M. (2018). Silicon fertilization as a sustainable approach to disease management of agricultural crops. *J. Plant. Prot. Res.* 58 (4): 317–320.

Shen, Y. (2017). Rice husk silica derived nanomaterials for sustainable applications. *Renew. Sustain. Energy Rev. 80*: 453–466.

Shen, X., Zhou, Y., Duan, L. et al. (2010). Silicon effects on photosynthesis and antioxidant parameters of soybean seedlings under drought and ultraviolet-B radiation. *J. Plant Physiol.* 167 (15): 1248–1252.

Shi, Y., Zhang, Y., Han, W. et al. (2016). Silicon enhances water stress tolerance by improving root hydraulic conductance in *Solanum lycopersicum* L. *Front. Plant Sci.* 7: 196.

Shim, J., Shea, P.J., and Oh, B.T. (2014). Stabilization of heavy metals in mining site soil with silica extracted from corn cob. *Water Air Soil Pollut.* 225 (10): 2152.

Shivaraj, S.M., Mandlik, R., Bhat, J.A. et al. (2021). Outstanding questions on the beneficial role of silicon in crop plants. *Plant Cell Physiol.* 63 (1): 4–18.

Sidhu, J.K., Stout, M.J., Blouin, D.C., and Datnoff, L.E. (2013). Effect of silicon soil amendment on performance of sugarcane borer, *Diatraea saccharalis* (Lepidoptera: Crambidae) on rice. *Bull. Entomol. Res.* 103 (6): 656–664.

Sohail, M.I., Ur Rehman, M.Z., Rizwan, M. et al. (2020a). Effect of biochars, biogenic, and inorganic amendments on dissolution and kinetic release of phytoavailable silicon in texturally different soils under submerged conditions. *Arab. J. Geosci. 13*: 376.

Sohail, M.I., Ur Rehman, M.Z., Rizwan, M. et al. (2020b). Efficiency of various silicon rich amendments on growth and cadmium accumulation in field grown cereals and health risk assessment. *Chemosphere* 244: 125481.

Sommer, M., Kaczorek, D., Kuzyakov, Y., and Breuer, J. (2006). Silicon pools and fluxes in soils and landscapes – a review. *J. Plant Nutr. Soil Sci.* 169 (3): 310–329.

Sommer, M., Jochheim, H., Höhn, A. et al. (2013). Si cycling in a forest biogeosystem–the importance of transient state biogenic Si pools. *Biogeosciences* 10 (7): 4991–5007.

Song, A., Ning, D., Fan, F. et al. (2015). The potential for carbon bio-sequestration in China's paddy rice (*Oryza sativa* L.) as impacted by slag-based silicate fertilizer. *Sci. Rep.* 5: 17354.

Song, A., Xue, G., Cui, P. et al. (2016). The role of silicon in enhancing resistance to bacterial blight of hydroponic-and soil-cultured rice. *Sci. Rep.* 6: 24640.

Sonobe, K., Hattori, T., An, P. et al. (2010). Effect of silicon application on sorghum root responses to water stress. *J. Plant Nutr.* 34 (1): 71–82.

Sousa, R.T.X. and Korndorfer, G.H. (2010). Slag efficacy as a lime and silicon source for rice crops through the biological method. *J. Plant Nutr.* 33: 1103–1111.

Sousa, A., Saleh, A.M., Habeeb, T.H. et al. (2019). Silicon dioxide nanoparticles ameliorate the phytotoxic hazards of aluminum in maize grown on acidic soil. *Sci. Total Environ. 693*: 133636.

Suzuki, S., Ma, J.F., Yamamoto, N. et al. (2012). Silicon deficiency promotes lignin accumulation in rice. *Plant Biotech.* 29 (4): 391–394.

Telles, C.C., de Freitas, L.M., Junqueira, A.M.R., and de Mendonça, R.S. (2019). Silicon application as an auxiliary method to control diamondback moth in cabbage plants. *Hortic. Bras. 37* (4): 390–394.

Tibbitts, S.A. (2018). Effect of silicon on wheat growth and development in drought and salinity stress. Doctoral dissertation. Utah State University.

Tripathi, D.K., Singh, V.P., Kumar, D., and Chauhan, D.K. (2012). Impact of exogenous silicon addition on chromium uptake, growth, mineral elements, oxidative stress, antioxidant capacity, and leaf and root structures in rice seedlings exposed to hexavalent chromium. *Acta Physiol. Plant.* 34 (1): 279–289.

Tripathi, D.K., Singh, V.P., Prasad, S.M. et al. (2015a). Silicon nanoparticles (SiNp) alleviate chromium (VI) phytotoxicity in *Pisum sativum* (L.) seedlings. *Plant Physiol. Biochem.* 96: 189–198.

Tripathi, D.K., Singh, V.P., Prasad, S.M. et al. (2015b). Silicon-mediated alleviation of Cr (VI) toxicity in wheat seedlings as evidenced by chlorophyll florescence, laser induced breakdown spectroscopy and anatomical changes. *Ecotoxicol. Environ. Saf.* 113: 133–144.

Tripathi, D.K., Shweta, S.S., Yadav, V. et al. (2017). Silicon: a potential element to combat adverse impact of UV-B in plants. In: *UV-B Radiation: From Environmental Stressor to Regulator of Plant Growth*, vol. 1: 175–195. Wiley Publishers.

Tripathi, D.K., Singh, V.P., Lux, A., and Vaculik, M. (2020). Silicon in plant biology: from past to present, and future challenges. *J. Exp. Bot.* 71 (21): 6699–6702.

Tubana, B.S., Babu, T., and Datnoff, L.E. (2016). A review of silicon in soils and plants and its role in US agriculture: history and future perspectives. *Soil Sci.* 181: 393–411.

Tuna, A.L., Kaya, C., Higgs, D. et al. (2008). Silicon improves salinity tolerance in wheat plants. *Environ. Exp. Bot.* 62 (1): 10–16.

Van Bockhaven, J., De Vleesschauwer, D., and Höfte, M. (2013). Towards establishing broad-spectrum disease resistance in plants: silicon leads the way. *J. Exp. Bot.* 64 (5): 1281–1293.

Vandevenne, F.I., Barão, L., Ronchi, B. et al. (2015). Silicon pools in human impacted soils of temperate zones. *Glob. Biogeochem. Cycles* 29 (9): 1439–1450.

Vasanthi, N., Saleena, L.M., and Raj, S.A. (2012a). Silicon in day today life. *World Appl. Sci. J.* 17: 1425–1440.

Vasanthi, N., Chandrasekeran, D., and Raj, S.A. (2012b). Phytosil as an alternative carrier to talc for biocontrol agents. In: *Proc. Natl. Symp. Recent Adv. Bioinoculatns Techn.* Madurai: Agricultural College & Research Institute.

Verma, K.K., Liu, X.H., Wu, K.C. et al. (2019). The impact of silicon on photosynthetic and biochemical responses of sugarcane under different soil moisture levels. *Silicon* 12 (1355–1367): 1–13.

Verma, K.K., Singh, P., Song, X.P. et al. (2020). Mitigating climate change for sugarcane improvement: role of silicon in alleviating abiotic stresses. *Sugar Tech.* 22 (5): 741–749.

Virta, R.L. (2004). Wollastonite – U.S. geological survey. *Miner. Yearb.* 82: 1–3.

Wang, S., Liu, P., Chen, D. et al. (2015). Silicon enhanced salt tolerance by improving the root water uptake and decreasing the ion toxicity in cucumber. *Front. Plant Sci.* 6: 759.

Wang, M., Gao, L., Dong, S. et al. (2017). Role of silicon on plant–pathogen interactions. *Front. Plant Sci.* 8: 701.

Wang, M., Wang, J.J., and Wang, X. (2018). Effect of KOH-enhanced biochar on increasing soil plant-available silicon. *Geoderma* 321: 22–31.

Wang, Y., Xiao, X., Xu, Y., and Chen, B. (2019). Environmental effects of silicon within Biochar (Sichar) and carbon–silicon coupling mechanisms: a critical review. *Environ. Sci. Technol. 53* (23): 13570–13582.

Wang, J., Xue, R., Ju, X. et al. (2020). Silicon-mediated multiple interactions: simultaneous induction of rice defense and inhibition of larval performance and insecticide tolerance of *Chilo suppressalis* by sodium silicate. *Ecol. Evol.* 10 (11): 4816–4827.

Xie, Z., Song, R., Shao, H. et al. (2015). Silicon improves maize photosynthesis in saline-alkaline soils. *Sci. World J.* 2015: 1–6.

Xu, L., Islam, F., Ali, B. et al. (2017). Silicon and water-deficit stress differentially modulate physiology and ultrastructure in wheat (*Triticum aestivum* L.). *3 Biotech.* 7 (4): 273.

Xu, Y., Hu, X., Kundu, S. et al. (2019). Silicon-based sensors for biomedical applications: a review. *Sensors* 19 (13): 2908.

Yamaji, N., Chiba, Y., Mitani-Ueno, N., and Ma, J.F. (2012). Functional characterization of a silicon transporter gene implicated in silicon distribution in barley. *Plant Physiol.* 160 (3): 1491–1497.

Yang, L., Han, Y., Li, P. et al. (2017). Silicon amendment is involved in the induction of plant defense responses to a phloem feeder. *Sci. Rep.* 7 (1): 1–9.

Yao, A., Wang, Y., Ling, X. et al. (2017). Effects of an iron-silicon material, a synthetic zeolite and an alkaline clay on vegetable uptake of As and Cd from a polluted agricultural soil and proposed remediation mechanisms. *Environ. Geochem. Hlth.* 39 (2): 353–367.

Yavaş, I. and Aydın, U.N.A.Y. (2017). The role of silicon under biotic and abiotic stress conditions. *Türkiye Tarımsal Araştırmalar Dergisi* 4 (2): 204–209.

Yin, L., Wang, S., Li, J. et al. (2013). Application of silicon improves salt tolerance through ameliorating osmotic and ionic stresses in the seedling of *Sorghum bicolor*. *Acta Physiol. Plant.* 35 (11): 3099–3107.

Yin, L., Wang, S., Tanaka, K. et al. (2016). Silicon-mediated changes in polyamines participate in silicon-induced salt tolerance in *Sorghum bicolor* L. *Plant Cell Environ.* 39 (2): 245–258.

Zamani, G.R., Shaabani, J., and Izanloo, A. (2017). Silicon effects on the growth and yield of chickpea under salinity stress. *Int. J. Agric. Biol. 19*: 1475–1482.

Zanetti, L.V., Milanez, C.R.D., Gama, V.N. et al. (2016). Leaf application of silicon in young cacao plants subjected to water deficit. *Pesqi Agropecu Bras 51* (3): 215–223.

Zargar, S.M., Mahajan, R., Bhat, J.A. et al. (2019). Role of silicon in plant stress tolerance: opportunities to achieve a sustainable cropping system. *3 Biotech.* 9 (3): 73.

Zhang, M., Liang, Y., and Chu, G. (2017a). Applying silicate fertilizer increases both yield and quality of table grape (*Vitis vinifera* L.) grown on calcareous grey desert soil. *Sci. Hortic.* 225: 757–763.

Zhang, W., Xie, Z., Lang, D. et al. (2017b). Beneficial effects of silicon on abiotic stress tolerance in legumes. *J. Plant Nutr.* 40 (15): 2224–2236.

Zhou, X., Shen, Y., Fu, X., and Wu, F. (2018). Application of sodium silicate enhances cucumber resistance to *Fusarium wilt* and alters soil microbial communities. *Front. Plant Sci. 9*: 624.

Zhu, Y. and Gong, H. (2014). Beneficial effects of silicon on salt and drought tolerance in plants. *Agron. Sustain. Dev.* 34 (2): 455–472.

Zhu, J., Liang, Y.C., Ding, Y.F., and Li, Z.J. (2006). Effects of silicon on photosynthesis and its related physiological parameters in two winter wheat cultivars under cold stress. *Sci. Agric. J.* 39: 1780–1788.

Zhu, Y.X., Xu, X.B., Hu, Y.H. et al. (2015). Silicon improves salt tolerance by increasing root water uptake in *Cucumis sativus* L. *Plant Cell Rep.* 34 (9): 1629–1646.

Zhu, D., Xue, B., Jiang, Y., and Wei, C. (2019). Using chemical experiments and plant uptake to prove the feasibility and stability of coal gasification fine slag as silicon fertilizer. *Environ. Sci. Pollut. Res.* 26 (6): 5925–5933.

Ziaee, M. and Ganji, Z. (2016). Insecticidal efficacy of silica nanoparticles against *Rhyzopertha dominica* F. and *Tribolium confusum* Jacquelin du Val. *J. Plant. Prot. Res.* 56 (3): 250–256.

11

Silicon-Mediated Alleviation of Heavy Metal Stress in Plants

Sana Rana[1], Muhammad Zia ur Rehman[1], Muhammad Umair[1], Muhammad Ashar Ayub[2], and Muhammad Arif[3]

[1] *Institute of Soil and Environmental Sciences, Faculty of Agriculture, University of Agriculture, Faisalabad, Pakistan*
[2] *Institute of Agro-Industry and Environment, The Islamia University of Bahawalpur, Punjab, Pakistan*
[3] *Muhammad Nawaz Sharif University of Agriculture, Multan*

Introduction

Heavy metals (HMs) are elements with specific gravity >5 g cm^{-3}, and some of these are toxic to human and plant health (Ali and Khan 2018). In present era of rapid urbanization and industrialization, pollution by HMs has taken a leap in contaminating air, water, and soil equally. Industrial activities along with other anthropogenic processes are releasing a significant amount of HMs like copper (Cu), lead (Pb), arsenic (As), manganese (Mn), cadmium (Cd), chromium (Cr), aluminium (Al), and mercury (Hg) in ecosystem (Duffus 2002; He et al. 2005; Tripathi et al. 2021). Major sources of HM pollution in environment are industrial effluents, weathering of rocks and soil minerals, pharmaceutical industries, metallurgic work, and mining (Fergusson 1990; Bradl 2005; He et al. 2005) leading to huge accumulation in agro-ecosystem (Herawati et al. 2000; Azhar et al. 2019; Sohail et al. 2020). Natural sources of these HMs in ecosystem include volcanic eruptions, erosion, and weathering of parent rocks and minerals. In agro-ecosystem, input sources of these metals are agro-chemicals and raw city effluents being used for irrigation (Vamerali et al. 2010; Singh et al. 2011; Azhar et al. 2019; Naccarato et al. 2020; Sohail et al. 2020). Accumulation of these metals in soil above permissible limits can be a major concern for food security and health risks (Nagajyoti et al. 2010; Tchounwou et al. 2012) as these metal ions are very persistent in soil and are highly bioavailable (Bolan et al. 2014). HMs like Zn, Cu, Fe, Mn, Mo, and Ni are beneficial elements for plant growth (WHO/FAO/IAEA 1996), but their higher concentrations in soil can also become toxic for plants (Tchounwou et al. 2012). While other metallic ions like Pb, Al, Hg, As, Cd, and Cr are non-essential HM ions with non-biological function and are potentially toxic to plant health (Chang et al. 1996; Tripathi et al. 2015). The higher contents of both types of these metallic ions are toxic for proper plant growth and can contaminate food supplies, thus having to be managed carefully (Tchounwou et al. 2012). Excessive uptake of these metallic ions and their translocation in plant's body may result in oxidative stresses (Mithöfer et al. 2004), modulation of

Beneficial Chemical Elements of Plants: Recent Developments and Future Prospects, First Edition. Edited by Sangeeta Pandey, Durgesh Kumar Tripathi, Vijay Pratap Singh, Shivesh Sharma, and Devendra Kumar Chauhan.

antioxidant defence (Srivastava et al. 2004), electron harvesting and transport chain disturbance (Qadir et al. 2004), and disturbance in mineral element homeostasis in plants (Dong et al. 2006) that disrupt plant growth and functioning (Rizwan et al. 2016a). Major symptoms of toxicity of HMs in plants involve dark green and wilted leave, necrosis, and stunted growth (Bhattacharyya et al. 2008) as well as potential necrosis and disturbed root growth (Guo et al. 2008).

A wide range of techniques have been proposed for tackling of HM stress in plants like addition of some inorganic and organic amendments (Bolan and Duraisamy 2003; Azhar et al. 2019; Sohail et al. 2020), silicon-rich amendments (Sohail et al. 2020) and silicon sources (Rizwan et al. 2016b; Rehman et al. 2019), and phytoremediation (Shackira and Puthur 2019). Role of Si in this regard in very prominent (Sohail et al. 2020) as it is present in Earth's crust abundantly (Kovda 1973), but very less bioavailable fraction is present in soil (Sommer et al. 2006). Plants take up Si is luxury, and its content in plants' body can be up to 10% (Hodson et al. 2005) as it shows a significant part in strengthening plant's integrity and stress tolerance (Shi et al. 2014; Rizwan et al. 2015; Khan et al. 2019; Verma et al. 2019). Application of Si in soil mastics has been associated with decreased metal availability (Rehman et al. 2019) via various mechanisms like chelation, pH modulation, and complexation (Dong et al. 2019), which ultimately help reduce metal accumulation in plant's body (Maghsoudi et al. 2019; Sousa et al. 2019; Sarkar et al. 2020). This chapter reviews sources and effects of HMs in plants and potential part of silicon in stress management of HMs.

Heavy Metal (HM) Sources in Agro-ecosystem

Toxic HMs like Pb, As, Mn, Cd, Cr, Al, and Hg are potentially being accumulated in agro-ecosystem (Duffus 2002; He et al. 2005; Deshmukh et al. 2020) via various natural in addition to anthropogenic sources (Vamerali et al. 2010; Singh et al. 2011; Azhar et al. 2019; Naccarato et al. 2020; Sohail et al. 2020). Among various anthropogenic sources, use of agro-chemicals and contaminated wastewater are prominent in agro-ecosystem (Vamerali et al. 2010; Singh et al. 2011; Mazurek et al. 2017; Azhar et al. 2019).

The Response of Plants Towards HM Stress

HMs (toxic ones) are potential stress inducers in plants' body and disturb normal functioning and physiology of plants (Mithöfer et al. 2004; Rahman et al. 2010; Adrees et al. 2015a). Upon uptake, HMs induce oxidative stress in plants that can cause cellular damage to counter which plants produce antioxidants and osmo-protectants (Srivastava et al. 2004). Once HMs are taken up by plants, plants try to avoid the potential toxicity via oxidative stress management, intercellular and intracellular binding and chelation of HMs, production of stress-suppressing proteins, and other cellular mechanisms (Manara 2012). The HMs are very mobile in soil matrices and can be taken up by plants via roots and then make their way towards aerial tissues (Wang et al. 2007; Azhar et al. 2019) where they cause toxicity (Sharma and Dietz 2009, Feng et al. 2010). In plant roots, HMs also cause toxicity leading to disturbed root structure, decreased root growth, and damage to root cells and membranes (Seregin et al. 2004). Once taken up in plants, major outcomes of HM toxicity in plant tissues are disturbance in gaseous exchange and CO_2 fixation, less chlorophyll production, and more ROS production (Viehweger and Geipel 2010; Seneviratne et al. 2019) causing disruption in thylakoid membrane and organelle responses (Kim et al. 2014). Upon onset of ROS production, antioxidants are responsible for saving plant cell from oxidative damage, but HM stress also disturbs enzymatic processes via lipid peroxidation overall, which causes decrease in plant development and physiology (Tchounwou et al. 2012) like reported in literature (Gill et al. 2011; Maghsoudi et al. 2019; Sarkar et al. 2020). This kind of disruption in the growth and product quality is of major concern in modern agriculture (Nagajyoti et al. 2010; Tchounwou et al. 2012).

Sources of Silicon in Soil

Part of Si in the plant's stress management is unique and significant (Wang et al. 2004; Broadley et al. 2011; Liang et al. 2015b; Ning et al. 2016; Rasoolizadeh et al. 2018), but its availability in soil is contrasting to its available contents (Datnoff et al. 2001; Broadley et al. 2011). Earth's crust and soil are rich pool of silicon (Haynes 2017), but most of it is fixed in solid and amorphous phase (Liang et al. 2015a, b; Zhu and Gong 2014) and limiting amount is present in liquid pool (Sauer et al. 2006). Among other natural sources of Si, weathering of parent minerals (Datnoff et al. 2001), plant biomass (Frings et al. 2014), and indigenous microorganisms are important (Puppe et al. 2015). Besides natural sources, a wide range of chemical and artificial sources of Si are also present involving slag (Haynes 2014; Liang et al. 2015a, b) and other silicate minerals (Gascho 2001). Bioavailable fraction of Si in soil comprises monosilicic acids that have very low concentration in soil solution (Hodson et al. 2005) and are much needed by plants (Gocke et al. 2013) especially under their stressed phase (Wang et al. 2004). Industrial use of Si has created many by-products having significant Si contents that can be applied in agro-ecosystems like Ca/Mg-silicates, dolomite, and slag (Savant et al. 1997; Guntzer et al. 2012; Qayyum et al. 2017), out of which slag is very important, as it being a by-product of metallurgic industry it is present in abundance (Epstein 2001). Use of silicate minerals like potassium silicate, sodium silicate, and calcium silicate has been reported in literature to control HM stress (Berthelsen et al. 2001; Ning et al. 2016; Rasoolizadeh et al. 2018; Rehman et al. 2019). Wollassonite ($CaSiO_3$) is also a Ca-based Si mineral that is abundantly being used as a Si source. Application of any of these sources can enhance Si uptake and HM stress management response of field crops (Nanayakkara et al. 2008; Ali et al. 2019; Sohail et al. 2020).

Role of Silicon in HM Stress Management

Silicon is termed as beneficial element for plants (Nanayakkara et al. 2008; Rasoolizadeh et al. 2018) especially under stressful environment (Maghsoudi et al. 2019; Rehman et al. 2019; Sousa et al. 2019). The potential role of Si in plant growth management under abiotic and biotic stresses is evident in literature (Fauteux et al. 2005;

Hernandez-Apaolaza 2014; Zhu and Gong 2014; Adrees et al. 2015b; Rizwan et al. 2015; Coskun et al. 2016; Imtiaz et al. 2016; Debona et al. 2017; Roychoudhury and Tripathi 2020), and its uptake by roots and accumulation in plants is very important in this regard (Epstein 1999; Fawe et al. 2001; Ma and Yamaji 2006; Exley 2015). Si can help plants in nutritional homeostasis (Hernandez-Apaolaza 2014), decreased metal uptake (Rehmane et al. 2019), strengthen cell wall via silicification and lignification (Exley 2015; Keller et al. 2015), and uplifting antioxidant activity in plants (Abbas et al. 2015; Khan et al. 2016). As a primary action, Si makes silicate complexes with metals in soil thus decreasing its bioavailable fraction (Adrees et al. 2015b; Imtiaz et al. 2016; Etesami 2018; Etesami and Jeong 2018), creating apoplastic barrier for HM translocation (Mali and Aery 2008; Detmann et al. 2012) and physiological strengthening (Anwaar et al. 2015; Khan et al. 2016). Silicon helps plant strengthen its cell wall and helps quench ROS (Abbas et al. 2015) and hydroxyl radical (Dragišić Maksimović et al. 2012). All of these mechanisms ultimately result in decreased metal uptake by plants under Si fertilizer application (Rizwan et al. 2012; Maghsoudi et al. 2019; Sousa et al. 2019).

Silicon Role in Plant Nutrition

Plant nutrition is responsible for proper growth and functioning that gets disturbed when plant is exposed to HM stress (Bienert et al. 2008; Rizwan et al. 2012). Silicon application can help plants uplift C and nutrient assimilation, thus improving growth (Tuna et al. 2008; Ashraf et al. 2010; Mehrabanjoubani et al. 2015; Fan et al. 2016; Hussain et al. 2019). Main reason behind role of Si in boasting nutrient acquisition is its ability for root strengthening and elongation (Isa et al. 2010), which can help manage micronutrient (Hernandez-Apaolaza 2014) and macronutrients like P (Eneji et al. 2008; Wang et al. 2015; Agostinho et al. 2017).

Silicon-Mediated HM Management Mechanisms

Plants tend to survive, avoid, or tolerate HM stress via various mechanisms (Colzi et al. 2011; Rehman et al. 2015) including metal ions compartmentation in vacuole (Tong et al. 2004) or cell wall (Zhang et al. 2007), managing the production of cellular proteins (Wollgiehn and Neumann 1999), and metal chelation (Hall et al. 2000). Use of Si for plants can help uplift all these mechanisms (Liang et al. 2007; Li et al. 2012; Rizwan et al. 2012; Adrees et al. 2015b). Si addition ultimately lead to reduced HMs bioavailability in soil, thus it will help plants grow better and accumulate less metals in their body (Ali et al. 2013; Hu et al. 2013; Tripathi et al. 2015).

Reduction of HM Uptake

A major mechanism of Si-mediated HM stress mitigation is its role in decreasing soil's bio-available metals (Rehman et al. 2019) and HM translocation (Liang et al. 2007; Wang et al. 2007; Zhang et al. 2008). This decreased metal ions' bioavailability and uptake by plant roots is associated with Si effect on soil properties (Liang et al. 2007; Ding et al. 2013) helping decrease metal mobility in maize (Dresler et al. 2015; Sousa et al. 2019), peanut (Shi et al. 2010), rice (Rehman et al. 2019), wheat (Maghsoudi et al. 2019), cotton

(Farooq et al. 2013), and Chinese cabbage (*Brassica chinensis* L.) (Song et al. 2009). Silicon fertilizer application also helps plants in net translocation of these metals, which can help them compartmentalize these metals in non-edible portions (Zhang et al. 2007; Hussain et al. 2015; Sahebi et al. 2015; Naeem et al. 2018). Application of silicates (specifically alkaline) can boast metal chelation and regulate translocation of metals (Liang et al. 2005, 2007)

Modification of Rhizosphere Chemistry/Making Si Complexes with Metals

Some ligands can chelate (form complexes) with soil metals at the phytotoxic level, making them 'inaccessible' and less harmful to plant growth. The Si-mediated removal of HMs primarily involves chelating metals by flavonoid phenolics or organic acids. At cellular and molecular levels, plants minimize the concentration and symplasmic uptake of available elements by the creation and storage of insoluble complexes between Si and metal(loid)s within cell walls. Si involves alterations to chemical conditions in soil, e.g. soil pH alteration, metal speciation, or reduced metal(loids) availability due to Si co-precipitation (Bhat et al. 2019). Si and Fe exclusively interconnected dynamics within the rhizosphere on the basis of elements activated by organic ligands (Gattullo et al. 2016). The two distinct groups of protein chelators abundant with cysteine are the phytochelatins (PCs) and the metallothionines (MTs) (Yadav 2010). Silicon application can help in fixation of metals in organic or crystalline phase such as use of Si-rich amendment (da Cunha and do Nascimento 2009; Sohail et al. 2020) mitigating toxic effects of HMs in plants (Kollárová et al. 2019).

Stimulation of Antioxidants

Uptake of metals in plants trigger oxidative stress (ROS) production, which damages (Lucas et al. 2011; Qin et al. 2011; Tripathi et al. 2020) and adversely affects plant functions and oxidative balance (Adrees et al. 2015a). Plants tend to produce antioxidants to counter these ROS, and Si application can boast this antioxidant production (Bharwana et al. 2013; Farooq et al. 2013; Anwaar et al. 2015). Silicon also helps to reduce ROS (MDA, H_2O_2) contents in plant cells under exposure of HMs (Song et al. 2009; Shi et al. 2010; Kim et al. 2014; Wang et al. 2016), which along with boasted antioxidant defence stimulates plant metabolism (Suzuki et al. 2012; Ehsan et al. 2014; Gill et al. 2015; Adrees et al. 2015b). Silicon application can help plants uplift enzymatic (Abbas et al. 2015; Khan et al. 2016) and non-enzymatic antioxidants under stress conditions (Gupta et al. 2009; Chanchal Malhotra et al. 2016). Regulation of these kinds of cellular and molecular responses by plants (uplifted by Si application) helps plants escape from HM stress (Khandekar and Leisner 2011; Rizwan et al. 2012).

Help in Compartmentation of HM Inside Plants

Plants possess homeostatic mechanisms in various cellular compartments to sustain the required amount of ions of essential metals. Metal(loids) homeostatic mechanisms involve processes of absorption, transport, chelation, and sequestration (Imtiaz et al. 2016). Transporters have a fundamental role in facilitating the distribution of metal ions into the root's cellular and intracellular spaces. Translocation of metal ions in plant tissues is also controlled by different families of metal transporters with substrate specificities, spatial or temporal signalling pathways, and functional characterization (Adrees et al. 2015a; Shivaraj et al. 2021). HMs sequester metabolically inside cells and less active cell compartments by

Si-mediated HM sequestration (Zhang et al. 2018). Rhizodermis traps ions as they pass by radial transportation via cortical tissues through the central cylinders. Intercellular and intracellular compartmentation of HMs involves deposition in vacuole (Tong et al. 2004) or cell wall (Zhang et al. 2007).

Gene Expression Modification

Plants respond to HM stress via gene expression (Kim et al. 2014) as various physiological processes are directly in control of metal transporter regulating genes (Khanna et al. 2019). Initiation of gene-mediated stress response in plants involves stimuli by stress-associated proteins, activation of transporter controlling genes, stimulated synthesis of chelating agents like phytochelatins (PCs) and MTs, and thiol/non-thiol compounds production (Sharma and Dietz 2006; Roychoudhury and Tripathi 2019), while silicon is central contributor in controlling all of these mechanisms (Chen et al. 2019). Silicon-mediated gene regulatory mechanism (involved in biosynthesis of phytochelatins) is responsible for decreased metal ions uptake and accumulation (Sasaki et al. 2012; Kim et al. 2014; Ma and Yamaji 2015; Pontigo et al. 2017). In another response, Si has found to be controlling expression of gene responsible for transported proteins (Ma et al. 2007) like in Poaceae family (Deshmukh et al. 2015; Ma and Yamaji 2015). In a study conducted by Cao et al. (2020) Si fertilizer application upregulated production of OVP1 leading to decreased Cd loading in xylem. In another work conducted by Khan and Gupta (2018), it was observed that Si upregulated transporting genes (Lsi1, Lsi2, and Lsi6) as well as increased uptake and accumulation of macronutrients. Silicon also participates actively in the regulation of gene expression of antioxidant machinery, but these mechanisms need to understand better.

Structural and Physiological Modification

HM-mediated toxic effect comprising oxidative disturbance is a leading cause of functional disturbance in plants, which can be reversed by application of Si as reported for mesophyll thickness and tissue diameters in maize (da Cunha and do Nascimento 2009), rapeseed, Indian mustards (Vaculík et al. 2012), and rice tissues (Singh et al. 2011). Among other developments, role of Si in the development of Casparian bands, strengthening of endodermal cell wall, and upregulation of apoplastic defence in crop plants were observed under HM stress (Vaculík et al. 2009, 2012; Lukačová et al. 2013). Besides cellular effects, Kim et al. (2014) reported that application of Si reduced Cu-initiated necrosis and leaf damages in rice plant. Silicon supplementation also enhances gas and light use efficiency and photosynthetic potential in plants, thus further assisting in mitigation of HM stress (Adrees et al. 2015b).

Exogenous Application of Silicon to Manage HM Toxicity

Exogenous Si supplementation is found to be very sustainable in diminishing HM harmfulness via various physiological and molecular mechanisms (Figure 11.1, Table 11.1). Major effect of exogenous applied Si involves net decrease in soil metal contents and assimilation (Zhang et al. 2008; Kaya et al. 2009; Song et al. 2009; Shi et al. 2010; Tubana et al. 2016), enhanced growth and productivity (Ma and Yamaji 2006), upregulation of antioxidants

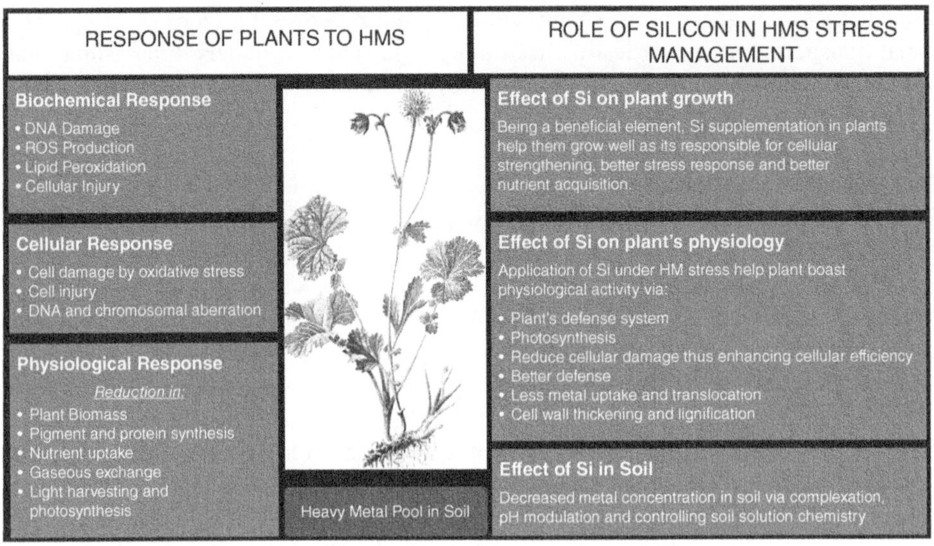

Figure 11.1 Role of silicon in management of heavy metals contaminated soils.

defence (Bharwana et al. 2013; Farooq et al. 2013; Anwaar et al. 2015), and improved plant physiology (Rodrigues et al. 2003; Adrees et al. 2015a, b). Exogenous sources of Si and their effect on plants are explained as follows.

Silicon Fertilizer

A range of criteria have been addressed for silicon fertilizers before use such as solubility, appropriate physical properties (i.e. large surface area), and contamination. In Japan industrial slag was being used as silicon fertilizers in 1950 (Guntzer et al. 2012). The most important Si-fertilizer products used are potassium silicate and calcium metasilicate, usually referred to as 'wollastonite'. The others are magnesium silicate and basalt dust and silicate slag. Silicate slags have high levels of total Si but low proportions of readily soluble Si. Most silicate fertilizers have the advantage of being comprised of certain other nutrients like Ca^{2+} (Almeida et al. 2009).The Si bioavailability depends on its particle size, surface area, and chemical composition of different silicon materials. For example, calcium metasilicates are much more soluble than calcium orthosilicate and are easily accessible for plants (de Tombeur et al. 2020). Diatomaceous rocks are also natural source of readily soluble Si because of their amorphous nature. Silicon application of wollastonite or slag by foliar spray has been most effective in raising Si content of rice (Agostinho et al. 2017).

Biogenic Si Sources (Organic Amendments Enriched in Si)

Metal stress in the soil-plant system can be mitigated by using highly soluble, low-cost, environmentally sustainable Si-enriched organic materials, biogenic silica, and industrial waste, which can be an acceptable approach rather than using expensive and insoluble commercial materials (Sohail et al. 2020; Dhiman et al. 2021). The biogenic pool primarily is composed of amorphous (plant-driven) phytogenic silica, but also minor zoogenic,

Table 11.1 Role of silicon in management of heavy metals contaminated soils.

Mode of application	Si application rate	Heavy metals	Concentration of heavy metal ions	Growth medium	Crop/plant	Remarks	References
Si (K_2SiO_3) supplementation	1 mM per	Cu	30 mg L^{-1}	Hydroponic	Durum wheat (*Triticumturgidum* L.)	Reduced Cu uptake and root to shoot translocation and improved growth	Keller et al. (2015)
Si-potash fertilizer	900 kg ha^{-1}	As and Cd	30 mg Kg^{-1} As and 0.4 mg Kg^{-1} Cd	Soil	Rice (*Oryza sativa* L.)	Si fertilizer improved growth of rice plants and reduced grain Cd and As	Wang et al. (2015)
Organosilicon Si fertilizer	1.5 mM	Cd	5 mg L^{-1} Cd	Soil	*Brassica chinensis* L.	Enhanced Cd and Pb-stressed wheat growth and photosynthetic activity, while decreased MDA and H_2O_2 contents	Song et al. (2009)
Silicon fertilizers SiO_2	500 mg Kg^{-1}	Hg	3 mg Kg^{-1}	Soil	Garlic plant (*Allium sativum*)	Si relieved metal stress, stimulated the antioxidant system, made complexes with Hg, and reduced the uptake and translocation of Hg	Hu et al. (2020)
Si fertilizer	5 mM and 0.08 mM	Cd	10 M and 1 M $Cd(NO_3)_2$	Hydroponic	Maize (*Zea mays* L)	Reduced the number of damaged chloroplasts and increased production of chelates in roots	Lukačová et al. (2013)
Foliar Si application	5 μM	Cd	5 μM	Soil	Chinese flowering cabbage (*Brassica rapa*)	Decreased absorption and translocation of Cd. Improved the antioxidant SOD, CAT, and APX in plant tissues	Wu et al. (2018)
Si fertilizer	50 μM	Cr	1 mM	Soil	Wheat (*Triticumaestivum* L.)	Si-enhanced non-enzymatic scavenging and proline contents and decreased Cr-induced oxidative stress	Sarkar et al. (2020)
Sodium silicate fertilizer	1 mM	Cd	100 mM $CdCl_2$	Soil	Cucumber (*Cucumissativus* L.)	Si protects towards damage to photosynthetic apparatus in Cd stress	Feng et al. (2010)
Sodium silicate fertilizer	1 mM	Mn	600 μM $MnSO_4$	Soil	Cucumber (*Cucumissativus* L.	Reduced H_2O_2 and Mn lipid peroxidation activity. Improved enzymes linked to the ascorbate-glutathione cycle, including ascorbate peroxidase (APX) and dehydroascorbate (DHAR)	Feng et al. (2009)

(*Continued*)

Table 11.1 (Continued)

Mode of application	Si application rate	Heavy metals	Concentration of heavy metal ions	Growth medium	Crop/plant	Remarks	References
Si nanoparticles	—	Cd, Pb, and Zn	7.38, 45.76, and 41.15 mg Kg^{-1}, respectively	Soil	Wheat (*Triticumaestivum* L.)	Increased growth and photosynthetic activity of wheat, oxidative stress, and Cd concentration decreased	Ali et al. (2019)
Si fertilizer K_2SiO_3	600 kg ha^{-1}	Cd	35 mg kg^{-1}	Soil	Rice crop (*Oryza sativa* L.)	Silicon improved plant growth and significantly reduced Cd accumulation, translocation factor, and bioaccumulation factor	Rehman et al. (2019)
Silicon dioxide nanoparticle	4 mg kg^{-1}	Al	25 mg kg^{-1}	Soil	Maize plants (*Zea mays* L)	SNPs improved the concentration of organic acids and metal detoxification (i.e. glutathione-S-transferase activity) in plants as protective mechanism against Al toxicity	Sousa et al. (2019)
Sodium silicate foliar spray ($Na_2SiO_3{\cdot}9H_2O$)	198 mg kg^{-1}	As	4.62 mg kg^{-1}	Soil	Rice plants (*Oryza sativa* L.)	Si spray was helpful in order to reduce inorganic As in rice grain at tillering stages. Spraying Si reduced As in rice grain by Si transporter regulations	Zhang et al. (2020)
Silicate	1 mM	Cr	50 μM Cr_2O_3	Hydroponic	Wheat (*Triticumaestivum* L.)	Adding Si in Cr-stressed plants greatly improved morphophysiological properties, total proteins, and membrane stability compared to Cr-stressed plants	Sarkar et al. (2020)

microbial, and nano silica components are available. In natural environments the pool of biogenic Si typically exceeds annual phyto-absorption in many orders of magnitude in the soil. The Si pool supplies the plant with Si (as silicic acid), whereas the litter fall fills the pool with freshly formed phytolite Si (Haynes 2017). Biochar (BC) is proven to have several impacts on agriculture as a healthy and sustainable future technology (Xu et al. 2017). BCs are recognized worldwide as effective fertilizer and soil modifications in various media for HM remediation (Abbas et al. 2017). Despite the BC precursor, Si is a vital inorganic element present in BC (Abbas et al. 2017). Recently, the word 'sichar' was suggested to present a BC with an amorphous Si of a minimum of 5.0 g per kilogram of BC. Therefore, BC from Si-rich feedstocks, such as residue of rice, may be a prospective amorphous source of Si to cope with metal toxicity (Wang et al. 2019). Silicon needs to be considered in BC for: (i) Si is predominantly a significant BC inorganic component, (ii) Si is a silicophilic plant nutrient, (iii) almost all of the plant tissues provide sufficient fraction of Si, and (iv) Si-rich crop straws and other residues provide a major material for producing BC (Wang et al. 2019). Methyl silanols are common waste products frequently introduced in the environment via industrial and municipal waste streams containing silicon. These silicon degradation products are more water soluble than silicic acid (Fauteux et al. 2006). The BCs produced from phytoliths-rich biomass, e.g. bamboo and rice, have high releasing capacity of Si (Li et al. 2015).

A vast number of Si-rich organic compounds containing excess phytogenic/amorphous Si can be used in preparing Si-rich BCs from feedstock. The release of Si from wollastonite and *Miscanthus giganteus* BC was evaluated by Li et al. (2015). Animal manure is usually rich in Si in rice growing areas (Babaso and Sharanagouda 2017). Rice residues are also among the feedstocks commonly used for bio-derived materials such as BCs, silica gels, and bioethanols. Rice straw and rice husk are amongst the major agricultural residues. In rice residue ash, the key element present is Si (75–75%) along Ca, Mg, P, Al, Fa, K, and Na. The amorphous silica source (opal or plantolite) ($SiO_2 \cdot nH_2O$) is rich in rice tissue, providing a significant and sustainable PASi (Sohail et al. 2019a). Rice straw is beneficial for many plants, since rice straw is a slow-release and durable silica fertilizer. It can be an even slow dissolution fertilizer, and only one application can be useful for several years (Seyfferth et al. 2016). Many crop residues BC have been reported for biogenic Si like coffee BC, miscanthus BC, wood BC (Houben et al. 2014), saw dust ash BC, bagasse ash BC, rice husk biochar (RHA-BC) (Suksabye et al. 2016), rice straw BC, bamboo BC (Lu et al. 2014), wheat straw BC, slag (Bian et al. 2016), corn straw BC (Chi et al. 2017), rice husk, and rice husk ash (Seyfferth et al. 2016).

Silicon Nanoparticles

Nanoparticles and nanotechnology are a very diverse field of study and have huge potential for managing plant production under stress environments (Ahmad et al. 2018; Sohail et al. 2019b; Ayub et al. 2019). Due to diverse action and scope, nanofertilizer has gained a nice pace in modern day agriculture. Use of Si nanoparticles is also reported in literature affecting crop growth and mitigating biotic and abiotic stresses. Application of Si-NPs have shown a significant effect on increased metabolism, improved yield, and decreased HM toxicity in maize (Suriyaprabha et al. 2012; Sousa et al. 2019) and rice (Cui et al. 2017) crops. Among Si-NPs, silica nanoparticle is also important as they are penetrating in cell and can be used for nutrient and chemical delivery into cell (Torney et al. 2007) and silica NPs' application can also enhance plant growth (Janmohammadi et al. 2016), better fertilizer use (Hossain et al. 2008; Wanyika et al. 2012), and better production (Hong et al. 2013).

Summary

Accumulation of HMs in agricultural soils above permissible limits is of great concern regarding their adverse effects on soil ecosystem, plant metabolic functions, food safety, and human health. Crops grown on such soil get contaminated with these HMs giving rise to health risks. Therefore, it is necessary to take some remedial measures to combat their hazards, and application of silicon in such soil is a viable option. Silicon being a beneficial element has several metabolic and physiological functions in plants. Natural Si is not much available for plants uptake, so there is a need of silicon fertilizers to encounter the plant supplies. Exogenous application of Si amendments is suitable for plant growth in HM soil and its remediation. Several roles of Si have been described during HM contaminations like reduced uptake of HMs, less translocation, and enhanced plant growth. A proper understanding regarding Si application rates and time can be helpful in harvesting the benefits of Si fertilizers.

References

Abbas, T., Balal, R.M., Shahid, M.A. et al. (2015). Silicon-induced alleviation of NaCl toxicity in okra (*Abelmoschus esculentus*) is associated with enhanced photosynthesis, osmoprotectants and antioxidant metabolism. *Acta physiol plant* 37 (6): 1–13.

Abbas, T., Rizwan, M., Ali, S. et al. (2017). Effect of biochar on cadmium bioavailability and uptake in wheat (*Triticum aestivum* L.) grown in a soil with aged contamination. *Ecotoxicol Environ Saf* 140: 37–47.

Adrees, M., Ali, S., Rizwan, M. et al. (2015a). The effect of excess copper on growth and physiology of important food crops: a review. *Environ Sci Pollut Res* 22: 8148–8162.

Adrees, M., Ali, S., Rizwan, M. et al. (2015b). Mechanisms of silicon-mediated alleviation of heavy metal toxicity in plants: a review. *Ecotoxicol Environ Saf* 119: 186–197.

Agostinho, F.B., Tubana, B.S., Martins, M.S., and Datnoff, L.E. (2017). Effect of different silicon sources on yield and silicon uptake of rice grown under varying phosphorus rates. *Plants* 6: 35.

Ahmad, H.R., Rehman, M.Z., Sohail, M.I. et al. (2018). Effects of rare earth oxide nanoparticles on plants. In: *Nanomaterials in Plants, Algae, and Microorganisms* (ed. D.K. Tripathi, P. Ahmad and N.K. Dubey), 239–275. Academic Press.

Ali, H. and Khan, E. (2018). What are heavy metals? Long-standing controversy over the scientific use of the term 'heavy metals'–proposal of a comprehensive definition. *Toxicol Environ Chem* 100: 6–19.

Ali, S., Farooq, M., Yasmeen, T. et al. (2013). The influence of silicon on barley growth, photosynthesis and ultra-structure under chromium stress. *Ecotoxicol. Environ. Saf.* 89: 66–72.

Ali S, Rizwan M, Hussain A, Zia ur Rehman M, *et al.* (2019) Silicon nanoparticles enhanced the growth and reduced the cadmium accumulation in grains of wheat (*Triticum aestivum* L.). *Plant Physiol. Biochem.* 140, 1–8. doi:https://doi.org/10.1016/j.plaphy.2019.04.041.

Almeida, G.D., Pratissoli, D., Zanuncio, J.C. et al. (2009). Calcium silicate and organic mineral fertilizer increase the resistance of tomato plants to *Frankliniella schultzei. Phytoparasitica* 37: 225–230.

Anwaar, S.A., Ali, S., Ishaque, W. et al. (2015). Silicon (Si) alleviates cotton (*Gossypium hirsutum* L.) from zinc (Zn) toxicity stress by limiting Zn uptake and oxidative damage. *Environ Sci Pollut Res* 22: 3441–3450.

Ashraf, M., Afzal, M., Ahmed, R. et al. (2010). Alleviation of detrimental effects of NaCl by silicon nutrition in salt-sensitive and salt-tolerant genotypes of sugarcane (*Saccharum officinarum* L.). *Plant Soil* 326: 381–391.

Ayub, M.A., Sohail, M.I., Umair, M. et al. (2019). Cerium oxide nanoparticles: advances in synthesis, prospects and application in agro-ecosystem. In: *Comprehensive Analytical Chemistry*, vol. 87 (ed. S.K. Verma and A.K. Das), 209–250. Elsevier.

Azhar, M., ur Rehman, M.Z., Ali, S. et al. (2019). Comparative effectiveness of different biochars and conventional organic materials on growth, photosynthesis and cadmium accumulation in cereals. *Chemosphere* 227: 72–81.

Babaso, P.N. and Sharanagouda, H. (2017). Rice husk and its applications: review. *Int J Curr Microbiol Appl Sci* 6: 1144–1156.

Berthelsen, S., Noble, A.D., and Garside, A.L. (2001). Silicon research down under: past, present, and future. In: *Studies in Plant Science* (ed. L.E. Datnoff, G.H. Snyder and G.H. Korndörfr), 241–255. Elsevier.

Bharwana, S.A., Ali, S., Farooq, M.A. et al. (2013). Alleviation of lead toxicity by silicon is related to elevated photosynthesis, antioxidant enzymes suppressed lead uptake and oxidative stress in cotton. *J Bioremed Biodeg* 4: 10-4172.

Bhat, J.A., Shivaraj, S.M., Singh, P. et al. (2019). Role of silicon in mitigation of heavy metal stresses in crop plants. *Plants* 8: 71.

Bhattacharyya, P., Chakrabarti, K., Chakraborty, A. et al. (2008). Fractionation and bioavailability of Pb in municipal solid waste compost and Pb uptake by rice straw and grain under submerged condition in amended soil. *Geosci J* 12: 41–45.

Bian, R., Li, L., Bao, D. et al. (2016). Cd immobilization in a contaminated rice paddy by inorganic stabilizers of calcium hydroxide and silicon slag and by organic stabilizer of biochar. *Environ Sci Pollut Res* 23: 10028–10036.

Bienert, G.P., Schüssler, M.D., and Jahn, T.P. (2008). Metalloids: essential, beneficial or toxic? Major intrinsic proteins sort it out. *Trends Biochem Sci* 33: 20–26.

Bolan, N.S. and Duraisamy, V.P. (2003). Role of inorganic and organic soil amendments on immobilisation and phytoavailability of heavy metals: a review involving specific case studies. *Soil Res* 41: 533–555.

Bolan, N., Kunhikrishnan, A., Thangarajan, R. et al. (2014). Remediation of heavy metal (loids) contaminated soils–to mobilize or to immobilize. *J Hazard Mater* 266: 141–166.

Bradl, H. (2005). *Heavy Metals in the Environment: Origin, Interaction and Remediation.* Elsevier.

Broadley, M., Brown, P., Cakmak, I. et al. (2011). Beneficial elements. In: *Marschner's Mineral Nutrition of Higher Plants*, 3e (ed. P. Marschner), 257–261. Amsterdam: Elsevier.
Cao, F., Dai, H., Hao, P.F., and Wu, F. (2020). Silicon regulates the expression of vacuolar H+-pyrophosphatase 1 and decreases cadmium accumulation in rice (*Oryza sativa* L.). *Chemosphere* 240: 124907.
Chanchal Malhotra, C., Kapoor, R., and Ganjewala, D. (2016). Alleviation of abiotic and biotic stresses in plants by silicon supplementation. *Scientia 13*: 59–73.
Chang, L.W., Magos, L., and Suzuki, T. (1996). *Toxicology of Metals*. Taylor & Francis US.
Chen, D., Chen, D., Xue, R. et al. (2019). Effects of boron, silicon and their interactions on cadmium accumulation and toxicity in rice plants. *J Hazard Mater* 367: 447–455.
Chi, T., Zuo, J., and Liu, F. (2017). Performance and mechanism for cadmium and lead adsorption from water and soil by corn straw biochar. *Front Environ Sci Eng* 11: 15.
Colzi, I., Doumett, S., Del Bubba, M. et al. (2011). On the role of the cell wall in the phenomenon of copper tolerance in *Silene paradoxa L. Environ Exp Bot 72*: 77–83.
Coskun, D., Britto, D.T., Huynh, W.Q., and Kronzucker, H.J. (2016). The role of silicon in higher plants under salinity and drought stress. *Front Plant Sci* 7: 1072.
Cui, J., Liu, T., Li, F. et al. (2017). Silica nanoparticles alleviate cadmium toxicity in rice cells: mechanisms and size effects. *Environ Pollut* 228: 363–369.
da Cunha, K.P.V. and do Nascimento, C.W.A. (2009). Silicon effects on metal tolerance and structural changes in maize (*Zea mays* L.) grown on a cadmium and zinc enriched soil. *Water Air Soil Pollut* 197: 323.
Datnoff, L.E., Snyder, G.H., and Korndörfer, G.H. (2001). *Silicon in Agriculture*. Elsevier.
Debona, D., Rodrigues, F.A., and Datnoff, L.E. (2017). Silicon's role in abiotic and biotic plant stresses. *Annu Rev Phytopathol* 55: 85–107.
Deshmukh, R.K., Vivancos, J., Ramakrishnan, G. et al. (2015). A precise spacing between the NPA domains of aquaporins is essential for silicon permeability in plants. *Plant J* 83: 489–500.
Deshmukh, R., Tripathi, D.K., and Guerriero, G. (2020). *Metalloids in Plants: Advances and Future Prospects*, vol. 1, 1–520. John Wiley & Sons.
Detmann, K.C., Araújo, W.L., Martins, S.C. et al. (2012). Silicon nutrition increases grain yield, which, in turn, exerts a feed-forward stimulation of photosynthetic rates via enhanced mesophyll conductance and alters primary metabolism in rice. *New Phytol* 196: 752–762.
Dhiman, P., Rajora, N., Bhardwaj, S. et al. (2021). Fascinating role of silicon to combat salinity stress in plants: an updated overview. *Plant Physiology and Biochemistry.* 162: 110–123.
Ding, X., Zhang, S., Li, S. et al. (2013). Silicon mediated the detoxification of Cr on Pakchoi (*Brassica Chinensis* L.) in Cr-contaminated soil. *Proced Environ Sci* 18: 58–67.
Dong, J., Wu, F., and Zhang, G. (2006). Influence of cadmium on antioxidant capacity and four microelement concentrations in tomato seedlings (*Lycopersicon esculentum*). *Chemosphere* 64: 1659–1666.
Dong, Q., Fang, J., Huang, F., and Cai, K. (2019). Silicon amendment reduces soil Cd availability and Cd uptake of two Pennisetum species. *Int J Environ Res Public Health* 16: 1624.
Dragišić Maksimović, J., Mojović, M., Maksimović, V. et al. (2012). Silicon ameliorates manganese toxicity in cucumber by decreasing hydroxyl radical accumulation in the leaf apoplast. *J Exp Bot 63*: 2411–2420.

Dresler, S., Wójcik, M., Bednarek, W. et al. (2015). The effect of silicon on maize growth under cadmium stress. *Russian J Plant Physiol* 62: 86–92.

Duffus, J.H. (2002). "Heavy metals" a meaningless term? (IUPAC Technical Report). *Pure Appl. Chem* 74: 793–807.

Ehsan, S., Ali, S., Noureen, S. et al. (2014). Citric acid assisted phytoremediation of cadmium by *Brassica napus* L. *Ecotoxicol Environ Saf* 106: 164–172.

Eneji, A.E., Inanaga, S., Muranaka, S. et al. (2008). Growth and nutrient use in four grasses under drought stress as mediated by silicon fertilizers. *J Plant Nutr 31*: 355–365.

Epstein, E. (1999). Silicon. *Annu Rev Plant Biol 50*: 641–664.

Epstein, E. (2001). Silicon in plants: facts vs. concepts. In: *Studies in Plant Science*, vol. 8 (ed. L.E. Datnoff, G.H. Snyder and G.H. Korndörfer), 1–15. Elsevier.

Etesami, H. (2018). Can interaction between silicon and plant growth promoting rhizobacteria benefit in alleviating abiotic and biotic stresses in crop plants. *Agric Ecosyst Environ* 253: 98–112.

Etesami, H. and Jeong, B.R. (2018). Silicon (Si): Review and future prospects on the action mechanisms in alleviating biotic and abiotic stresses in plants. *Ecotoxicol Environ Saf* 147: 881–896.

Exley, C. (2015). A possible mechanism of biological silicification in plants. *Front Plant Sci* 6: 853.

Fan, X., Wen, X., Huang, F. et al. (2016). Effects of silicon on morphology, ultrastructure and exudates of rice root under heavy metal stress. *Acta Physiol. Plant* 38: 197.

Farooq, M.A., Ali, S., Hameed, A. et al. (2013). Alleviation of cadmium toxicity by silicon is related to elevated photosynthesis, antioxidant enzymes; suppressed cadmium uptake and oxidative stress in cotton. *Ecotoxicol Environ Saf 96*: 242–249.

Fauteux, F., Rémus-Borel, W., Menzies, J.G., and Bélanger, R.R. (2005). Silicon and plant disease resistance against pathogenic fungi. *FEMS Microbiol Lett* 249: 1–6.

Fauteux, F., Chain, F., Belzile, F. et al. (2006). The protective role of silicon in the Arabidopsis-powdery mildew pathosystem. *PNAS 103*: 17554–17559.

Fawe, A., Menzies, J.G., Chérif, M., and Bélanger, R.R. (2001). Silicon and disease resistance in dicotyledons. In: *Studies in Plant Science*, vol. 8 (ed. L.E. Datnoff, G.H. Snyder and G.H. Korndörfer), 159–169. Elsevier.

Feng, J.P., Shi, Q.H., and Wang, X.F. (2009). Effects of exogenous silicon on photosynthetic capacity and antioxidant enzyme activities in chloroplast of cucumber seedlings under excess manganese. *Agr Sci China* 8: 40–50.

Feng, J., Shi, Q., Wang, X. et al. (2010). Silicon supplementation ameliorated the inhibition of photosynthesis and nitrate metabolism by cadmium (Cd) toxicity in *Cucumis sativus* L. *Sci Hortic* 123: 521–530.

Fergusson, J.E. (1990). The heavy elements: chemistry, environmental impact adn health effects. Fergusson (No. 628.53 F4.).

Frings, P.J., Clymans, W., Jeppesen, E. et al. (2014). Lack of steady-state in the global biogeochemical Si cycle: emerging evidence from lake Si sequestration. *Biogeochem* 117: 255–277.

Gascho, G.J. (2001). Silicon sources for agriculture. In: *Studies in Plant Science*, vol. 8 (ed. L.E. Datnoff, G.H. Snyder and G.H. Korndörfer), 197–207. Elsevier.

Gattullo, C.E., Allegretta, I., Medici, L. et al. (2016). Silicon dynamics in the rhizosphere: connections with iron mobilization. *J Soil Sci Plant Nutr* 179: 409–417.

Gill, S.S., Khan, N.A., Anjum, N.A., and Tuteja, N. (2011). Amelioration of cadmium stress in crop plants by nutrients management: morphological, physiological and biochemical aspects. *Plant Stress 5*: 1–23.

Gill, R.A., Zang, L., and Ali, B. (2015). Chromium-induced physio-chemical and ultrastructural changes in four cultivars of *Brassica napus* L. *Chemosphere 120*: 154–164.

Gocke, M., Liang, W., Sommer, M., and Kuzyakov, Y. (2013). Silicon uptake by wheat: effects of Si pools and pH. *J Plant Nutr Soil Sci* 176: 551–560.

Guntzer, F., Keller, C., and Meunier, J.D. (2012). Benefits of plant silicon for crops: a review. *Agron Sustain Dev* 32: 201–213.

Guo, J., Dai, X., Xu, W., and Ma, M. (2008). Overexpressing GSH1 and AsPCS1 simultaneously increases the tolerance and accumulation of cadmium and arsenic in *Arabidopsis thaliana*. *Chemosphere* 72: 1020–1026.

Gupta, D.K., Nicoloso, F.T., and Schetinger, M.R.C. (2009). Antioxidant defense mechanism in hydroponically grown *Zea mays* seedlings under moderate lead stress. *J Hazard Mater* 172: 479–484.

Hall, I.H., Lackey, C.B., Kistler, T.D. et al. (2000). The cytotoxicity of symmetrical and unsymmetrical bis (thiosemicarbazones) and their metal complexes in murine and human tumor cells. *Arch Pharm 333*: 217–225.

Haynes, R.J. (2014). A contemporary overview of silicon availability in agricultural soils. *J Plant Nut Soil Sci* 177: 831–844.

Haynes, R.J. (2017). The nature of biogenic Si and its potential role in Si supply in agricultural soils. *Agric Ecosyst Environ* 245: 100–111.

He, Z.L., Yang, X.E., and Stoffella, P.J. (2005). Trace elements in agroecosystems and impacts on the environment. *J Trace Elem Med Biol* 19: 125–140.

Herawati, N., Suzuki, S., Hayashi, K. et al. (2000). Cadmium, copper, and zinc levels in rice and soil of Japan, Indonesia, and China by soil type. *B Environ Contam Tox* 64: 33–39.

Hernandez-Apaolaza, L. (2014). Can silicon partially alleviate micronutrient deficiency in plants? A review. *Planta* 240: 447–458.

Hodson, M.J., White, P.J., Mead, A., and Broadley, M.R. (2005). Phylogenetic variation in the silicon composition of plants. *Ann Bot 96*: 1027–1046.

Hong, J., Peralta-Videa, J.R., and Gardea-Torresdey, J.L. (2013). Nanomaterials in agricultural production: benefits and possible threats. In: *Sustainable Nanotechnology and the Environment: Advances and Achievements*, Am Chem Soc, 73–90.

Hossain, K.Z., Monreal, C.M., and Sayari, A. (2008). Adsorption of urease on PE-MCM-41 and its catalytic effect on hydrolysis of urea. *Colloid Surface B* 62: 42–50.

Houben, D., Sonnet, P., and Cornelis, J.T. (2014). Biochar from Miscanthus: a potential silicon fertilizer. *Plant Soil 374*: 871–882.

Hu, B., Cui, R., Li, J. et al. (2013). Occurrence and distribution of heavy metals in surface sediments of the Changhua River Estuary and adjacent shelf (Hainan Island). *Mar Pollut Bull 76*: 400–405.

Hu, Y., Wang, Y., Liang, Y. et al. (2020). Silicon alleviates mercury toxicity in garlic plants. *J Plant Nutr* 43 (16): 1–10.

Hussain, I., Ashraf, M.A., Rasheed, R. et al. (2015). Exogenous application of silicon at the boot stage decreases accumulation of cadmium in wheat (*Triticum aestivum* L.) grains. *Braz J Bot* 38: 223–234.

Hussain, A., Rizwan, M., Ali, Q., and Ali, S. (2019). Seed priming with silicon nanoparticles improved the biomass and yield while reduced the oxidative stress and cadmium concentration in wheat grains. *Environ Sci Pollut Res* 26: 7579–7588.

Imtiaz, M., Rizwan, M.S., Mushtaq, M.A. et al. (2016). Silicon occurrence, uptake, transport and mechanisms of heavy metals, minerals and salinity enhanced tolerance in plants with future prospects: a review. *J Environ Manage* 183: 521–529.

Isa, M., Bai, S., Yokoyama, T. et al. (2010). Silicon enhances growth independent of silica deposition in a low-silica rice mutant, lsi1. *Plant Soil* 331: 361–375.

Janmohammadi, M., Amanzadeh, T., Sabaghnia, N., and Ion, V. (2016). Effect of nano-silicon foliar application on safflower growth under organic and inorganic fertilizer regimes. *Botanica* 22: 53–64.

Kaya, C., Tuna, A.L., Sonmez, O. et al. (2009). Mitigation effects of silicon on maize plants grown at high zinc. *J Plant Nutr* 32: 1788–1798.

Keller, C., Rizwan, M., Davidian, J.C. et al. (2015). Effect of silicon on wheat seedlings (*Triticum turgidum* L.) grown in hydroponics and exposed to 0 to 30 μM Cu. *Planta* 241: 847–860.

Khan, E. and Gupta, M. (2018). Arsenic–silicon priming of rice (*Oryza sativa* L.) seeds influence mineral nutrient uptake and biochemical responses through modulation of Lsi-1, Lsi-2, Lsi-6 and nutrient transporter genes. *Sci Rep* 8: 1–16.

Khan, W.U.D., Aziz, T., Hussain, I. et al. (2016). Silicon: beneficial nutrient for maize crop to enhance photochemical efficiency of photosystem II under salt stress. *Arch Agron Soil Sci* 63 (5): 599–611.

Khan, A., Khan, A.L., Muneer, S. et al. (2019). Silicon and salinity: Cross-talk in crop mediated stress tolerance mechanisms. *Frontiers in Plant Sci* 10: 1429.

Khandekar, S. and Leisner, S. (2011). Soluble silicon modulates expression of *Arabidopsis thaliana* genes involved in copper stress. *J Plant Physiol 168*: 699–705.

Khanna, K., Jamwal, V.L., Gandhi, S.G. et al. (2019). Metal resistant PGPR lowered Cd uptake and expression of metal transporter genes with improved growth and photosynthetic pigments in *Lycopersicon esculentum* under metal toxicity. *Sci Rep* 9: 1–14.

Kim, Y.H., Khan, A.L., Kim, D.H. et al. (2014). Silicon mitigates heavy metal stress by regulating P-type heavy metal ATPases, Oryza sativa low silicon genes, and endogenous phytohormones. *BMC Plant Biol* 14: 1–13.

Kollárová, K., Kusa, Z., Vatehova-Vivodova, Z., and Lišková, D. (2019). The response of maize protoplasts to cadmium stress mitigated by silicon. *Ecotoxicol Environ Saf* 170: 488–494.

Kovda, V.A. (1973). The bases of learning about soils. *Nauka* 2: 377–428.

Li, Z.C., Fu, Y., Wu, D. et al. (2012). Silicate-mediated alleviation of Pb toxicity in banana grown in Pb-contaminated soil. *Biol Trace Elem Res* 145: 101–108.

Li, Z., Dufour, N., Delvaux, B., and Cornelis, J.T. (2015). Biochar supplying plant-available Si in soils with contrasting weathering degrees. In SSSB 2015.

Liang, Y., Wong, J.W.C., and Wei, L. (2005). Silicon-mediated enhancement of cadmium tolerance in maize (*Zea mays* L.) grown in cadmium contaminated soil. *Chemosphere 58*: 475–483.

Liang, Y., Sun, W., Zhu, Y.G., and Christie, P. (2007). Mechanisms of silicon-mediated alleviation of abiotic stresses in higher plants: a review. *Environ Pollution* 147: 422–428.

Liang, Y., Nikolic, M., BélangerR, G.H., and Song, A. (2015a). Effect of silicon on crop growth, yield and quality. In: *Silicon in Agric* (ed. L.E. Datnoff, G.H. Snyder and G.H. Korndörfer), 209–223. Dordrecht: Springer.

Liang, Y., Nikolic, M., Bélanger, R. et al. (2015b). *Silicon in Agriculture*. Dordrecht: *Springer*.

Lu, K., Yang, X., Shen, J. et al. (2014). Effect of bamboo and rice straw biochars on the bioavailability of Cd, Cu, Pb and Zn to *Sedum plumbizincicola*. *Agric Ecosyst Environ* 191: 124–132.

Lucas, S., Durmaz, E., Akpınar, B.A., and Budak, H. (2011). The drought response displayed by a DRE-binding protein from *Triticum dicoccoides*. *Plant Physiol Biochem* 49: 346–351.

Lukačová, Z., Švubová, R., Kohanová, J., and Lux, A. (2013). Silicon mitigates the Cd toxicity in maize in relation to cadmium translocation, cell distribution, antioxidant enzymes stimulation and enhanced endodermal apoplasmic barrier development. *Plant Growth Regul* 70: 89–103.

Ma, J.F. and Yamaji, N. (2006). Silicon uptake and accumulation in higher plants. *Trends Plant Sci* 11: 392–397.

Ma, J.F. and Yamaji, N. (2015). A cooperative system of silicon transport in plants. *Trends Plant Sci 20*: 435–442.

Ma, J.F., Yamaji, N., Tamai, K., and Mitani, N. (2007). Genotypic difference in silicon uptake and expression of silicon transporter genes in rice. *Plant Physiol 145*: 919–924.

Maghsoudi, K., Arvin, M.J., and Ashraf, M. (2019). Mitigation of arsenic toxicity in wheat by the exogenously applied salicylic acid, 24-epi-brassinolide and silicon. *J Soil Sci Plant Nutr* 20 (2): 1–12.

Mali, M. and Aery, N.C. (2008). Influence of silicon on growth, relative water contents and uptake of silicon, calcium and potassium in wheat grown in nutrient solution. *J Plant Nutr 31*: 1867–1876.

Manara, A. (2012). Plant responses to heavy metal toxicity. In: *Plants and Heavy Metals* (ed. A. Furini), 27–53. Dordrecht: Springer.

Mazurek, R., Kowalska, J., Gąsiorek, M. et al. (2017). Assessment of heavy metals contamination in surface layers of Roztocze National Park forest soils (SE Poland) by indices of pollution. *Chemosphere 168*: 839–850.

Mehrabanjoubani, P., Abdolzadeh, A., Sadeghipour, H.R., and Aghdasi, M. (2015). Silicon affects transcellular and apoplastic uptake of some nutrients in plants. *Pedosphere 25*: 192–201.

Mithöfer, A., Schulze, B., and Boland, W. (2004). Biotic and heavy metal stress response in plants: evidence for common signals. *FEBS Letters* 566: 1–5.

Naccarato, A., Tassone, A., Cavaliere, F. et al. (2020). Agrochemical treatments as a source of heavy metals and rare earth elements in agricultural soils and bioaccumulation in ground beetles. *Sci Total Environ* 141438.

Naeem, A., Rehman, M.Z., Akhtar, T. et al. (2018). Silicon nutrition lowers cadmium content of wheat cultivars by regulating transpiration rate and activity of antioxidant enzymes. *Environ Pollut 242*: 126–135.

Nagajyoti, P.C., Lee, K.D., and Sreekanth, T.V.M. (2010). Heavy metals, occurrence and toxicity for plants: a review. *Environ Chem Lett* 8: 199–216.

Nanayakkara, U.N., Uddin, W., and Datnoff, L.E. (2008). Application of silicon sources increases silicon accumulation in perennial ryegrass turf on two soil types. *Plant Soil* 303: 83–94.

Ning, D., Liang, Y., Liu, Z. et al. (2016). Impacts of steel-slag-based silicate fertilizer on soil acidity and silicon availability and metals-immobilization in a paddy soil. *PLoS One* 11: e0168163.

Pontigo, S., Godoy, K., and Jiménez, H. (2017). Silicon-mediated alleviation of aluminum toxicity by modulation of Al/Si uptake and antioxidant performance in ryegrass plants. *Front Plant Sci* 8: 642.

Puppe, D., Ehrmann, O., Kaczorek, D. et al. (2015). The protozoic Si pool in temperate forest ecosystems—quantification, abiotic controls and interactions with earthworms. *Geoderma* 243: 196–204.

Qadir, S., Qureshi, M.I., Javed, S., and Abdin, M.Z. (2004). Genotypic variation in phytoremediation potential of *Brassica juncea* cultivars exposed to Cd stress. *Plant Sci* 167: 1171–1181.

Qayyum, M.F., Rehman, M.Z., Ali, S. et al. (2017). Residual effects of monoammonium phosphate, gypsum and elemental sulfur on cadmium phytoavailability and translocation from soil to wheat in an effluent irrigated field. *Chemosphere 174*: 515–523.

Qin, F., Shinozaki, K., and Yamaguchi-Shinozaki, K. (2011). Achievements and challenges in understanding plant abiotic stress responses and tolerance. *Plant Cell Physiol* 52: 1569–1582.

Rahman, A.K.M.R., Hossain, S.M., and Akramuzzaman, M.M. (2010). Distribution of heavy metals in rice plant cultivated in industrial effluent receiving soil. *Environ Asia 3*: 15–19.

Rasoolizadeh, A., Labbé, C., Sonah, H. et al. (2018). Silicon protects soybean plants against Phytophthorasojae by interfering with effector-receptor expression. *BMC Plant Biol* 18: 97.

Rehman, M.Z., Sabir, M., and Nadeem, M. (2015). Remediating cadmium-contaminated soils by growing grain crops using inorganic amendments. In: *Soil Remediation and Plants: Prospects and Challenges*, 367–396.

Rehman, M.Z., Rizwan, M., Rauf, A. et al. (2019). Split application of silicon in cadmium (Cd) spiked alkaline soil plays a vital role in decreasing Cd accumulation in rice (*Oryza sativa* L.) grains. *Chemosphere* 226: 454–462.

Rizwan, M., Meunier, J.D., Miche, H., and Keller, C. (2012). Effect of silicon on reducing cadmium toxicity in durum wheat (*Triticum turgidum* L. cv. Claudio W.) grown in a soil with aged contamination. *J Hazard Mater* 209: 326–334.

Rizwan, M., Ali, S., Ibrahim, M. et al. (2015). Mechanisms of silicon-mediated alleviation of drought and salt stress in plants: a review. *Environ Sci Pollut Res* 22: 15416–15431.

Rizwan, M., Ali, S., Abbas, T. et al. (2016a). Cadmium minimization in wheat: a critical review. *Ecotoxicol Environ Saf* 130: 43–53.

Rizwan, M., Meunier, J., Davidian, D. et al. (2016b). Silicon alleviates Cd stress of wheat seedlings (*Triticum turgidum* L. cv. Claudio) grown in hydroponics. *Environ Sci Pollut Res* 23: 1414–1427.

Rodrigues, F.Á., Benhamou, N., Datnoff, L.E. et al. (2003). Ultrastructural and cytochemical aspects of silicon-mediated rice blast resistance. *Phytopathology* 93: 535–546.

Roychoudhury, A. and Tripathi, D.K. (2019). *Molecular Plant Abiotic Stress: Biology and Biotechnology*, vol. 1, 1–480. John Wiley & Sons.

Roychoudhury, A. and Tripathi, D.K. (2020). *Protective Chemical Agents in the Amelioration of Plant Abiotic Stress: Biochemical and Molecular Perspectives*, vol. 1, 1–704. John Wiley & Sons.

Sahebi, M., Hanafi, M. et al. (2015). Importance of silicon and mechanisms of biosilica formation in plants. *BioMed REs Int* 2015: 16p.

Sarkar, U., Tahura, S., Das, U. et al. (2020). Mitigation of chromium toxicity in wheat (Triticum aestivum L.) through silicon. *Gesunde Pflanz* 72 (3): 237–244.

Sasaki, A., Yamaji, N., Yokosho, K., and Ma, J.F. (2012). Nramp5 is a major transporter responsible for manganese and cadmium uptake in rice. *The Plant Cell* 24: 2155–2167.

Sauer, D., Saccone, L., Conley, D.J. et al. (2006). Review of methodologies for extracting plant-available and amorphous Si from soils and aquatic sediments. *Biogeochem* 80: 89–108.

Savant, N.K., Datnoff, L.E., and Snyder, G.H. (1997). Depletion of plant-available silicon in soils: a possible cause of declining rice yields. *Commun Soil Sci Plan 28*: 1245–1252.

Seneviratne, M., Rajakaruna, N., Rizwan, M. et al. (2019). Heavy metal-induced oxidative stress on seed germination and seedling development: a critical review. *Environ Geochem Health* 41: 1813–1831.

Seregin, I.V., Shpigun, L.K., and Ivanov, V.B. (2004). Distribution and toxic effects of cadmium and lead on maize roots. *Russ J Plant Phys 51*: 525–533.

Seyfferth, A.L., Morris, A.H., Gill, R. et al. (2016). Soil incorporation of silica-rich rice husk decreases inorganic arsenic in rice grain. *Journal of Agricultural and Food Chemistry* 64: 3760–3766.

Shackira, A.M. and Puthur, J.T. (2019). Phytostabilization of heavy metals: understanding of principles and practices. In: *Plant-Metal Interactions* (ed. S. Srivastava, A.K. Srivastava and P. Suprasanna), 263–282. Cham: Springer.

Sharma, S.S. and Dietz, K.J. (2006). The significance of amino acids and amino acid-derived molecules in plant responses and adaptation to heavy metal stress. *J Exp Bot 57*: 711–726.

Sharma, S.S. and Dietz, K.J. (2009). The relationship between metal toxicity and cellular redox imbalance. *Trends Plant Sci* 14: 43–50.

Shi, G.R., Cai, Q.S., Liu, C.F., and Wu, L. (2010). Silicon alleviates cadmium toxicity in peanut plants in relation to cadmium distribution and stimulation of antioxidative enzymes. *Plant Growth Regul* 61: 45–62.

Shi, Y., Zhang, Y., Yao, H. et al. (2014). Silicon improves seed germination and alleviates oxidative stress of bud seedlings in tomato under water deficit stress. *Plant Physiol Biochem* 78: 27–36.

Shivaraj SM, Mandlik R, Bhat JA, Raturi G, Elbaum R, Alexander L, Tripathi DK, Deshmukh R, Sonah H.2021. Outstanding questions on the beneficial role of silicon in crop plants. *Plant and Cell Physiology* doi: https://doi.org/10.1093/pcp/pcab145.

Singh, V.P., Tripathi, D.K., Kumar, D., and Chauhan, D.K. (2011). Influence of exogenous silicon addition on aluminium tolerance in rice seedlings. *Biol Trace Elem Res* 144: 1260–1274.

Sohail, M.I., ur Rehman, M.Z., Murtaza, G., and Wahid, M.A. (2019a). Chemical investigations of Si-rich organic and inorganic amendments and correlation analysis between different chemical composition and Si contents in amendments. *Arab J Geosci* 12: 47.

Sohail, M.I., Waris, A.A., Ayub, M.A. et al. (2019b). Environmental application of nanomaterials: A promise to sustainable future. In: *Comprehensive Analytical Chemistry*, vol. 87 (ed. S.K. Verma and A.K. Das), 1–54. Elsevier.

Sohail, M.I., ur Rehman, M.Z., Rizwan, M. et al. (2020). Efficiency of various silicon rich amendments on growth and cadmium accumulation in field grown cereals and health risk assessment. *Chemosphere* 244: 125481.

Sommer, M., Kaczorek, D., Kuzyakov, Y., and Breuer, J. (2006). Silicon pools and fluxes in soils and landscapes—a review. *J Plant Nutr Soil Sci* 169: 310–329.

Song, A., Li, Z., Zhang, J. et al. (2009). Silicon-enhanced resistance to cadmium toxicity in *Brassica chinensis* L. is attributed to Si-suppressed cadmium uptake and transport and Si-enhanced antioxidant defense capacity. *J Hazard Mater* 172: 74–83.

Sousa, A., Saleh, A.M., Habeeb, T.H. et al. (2019). Silicon dioxide nanoparticles ameliorate the phytotoxic hazards of aluminum in maize grown on acidic soil. *Sci Total Environ* 693: 133636.

Srivastava, S., Tripathi, R.D., and Dwivedi, U.N. (2004). Synthesis of phytochelatins and modulation of antioxidants in response to cadmium stress in *Cuscuta reflexa*–an angiospermic parasite. *J Plant Physiol* 161: 665–674.

Suksabye, P., Pimthong, A., Dhurakit, P. et al. (2016). Effect of biochars and microorganisms on cadmium accumulation in rice grains grown in Cd-contaminated soil. *Environ Sci Pollut Res* 23: 962–973.

Suriyaprabha, R., Karunakaran, G., Yuvakkumar, R. et al. (2012). Growth and physiological responses of maize (*Zea mays* L.) to porous silica nanoparticles in soil. *J Nanopart Res* 14: 1294.

Suzuki, S., Ma, J.F., Yamamoto, N. et al. (2012). Silicon deficiency promotes lignin accumulation in rice. *Plant Biotech 29*: 391–394.

Tchounwou, P.B., Yedjou, C.G., Patlolla, A.K., and Sutton, D.J. (2012). Heavy metal toxicity and the environment. In: *Molecular, Clinical and Environmental Toxicology* (ed. A. Luch), 133–164. Basel: Springer.

de Tombeur, F., Vander Linden, C., Cornélis, J.T. et al. (2020). Soil and climate affect foliar silicification patterns and silica-cellulose balance in sugarcane (*Saccharum officinarum*). *Plant Soil* 452 (1): 1–18.

Tong, Y.P., Kneer, R., and Zhu, Y.G. (2004). Vacuolar compartmentalization: a second-generation approach to engineering plants for phytoremediation. *Trends Plant Sci 9*: 7–9.

Torney, F., Trewyn, B.G., Lin, V.S.Y., and Wang, K. (2007). Mesoporous silica nanoparticles deliver DNA and chemicals into plants. *Nat Nanotechnol* 2: 295–300.

Tripathi, D.K., Singh, S., Singh, S. et al. (2015a). Micronutrients and their diverse role in agricultural crops: advances and future prospective. *Acta Physiologiae Plantarum* 37 (7): 1–4.

Tripathi, D., Singh, K. et al. (2015b). Silicon-mediated alleviation of Cr (VI) toxicity in wheat seedlings as evidenced by chlorophyll florescence, laser induced breakdown spectroscopy and anatomical changes. *Ecotoxicol Environ Saf* 113: 133–144.

Tripathi, D.K., Singh, V.P., Lux, A., and Vaculik, M. (2020). Silicon in plant biology: from past to present, and future challenges. *J Exp Bot* 71 (21): 6699–6702.

Tripathi, D.K., Singh, V.P., Ahmad, P. et al. (2021). Metalloids in plant biology: new avenues in their research. *Journal of Hazardous Materials* 422: 126738.

Tubana, B.S., Babu, T., and Datnoff, L.E. (2016). A review of silicon in soils and plants and its role in US agriculture: history and future perspectives. *Soil Sci* 181: 393–411.

Tuna, A., Kaya, C., Higgs, D. et al. (2008). Silicon improves salinity tolerance in wheat plants. *Environ Exp Bot* 62: 10–16.

Vaculík, M., Lux, A., Luxová, M. et al. (2009). Silicon mitigates cadmium inhibitory effects in young maize plants. *Environ Exp Bot* 67: 52–58.

Vaculík, M., Landberg, T., Greger, M. et al. (2012). Silicon modifies root anatomy, and uptake and subcellular distribution of cadmium in young maize plants. *Ann Bot 110*: 433–443.

Vamerali, T., Bandiera, M., and Mosca, G. (2010). Field crops for phytoremediation of metal-contaminated land. A review. *Environ Chem Lett 8*: 1–17.

Verma, K.K., Liu, X.H., Wu, K.C. et al. (2019). The impact of silicon on photosynthetic and biochemical responses of sugarcane under different soil moisture levels. *Silicon* 12 (6): 1–13.

Viehweger, K. and Geipel, G. (2010). Uranium accumulation and tolerance in *Arabidopsis halleri* under native versus hydroponic conditions. *Environ Exp Bot* 69: 39–46.

Wang, Y., Stass, A., Horst, W.J. et al. (2004). Apoplastic binding of aluminum is involved in silicon-induced amelioration of aluminum toxicity in maize. *Plant Physiol* 136: 3762–3770.

Wang, Y., Shi, J., Wang, H. et al. (2007). The influence of soil heavy metals pollution on soil microbial biomass, enzyme activity, and community composition near a copper smelter. *Ecotoxicol Environ Saf* 67: 75–81.

Wang, S., Wang, F., and Gao, S. (2015). Foliar application with nano-silicon alleviates Cd toxicity in rice seedlings. *Environ Sci Pollut Res 22*: 2837–2845.

Wang, H.Y., Wen, S.L., Chen, P. et al. (2016). Mitigation of cadmium and arsenic in rice grain by applying different silicon fertilizers in contaminated fields. *Environ Sci Pollut Res 23*: 3781–3788.

Wang, Y., Xiao, X., Xu, Y., and Chen, B. (2019). Environmental effects of silicon within biochar (sichar) and carbon–silicon coupling mechanisms: a critical review. *Environ Sci Technol* 53: 13570–13582.

Wanyika, H., Gatebe, E., Kioni, P. et al. (2012). Mesoporous silica nanoparticles carrier for urea: potential applications in agrochemical delivery systems. *Journal of Nanoscience and Nanotechnology* 12: 2221–2228.

WHO/FAO/IAEA (1996). *Trace Elements in Human Nutrition and Health*. Switzerland: Geneva: World Health Organization.

Wollgiehn, R. and Neumann, D. (1999). Metal stress response and tolerance of cultured cells from *Silene vulgaris* and *Lycopersicon peruvianum*: role of heat stress proteins. *J Plant Physiol* 154: 547–553.

Wu, Z., Xu, S., Shi, H. et al. (2018). Comparison of foliar silicon and selenium on cadmium absorption, compartmentation, translocation and the antioxidant system in Chinese flowering cabbage. *Ecotoxicol Environ Saf* 166: 157–164.

Xu, X., Zhao, Y., Sima, J. et al. (2017). Indispensable role of biochar-inherent mineral constituents in its environmental applications: a review. *Bioresour Technol* 241: 887–899.

Yadav, S.K. (2010). Heavy metals toxicity in plants: an overview on the role of glutathione and phytochelatins in heavy metal stress tolerance of plants. *S Afr J Bot 76*: 167–179.

Zhang, F.Q., Wang, Y.S., Lou, Z.P., and Dong, J.D. (2007). Effect of heavy metal stress on antioxidative enzymes and lipid peroxidation in leaves and roots of two mangrove plant seedlings (*Kandelia candel* and *Bruguiera gymnorrhiza*). *Chemosphere 67*: 44–50.

Zhang, Y., Zhang, H.W., Zhen-Cheng, S.U., and Zhang, C.G. (2008). Soil microbial characteristics under long-term heavy metal stress: a case study in Zhangshi wastewater irrigation area, Shenyang. *Pedosphere 18*: 1–10.

Zhang, X., Yang, H., and Cui, Z. (2018). Evaluation and analysis of soil migration and distribution characteristics of heavy metals in iron tailings. *J Clean Prod 172*: 475–480.

Zhang, S., Geng, L., Fan, L. et al. (2020). Spraying silicon to decrease inorganic arsenic accumulation in rice grain from arsenic-contaminated paddy soil. *Sci Total Environ* 704: 135239.

Zhu, Y. and Gong, H. (2014). Beneficial effects of silicon on salt and drought tolerance in plants. *Agron Sustain Dev* 34: 455–472.

12

How Does Sodium Content in Growing Media Affect the Chemical Content of Medicinal and Aromatic Plants? Two Sides of the Coin

Ahmet Metin Kumlay[1,], Muhittin Kulak[2,*], Mehmet Zeki Kocak[2], Ferdi Celikcan[3], and Mehmet Hakki Alma[4]*

[1] *Department of Field Crops, Faculty of Agriculture, Igdir University, Igdir, Türkiye*
[2] *Department of Herbal and Animal Production, Vocational School of Technical Sciences, Igdir University, Igdir, Türkiye*
[3] *Department of Organic Farming, College of Applied Science, Igdir University, Igdir, Türkiye*
[4] *Department of Forest Industry Engineering, Faculty of Forestry, Kahramanmaraş Sutcu Imam University, Kahramanmaraş, Türkiye*

All things are poisons, for there is nothing without poisonous qualities. It is only the dose which makes a thing poison (Paracelsus).

Introduction

Mineral nutrients are elements that are primarily taken from the soil in the form of inorganic ions. Although mineral nutrients are in a continuous cycle from one to another in all organisms, they mostly enter the biosphere through the root systems of plants. After absorption by the root system, the minerals are translocated to or accumulated into different parts of the plants for subsequent significant functions for the complex and proper plant systems (Taiz and Zeiger 2002; Tripathi et al. 2015). It is worthy to note herein that mineral nutrients are necessary and required not only for the proper and desired growth and development of the plant and in turn for higher crop yield but also for their protective properties in response to the biotic and abiotic stress factors (Marschner 1995). In this regard, it is important to keep the nutrients at a certain level in the functional organs of the plant in order to expand the tolerance or resilience of the plant.

Elements are required for proper and complete life cycle of the plant, ensuring successful growth and development. Based on their amount in plant tissues, those elements are classified as macronutrients and micronutrients (Pilon-Smits et al. 2009; Gómez-Merino and Trejo-Téllez 2018). Of those elements, elements that encourage growth but for particular taxa of the plants are called beneficial elements, out of which Al, Co, Na, Se, and Si are the most investigated elements (Pilon-Smits et al. 2009).

* Both authors equally contributed to the chapter

Beneficial Chemical Elements of Plants: Recent Developments and Future Prospects, First Edition. Edited by Sangeeta Pandey, Durgesh Kumar Tripathi, Vijay Pratap Singh, Shivesh Sharma, and Devendra Kumar Chauhan.

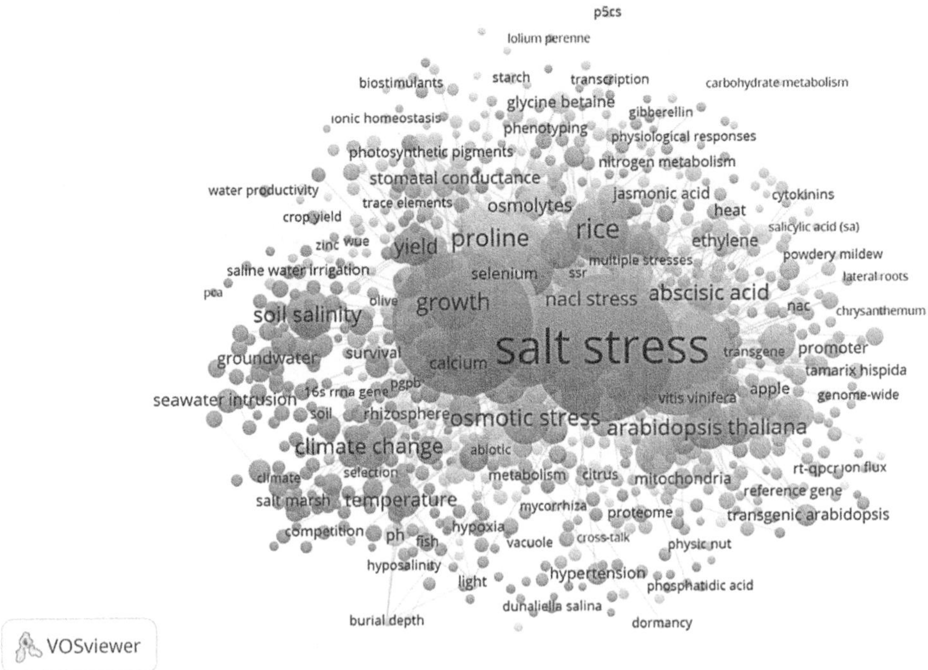

Figure 12.1 Keywords of the salinity-related studies in agricultural and biological sciences.

Up to now, great detailed reviews and research results have been reported with respect to the physiological, biochemical, and molecular responses of plants against high salinity in the growing media. After a search at SCOPUS with TITLE-ABS-KEY (salinity OR salt AND stress OR saline OR sodic) AND (LIMIT-TO (SUBJAREA, 'AGRI')), 39.236 documents came out (August 10th, 2020) and then the documents were limited to years between 2016 and 2020 and visualized (Figure 12.1). For the visualization, we herein used keyword analysis since keywords are the core of the scientific reports as indicators of the development and progress of the knowledge (He 1999). For the assessment, we set minimum occurrences of a keyword to be 5, visualizing 1324 keywords out of 21.225 keywords proposed along with the studies (Figure 12.1). Consequently, 15 clusters corresponding to the 1000 items were obtained. Studies are mainly concentrated on oxidative stress, osmoprotectants, enzymatic and non-enzymatic activities, and genomic, transcriptomic, and proteomic responses of the plants against higher salinity levels. Once compared with the published number of documents, there are limited studies regarding stress-metabolite interaction. For that reason, herewith the study, we concentrated on secondary metabolites accumulation, *viz.* alkaloids, terpenoids, and phenolics, in specifically medicinal and aromatic plants exposed to salinity stress manipulation without any discrimination regarding field or greenhouse studies.

What Kinds of Functions Have Been Attributed to Sodium for Proper Metabolism of the Plant?

According to essentiality criteria postulated by Arnon and Stout (1939), an essential nutrient for plants is completely required for the survival and reproduction of the plants. However, high yield or biomass production may not be classified as an important aspect for the nutrient essentiality. The elements such as Na, Se, and Si exhibit significant roles in promoting augmented production but may not be required for the plant for its survival (Subbarao et al. 2003). Nevertheless, an exception of a subgroup of C_4 plants, *viz. Atriplex spp., Kochia childsii*, millet, and a group of grasses, Na^+ at trace levels for some C_4 plant species are essential in order to forward particular mechanisms such as transport process and pyruvate uptake into the chloroplasts (Maathuis 2014). Although not essential for all terrestrial plants, Na^+ can be beneficial at its low levels, particularly when K^+ level is low (Maathuis 2014; Kaur et al. 2016).

What Kind of Perturbations Might Emerge in Case of Deficiency or Excessive Accumulation of Sodium in Growing Media and in Turn, in Plants?

Salinity is characterized with high levels of Na^+ and Cl^-, affecting about 800 millions of hectares of arable land worldwide (Munns 2005; Munns and Tester 2008; Gohari et al. 2020). The adverse effects of high levels of salt on seed germination, crop survival, growth, and productivity have been well-documented. The alterations in morphological, physiological, biochemical characteristics; ion toxicity; disruption in nutritional balance; and osmotic stress have been revealed in the plants exposed to the high levels of salinity, which in turn cause significant reductions in plant yield and production (Munns and Tester 2008; Parihar et al. 2015; Gohari et al. 2020; Sharifi and Bidabadi 2020).

Although the high levels of sodium damage most of the terrestrial plants and are not essential for many plant species, low concentration of Na^+ might be needed for normal growth and development of some plant species (Wakeel et al. 2011; Subbarao et al. 2003). Sodium has been essential for plants that use C_4 or CAM photosynthetic pathways (Ohnishi et al. 1990). Sodium is required for the regeneration of phosphoenolpyruvate from pyruvate, and phosphoenolpyruvate is used for fixing atmospheric carbon for photosynthesis by C_4 and CAM plants (Pilon-Smits et al. 2009). Due to low concentration or deficiency of Na^+, phosphoenolpyruvate cannot be regenerated from pyruvate in mesophyll chloroplasts. In C_4 plants (*Amaranthus tricolor* L. and *K. childsii*), lower activities of PSII were reported in mesophyll thylakoids of Na^+-deficient plants (Grof et al. 1989).

What Are the Major Mechanisms Associated with the Damage Caused by High Salinity?

As reported in previous detailed reviews and original researches, deleterious effects of salinity on relevant plant system have been associated with three different mechanisms (Levinsh 2006). Of those mechanisms, the first one concerns with the ion toxicity caused by overaccumulation of Na^+ and Cl^- ions in the cytoplasm, which in turn results in ionic imbalance. Second, in order to cope with ionic imbalance, an increased transport of the ions to vacuole occurs. Along with the compartmented ions into the vacuole, cellular osmotic potential must be coordinated and balanced according to the external water potential. In the case of not balanced water potential, damage to macromolecules including carbohydrates, lipids, proteins, and nucleic acids emerges. Third, combined effect of water deficit and ion toxicity disturbs proper metabolism and subsequently results in accumulation of reactive oxygen species, at a level beyond cellular capacity of oxidant detoxification for redox homeostasis maintenance (Mittler 2002; Levinsh 2006; Hasegawa 2013).

Compartmentalization of Sodium Through Plant Parts

Compartmentalization of Na^+ through plant cells is of significant and key mechanisms regarding salt tolerance and that process includes Na^+ absorption from soil to root, transport of Na^+ to leaf, intracellular compartmentalization of Na^+ in leaf, and finally secretion of Na^+ out of leaves (Peng et al. 2016). With the compartmentalization of Na^+ into the vacuoles, plants reduce the toxic concentration in the cytosol (Keisham et al. 2018) as high content of Na^+ hinders the cellular metabolism (Wakeel et al. 2011). In the case of translocation of Na^+ towards the aerial parts of the plant by Na^+-sensitive species, serious damages may emerge due to the specific ion toxicity (Greenway and Munns 1980; Wakeel et al. 2011).

However, we should herein highlight again that the response of Na^+ greatly varies among the plant species, which might be sorted into two categories (natrophilic and natrophobic) according to Na^+ nutrition. The categorization is dependent on the plant growth response to Na^+ and its translocation to the aboveground parts of the plant (Wakeel et al. 2011). Natrophilic plants accumulate large quantity of Na^+ in their leaves in comparison to other plants but have lower concentrations in their roots, whereas natrophobic species prefers to accumulate in the roots or lower aboveground tissues but a small amount of Na^+ is accumulated in leaf tissues (Smith et al. 1978).

Why Is the Sodium/Potassium Ratio Important for Plant Metabolism?

Two stresses on plant tissues by high levels of salinity have been postulated. The first has been noted as a water deficiency resulting from relatively high solute concentration of soil and the later has been associated with ion-specific stress that results from altered ratio of K^+/Na^+ and concentrations of Na^+ and Cl^- (Blumwald 2000). In order to combat with the stress, low concentration of Na^+ in cytosol and a high ratio of K^+/Na^+ are desired for reducing the influx of Na^+ into root cells, partitioning off Na^+ into the vacuoles and increasing the efflux from the root cells (Tester and Davenport 2003; Keisham et al. 2018).

It is well-known that Na^+ has antagonistic effect on K^+ uptake in plants. For that reason, excessive Na^+ content in media – in case of high level of salinity – significantly reduces K^+ content. The K-Na selectivity was reported to be correlated with salt tolerance in some plant species (Ashraf and Naqvi 1992). K^+ is of essential macronutrients that exhibit important vital roles in regulation of enzymatic activities, osmotic adjustments, turgor status, membrane potential, and pH homeostasis in cytosol (Khan et al. 2009; Barragán et al. 2012; Almeida et al. 2017). For that reason, in order to maintain the desired K^+/Na^+ ratio, preventing or restricting the excessive content and efflux of Na^+ and uptake of K^+ are of the important regulation in the cytosol (Khan et al. 2009) since a high cytosolic K^+/Na^+ ratio have been associated with plant survival against salt stress (Tahal et al. 2000; García Morales et al. 2012; Almeida et al. 2017).

How Do Priming or Osmo-Conditioning Seeds Using NaCl Solutions Imprint the Sequential Growth Performance or Stage of the Plants? An Approach Regarding Imprint Memory with Low Concentration versus Higher Subsequent Concentration of NaCl

Seed priming is a simple and cost-effective pre-sowing technique to improve subsequent growth and development of plants. The effectiveness of the technique for enhancing the tolerance against stressful conditions has been revealed (Wu et al. 2019). With the priming, seeds are hydrated to the point of pre-germination metabolic process but not letting seeds germinate (Farooq et al. 2019). In that context, many molecules, *viz.* salicylic acid (Wang et al. 2020a; Nazari et al. 2020; Bortolin et al. 2020; Mahmood-ur-Rehman et al. 2020), melatonin (Bahcesular et al. 2020; Khan et al. 2020; Azadbakht and Balouchi 2020), brassinosteroid (Soliman et al. 2020; Huang et al. 2020), jasmonate (Bali et al. 2020; Wang et al. 2020b), and 5-aminolevulinic acid (Sheteiwy et al. 2017; Kanto et al. 2015), have been used recently. In addition to signal molecules, low concentration of sodium chloride solutions (NaCl) have been also widely used for priming (Ashraf and Rauf 2001; Olave and Guzmán 2004; Mohammadi 2009; İşeri et al. 2014; Khan et al. 2009; Gebreegziabher and Qufa 2017; Yang et al. 2020).

According to the studies, priming with NaCl was effective in reducing the harmful impact of salinity through enhancing the percentage, index and speed of germination, vigour index, length of plumule and radicle length, as well as dry weight of the plants in comparison with those plants of control (Khan et al. 2009). The improvements in germination and seedling emergence were recorded after low salt seed priming. Stress tolerance and subsequent growth of plants were enhanced through coupling with modifications in expression levels of genes associated with growth and stress (Nakaune et al. 2012). The study by Mohammadi (2009) also reported high values of germination percentage, germination rate, and seedling dry weight after NaCl priming in canola (*Brassica napus* L.). The studies suggest that low sodium chloride priming might enhance seedling performance and stress tolerance, but we herein note that the studies regarding NaCl priming have remained limited with the growth and development performance of the plants. Forthcoming studies can be concentrated on NaCl priming, subsequent stress, and NaCl priming *x* subsequent stress interaction on secondary metabolites, as a possible elicitor role of priming with low concentration of NaCl priming.

What Are Medicinal and Aromatic Plants and Metabolites of Those Plants? How Do Those Metabolites Respond to Higher Content of Na in Media Regarding Total Content and Their Specific Compounds?

With an estimation number of 50 000–80 000 flowering plants, medicinal and aromatic plants are of significant plant groups of the plant diversity. Due to chemically active constituents available but organ-specified, ontogenetic-specified, or diurnal-specified active constituents, those plants trigger physiological responses when used for combating various ailments (Imadi et al. 2016). In addition to the beneficial physiological impact with respect to the human health, those plants have been used for many purposes, *viz*. pharmaceuticals, food additives, and cosmetics from prehistoric times to present (Iqbal 2013; Nayanabhirama 2016). It is well-known and documented in many original researches and reputed review reports that the efficacies of the plants depend on the quality and quantity of the secondary metabolites present. As mentioned herein, the chemical content and composition are not constant, responsive to any internal and external changes (Aghaei and Komatsu 2013; Akgunlu et al. 2016).

Plants produce a huge array of low molecular weight compounds with estimates ranging from 100 000 to 1 million (Fang et al. 2019). At the beginning of the 2000s, it was reported that the structures of approximately 50 000 metabolites have been elucidated (De Luca and St Pierre 2000). Those metabolites are traditionally classified into two major groups as primary and secondary metabolites. Of those metabolites, a few parts of the metabolites are derived from primary metabolic pathways, which are common to all organisms. The essential roles regarding plant growth and development for plants have been attributed to primary metabolites (Patra et al. 2013). The rest are deemed as secondary metabolites, which were clearly defined as a compound whose biosynthesis is confined into selected plant groups (Pichersky and Gang 2000). Furthermore, Cheynier et al. (2013) describes the secondary metabolites as organic compounds available in specialized cells that are not considered as fundamental and essential for photosynthetic and respiratory metabolism, but exhibiting significant roles for plant survival against stress factors. Not only restricted roles against abiotic stress factors, secondary metabolites have been well-documented to exhibit certain roles in plant defence against herbivory and interspecies defences (Fang et al. 2019). According to the chemical composition, secondary metabolites are composed of two major groups. The first group contains nitrogen-containing molecules (alkaloid), whilst the later includes nitrogen-deficient molecules (terpenoids and phenolics) (Patra et al. 2013).

The Growth, Development, and Yield are Adversely Affected Under High Sodium Concentration of Growing Media, but What Can We Say for Contents of Total Metabolites or Specific Compounds?

Alkaloids

Most of the plant secondary metabolites have nitrogen in the structure. Of those groups, alkaloids are highly interesting to humans due to their toxicity and medicinal properties, containing approximately more than 12 000 low molecular compounds in their family (Verma and Shukla 2015). Those molecules are mostly derived from amino acids such as tryptophan, tyrosine, phenylalanine, and lysine, as well as ornithine (De Luca and St Pierre 2000; Patra et al. 2013). Of those compounds, approximately 2500 compounds have been attributed to exhibit significant pharmacological activities such as analgesic, cough suppressant, and muscle relaxant (Verma and Shukla 2015). Not to confine properties for human and animals, alkaloids are well known to exhibit protective roles against microbial or herbivore attack and UV -radiation (Patra et al. 2013).

As dynamic structure of the plants and metabolites in the complex plant system, after being exposed to stressful conditions, the alkaloid concentration of the plants is frequently enhanced through its increased biosynthesis as a consequence of higher concentration of NADPH in plants exposed to stress (Yahyazadeh et al. 2018). However, in comparison to the terpenoid and phenolics, the studies with respect to alkaloid concentration and compositions under stress conditions are limited and alkaloid-bearing plants received less interest. Also, it is worthy to underline that pharmacologically interesting compounds are not obtained from plant material in large quantities, and for that reason, in order to enhance those compounds researchers have concentrated on genetic engineering like transferring plant genes to microorganisms (Heldt and Heldt 2015). For higher yield of alkaloids, higher Na^{+}-including media can be considered as an efficient and economical elicitor source (Bhat et al. 2008), and in that context, an elicitation role can be attributed to Na, a beneficial element that triggers the secondary metabolite formation (Ahmad et al. 2013).

In response to higher concentration of Na in growing media, higher alkaloid concentrations were recorded for plants under salt stress (Frischknecht and Baumann 1985; Brachet and Cosson 1986; Zhang et al. 2011; Jasmin Šutković et al. 2011). However, total alkaloid content was declined with higher Na in media (Idrees et al. 2011). Regarding total alkaloid measurements, alkaloid composition (more than 12000 elucidated structures) might be revealed in order to explain whether plants respond to treatments through yielding the total alkaloid content with the same composition or diversifying the alkaloid composition or both responses.

Terpenoids

Terpenoids, also known as isoprenoids due to being built up from isoprene, are the largest family and most structural class of plants specializing as secondary metabolites, with an estimate of >40000 different molecules (Aharoni et al. 2005; Grassmann 2005; Aharoni et al. 2006; Yazaki et al. 2017). Those molecules are classified according to their basis of carbon skeleton, viz. monoterpene, sesquiterpene, diterpene, triterpene, tetraterpene, and polyterpene (Grassmann 2005). Many significant functions have been attributed to those molecules for both plants and humans. Those have important roles for plant survival against biotic and abiotic environmental factors and also have biological properties beneficial for human health (Aharoni et al. 2005). Plants exhibit flexibility to environmental stimuli by changing in yield, content, or diversity of the secondary metabolites. Along with the report, increases in essential oil content (Gohari et al. 2020; Gohari et al. 2020a; Sharifi and Bidabadi 2020; de Azevedo Neto et al. 2000; Estaji et al. 2018), increases in essential oil yield (de Azevedo Neto et al. 2000; Pankaj et al. 2019), decreases in essential oil content (Gadallah et al. 2020; Sany et al. 2020; Hegazy et al. 2019; Keramati et al. 2016), and decreases in essential oil yield (Gadallah et al. 2020; Sany et al. 2020; de Azevedo Neto et al. 2000; Hegazy et al. 2019; Sarmoum et al. 2019; Keramati et al. 2016) have been reported. Furthermore, no significant changes have been documented but up to a certain threshold. Below or above the threshold, significant increases or decreases were recorded (Talebi et al. 2018; Yadav et al. 2017; Ahmed et al. 2017) (Table 12.1).

With respect to the individual components of essential oil, linalool did not significantly change at 40 mM NaCl in *Ocimum basilicum* (Farsaraei et al. 2020). The content of the

Table 12.1 Changes in total yield or content of essential oil in those plants exposed to higher levels of salinity.

Plant species	Secondary metabolites	Changes	Authors
D. moldavica	Essential oil content	Increase at 50 and 100 mM NaCl	Gohari et al. (2020)
O. basilicum	Essential oil content	No change	Farsaraei et al. (2020)
S. nemorosa	Essential oil content	Increase	Sharifi and Bidabadi (2020)
P. graveolens	Essential oil content	Decrease	Gadallah et al. (2020)
P. graveolens	Essential oil yield	Decrease	Gadallah et al. (2020)
O. basilicum	Essential oil content	Increase at 50 mM NaCl	Gohari et al. (2020a)
O. basilicum	Essential oil content	Increase at 50 mM NaCl	Gohari et al. (2020b)
H. officinalis	Essential oil yield	Decrease at 50, 75, and 100 mM NaCl	Ben Hamida et al. (2020)
P. amboinicus	Essential oil content	Increase at 5, 10, and 15 dSm^{-1}	Sany et al. (2020)
P. amboinicus	Essential oil yield	Decrease at 5, 10, and 15 dSm^{-1}	Sany et al. (2020)
O. basilicum (TFA)	Essential oil content	Decrease at 80 mM NaCl	de Azevedo Neto et al. (2000)
O. basilicum (TFA)	Essential oil yield	Decrease at 80 mM NaCl	de Azevedo Neto et al. (2000)
O. basilicum (GDM)	Essential oil content	Increase at 80 mM NaCl	de Azevedo Neto et al. (2000)
O. basilicum (GDM)	Essential oil yield	Increase at 80 mM NaCl	de Azevedo Neto et al. (2000)
S. montana	Essential oil content	Decrease at 4 dSm^{-1}	Hegazy et al. (2019)
S. montana	Essential oil yield	Decrease at 4 dSm^{-1}	Hegazy et al. (2019)
R. officinalis L.	Essential oil yield	Decrease at 4.2 g of NaCl/L concentration	Sarmoum et al. (2019)
C. martinii	Essential oil yield	Increase at salted soils	Pankaj et al. (2019)
R. officinalis	Essential oil content	Increase at 2.5, 5, 7.5, 10, and 12.5 NaCl g/L	Bidgoli et al. (2019)
O. basilicum (GDM)	Essential oil content	Increase at 80 mM NaCl	Silva et al. (2019)

Table 12.1 (Continued)

Plant species	Secondary metabolites	Changes	Authors
O. basilicum (GDM)	Essential oil yield	Increase at 80 mM NaCl	Silva et al. (2019)
M. piperita	Essential oil content	Decrease at 50 and 100 but no change 150 mM NaCl	Khanam and Mohammad (2018)
M. piperita	Essential oil yield	Decrease at 50, 100, 150 mM NaCl	Khanam and Mohammad (2018)
S. hortensis (all accessions)	Essential oil content	Increase at 50 mM NaCl	Estaji et al. (2018)
O. basilicum (Cv. Rubi)	Essential oil content	No change at 30, decrease at 60, and increase at 90 mM NaCl	Talebi et al. (2018)
O. basilicum (Cv. Genove)	Essential oil content	No change at 30 and 90 mM but increase at 60 mM	Talebi et al. (2018)
R. officinalis	Essential oil content	Increases at 100, 200, and 300 mM NaCl	El-Esawi et al. (2017)
O. basilicum	Essential oil content	Increase at 30 mM NaCl	Bernstein et al. (2017)
C. cyminum	Essential oil yield	Increase at 25, 50, and 75 mmol NaCl	Rebey et al. (2017)
N. cataria	Essential oil content	Increase at 1200, 2400, and 3600 mg NaCl kg^{-1} soil	Mohammadi et al. (2017)
A. annua (EVT)	Essential oil yield	Increase at 50 and 100, but decrease at 200 mM NaCl	Yadav et al. (2017)
A. annua (MVS)	Essential oil yield	Increase at 50 and 100, but no change at 200 mM NaCl	Yadav et al. (2017)
A. annua (LVS)	Essential oil yield	No change at 50, increase at 150, decrease at 200 mM NaCl	Yadav et al. (2017)
A. annua (FBS)	Essential oil yield	Increase at 50 and 100 but decrease at 200 mM NaCl	Yadav et al. (2017)
M. officinalis	Essential oil content	Increase at 1.6, 3.1, and 6.3 dSm^{-1}	Ahmed et al. (2017)
M. officinalis	Essential oil yield	No change at 1.6 but decrease at 3.1 and 6.3 dSm^{-1}	Ahmed et al. (2017)
O. basilicum	Essential oil content	Decrease at 3, 6, and 9 dS m^{-1} NaCl	Keramati et al. (2016)
O. basilicum	Essential oil yield	Decrease at 3, 6, and 9 dS m^{-1} NaCl	Keramati et al. (2016)
R. officinalis	Essential oil content	No change at 2, 4, 6, and 8 dS m^{-1} NaCl	Bahonar et al. (2016)

EVT, Early vegetative stage; MVS, Mid-vegetative stage; LVS, Late vegetative stage; FBS, Full bloom stage; TFA, *Toscano folha de alface*; GDM, *Gennaro de menta*; Cv, Cultivar.

compound decreased at 100 mM NaCl in *O. basilicum* (Bahcesular et al. 2020) and *Thymus vulgaris* (Zrig et al. 2019). In cultivar Rubi of *O. basilicum*, linalool content increased at 30 mM, decreased at 60, and did not change at 90 mM NaCl, but in cultivar Genove of *O. basilicum*, linalool content increased at 30, 60, and 90 mM NaCl (Talebi et al. 2018). Gohari et al. (2020a) reported that linalool increased at 50 mM NaCl but did not change at 100 mM NaCl in *O. basilicum.* Moreover, 1,8 cineol content decreased at 100 mM NaCl in *O. basilicum* (Bahcesular et al. 2020). The content increased at 25, 50, 75, and 100 mM NaCl in *Salvia mirzayanii* (Valifard et al. 2019). However, Chrysargyris et al. (2019a) recorded no significant changes at 150 mM NaCl in *Mentha spicata* (Table 12.2).

According to the bibliometric analysis of salt stress and essential oil, Apiaceae and Lamiaceae family were the most studied plant groups, which are characterized with strong essence and aromatic properties. *O. basilicum, Salvia spp., Mentha spp., Matricaria chamomile, Coriandrum sativum, Origanum majorana,* and *Rosmarinus officinalis* are mostly examined for their essential oil profile in response to salinity stress. Along with the studies, in addition to the basic physiological measurements combined with antioxidant enzyme activities, changes in essential oil yield and composition have been monitored as a response to higher levels of sodium growing media. Furthermore, among the plant growth regulators, salicylic acid, brassinosteroid, chitosan, and spermidine are widely used for salt stress mitigation (Figure 12.2).

Phenolics

Phenolic compounds with an aromatic ring with a hydroxyl group attached are the secondary metabolites ubiquitously present in the plant kingdom. The phenolics comprise over 9000 compounds that are different in their chemical structure and subsequently in their biological properties (Crozier et al. 2006; Waśkiewicz et al. 2013). This group is not common in bacteria, fungi, and algae, but a limited number of phenolics like flavonoids are produced in bryophytes (Cheynier et al. 2013). Essential roles for the plant defence mechanisms in response to stress conditions have been attributed to phenolics like other natural compounds (Sakihama et al. 2002; Michalak 2006; Sharma et al. 2012; Caretto et al. 2015).

Once compared to alkaloids and terpenoids, phenolics have been more extensively investigated (Tables 12.3 and 12.4). Stress protectant functions as potent antioxidant compounds against reactive oxygen species have been attributed to phenolics in plants grown under high levels of salinity. In addition to the increases in total phenolics (Bistgani et al. 2019; Ghassemi-Golezani et al. 2020; Chrysargyris et al. 2019b; Al-Garni et al. 2019; Javed and Gürel 2019; Sarker and Oba 2018; Estaji et al. 2018) and significant decreases regarding total content of phenolics (Bahcesular et al. 2020; Astaneh et al. 2018), we cannot illustrate a common trend for individual phenolic fractions. For example, caffeic acid content increased at 25 mM NaCl but decreased at 50 and 75 mM NaCl in *Echinacea purpurea* (Khorasaninejad and Hemmati 2020). The compound did not exhibit significant changes at 25 but increased at 50 and 100 mM NaCl in *A. tricolor* (Sarker and Oba 2018). On the other hand, the content of the compound significantly declined with the addition of 100 mM NaCl in *O. basilicum* (Bahcesular et al. 2020). Of the prominent phenolics, chlorogenic acid content did not significantly respond to 50 and 75 mM NaCl but exhibited a significant

Table 12.2 Changes in major essential oil components in those plants exposed higher levels of salinity.

Plant species	Major component	Changes	Authors
D. moldavica	Z-citral	Increase at 100 mM but decrease at 50 mM NaCl	Gohari et al. (2020)
D. moldavica	Geranial (E-citral)	Decrease at 50 mM but no change at 100 mM NaCl	Gohari et al. (2020)
D. moldavica	Geranyl acetate	Increase at 50 mM and 100 mM NaCl	Gohari et al. (2020)
O. basilicum	Linalool	No change at 40 mM NaCl	Farsaraei et al. (2020)
O. basilicum	Eugenol	Decrease at 40 mM NaCl	Farsaraei et al. (2020)
O. basilicum	tau-Muurolol	No change at 40 mM NaCl	Farsaraei et al. (2020)
O. basilicum	α-Cadinol	Decrease at 40 mM NaCl	Farsaraei et al. (2020)
O. basilicum	Linalool	Increase at 50 mM but no change at 100 mM NaCl	Gohari et al. (2020a)
O. basilicum	Camphor	Increase at 50 and 100 mM NaCl	Gohari et al. (2020a)
O. basilicum	Estragole	Increase at 50 and 100 mM NaCl	Gohari et al. (2020a)
O. basilicum	a-Bergamotene	Increase at 100 mM but decrease at 50 Mm NaCl	Gohari et al. (2020a)
O. basilicum	Germacrene D	Decrease at 50 and 100 mM NaCl	Gohari et al. (2020a)
O. basilicum	Methyl chavicol	Decrease at 50 and 100 mM NaCl	Gohari et al. (2020a)
O. basilicum	1,8 cineol	Decrease at 100 mM NaCl	Bahcesular et al. (2020)
O. basilicum	Linalool	Decrease at 100 mM NaCl	Bahcesular et al. (2020)
O. basilicum	Methlyeugenol	Increase at 100 mM NaCl	Bahcesular et al. (2020)
O. basilicum	Eugenol	Decrease at 100 mM NaCl	Bahcesular et al. (2020)
H. officinalis	Pinocarvone	Decrease at 50, 75, and 100 mM NaCl	Ben Hamida et al. (2020)
H. officinalis	Isopinocamphone	Decrease at 50, 75, and 100 mM NaCl	Ben Hamida et al. (2020)
H. officinalis	Elemol	Increase at 50, 75, and 100 mM NaCl	Ben Hamida et al. (2020)

(*Continued*)

Table 12.2 (Continued)

Plant species	Major component	Changes	Authors
P. amboinicus	α-terpinene	Decrease at 5 dSm^{-1} but increase at 10 and 15 dSm^{-1}	Sany et al. (2020)
P. amboinicus	γ-terpinene	Decrease at 5, 10, and 15 dSm^{-1}	Sany et al. (2020)
P. amboinicus	Thymol	Increase at 5, 10, and 15 dSm^{-1}	Sany et al. (2020)
P. amboinicus	Trans-caryophyllene	Increase at 5 dSm^{-1} but decrease at 10 and 15 dSm^{-1}	Sany et al. (2020)
S. montana	(−)Camphor	Not detected at 4 dSm^{-1}	Hegazy et al. (2019)
S. montana	trans-β-Bergamotene	Not detected at 4 dSm^{-1}	Hegazy et al. (2019)
T. vulgaris	Camphene	Increase at 100 mM NaCl	Zrig et al. (2019)
T. vulgaris	p-cymene	No change at 100 mM NaCl	Zrig et al. (2019)
T. vulgaris	Myrcene	Increase at 100 mM NaCl	Zrig et al. (2019)
T. vulgaris	Thymol	Decrease at 100 mM NaCl	Zrig et al. (2019)
T. vulgaris	Linalool	Decrease at 100 mM NaCl	Zrig et al. (2019)
T. vulgaris	Borneol	Decrease at 100 mM NaCl	Zrig et al. (2019)
S. mirzayanii	α-terpinyl acetate	Increase at 25 and 50 Mm but no change at 75 and 100 mM NaCl	Valifard et al. (2019)
S. mirzayanii	1,8-Cineole	Increase at 25, 50, 75, and 100 mM NaCl	Valifard et al. (2019)
S. mirzayanii	Linalyl acetate	Decrease at 25 and 50 Mm but increase at 75 and 100 mM NaCl	Valifard et al. (2019)
S. mirzayanii	Bicyclogermacrene	Decrease at 25, 50, 75, and 100 mM NaCl	Valifard et al. (2019)
M. spicata	Carvone	Decrease at 150 mM NaCl	Chrysargyris et al. (2019a)
M. spicata	Germacrene D	No change at 150 mM NaCl	Chrysargyris et al. (2019b)
M. spicata	Limonene	Increase at 150 mM NaCl	Chrysargyris et al. (2019a)
M. spicata	1,8-Cineole	No change at 150 mM NaCl	Chrysargyris et al. (2019b)
C. martinii	Myrcene	Decrease at salted soils	Pankaj et al. (2019)
C. martinii	Geraniol	Decrease at salted soils	Pankaj et al. (2019)
C. martinii	Geranyl acetate	Increase at salted soils	Pankaj et al. (2019)
M. piperita	Menthol	Decrease at 50, 100, and 150 mM NaCl	Khanam and Mohammad (2018)
S. hortensis (all accessions)	β-Pinene	Increase at 50 mM NaCl	Estaji et al. (2018)
S. hortensis (all accessions)	Myrcene	Decrease at 50 mM NaCl	Estaji et al. (2018)

Table 12.2 (Continued)

Plant species	Major component	Changes	Authors
S. hortensis (all accessions)	p-Cymene	Increase at 50 mM NaCl (except accession Kahnu)	Estaji et al. (2018)
O. basilicum (cv. Rubi)	Linalool	Increase at 30 mM, decrease at 60, and no change at 90 mM NaCl	Talebi et al. (2018)
O. basilicum (cv. Genove)	Linalool	Increase at 30, 60, and 90 mM NaCl	Talebi et al. (2018)
O. basilicum (cv. Rubi)	α-Bergamotene ⟨Z⟩	Decrease at 30, 60, and 90 mM NaCl	Talebi et al. (2018)
O. basilicum (cv. Genove)	α-Bergamotene ⟨Z⟩	Decrease at 30, 60, and 90 mM NaCl	Talebi et al. (2018)
O. basilicum (cv. Rubi)	α-Elemene	Decrease at 30, 60, and 90 mM NaCl	Talebi et al. (2018)
O. basilicum (cv. Genove)	α-Elemene	Increase at 30 mM but disappeared at 60 and 90 mM NaCl	Talebi et al. (2018)
O. basilicum (cv. Rubi)	α-Cadinol	Decrease at 30, 60, and 90 mM NaCl	Talebi et al. (2018)
O. basilicum (cv. Genove)	α-Cadinol	Increase at 30, 60, and 90 mM NaCl	Talebi et al. (2018)

EVT, Early vegetative stage; MVS, Mid-vegetative stage; LVS, Late vegetative stage; FBS, Full bloom stage; TFA, *Toscano folha de alface*; GDM, *Gennaro de menta*; Cv, Cultivar.

Table 12.2 (continued) Change in individual components of essential oil of different plants exposed to higher salinity.

Plant species	Major component	Changes	Authors
M. officinalis	Myrcene	Increase at 50, 100, and 200 mM but decrease at 150 Mm NaCl	Bonacina et al. (2017)
M. officinalis	Citronellal	Decrease at 50, 100, and 150 mM but increase at 200 mM NaCl	Bonacina et al. (2017)
M. officinalis	Citronellol	Increase at 50 and 200 mM but decrease at 100 and 150 mM NaCl	Bonacina et al. (2017)
M. officinalis	Neral	Decrease at 50, 100, 150, and 200 mM NaCl	Bonacina et al. (2017)
M. officinalis	α-Citral	Decrease at 50, 100, 150, and 200 mM NaCl	Bonacina et al. (2017)
M. officinalis	Trans-caryophyllene	Increase at 50, 100, 150, and 200 mM NaCl	Bonacina et al. (2017)

(Continued)

Table 12.2 (Continued)

Plant species	Major component	Changes	Authors
R. officinalis	α-Pinene	Decrease at 100, 200, and 300 ppm NaCl	El-Esawi et al. (2017)
R. officinalis	β-Pinene	Decrease at 100, 200, and 300 ppm NaCl	El-Esawi et al. (2017)
R. officinalis	1,8-Cineole	No change at 100 and 200 but slight decrease at 300 ppm NaCl	El-Esawi et al. (2017)
R. officinalis	Linalool	Increase at 100, 200, and 300 ppm NaCl	El-Esawi et al. (2017)
R. officinalis	Camphor	Increase at 100, 200, and 300 ppm NaCl	El-Esawi et al. (2017)
R. officinalis	Borneol	Increase at 100, 200, and 300 ppm NaCl	El-Esawi et al. (2017)
C. cyminum	β-Pinene	Increase at 25, 50, and 75 mmol NaCl	Rebey et al. (2017)
C. cyminum	γ-Terpinene	Increase at 25, 50, and 75 mmol NaCl	Rebey et al. (2017)
C. cyminum	1-Phenyl-1,2 ethanediol	Decrease at 25, 50, and 75 mmol NaCl	Rebey et al. (2017)
C. cyminum	Cuminaldehyde	Decrease at 25 and 50 but increase at 75 mmol NaCl	Rebey et al. (2017)
M. communis	α-Pinene	Increase at 2, 4, and 6 dS/m	Vafadar Shoshtari et al. (2017)
M. communis	1,8-Cineole	Increase at 2 and 4 but decrease at 6 dS/m	Vafadar Shoshtari et al. (2017)
M. communis	Linalool	Decrease at 2, 4, and 6 dS/m	Vafadar Shoshtari et al. (2017)
M. communis	Linalyl acetate	Increase at 2, 4, and 6 dS/m	Vafadar Shoshtari et al. (2017)
A. annua (EVT)	Camphor	Increase at 50, 100, and 200 mM NaCl	Yadav et al. (2017)
A. annua (EVT)	Trans-chrysanthenylacetate	Increase at 50, 100, and 200 mM NaCl	Yadav et al. (2017)
A. annua (EVT)	α-Copaene	Increase at 50, 100, and 200 mM NaCl	Yadav et al. (2017)
A. annua (EVT)	Isocycloseychellene	Increase at 50, 100, and 200 mM NaCl	Yadav et al. (2017)
A. annua (EVT)	4,5-Dehydroisolongifolene	Increase at 50 and 100 but decrease at 200 mM NaCl	Yadav et al. (2017)
A. annua (LVT)	Camphor	Increase at 50, 100, and 200 mM NaCl	Yadav et al. (2017)

Table 12.2 (Continued)

Plant species	Major component	Changes	Authors
A. annua (LVS)	Trans-chrysanthenylacetate	Increase at 50, 100, and 200 mM NaCl	Yadav et al. (2017)
A. annua (LVS)	α-Copaene	Increase at 50, 100, and 200 mM NaCl	Yadav et al. (2017)
A. annua (LVS)	Isocycloseychellene	Increase at 50, 100, and 200 mM NaCl	Yadav et al. (2017)
A. annua (LVS)	4,5-Dehydroisolongifolene	Increase at 50, 100, and 200 mM NaCl	Yadav et al. (2017)
A. annua (FBS)	Camphor	Increase at 50, 100, and 200 mM NaCl	Yadav et al. (2017)
A. annua (FBS)	Trans-chrysanthenylacetate	Increase at 50, 100, and 200 mM NaCl	Yadav et al. (2017)
A. annua (FBS)	α-Copaene	Increase at 50, 100, and 200 mM NaCl	Yadav et al. (2017)
A. annua (FBS)	Isocycloseychellene	Decrease at 50, disappeared at 100, and increased at 200 mM NaCl	Yadav et al. (2017)
A. annua (FBS)	4,5-Dehydroisolongifolene	Not detected at both control and experimental groups	Yadav et al. (2017)
M. officinalis	Trans-verbenol	Decrease at 1.6, 3.1, and 6.3 dSm^{-1}	Ahmed et al. (2017
M. officinalis	Citral	No change at 1.6, 3.1, and 6.3 dSm^{-1}	Ahmed et al. (2017)
M. officinalis	Geranyl acetate	No change at 1.6, 3.1, and 6.3 dSm^{-1}	Ahmed et al. (2017)
M. officinalis	Caryophyllene oxide	Slight increase at 1.6, 3.1, and 6.3 dSm^{-1}	Ahmed et al. (2017)
R. officinalis	α-Pinene	Increase at 2 but no change at 4, 6, and 8 dS m^{-1} NaCl	Bahonar et al. (2016)
R. officinalis	Camphene	Increase at 2, 4, 6, and 8 dS m^{-1} NaCl	Bahonar et al. (2016)
R. officinalis	β-Pinene	Decrease at 2, 4, 6, and 8 dS m^{-1} NaCl	Bahonar et al. (2016)
R. officinalis	Myrecene	Decrease at 2 but increase at 4, 6, and 8 dS m^{-1} NaCl	Bahonar et al. (2016)
R. officinalis	Limonene	Increase at 2, 4, 6, and 8 dS m^{-1} NaCl	Bahonar et al. (2016)
R. officinalis	1,8-Cineole	No change at 2, 4, 6, and 8 dS m^{-1} NaCl	Bahonar et al. (2016)

EVT, Early vegetative stage; MVS, Mid-vegetative stage; LVS, Late vegetative stage; FBS, Full bloom stage.

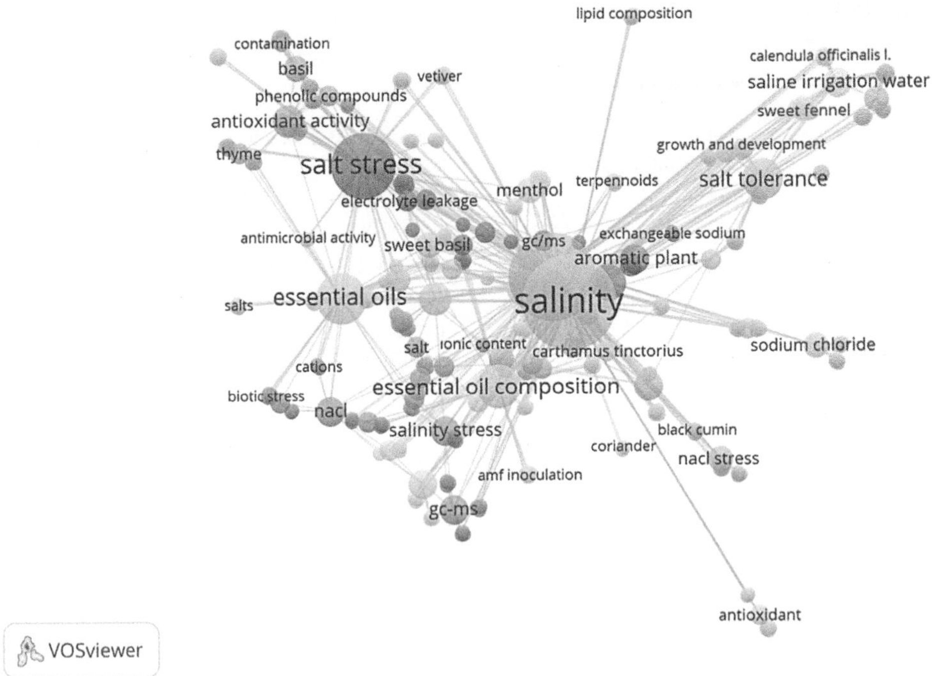

Figure 12.2 Visualization of the studies regarding salinity stress and essential oil.

reduction at 25 mM NaCl in *E. purpurea* (Khorasaninejad and Hemmati 2020). In *T. vulgaris*, chlorogenic acid decreased at 30, 60, and 90 mM NaCl, but the compound did not change at 30 mM NaCl, whereas it increased at 60 and 90 mM NaCl in *Thymus daenensis* (Bistgani et al. 2019). Furthermore, the content increased with a supply of 25, 50, and 100 mM NaCl in the media as reported by Sarker and Oba (2018).

In this context, regarding secondary metabolites, we accordingly underline that every element, not only beneficial elements, possesses a certain threshold that is nutrient or toxicant for the plant system. As we can see from the reports, the threshold is not only associated with concentration/doses of NaCl, also being related with plant species and its cultivars, population, or genotypes; timing and duration of NaCl addition to the growing media; developmental stages of the plants; source of Na^+ (NaCl, Na_2SO_4, etc.); mode of NaCl treatment, *viz.* priming or application to rooting media; interaction with the biotic and abiotic factors around the plants such as growing conditions (field, greenhouse, or growth chamber); as well as soil types of growing media.

Bibliometric analysis revealed that, in comparison to the studies regarding terpenoids, phenolic studies are more concentrated on iconic plant species essential for nutrition and economic values such as wheat, barley, strawberry, eggplant, maize, bean, chickpea, and grape. Of the group of medicinal and aromatic plants, *R. officinalis, O. basilicum, Dracocephalum kotschyi, Salvia spp., Mentha spp.,* and *Nigella sativa* are of the plant species. Furthermore, salicylic acid, jasmonic acid, 24-epibrassinolide, melatonin, glycinebetaine, and γ-aminobutyric acid have been used for stress-reducing agents. In those reports, in comparison to the reports of alkaloids and terpenoids, more detailed studies have been carried out in order to reveal the mechanisms of biosynthesis, viz. gene expression levels, phenylalanine ammonia lyase activities, and other well-known antioxidant enzymes (Figure 12.3).

Table 12.3 Changes in total phenolic and flavonoid content in those plants exposed to higher levels of salinity.

Plant species	Secondary metabolites	Changes	Authors
S. ramosissima	Total phenolic content	Increase at 110 and 200 mM, decrease at 275 and 350 mM, no change at 465 mM NaCl	Lima et al. (2020)
S. ramosissima	Total flavonoid content	Increase at 200 mM but decrease at 110, 275, 350, and 465 mM NaCl	Lima et al. (2020)
D. moldavica	Total phenolic content	Increase at 50 and 100 mM for both shoot and root	Moradbeygi et al. (2020)
D. moldavica	Total flavonoid content	Increase at 50 and 100 mM for both shoot and root	Moradbeygi et al. (2020)
D. kotschyi	Total phenolic content	Increase at 75 mM	Vafadar et al. (2020a)
D. kotschyi	Total flavonoid content	Increase at 75 mM	Vafadar et al. (2020a)
B. nigra	Total phenolic content	Increase at 4, 8, and 12 dS m^{-1} NaCl	Ghassemi-Golezani et al. (2020)
B. nigra	Total flavonoid content	Increase at 4, 8, and 12 dS m^{-1} NaCl	Ghassemi-Golezani et al. (2020)
R. officinalis	Total phenolic content	No significant changes at 75 and 150 but decrease 225 mM NaCl	Hassanpouraghdam et al. (2019)
R. officinalis	Total flavonoid content	No significant changes at 75, 150, and 225 mM NaCl	Hassanpouraghdam et al. (2019)
V. officinalis	Total phenolic content	Increase at 5 and 10 dS m^{-1} NaCl	Amanifar and Toghranegar (2020)
E. purpurea	Total phenolic content	No change at 25 but increase at 50 and 75 mM NaCl	Khorasaninejad and Hemmati (2020)
E. purpurea	Total flavonoid content	No change at 50 but increase at 25 and 75 mM NaCl	Khorasaninejad and Hemmati (2020)
O. basilicum	Total phenolic content	Decrease at 100 mM NaCl	Bahcesular et al. (2020)
O. basilicum	Total flavonoid content	Decrease at 100 mM NaCl	Bahcesular et al. (2020)
M. officinalis	Total phenolic content	Increase at 100 mM NaCl	Safari et al. (2020)
M. officinalis	Total flavonoid content	Increase at 100 mM NaCl	Safari et al. (2020

(*Continued*)

Table 12.3 (Continued)

Plant species	Secondary metabolites	Changes	Authors
A. tricolor	Total phenolic content	Increase at 100 mM NaCl	Hoang et al. (2020)
A. tricolor	Total flavonoid content	Increase at 100 mM NaCl	Hoang et al. (2020
T. vulgaris	Total phenolic content	Increase at 30, 60, and 90 mM NaCl	Bistgani et al. (2019)
T. vulgaris	Total flavonoid content	Increase at 30, 60, and 90 mM NaCl	Bistgani et al. (2019)
T. daenensis	Total phenolic content	Increase at 30, 60, and 90 mM NaCl	Bistgani et al. (2019)
T. daenensis	Total flavonoid content	Increase at 30, 60, and 90 mM NaCl	Bistgani et al. (2019)
Mentha spicata	Total phenolic content	Increase at 25, 50, and 100 mM NaCl	Chrysargyris et al. (2019b)
C. sativum	Total phenolic content	Increase at 75 mM NaCl	Al-Garni et al. (2019)
S. rebaudiana	Total phenolic content	Increase at 100 mM but decrease at 200 mM NaCl	Javed and Gürel (2019)
S. rebaudiana	Total flavonoid content	Increase at 100 mM but decrease at 200 mM NaCl	Javed and Gürel (2019)
A. tricolor	Total phenolic content	Increase at 25, 50, and 100 mM NaCl	Sarker and Oba (2018)
A. tricolor	Total flavonoid content	Increase at 25, 50, and 100 mM NaCl	Sarker and Oba (2018)
A. sativum	Total phenolic content	Decrease at 30, 60, and 90 mM NaCl	Astaneh et al. (2018)
S. hortensis	Total phenolic content	Increase at 50 mM NaCl for all accessions	Estaji et al. (2018)
P. granatum	Total phenolic content	No change at 125 and 250 mM NaCl after 10 days harvest but decrease at 250 mM after 20 days	Urbinati et al. (2018)

Table 12.4 Changes in major phenolic components in those plants exposed higher levels of salinity.

Plant species	Secondary metabolites	Changes	Authors
E. purpurea	Caffeic acid	Increase at 25 but decrease at 50 and 75 mM NaCl	Khorasaninejad and Hemmati (2020)
E. purpurea	Chlorogenic acid	No change at 50 and 75 but decrease at 25 mM NaCl	Khorasaninejad and Hemmati (2020)
O. basilicum	Caffeic acid	Decrease at 100 mM NaCl	Bahcesular et al. (2020)
O. basilicum	Cichoric acid	Decrease at 100 mM NaCl	Bahcesular et al. (2020)
O. basilicum	Rosmarinic acid	Decrease at 100 mM NaCl (not detected)	Bahcesular et al. (2020)
D. kotschyi	Rosmarinic acid	Increase at 25, 50, 75, and 100 mM NaCl	Vafadar et al. (2020b)
D. kotschyi	Luteolin	Increase at 25, 50, 75, and 100 mM NaCl	Vafadar et al. (2020b)
D. kotschyi	Apigenin	No change at 25 but increase at 50, 75, and 100 mM NaCl	Vafadar et al. (2020b)
T. vulgaris	Gallic acid	Increase at 30, 60, and 90 mM NaCl	Bistgani et al. (2019)
T. vulgaris	Chlorogenic acid	Decrease at 30, 60, and 90 mM NaCl	Bistgani et al. (2019)
T. vulgaris	Cinnamic acid	Increase at 30, 60, and 90 mM NaCl	Bistgani et al. (2019)
T. vulgaris	Rosmarinic acid	Increase at 30, 60, and 90 mM NaCl	Bistgani et al. (2019)
T. vulgaris	Rutin	No change at 30 but increase at 60 and 90 mM NaCl	Bistgani et al. (2019)
T. vulgaris	Quercetin	Increase at 30, 60, and 90 mM NaCl	Bistgani et al. (2019)
T. vulgaris	Luteolin	Increase at 30, 60, and 90 mM NaCl	Bistgani et al. (2019)
T. daenensis	Gallic acid	Increase at 30, 60, and 90 mM NaCl	Bistgani et al. (2019)
T. daenensis	Chlorogenic acid	No change at 30 but increase at 60 and 90 mM NaCl	Bistgani et al. (2019)
T. daenensis	Cinnamic acid	Increase at 30, 60, and 90 mM NaCl	Bistgani et al. (2019)
T. daenensis	Rosmarinic acid	Increase at 30, 60, and 90 mM NaCl	Bistgani et al. (2019)
T. daenensis	Rutin	Increase at 30, 60, and 90 mM NaCl	Bistgani et al. (2019)
T. daenensis	Quercetin	Increase at 30, 60, and 90 mM NaCl	Bistgani et al. (2019)

(*Continued*)

Table 12.4 (Continued)

Plant species	Secondary metabolites	Changes	Authors
T. daenensis	Luteolin	No change at 30 but increase at 60 and 90 mM NaCl	Bistgani et al. (2019)
A. tricolor	Gallic acid	No change at 25 but increase at 50 and 100 mM NaCl	Sarker and Oba (2018)
A. tricolor	Vanilic acid	No change at 25 but increase at 50 and 100 mM NaCl	Sarker and Oba (2018)
A. tricolor	Syringic acid	Decrease at 25 and 50 but increase at 100 mM NaCl	Sarker and Oba (2018)
A. tricolor	p-Hydroxybenzoic acid	No change at 25 but increase at 50 and 100 mM NaCl	Sarker and Oba (2018)
A. tricolor	Salicylic acid	Increase at 25, 50, and 100 mM NaCl	Sarker and Oba (2018)
A. tricolor	Ellagic acid	Increase at 25, 50, and 100 mM NaCl	Sarker and Oba (2018)
A. tricolor	Caffeic acid	No change at 25 but increase at 50 and 100 mM NaCl	Sarker and Oba (2018)
A. tricolor	Chlorogenic acid	Increase at 25, 50, and 100 mM NaCl	Sarker and Oba (2018)
A. tricolor	p-Coumaric acid	Increase at 25, 50, and 100 mM NaCl	Sarker and Oba (2018)
A. tricolor	Ferulic acid	No change at 25 but increase at 50 and 100 mM NaCl	Sarker and Oba (2018)
A. tricolor	m-Coumaric acid	No change at 25 but increase at 50 and 100 mM NaCl	Sarker and Oba (2018)
A. tricolor	Sinapic acid	No change at 25 but increase at 50 and 100 mM NaCl	Sarker and Oba (2018)
A. tricolor	Trans-cinnamic acid	No change at 25 but increase at 50 and 100 mM NaCl	Sarker and Oba (2018)
A. tricolor	Rutin	Increase at 25, 50, and 100 mM NaCl	Sarker and Oba (2018)
P. granatum	Ellagic acid	Increase after 10 days harvest at 125 but decreases after 20 days at both salinity	Urbinati et al. (2018)

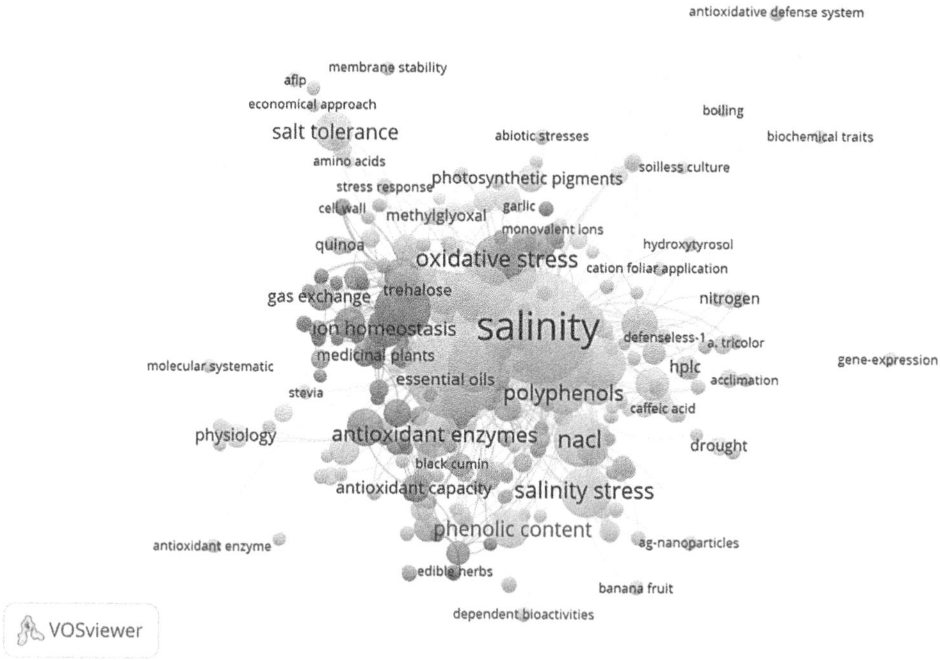

Figure 12.3 Visualization of the studies regarding salinity stress and phenolics.

What Kinds of Explanations Have Been Postulated for Changes Concerned with Defence-Related Metabolites in Those Plants Exposed to Higher Levels of Sodium in Growing Media?

In general, oil production decreased with the stress but the content increased. Those responses have been explained on the basis of declined plant metabolism, but decline in the primary metabolites as a consequence of high level of salinity resulted in high content of essential oil. The stress factor caused intermediary products to be used for biosynthesis of secondary metabolites (Sarmoum et al. 2019). In detailed review study by Caretto et al. (2015), the explanation has been done on carbon fluxes among primary metabolism and phenolic biosynthesis pathway. As a response to the limited resources available under stressful conditions, the carbon fluxes are preferences of the plants as a consequence of decision regarding allocation of the resources into stress defence functioning metabolites (Caretto et al. 2015).

Do Lower or Higher Concentration of the Sodium Favour Metabolites?

Since the higher levels of salinity impose agricultural production and also significant issues for other livings rather than plants, most of the studies have been on understanding the mechanism of plants against high level of sodium content in growing media.

There are numerous reports, which were about ~40 000 documents according to the Scopus results, but there is a lacking number of studies regarding sodium-deficit impacts. In study by Brownell and Nicholas (1967) – a culture media study, less phycocyanin was noted in sodium-deficient cultures in comparison to the adequate supply of sodium, but the chlorophyll content did not differ between the treatments. In our view, performing studies on the impacts of sodium deficit on secondary metabolites seem to be a dilemma, and necessity of the studies on that context remains questionable in the situation that salinity is a rising problem day by day.

Two Sides of the Coin: Is a Third Probability Possible for Plant Production Versus Secondary Metabolite Production?

Based on the reports, nutritional stress, or herein better to name high level of salinity might promote the biosynthesis of secondary metabolites through induction of supply pathways from primary to phenolic secondary metabolite production (Caretto et al. 2015). However, the adverse impacts of high levels of salinity on plant growth and subsequently production has been well-known and reported in many plant species.

On the study by Lattanzio et al. (2009), shoot weight of oregano (*Origanum vulgare* L.) was negatively correlated with total phenolic content. That response has been explained with carbon fluxes between primary and secondary metabolism. In the case of restriction of photosynthesis, excess of carbon skeletons are used for production of defence-related phenolics (Lattanzio et al. 2009).

However, the high-yielding genotype of *Phaseolus vulgaris* L. under salt stress included higher levels of phenolic, and low-yielding genotype had lower total phenolics. That situation has been explained as a protectant role of phenolics against high salinity, which provided protection for genotype and contributed to the genotype for high yield (Taïbi et al. 2016).

Based on the studies, selecting the genotype, cultivar, or populations of the plant species is possible (as a third possibility) in accordance with high agricultural yield coupled with high content of secondary metabolites, which are of great interest in industrial, pharmaceutical, and medicinal sector. We should not neglect that yielding and diversifying might not act in parallel. High agricultural yielding might not mean qualified 'high' yielding secondary metabolites since the effectiveness of metabolites are associated with their chemical diversity. In that context, for the selection studies, individual secondary metabolite fractions might be better to be profiled using chromatographic approaches.

Conclusion

We herewith mainly concentrate on the dilemma that plants face regarding Na^+ metabolism. Whilst the uptake and subsequently intracellular compartmentalization of Na^+ are required in order to build osmotic potential and regulate absorption of water and finally sustain the turgid status of the cell, the higher concentration of Na^+ might be toxic (Pardo and Quintero 2002) unless translocation and compartmentalization are not sufficient. In accordance with '*All things are poisons, for there is nothing without poisonous qualities. It is only the dose which makes a thing poison*' principle famously attributed to Theophrastus Bombastus von Hohenheim (*Paracelsus*), selecting the appropriate 'dose' of NaCl but coupled with appropriate plant species, timing, duration, frequency, and mode of treatments of NaCl might favour regarding high yield or not affecting yield and qualified high yield of individual of metabolites and their individual fractions is of the fundamental of elements to be beneficial or toxicant. We herein should emphasize that a beneficial 'dose' can be toxicant for other plant species and vice versa.

References

Aghaei, K. and Komatsu, S. (2013). Crop and medicinal plants proteomics in response to salt stress. *Front Plant Sci* 4: 1–9.

Aharoni, A., Jongsma, M.A., and Bouwmeester, H.J. (2005). Volatile science? Metabolic engineering of terpenoids in plants. *Trends Plant Sci* 10 (12): 594–602.

Aharoni, A., Jongsma, M.A., Kim, T.Y. et al. (2006). Metabolic engineering of terpenoid biosynthesis in plants. *Phytochem Rev* 5 (1): 49–58.

Ahmad, S., Garg, M., Tamboli, E.T. et al. (2013). in vitro production of alkaloids: factors, approaches, challenges and prospects. *Pharmacogn Rev* 7 (13): 27.

Ahmed, A.M.A., Talaat, I.M., and Khalid, K.A. (2017). Citric acid affects *Melissa officinalis* L. essential oil under saline soil. *Asian J Crop Sci* 9: 40–49.

Akgunlu, S., Sekeroglu, N., Koca-Caliskan, U. et al. (2016). Research on selected wild edible vegetables: mineral content and antimicrobial potentials. *Annals of Phytomedicine-An International Journal* 5 (2): 50–57.

Al-Garni, S.M., Khan, M.M.A., and Bahieldin, A. (2019). Plant growth-promoting bacteria and silicon fertilizer enhance plant growth and salinity tolerance in *Coriandrum sativum*. *J Plant Interact* 14 (1): 386–396.

Almeida, D.M., Oliveira, M.M., and Saibo, N.J. (2017). Regulation of Na^+ and K^+ homeostasis in plants: towards improved salt stress tolerance in crop plants. *Genet Mol Biol* 40 (1): 326–345.

Amanifar, S. and Toghranegar, Z. (2020). The efficiency of arbuscular mycorrhiza for improving tolerance of *Valeriana officinalis* L. and enhancing valerenic acid accumulation under salinity stress. *Ind Crops. Prod* 147: 112234.

Arnon, D.I. and Stout, P.R. (1939). The essentiality of certain elements in minute quantity for plants with special reference to copper. *Plant Physiol* 14: 371–375.

Ashraf, M. and Naqvi, M.I. (1992). Growth and ion uptake of four Brassica species as affected by Na/Ca ratio in saline sand culture. *Z Pflanz Bodenkunde* 155 (2): 101–108.

Ashraf, M. and Rauf, H. (2001). Inducing salt tolerance in maize (*Zea mays* L.) through seed priming with chloride salts: growth and ion transport at early growth stages. *Acta Physiol. Plant* 23 (4): 407–414.

Astaneh, R.K., Bolandnazar, S., Nahandi, F.Z., and Oustan, S. (2018). Effect of selenium application on phenylalanine ammonia-lyase (PAL) activity, phenol leakage and total phenolic content in garlic (*Allium sativum L.*) under NaCl stress. *Inf Process Agric* 5 (3): 339–344.

Azadbakht, M. and Balouchi, H. (2020). The effect of melatonin and hydropriming on some physiological characteristics of Chickpea (*Cicer arietinum* L.) seed and seedling under salinity stress. *J Plant ProFunct* 9 (35): 347–358.

de Azevedo Neto, A.D., Menezes, R.V., Gheyi, H.R. et al. (2000). The cell and developmental biology of alkaloid biosynthesis. *Trends Plant Sci* 5: 168–173.

Bahcesular, B., Yildirim, E.D., Karaçocuk, M. et al. (2020). Seed priming with melatonin effects on growth, essential oil compounds and antioxidant activity of basil (*Ocimum basilicum* L.) under salinity stress. *Ind Crops Prod* 146: 112165.

Bahonar, A., Mehrafarin, A., Abdousi, V. et al. (2016). Quantitative and qualitative changes of rosemary (*Rosemarinus officinalis* L.) in response to mycorrhizal fungi (*Glomus intraradices*) inoculation under saline environments. *J Med Plant Res* 1 (57): 25–37.

Bali, S., Kaur, P., Jamwal, V.L. et al. (2020). Seed priming with jasmonic acid counteracts root knot nematode infection in tomato by modulating the activity and expression of antioxidative enzymes. *Biomolecules* 10 (1): 98.

Barragán, V., Leidi, E.O., Andrés, Z. et al. (2012). Ion exchangers NHX1 and NHX2 mediate active potassium uptake into vacuoles to regulate cell turgor and stomatal function in *Arabidopsis. The Plant Cell* 24 (3): 1127–1142.

Ben Hamida, N., Martínez-Díaz, R.A., Hela, M. et al. (2020). Effect of salinity on the antiparasitic activity of hyssop essential oil. *J Essent Oil Res* 32 (1): 69–78.

Bernstein, N., Dudai, N., and Gorbatsevich, E. (2017). Salinity stress does not affect root uptake, dissemination and persistence of salmonella in sweet-basil (*Ocimum basilicum*L.). *Front. Plant Sci* 8: 675.

Bhat, M.A., Ahmad, S., Aslam, J., and Mujib, A. (2008). Salinity stress enhances production of solasodine in *Solanum nigrum* L. *Chem Pharm Bull* 56 (1): 17–21.

Bidgoli, R.D., Azarnezhad, N., Akhbari, M., and Ghorbani, M. (2019). Salinity stress and PGPR effects on essential oil changes in *Rosmarinus officinalis* L. *Agric Food Secur* 8 (1): 1–7.

Bistgani, Z.E., Hashemi, M., DaCosta, M. et al. (2019). Effect of salinity stress on the physiological characteristics, phenolic compounds and antioxidant activity of *Thymus vulgaris* L. and *Thymus daenensis* Celak. *Ind Crops Prod* 135: 311–320.

Blumwald, E. (2000). Sodium transport and salt tolerance in plants. *Curr Opin Cell Biol* 12 (4): 431–434.

Bonacina, C., Trevizan, C.B., Stracieri, J. et al. (2017). Changes in growth, oxidative metabolism and essential oil composition of lemon balm (*Melissa officinalis* L.) subjected to salt stress. *Aust. J Crop Sci* 11 (12): 1665.

Bortolin, G.S., Teixeira, S.B., de Mesquita, P.R. et al. (2020). Seed priming with salicylic acid minimizes oxidative effects of *Aluminum* on *Trifolium* seedlings. *J Soil Sci Plant Nutr* 20 (4): 1–10.

Brachet, J. and Cosson, L. (1986). Changes in the total alkaloid content of *Datura innoxia* Mill. subjected to salt stress. *J Exp Bot* 37 (5): 650–656.

Brownell, P.F. and Nicholas, D.J.D. (1967). Some effects of sodium on nitrate assimilation and N2 fixation in *Anabaena cylindrica. Plant Physiol* 42 (7): 915–921.

Caretto, S., LinsalataV, C.G., Mita, G., and Lattanzio, V. (2015). Carbon fluxes between primary metabolism and phenolic pathway in plant tissues under stress. *J Mol Sci* 16 (11): 26378–26394.

Cheynier, V., Comte, G., Davies, K.M. et al. (2013). Plant phenolics: recent advances on their biosynthesis, genetics, and ecophysiology. *Plant Physiol Biochem* 72: 1–20.

Chrysargyris, A., Papakyriakou, E., Petropoulos, S.A., and Tzortzakis, N. (2019a). The combined and single effect of salinity and copper stress on growth and quality of *Mentha spicata* plants. *J Hazard Mater* 368: 584–593.

Chrysargyris, A., Solomou, M., Petropoulos, S.A., and Tzortzakis, N. (2019b). Physiological and biochemical attributes of *Mentha spicata* when subjected to saline conditions and cation foliar application. *J Plant Physiol* 232: 27–38.

Crozier, A., Jaganath, I.B., and Clifford, M.N. (2006). Phenols, polyphenols and tannins: an overview. In: *Plant Secondary Metabolites: Occurrence, Structure and Role in the Human Diet* (ed. A. Crozier and H. Ashihara), 1–24. Oxford: *Blackwell.*

De Luca, V. and St Pierre, B. (2000). The cell and developmental biology of alkaloid biosynthesis. *Trends Plant Sci* 5 (4): 168–173.

El-Esawi, M.A., Elansary, H.O., El-Shanhorey, N.A. et al. (2017). Salicylic acid-regulated antioxidant mechanisms and gene expression enhance rosemary performance under saline conditions. *Front Plant Sci* 8: 716.

Estaji, A., Roosta, H.R., Rezaei, S.A. et al. (2018). Morphological, physiological and phytochemical response of different *Satureja hortensis* L. accessions to salinity in a greenhouse experiment. *J Appl Res Med Aroma* 10: 25–33.

Fang, C., Fernie, A.R., and Luo, J. (2019). Exploring the diversity of plant metabolism. *Trends Plant Sci* 24 (1): 83–98.

Farooq, M., Usman, M., Nadeem, F. et al. (2019). Seed priming in field crops: potential benefits, adoption and challenges. *Crop Pasture Sci* 70 (9): 731–771.

Farsaraei, S., Moghaddam, M., and Pirbalouti, A.G. (2020). Changes in growth and essential oil composition of sweet basil in response of salinity stress and superabsorbents application. *Sci Hortic* 271: 109465.

Frischknecht, P.M. and Baumann, T.W. (1985). Stress induced formation of purine alkaloids in plant tissue culture of *Coffea arabica*. *Phytochemistry* 24 (10): 2255–2257.

Gadallah, L.S., El-Ghadban, M.E.M., Soliman, A.S. et al. (2020). Effect of foliar application of anti-salinity substances on geranium (*Pelargonium graveolens*) plants under salt stress. *J Anim Plant Sci* 30 (3): 723–729.

García Morales, S., Trejo-Téllez, L.I., Gómez Merino, F.C. et al. (2012). Growth, photosynthetic activity, and potassium and sodium concentration in rice plants under salt stress. *Acta Sci Agron* 34 (3): 317–324.

Gebreegziabher, B.G. and Qufa, C.A. (2017). Plant physiological stimulation by seeds salt priming in maize (*Zea mays*): prospect for salt tolerance. *Afr J Biotechnol* 16 (5): 209–223.

Ghassemi-Golezani, K., Hassanzadeh, N., Shakiba, M.R., and Esmaeilpour, B. (2020). Exogenous salicylic acid and 24-epi-brassinolide improve antioxidant capacity and secondary metabolites of *Brassica nigra*. *Biocatal Agric. Biotechnol* 26: 101636.

Gohari, G., Mohammadi, A., Akbari, A. et al. (2020). Titanium dioxide nanoparticles (TiO_2 NPs) promote growth and ameliorate salinity stress effects on essential oil profile and biochemical attributes of *Dracocephalum moldavica*. *Sci Rep* 10 (1): 1–14.

Gohari, G., Safai, F., Panahirad, S. et al. (2020a). Modified multiwall carbon nanotubes display either phytotoxic or growth promoting and stress protecting activity in *Ocimum basilicum* L. in a concentration-dependent manner. *Chemosphere* 249: 126171.

Gohari, G., Alavi, Z., Esfandiari, E. et al. (2020b). Interaction between hydrogen peroxide and sodium nitroprusside following chemical priming of *Ocimum basilicum* L. against salt stress. *Physiol Plant* 168 (2): 361–373.

Gómez-Merino, F.C. and Trejo-Téllez, L.I. (2018). The role of beneficial elements in triggering adaptive responses to environmental stressors and improving plant performance. In: *Biotic and Abiotic Stress Tolerance in Plants* (ed. S. Vats), 137–172. Singapore: *Springer*.

Grassmann, J. (2005). Terpenoids as plant antioxidants. *Vitam Horm* 72: 505–535.

Greenway, H. and Munns, R. (1980). Mechanisms of salt tolerance in nonhalophytes. *Annu Rev Plant Physiol 31* (1): 149–190.

Grof, P.L.G., Johnston, M., and Brownell, P.F. (1989). Effect of Na nutrition on ultra-structure of chloroplasts of C4 plants. *Plant Physiol* 89: 539–543.

Hasegawa, P.M. (2013). Sodium (Na^+) homeostasis and salt tolerance of plants. *Environ Exp Bot* 92: 19–31.

Hassanpouraghdam, M.B., Mehrabani, L.V., and Tzortzakis, N. (2019). Foliar application of nano-zinc and iron affects physiological attributes of *Rosmarinus officinalis* and quietens NaCl salinity depression. *J Soil Sci Plant Nutr* 20 (2): 1–11.

He, Q. (1999). Knowledge discovery through co-word analysis. *Libr Trends* 48: 133–159.

Hegazy, M.H., Sabra, A.S., Alharbi, B.M. et al. (2019). Ameliorative effects of supplemental nutrition on growth and essential oil yield of saline irrigated *Satureja montana*. *J Essent Oil-Bear Plants* 22 (5): 1218–1227.

Heldt, H.W. and Heldt, F. (2015). *Pflanzenbiochemie*. Springer Berlin Heidelberg.

Hoang, H.L., de Guzman, C.C., Cadiz, N.M. et al. (2020). Salicylic acid and calcium signaling induce physiological and phytochemical changes to improve salinity tolerance in red Amaranth (*Amaranthus tricolor* L.). *J Soil Sci Plant Nutr* 20: 1759–1769. https://doi.org/10.1007/s42729-020-00248-4.

Huang, L., Zhang, L., Zeng, R. et al. (2020). Brassinosteroid priming improves peanut drought tolerance via eliminating inhibition on genes in photosynthesis and hormone signaling. *Genes* 11 (8): 919.

Idrees, M., Naeem, M., Aftab, T., and Khan, M.M.A. (2011). Salicylic acid mitigates salinity stress by improving antioxidant defence system and enhances vincristine and vinblastine alkaloids production in periwinkle [*Catharanthus roseus* (L.) G. Don]. *Acta Physiol Plant* 33 (3): 987–999.

Imadi, S.R., Kazi, A.G., Hashem, A. et al. (2016). Medicinal plants under abiotic stress: An overview. In: *Plant-Environment Interaction: Responses and Approaches to Mitigate Stress*, 1e (ed. M.M. Azooz and P. Ahmad). John Wiley & Sons.

Iqbal, M. (2013). Invited commentary: from medicine to phytomedicine. *Ann Phytomed* 2: 1–2.

İşeri, Ö.D., Sahin, F.I., and Haberal, M. (2014). Sodium chloride priming improves salinity response of tomato at seedling stage. *J Plant Nutr* 37 (3): 374–392.

Jasmin Šutković, J., Ler, D., and Gawwad, M.R.A. (2011). Vitro production of solasodine alkaloid in *solanum nigrum* under salinity stress. *J Phytol* 3 (1): 43–49.

Javed, R. and Gürel, E. (2019). Salt stress by NaCl alters the physiology and biochemistry of tissue culture-grown *Stevia rebaudiana* Bertoni. *Turk J Agric For* 43 (1): 11–20.

Kanto, U., Jutamanee, K., Osotsapar, Y. et al. (2015). Promotive effect of priming with 5-aminolevulinic acid on seed germination capacity, seedling growth and antioxidant enzyme activity in rice subjected to accelerated ageing treatment. *Plant Prod Sci* 18 (4): 443–454.

Kaur, S., Kaur, N., Siddique, K.H., and Nayyar, H. (2016). Beneficial elements for agricultural crops and their functional relevance in defence against stresses. *Arch Agron Soil Sci* 62 (7): 905–920.

Keisham, M., Mukherjee, S., and Bhatla, S.C. (2018). Mechanisms of sodium transport in plants—progresses and challenges. *Int J Mol Sci* 19 (3): 647.

Keramati, S., Pirdashti, H., Babaeizad, V., and Dehestani, A. (2016). Essential oil composition of sweet basil (*Ocimum basilicum* L.) in symbiotic relationship with *Piriformospora indica* and paclobutrazol application under salt stress. *Acta Biol Hung* 67 (4): 412–423.

Khan, H.A., Ayub, C.M., Pervez, M.A. et al. (2009). Effect of seed priming with NaCl on salinity tolerance of hot pepper (*Capsicum annuum* L.) at seedling stage. *Soil Environ* 28 (1): 81–87.

Khan, M.N., Khan, Z., Luo, T. et al. (2020). Seed priming with gibberellic acid and melatonin in rapeseed: Consequences for improving yield and seed quality under drought and non-stress conditions. *Ind Crops Prod* 156: 112850.

Khanam, D. and Mohammad, F. (2018). Plant growth regulators ameliorate the ill effect of salt stress through improved growth, photosynthesis, antioxidant system, yield and quality attributes in *Mentha piperita* L. *Acta Physiol Plant* 40 (11): 188.

Khorasaninejad, S. and Hemmati, K. (2020). Effects of silicon on some phytochemical traits of purple coneflower (*Echinacea purpurea* L.) under salinity. *Sci Hortic* 264: 108954.

Lattanzio, V., Cardinali, A., Ruta, C. et al. (2009). Relationship of secondary metabolism to growth in oregano (*Origanum vulgare* L.) shoot cultures under nutritional stress. *Environ Exp Bot* 65 (1): 54–62.

Levinsh, G. (2006). Biological basis of biological diversity: physiological adaptations of plants to heterogeneous habitats along a sea coast. *Acta Univ Latv* 710: 53–79.

Lima, A.R., Castañeda-Loaiza, V., Salazar, M. et al. (2020). Influence of cultivation salinity in the nutritional composition, antioxidant capacity and microbial quality of *Salicornia ramosissima* commercially produced in soilless systems. *Food Chem* 333: 127525.

Maathuis, F.J. (2014). Sodium in plants: perception, signalling, and regulation of sodium fluxes. *J Exp Bot* 65 (3): 849–858.

Mahmood-ur-Rehman, M., Amjad, M., Ziaf, K., and Ahmad, R. (2020). Seed priming with salicylic acid improve seed germination and physiological responses of carrot seeds. *Pak J Agric Sci* 57 (2).

Marschner, H. (1995). *Mineral Nutrition of High Plants*, 2e, 889. London: *London Academic Press.*

Michalak, A. (2006). Phenolic compounds and their antioxidant activity in plants growing under heavy metal stress. *Pol J Environ Stud* 15 (4): 523–530.

Mittler, R. (2002). Oxidative stress, antioxidants and stress tolerance. *Trends Plant Sci* 7: 405–410.

Mohammadi, G.R. (2009). The influence of NaCl priming on seed germination and seedling growth of canola (*Brassica napus* L.) under salinity conditions. *American-Eurasian J Agric Environ Sci* 5 (5): 696–700.

Mohammadi, S., Saharkhiz, M.J., and Javanmardi, J. (2017). Evaluation of interaction effects of spermidine and salinity on physiological and morphology trait of catnip (*Nepeta cataria* L.). *Z Arznei-Gewurzpfla* 22 (3): 104–109.

Moradbeygi, H., Jamei, R., Heidari, R., and Darvishzadeh, R. (2020). Investigating the enzymatic and non-enzymatic antioxidant defense by applying iron oxide nanoparticles in *Dracocephalum moldavica* L. plant under salinity stress. *Sci Hortic* 272: 109537.

Munns, R. (2005). Genes and salt tolerance: Bringing them together. *New Phytol* 167: 645–663.

Munns, R. and Tester, M. (2008). Mechanisms of salinity tolerance. *Annu Rev Plant Biol* 59: 651–681.

Nakaune, M., Tsukazawa, K., Uga, H. et al. (2012). Low sodium chloride priming increases seedling vigor and stress tolerance to *Ralstonia solanacearum* in tomato. *Plant Biotechnol* 29: 9–18.

Nayanabhirama, U. (2016). Invited commentary: status on herbal drugs and their future perspectives. *Ann Phytomed* 5: 1–3.

Nazari, R., Parsa, S., Tavakkol Afshari, R. et al. (2020). Salicylic acid priming before and after accelerated aging process increases seedling vigor in aged soybean seed. *J Crop Improv* 34 (2): 218–237.

Ohnishi, J.-I., Flugge, U.-I., Heldt, H.W., and Kanai, R. (1990). Involvement of Na^+ in active uptake of pyruvate in mesophyll chloroplasts of some C4 plants. *Plant Physiol* 94: 950–959.

Olave, J. and Guzmán, M. (2004). Effects of N-form and saline priming on germination and vegetative growth of Galia-type melon (Cucumis melo L. cv. Primal) under salinity. *VII International Symposium on Protected Cultivation in Mild Winter Climates: Production Pest Management and Global Competition* 659, Kissimmee, Florida, USA (pp. 253–260).

de Oliveira, R.M. (2019). Salt-induced changes in solutes, pigments and essential oil of two basil (*Ocimum basilicum* L.) genotypes under hydroponic cultivation. *Aust. J Crop Sci* 13 (11): 1856.

Pankaj, U., Verma, R.S., Yadav, A., and Verma, R.K. (2019). Effect of arbuscular mycorrhizae species on essential oil yield and chemical composition of palmarosa (*Cymbopogon martinii*) varieties grown under salinity stress soil. *J Essent Oil Res* 31 (2): 145–153.

Pardo, J.M. and Quintero, F.J. (2002). Plants and sodium ions: keeping company with the enemy. *Genome Biol* 3 (6): 1–4.

Parihar, P., Singh, S., Singh, R. et al. (2015). Effect of salinity stress on plants and its tolerance strategies: a review. *Environ Sci Pollut Res* 22 (6): 4056–4075.

Patra, B., Schluttenhofer, C., Wu, Y. et al. (2013). Transcriptional regulation of secondary metabolite biosynthesis in plants. *BBA-Gene Regul Mech* 1829 (11): 1236–1247.

Peng, Z., He, S., Sun, J. et al. (2016). Na+ compartmentalization related to salinity stress tolerance in upland cotton (*Gossypium hirsutum*) seedlings. *Sci Rep* 6: 34548.

Pichersky, E. and Gang, D.R. (2000). Genetics and biochemistry of secondary metabolites in plants: an evolutionary perspective. *Trends Plant Sci* 5 (10): 439–445.

Pilon-Smits, E.A., Quinn, C.F., Tapken, W. et al. (2009). Physiological functions of beneficial elements. *Curr Opin Plant Biol* 12 (3): 267–274.

Rebey, I.B., Bourgou, S., Rahali, F.Z. et al. (2017). Relation between salt tolerance and biochemical changes in cumin (*Cuminum cyminum* L.) seeds. *J Food Drug Anal* 25 (2): 391–402.

Safari, F., Akramian, M., and Salehi-Arjmand, H. (2020). Physiochemical and molecular responses of salt-stressed lemon balm (*Melissa officinalis* L.) to exogenous protectants. *Acta Physiol Plant* 42 (2): 27.

Sakihama, Y., Cohen, M.F., Grace, S.C., and Yamasaki, H. (2002). Plant phenolic antioxidant and prooxidant activities: phenolics-induced oxidative damage mediated by metals in plants. *Toxicology* 177 (1): 67–80.

Sany, H., Alharbi, B.M., Khalid, A. et al. (2020). Effects of NaCl on growth, essential oil and chemical composition of *Plectranthus amboinicus*. *Plant Arch* 20 (1): 2471–2477.

Sarker, U. and Oba, S. (2018). Augmentation of leaf color parameters, pigments, vitamins, phenolic acids, flavonoids and antioxidant activity in selected *Amaranthus tricolor* under salinity stress. *Sci Rep* 8 (1): 1–9.

Sarmoum, R., Haid, S., Biche, M. et al. (2019). Effect of salinity and water stress on the essential oil components of rosemary (*Rosmarinus officinalis* L.). *Agronomy* 9 (5): 214.

Sharifi, P. and Bidabadi, S.S. (2020). Strigolactone could enhances gas-exchange through augmented antioxidant defense system in *Salvia nemorosa* L. plants subjected to saline conditions stress. *Ind Crops. Prod* 151: 112460.

Sharma, P., Jha, A.B., Dubey, R.S., and Pessarakli, M. (2012). Reactive oxygen species, oxidative damage, and antioxidative defense mechanism in plants under stressful conditions. *J Bot* 2012: 217037. https://doi.org/10.1155/2012/217037.

Sheteiwy, M., Shen, H., Xu, J. et al. (2017). Seed polyamines metabolism induced by seed priming with spermidine and 5-aminolevulinic acid for chilling tolerance improvement in rice (*Oryza sativa* L.) seedlings. *Environ Exp Bot* 137: 58–72.

Silva, H., Neto, A.A., Menezes, R. et al. (2019). Use of hydrogen peroxide in acclimation of basil (*Ocimum basilicum* L.) to salt stress. *Turk J Bot* 43 (2): 208–217.

Smith, G.S., Middleton, K.R., and Edmonds, A.S. (1978). A classification of pasture and fodder plants according to their ability to translocate sodium from their roots into aerial parts. *New Zeal J Exp Agr* 6 (3): 183–188.

Soliman, M., Elkelish, A., Souad, T. et al. (2020). Brassinosteroid seed priming with nitrogen supplementation improves salt tolerance in soybean. *Physiol Mol Biol Plants* 26 (3): 501–511.

Subbarao, G.V., Ito, O., Berry, W.L., and Wheeler, R.M. (2003). Sodium a functional plant nutrient. *Crit Rev Plant Sci* 22 (5): 391–416.

Tahal, R., Mills, D., Heimer, Y., and Tal, M. (2000). The relation between low K^+/Na^+ ratio and salt-tolerance in the wild tomato species *Lycopersicon pennellii*. *J Plant Physiol* 157 (1): 59–64.

Taïbi, K., Taïbi, F., Abderrahim, L.A. et al. (2016). Effect of salt stress on growth, chlorophyll content, lipid peroxidation and antioxidant defence systems in *Phaseolus vulgaris* L. *S Afr J Bot* 105: 306–312.

Taiz, L. and Zeiger, E. (2002). *Plant Physiol*, 3e. England: Sinauer Associates.

Talebi, M., Moghaddam, M., and Pirbalouti, A.G. (2018). Methyl jasmonate effects on volatile oil compounds and antioxidant activity of leaf extract of two basil cultivars under salinity stress. *Acta Physiol Plant* 40 (2): 34.

Tester, M. and Davenport, R. (2003). Na+ tolerance and Na+ transport in higher plants. *Ann Bot* 91 (5): 503–527.

Tripathi, D.K., Singh, S., Singh, S. et al. (2015). Micronutrients and their diverse role in agricultural crops: advances and future prospective. *Acta Physiologiae Plantarum.* 37 (7): 1–4.

Urbinati, G., Nota, P., Frattarelli, A. et al. (2018). Morpho-physiological and antioxidant response to NaCl-induced stress in in vitro shoots of pomegranate (*Punica granatum* L.). *Acta Physiol Plant* 40 (8): 151.

Vafadar Shoshtari, Z., Rahimmalek, M., Sabzalian, M.R., and Hosseini, H. (2017). Essential oil and bioactive compounds variation in myrtle (*Myrtus communis* L.) as affected by seasonal variation and salt stress. *Chem. Biodiversity* 14 (4): e1600365.

Vafadar, F., Amooaghaie, R., Ehsanzadeh, P. et al. (2020a). Melatonin and calcium modulate the production of rosmarinic acid, luteolin, and apigenin in *Dracocephalum kotschyi* under salinity stress. *Phytochemistry* 177: 112422.

Vafadar, F., Amooaghaie, R., Ehsanzadeh, P., and Ghanadian, M. (2020b). Salinity stress alters ion homeostasis, antioxidant activities and the production of rosmarinic acid, luteolin and apigenin in *Dracocephalum kotschyi* Boiss. *Biologia* 75 (12): 2147–2158.

Valifard, M., Mohsenzadeh, S., Kholdebarin, B. et al. (2019). Effect of salt stress on terpenoid biosynthesis in *Salvia mirzayanii*: From gene to metabolite. *J Hortic Sci Biotech* 94 (3): 389–399.

Verma, N. and Shukla, S. (2015). Impact of various factors responsible for fluctuation in plant secondary metabolites. *J Appl Res Med Aroma* 2 (4): 105–113.

Wakeel, A., Farooq, M., Qadir, M., and Schubert, S. (2011). Potassium substitution by sodium in plants. *Crit Rev Plant Sci* 30 (4): 401–413.

Wang, W., Wang, X., Zhang, J. et al. (2020a). Salicylic acid and cold priming induce late-spring freezing tolerance by maintaining cellular redox homeostasis and protecting photosynthetic apparatus in wheat. *Plant Growth Regul* 90 (1): 109–121.

Wang, X., Li, Q., Xie, J. et al. (2020b). Abscisic acid and jasmonic acid are involved in drought priming-induced tolerance to drought in wheat. *J Crop Prod* https://doi.org/10.1016/j.cj.2020.06.002.

Waśkiewicz, A., Muzolf-Panek, M., and Goliński, P. (2013). Phenolic content changes in plants under salt stress. In: *Ecophysiology and Responses of Plants Under Salt Stress* (ed. P. Ahmad, M.M. Azooz and M.N.V. Prasad), 283–314. New York, NY: *Springer*.

Wu, L., Huo, W., Yao, D., and Li, M. (2019). Effects of solid matrix priming (SMP) and salt stress on broccoli and cauliflower seed germination and early seedling growth. *Sci Hortic* 255: 161–168.

Yadav, R.K., Sangwan, R.S., Srivastava, A.K., and Sangwan, N.S. (2017). Prolonged exposure to salt stress affects specialized metabolites-artemisinin and essential oil accumulation in *Artemisia annua* L.: metabolic acclimation in preferential favour of enhanced terpenoid accumulation accompanying vegetative to reproductive phase transition. *Protoplasma* 254 (1): 505–522.

Yahyazadeh, M., Meinen, R., Hänsch, R. et al. (2018). Impact of drought and salt stress on the biosynthesis of alkaloids in *Chelidonium majus* L. *Phytochemistry* 152: 204–212.

Yang, J.S., Dai, Y., Liu, Y. et al. (2020). Reduced cadmium accumulation in tobacco by sodium chloride priming. *Environ Sci Pollut Res* https://doi.org/10.1007/s11356-020-09134-z.

Yazaki, K., Arimura, G.I., and Ohnishi, T. (2017). 'Hidden'terpenoids in plants: their biosynthesis, localization and ecological roles. *Plant Cell Physiol* 58 (10): 1615–1621.

Zhang, X.X., Li, C.J., and Nan, Z.B. (2011). Effects of salt and drought stress on alkaloid production in endophyte-infected drunken horse grass (*Achnatherum inebrians*). *Biochem Syst Ecol* 39 (4–6): 471–476.

Zrig, A., Ferreira, J.F., Hamouda, F. et al. (2019). The impact of foliar fertilizers on growth and biochemical responses of *Thymus vulgaris* to salinity stress. *Arid Land Res Manag* 33 (3): 297–320.

13

Sodium and Abiotic Stress Tolerance in Plants

Misbah Naz[1], Muhammad Imran Ghani[2], Muhammad Jawaad Atif[3], Muhammad Ammar Raza[4], Sarah Bouzroud[5], Muhammad Rahil Afzal[6], Muhammad Riaz[7], Maratab Ali[8], Muhammad Tariq[9], and Xiaorong Fan[1,10]

[1] *State Key Laboratory of Crop Genetics and Germplasm Enhancement, Nanjing Agricultural University, Nanjing, China*
[2] *College of Natural Resources and Environment, Northwest A&F University, Yangling, Shaanxi, China*
[3] *College of Horticulture, Northwest A&F University, Yangling, Shaanxi, China*
[4] *College of Food Science and Biotechnology, Key Laboratory of Fruits and Vegetables Postharvest and Processing Technology Research of Zhejiang Province, Zhejiang Gongshang University, Hangzhou, China*
[5] *Laboratoire de Biotechnologie et Physiologie Végétales, Centre de Biotechnologie Végétale et Microbienne Biodiversité et Environnement, Faculté des Sciences, Université Mohammed V de Rabat, Rabat, Morocco*
[6] *Faculty of Life Sciences, Institute of Environmental and Agricultural Sciences, University of Okara, Okara, Pakistan*
[7] *School of Agriculture and Biology, Shanghai Jiao Tong University, Minhang, Shanghai, China*
[8] *College of Food Science and Biotechnology, Key Laboratory of Fruits and Vegetables Postharvest and Processing Technology Research of Zhejiang Province, Zhejiang Gongshang University, Hangzhou, PR China*
[9] *Department of Pharmacology, Lahore Pharmacy Collage, Lahore, Pakistan*
[10] *Key Laboratory of Plant Nutrition and Fertilization in Lower-Middle Reaches of the Yangtze River, Ministry of Agriculture, Nanjing Agricultural University, Nanjing, China*

Introduction

The majority of salts existing in irrigation water are sulphates, chlorides, carbonates, and bicarbonates of calcium, magnesium, potassium, and sodium. The sodium content in irrigated water is referred to as alkalinity. Irrigation with water containing excessive sodium has adverse effects on soil structure and makes plant growth difficult (Choudhary 2017). High salinity and soda water quality can lead to irrigation problems, depending on the type and quantity of salt, the type of soil irrigated, the particular plant species and growth stage, and the quantity of water that may flow through the root zone (Shrivastava and Kumar 2015). Na^+ is a rich element, accounting for about 3% of the Earth's crust. In addition, it exists in nearly all surface and underground water sources, and it is, of course, abundant in the oceans, where it can reach more than 5% (w/w) (Maathuis et al. 2014). Therefore, it is not surprising that most plants are exposed to Na^+ at any stage of their life cycle; nevertheless, the level of Na^+ encountered may vary greatly (Zörb et al. 2019). Enormous studies have documented that salt stress influences the metabolic process of plants, triggers nutritional imbalance, alters levels of growth regulators, prevents photosynthesis and protein

Beneficial Chemical Elements of Plants: Recent Developments and Future Prospects, First Edition. Edited by Sangeeta Pandey, Durgesh Kumar Tripathi, Vijay Pratap Singh, Shivesh Sharma, and Devendra Kumar Chauhan.

synthesis, and all of this leads to a decline in plant growth (Carillo et al. 2011; Reddy et al. 2017). Therefore, abiotic stresses such as drought, salinity, and extreme temperature (high/low) become a significant challenge to crop production and lead to large-scale yield loss (Yadav et al. 2011, Roychoudhury and tripathi 2019). Although salt can improve soil structure, it can also hamper plant growth and crop yield (Khaled and Fawy 2011). Salt stress caused by sodium chloride (NaCl) and magnesium sulphate (Mg_2SO_4) at a concentration of 100 mm was found to impair cotton seed germination and seedling root and bud growth. The phenomenon of reducing plant germination and seedling development by the increase of salinity concentration is described in *Carthamus tinctorius* L. (da Silva et al. 2008; Aydin et al. 2012), wheat (*Triticum aestivum*) (Egamberdieva 2009), peanut (*Arachis hyphaea* L) (Mensah and Ihenyen 2009), chickpea (*Cicer arietinum*) (Egamberdieva et al. 2016), and other medicinal plants, such as *Ochradenus baccatus* (Del.) and *Ephedra alata* (Khaled and Fawy 2011). Salinity prevents the production of plant hormones such as auxin and cytokinins in plants (Fahad et al. 2015; Mishra et al. 2018). According to Egamberdieva et al. (2018), salt disrupts plant hormone balance. Therefore, the dynamic balance of hormones under salt stress may be one of the mechanisms of plant hormone-induced salt tolerance. Exogenous application of plant hormones (Javid et al. 2011), such as gibberellin (Kim et al. 2006), auxin (Egamberdieva et al. 2016), and cytokinin (Dobrev and Vankova 2012) alleviate salt stress and promote root and shoot growth of plants under salt stress. Plant hormone content in plants may also be influenced by root-associated microorganisms (Egamberdieva et al. 2016). In an earlier study, SAHARAN and NEHRA (2011) reported that the contents of indoleacetic acid (IAA) and GA3 in maize (*Zea mays*) seedlings inoculated with PGPR strain *Azospirillum lipoferum* were relatively high. Inoculating chickpea with *Bradyrhizobium* could increase IAA content in leaves and promote root growth (García et al. 2017). Root-associated bacteria that are produced by plant hormones can colonize the plant by effectively colonizing plant roots, providing more IAA to the rhizosphere, and promoting the development of lateral roots and root hairs (Khan et al. 2016). From these studies, it can be inferred that the existence of PGPR strains that produce plant hormones can influence the metabolism of endogenous plant hormones, raising the root surface and consequently, enhance the absorption of key nutrients (Esitken, 2011).

Rice is a typical plant for which salt stress is an important factor restricting its growth. Soil salinity is a key obstacle on global rice production, particularly in coastal areas (Hasegawa 2013). The sensitivity or tolerance of rice to high salt stress results from interaction of various stress response genes, which often interact with other components of stress signal transduction pathway. By introducing salt-tolerant genes, salt-tolerant varieties can be cultivated through marker-assisted selection or genetic engineering (Jamil et al. 2011).

According to Reddy et al. (2017), the mechanisms and genes involved in transferring salt tolerance to high-yield rice varieties were updated. Reddy et al. (2017) emphasized the importance of incorporating phenotype, metabolic profiling, genomics, and phenomics into transgenic and breeding methods to produce high-yielding and salt-tolerant rice varieties. Crop yield optimization is dependent on the optimum supply of water, mineral nutrition, organic small molecules, protein, and hormone from the root system through xylem (Pérez-Alfocea et al. 2011). Soil drying and salinization have changed these xylem fluxes, with new omic approaches offering distinctive ways to understand the complex nature of these reactions. Although the absolute xylem concentration of any component depends on the genotype and xylem sap sampling method, the study of relative changes in concentration reveals some conservative behaviours (Lipiec et al. 2013). Generally, these stresses increase the concentration of abscisic acid (ABA), a plant hormone in xylem, which limits the water loss of crops, but decreases the concentration of some cytokinins, which can stimulate the expansion growth and prevent the premature senescence of leaves. To further understand the ion and biophysical changes in the rhizosphere environment, it is necessary to increase the xylem concentration of ethylene precursor 1-aminocyclopropane-1-carboxylic acid (ACC) (Yang et al. 2013).

The interaction of these plant hormones with plant nutrient status and xylem nutrient transfer can be a key factor in regulating plant response to the environment. Xylem proteomics is a new field, which will help to understand the mechanism of plant stress adaptation. Using omics to support rootstock-mediated plant improvement may increase crop yield in arid or saline land (Benková and Hejátko 2009).

Relationship Between Salinity and Plant

Salt is a major environmental stress that exerts ionic toxicity and osmotic stress on plants, leading to nutritional imbalance, oxidative damage, and plant growth and crop yield limitation (Kamran et al. 2020). In many parts of the world, salinity is regarded as a significant factor influencing crop productivity and agricultural sustainability because it decreases the value and efficiency of affected soil. Salinity is most common in arid and semi-arid environments. (Dawood et al. 2020; Dhiman et al. 2021). If the sediment is insufficient to leach the excessive soluble salt into the root zone. In irrigated agriculture, salt problems also occur, especially when low- quality water is used for irrigation (Läuchli and Grattan 2007). The shortage of water resources in most countries in arid and semi-arid areas has led many farmers to use low-quality water for irrigation. Considerable quantity of such marginal water is available and can be efficiently used for irrigation under proper irrigation management (Chanwala et al. 2020). The ability of crops to grow successfully and accumulate a high concentration of salt in tissues is different under saline alkali conditions. Increasing the concentration of soluble salts in soil solutions appears to raise their osmotic pressure and/or leads to individual ionic toxicity (Iqbal et al. 2020). This results in reduced water and nutrient uptake by plants (Farooqi et al. 2020). It is a simple technique to grow plants in hydroponic solution with salt addition and strictly control the root environment to evaluate the plant response to salt (Munns et al. 2020). Through this technology, most of the complexity and interference caused by soil and environmental factors are eliminated, and better control of the experiment is realized (Kamran et al. 2020). The effects of salinity on plant growth have been extensively studied under different nitrogen conditions (Kaya et al. 2020). Under high salinity, the mechanism of plant growth inhibition is still unclear (Xiong et al. 2020). Salinity can alter nutrient uptake by antagonizing essential nutrients (Samaddar et al. 2019). Nutrient imbalance caused by antagonistic and synergistic effects in halophyte medium also affects nutrient uptake and reduces plant growth (Minhas et al. 2020).

Among the proteomic measurements of xylem sap, the role of ionomics in mediating salt stress response and the role of xylem hormone status in mediating dry soil response are of great concern (Frukh et al. 2019). More and more people realize that the interaction between ionic and hormonal status is becoming highly essential for the regulation of physiological reactions. Under suboptimal (resource-constrained) states (El-Esawi et al. 2020), modification of xylem sap structure (e.g. by grafting plants on selected rootstocks) has a beneficial impact on growth and yield, with no penalty under optimal (non-resource-constrained) conditions. While root-derived cytokinins (CK), ABA, and ACC seem to be vital to controlling shoot efficiency under abiotic stress, a better understanding of their physiological and genetic determinants and their regulation would help rootstock genetic improvement (Agarwal and Khurana 2020). The state of the rootstock hormone is also essential for deciding resistance to biological stress, which is another ideal agronomic trait (Colmenero-Flores et al. 2020). A crucial element in plant salt resistance is the equilibrium between stress-induced polyamine production and reactive oxygen species (ROS). Polyamines are molecules that accumulate in plants when exposed to salt stress and help to sustain intracellular ROS homeostasis (Xiong et al. 2020). Polyamines mediate salt stress response by regulating redox homeostasis. Recent studies are used to explore the two proposed roles of polyamines in ROS regulation as antioxidant molecules and origins of reactive oxygen synthesis (Satisha et al. 2020). Second, in the context of plant salt stress, the proposed functions of polyamines as ion transport regulators were discussed (Ahmad and Anjum 2020). Finally, we emphasize the direct relationship between polyamine accumulation and the activation of programmed cell death under stress. As a result, polyamines play multidimensional roles in regulating cellular signalling and metabolism under stress (Acosta-Motos et al. 2017). By focusing on how to regulate the accumulation and turnover of polyamines in the future, research in this field may provide new goals for the development of stress tolerance (Salehi-Lisar and Bakhshayeshan-Agdam 2020).

Salinity and Ideal Sustainable Agricultural System

For the improvement and maintenance of human health, a sustainable agricultural system is considered ideal, which can benefit the economical values of producers and consumers, and also protects the environment and production of enough food for the growing population of the globe (Farley and Smith 2020). Abiotic stresses in the environment are major constrains of production in the agriculture sector throughout the globe (Rolston 2020). Microorganisms that are associated with plants play significant part against abiotic stress resistance. These microorganisms are rhizosphere and endophytic bacteria and symbiotic fungi that operate through diverse mechanisms in the likes of osmotic trigger response, hormones, and nutrients that act as biological agents, resulting in induction of novel plant genes (Mercado-Blanco et al. 2018, Roychoudhury and tripathi 2019). Cultivation of stress-tolerant crops might be possible through genetic engineering and plant breeding techniques, which is a time-consuming and expensive process; however, inoculation of microbes for alleviation of stress might be a cost-effective and environment friendly option in a short period of time. Using existing clues, coordinated future research in this area is needed, especially in field assessment and potential biofertilizer application in stressed soils (Hardoim et al. 2019). Abiotic stress affects crop productivity worldwide. Understanding the molecular mechanisms of abiotic stress tolerance in plants is of great significance for developing stress-tolerant crops to maintain future crop productivity (Guerriero et al. 2018, Roychoudhury and tripathi 2019)). It is necessary to develop new rice varieties with higher and more stable yield in different climates, environments, and geographical locations. The availability of rice genome will help to find genes that can be used for breeding or transgenic. The ideal salt-tolerant varieties should have the agronomic advantages of high Na^+ tolerance, controlling Na^+ and N absorption, maintaining high K^+ absorption, good initial vigour, and high yield potential (Reddy et al. 2017).

Relationship Between Salinity and Sodicity and Soil

The relationship between soil salinity and its flocculation effect, alkalinity, and its dispersion effect affects whether the soil maintains aggregation or dispersion under different combinations of salinity and alkalinity (Basak et al. 2015). When low salinity irrigation water is applied to the soil through irrigation or rainfall, it will flow into the space between clay particles (micropores). If the salinity of applied water is lower than that of soil, clay particles will expand and disperse (Ezlit et al. 2010). In contrast, irrigation water with higher salinity than soil often leads to the aggregation of particles to maintain soil structure. Excessive soil salinity will lead to poor, uneven, slow growth, and low yield of crops, which depends on the degree of salinity (Bodale and Filipov 2019). The main function of excessive salt is to reduce the water that plants get, although some water still exists in the root zone. This is because the osmotic pressure in soil solution rises as salt content rises. Aside from the osmotic impact of salt in soil solution, excessive concentrations and absorption of individual ions can be harmful to plants and/or impede absorption of other important plant nutrients (Abd El-Samad 2016).

Salt Stress Effects on Plants

Salt is problematic for plants with increment of inorganic minerals in the environment, which resulted in osmotic and water stress and provides cheap osmotic agents that reduces osmotic potential and prevents water loss (Isayenkov and Maathuis 2019). Even though research has been carried out for decades, one of the most enigmatic issues of salt stress in plants is Na^+ and Cl^-, a mechanism of action. In conclusion, most of the species and conditions evidently observed that non-selective cation channels (NSCC) is the major pathway for Na^+ to flow into glycophytes in the roots that contain multiple channels from multiple families (Hasanuzzaman et al. 2013). This complication hampers the creation of detailed description of 'who does what' and how much. In monocotyledonous crops, high-affinity potassium (HKT) family members might promote uptake of Na^+. The use of mutants resulted in progression in uptake of Cl^-. It has been observed that ectoplasmic bypass of Na^+ and Cl^- is very crucial for aquatic plants that have special anatomy of roots like rice (Flowers et al. 2019). Stress tolerance improvement in crop plants is one of the solutions to cope with increased salinization of cultivated land areas. We urgently need further research to reveal the details of mechanism of Na^+ and Cl^- uptake. Toxicity profiles will be mapped at the tissue and cellular levels to enhance stress tolerance in crop plants. Important components are lack during perception, sensing, and signalling chains, especially at the start of the pathway, whereas better understandings of the roles of other minerals like K^+ also facilitate the alleviation of salt stress by regulation of absorption and distribution of mineral elements (Assaha et al. 2017). Plants are affected by salt when salt is sprayed on the stems, buds of deciduous woody plants, as well as leaves and needles of evergreen plants from past cars. Salt spray resulted in salt burns on buds, leaves, and twigs. Salt spray also resulted in dryness of bud scales and caused damage of developing leaves and flower tenders. Developing leaves and buds that are unprotected are often dried up and are destroyed by cold winter winds. Most of the time, damage caused by cold winter winds did not appear until the winter ends and in the start of spring. Browning of needles or leaves, death of buds, and death of branches on the side of plants facing road or sidewalk are typical signs of salt fog injury. Deciduous plants will not be damaged unless spring returns to growth (Miller et al. 2010). Dissolved salts in the runoffs also affected crop plants. Sodium and chloride ions are isolated if salt is dissolved in water. High level of dissolved sodium and chloride ions in soil can exchange additional mineral nutrients. Plants then consume chlorine and sodium rather than essential plant nutrients like phosphorus and potassium resulting in malnutrition (Beck et al. 2011). Chloride ions are capable of being transferred to leaves, where they interrupt photosynthesis and chlorophyll production. Chloride concentration may exceed toxic levels, resulting in leaf burn and death (Wahid et al. 2009). When saline snow is ploughed or shovelled into lawns and garden beds, rock salt may also cause massive damage. The salt in the soil can absorb water. Water stress and root dehydration were exacerbated as plant water absorption declined. This is known as physiological drought, and if not addressed, it may lead to a decline of plant development. The replacement of other mineral nutrients by sodium ions would also have an impact on soil fertility. When the drainage and ventilation decrease, the compaction degree will increase, which usually leads to the decrease of plant growth. Salt damage in the soil can be prolonged, and plant symptoms do not appear until next summer or even years after. Symptoms can also appear during hot and dry conditions (Jull 2009).

Management Strategies to Mitigate Salt Injury

High concentration of salt stress is considered to be one of the biggest threats to plants worldwide (Ejaz et al. 2020). Reducing the use of salt combines salt with other ingredients like sawdust, sand, or cinder, which may supply gravel for adhesion. De-icing materials using salts other than sodium chloride, which include magnesium chloride, calcium chloride, potassium calcium, magnesium, or chloride acetate, are costlier, but they can minimize damage (Bhattacharyya and Maier 2004). The application should be for sidewalks and roads, not landscape beds or lawns. When the snow melts, the discharge of saline runoff should be considered. Planting should be avoided in areas where natural flow of runoff occurs. Washing the soil with a lot of water can help to remove salt from well-drained soil. This might not be possible in poorly drained soil. Incorporating organic matter is an essential method for enhancing soil drainage of poorly drained soils (Dudley et al. 2008). Physical barriers like burlap, plastic, and wood might be used for protection of plants. Salt-tolerant plants might be planted close to roads, driveways, and sidewalks. Salt tolerance does not mean any harm. When selecting 'salt tolerant' plants, it is essential to remember that the degree of salt tolerance and damage depends on many factors, and plants within the same species have different salt tolerance. Salt tolerance may also vary with salt exposure. There are contradictory studies regarding salt tolerance in plant species. Soil types and climate fluctuations can lead to differences in plant responses in different regions (Tester and Davenport 2003). Plant salt stress is a kind of situation where excessive salt in soil solution leads to plant growth inhibition or death. Salt reduces or alters plant production by inhibiting or changing plant development, seed germination, photosynthesis, dry matter distribution, and yield, causing osmotic stress and ionic toxicity. Salinity and alkalinity are soil conditions that occur mainly in arid and semi-arid areas (Yadav et al. 2011). In wet areas, rainwater leaches salt from the soil, and salt problems are rare and transient. The ions contributing to soil salinity include Cl^-, HCO_3^-, $S0_4^{2-}$, Ca^{2+}, Mg^{2+}, and NO_3^- and few K^+ or Na^+. The concentration and proportion of salts of these ions vary greatly. They may be local, but more often they are taken to areas where irrigation water or drainage from adjacent areas is available (Raney 1960). In arid areas, natural drainage systems are often so underdeveloped that salt accumulates in inland basins rather than in the ocean. Saline soil contains soluble salts that are harmful to plant growth. The lower limit of saline soil is usually set as the conductivity of 4 mmho cm^{-1} in saturated soil extract.

It is well known that soil salt stress can inhibit plant growth (Shrivastava and Kumar 2015). Plants in the natural growth environment are colonized by intercellular and intracellular microorganisms (Douglas 2015). Rhizosphere microorganisms, especially beneficial bacteria and fungi, can improve the growth performance of plants under stress, thus directly and indirectly increasing the yield (Hayat et al. 2010). Some plant growth-promoting rhizosphere (PGPR) bacteria may directly stimulate plant growth and development by providing fixed nitrogen, plant hormones, iron isolated by bacterial iron carriers, and soluble phosphate to plants (Hayat et al. 2010). Others protect plants from indirect effects of soil-borne diseases, most of them are affected as a result of pathogenic fungi (Gao et al. 2010). Soil salinization is a major problem in agricultural production worldwide. Growth and production of crops was affected by high osmotic stress, malnutrition toxicity, and poor soil physical conditions grown on saline soil (Allam et al. 2018) (Figures 13.1–13.4).

Salt Sensitivity

Many crops are sensitive to salt. Corn, oranges, onions, lettuce, beans, and pecan are the most susceptible and wilt when grown in high salinity soil. Barley and cotton can bear medium level of salt; dates and beets are particularly salt tolerant. Agriculture's primary restricting factor is soil salinity (Lewis 2002). Various traits, such as osmotic tolerance, tissue tolerance, and ion exclusion, may be used extensively to increase salt tolerance. One effective strategy for addressing the salt issue is to grow salt-tolerant crops. 'Dehydrated' species need to be tolerant, but annual crops also need to be tolerant because salt remains in the soil during the fall of water table. Crops' tolerance to salt will also result in extra use of irrigation water with poor quality (Galvani 2006).

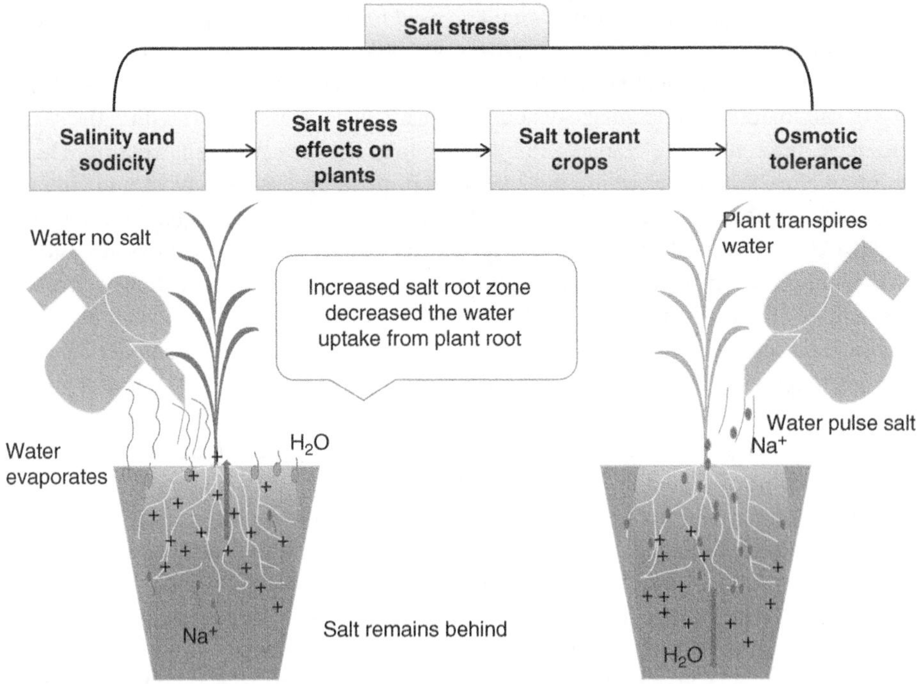

Figure 13.1 Plant growth under normal and salt stress condition.

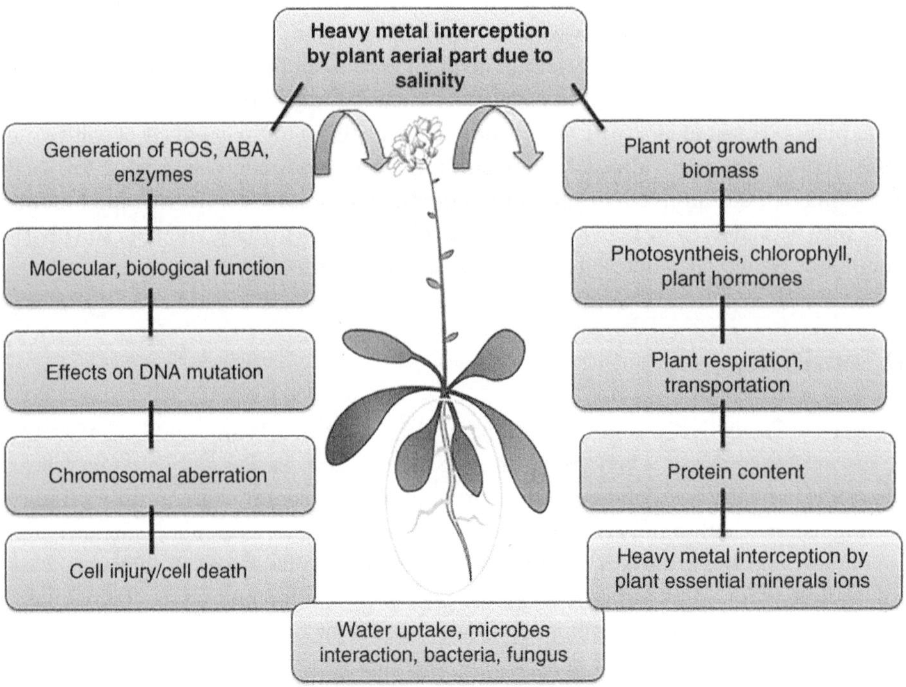

Figure 13.2 Plant performance under stress environments.

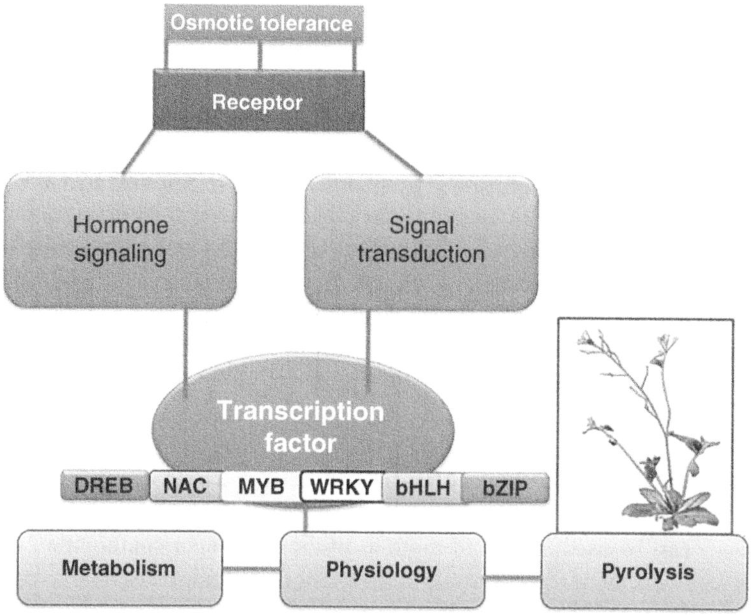

Figure 13.3 Plant osmotic stress and role of some mechanisms.

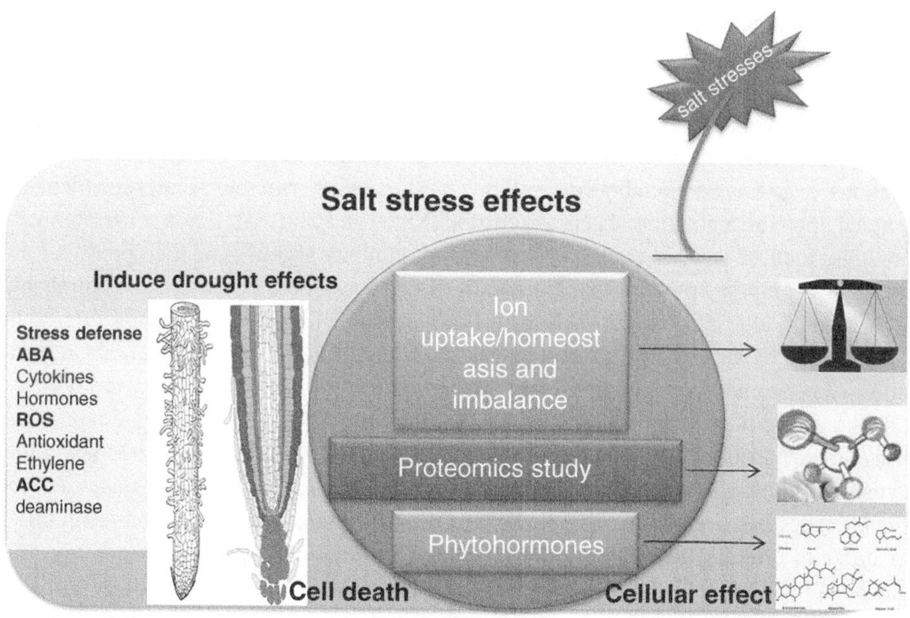

Figure 13.4 Salt stress effects in plant and interaction of some hormones to balance the plant defense system.

To improve salt tolerance in plants, it is necessary to understand the mechanism of salt restriction and tolerance at organelle and molecular level (Arzani 2008). Under the condition of salinity, the gene expression pattern changed, and the protein synthesis also changed qualitatively and quantitatively. Though, it is commonly believed that salt stress can lead to protein synthesis fluctuations, there are still some controversies about whether salinity can activate the specific genes involved in salt stress (Deinlein et al. 2014). Salt tolerance does not seem to be given by unique genes. When plants are subjected to abiotic stresses, many genes are turned on, leading to the increase of numerous metabolites and protein levels, some of which might be the reason for providing a protection against stresses (Bhatnagar-Mathur et al. 2008). Under environmental stress, determinations for increment of crop yield through transgenic methods have not achieved fruitful results, because the basic mechanism of plant stress tolerance remains to be fully understood.

Genetic Engineering and Salt-Tolerant Transgenic Plants

Modern biotechnology can be used more and more to insert the proper combination of genes into elite crop cultivars (Munns et al. 2002). Alterations in osmotic resistance would probably include long-distance signalling, cell cycle regulation, and methods requiring detecting signals from roots to the shoots. Given that crop plants are typically subjected to low salinity levels during the planting period, or salinity levels that begin low but increase towards the end of the planting period, genes encoding for osmotic tolerance traits have the capacity to have a substantially greater effect on crop salinity tolerance than those participated in ion exclusion (Christmann et al. 2013).

The osmotic stress was more important for the end output of these plants, which grew at low to moderate salinity than the ionic stress. Therefore, the detection of the osmotic resistance genes should be prioritized for enhancing salinity tolerance of plants grown in low to moderate saline (Sairam and Tyagi 2004). Molecular and cellular level responses include the adaptation of ion transfer (such as ion absorption, extrusion, and sequestration) and metabolic alterations (e.g. carbon metabolism and the production of suitable solutes), which are prompted upon the regulation of gene expression (Xiong and Zhu 2002). As a consequence of these stress-induced genes, two classes were known. Functional genes and their products: the first category comprises proteins that specifically protect against ecological stress. It is probable that these molecules safeguard the cell against dehydration. For instance, enzymes that aid in the production of various late embryogenesis-abundant proteins, osmoprotectants, detoxification enzymes, antifreeze proteins, and chaperones can provide protection. Additionally, regulatory genes and their products comprise a second category that regulates gene expression and signal transduction during the stress response. Transcription causes, protein kinases, and enzymes implicated in phosphoinositide synthesis are among them. Stress-spiepr DHYP inducible genes were used to enhance plant stress resistance through gene transfer techniques.

In order to understand not only the molecular processes of stress tolerance and the response of higher plants, the roles of stress-inducing genes are necessary to study, but also to increase crop stress resistance through genetic modification (Hasanuzzaman et al. 2013). This chapter highlights recent studies on transgenic plant technologies for enhancing environmental stress resistance by metabolic engineering and signalling pathway modulation.

Salinity resistance is a physiologically diverse trait that can be connected to a number of pathways. Furthermore, transcriptional modifications in the gene expression profile of core genes modulating plant ionic and oxidative homeostasis Rubidium hydroxide (RBOH), Alpha-hydroxy acids (AHAs), sodium/hydrogen antiporter (NHX), Gated or Guard cell Outward Rectifying K(+) channels (GORK), and Superoxide dismutases (SOD) were investigated to compare the influence of transcriptional and post-translational factors to the basic components of salinity tolerance (Adem et al. 2014). Sodium sequestration, K^+ preservation, and oxidative stress resistance all seemed to be vital for enhanced tissue tolerance (Ashraf 2009). Since these traits are very closely linked, it is believed that substantial advances in crop breeding for salinity tolerance can only be attained if these complementary traits are simultaneously targeted. The significance of post-translational changes in plant response to salt adaption (Reynolds and Tuberosa 2008).

Role of Sodium in Plants

As a result of the sodium and chlorine ions being isolated by the high concentration of salts in the water, the root system's ability to absorb water is diminished, which eventually causes drought stress and root dehydration. Physiological drought, also known as rising water stress and root dehydration, is a condition that, if not properly managed, can impair plant production. (Geise et al. 2012). Sodium is not considered a critical component for plant growth, but it may be used in a minute quantity to help the metabolism and synthesis of chlorophyll (Harb et al. 2010). When excessive salts are accumulated in the soil, it causes two main problems: first, soil becomes vulnerable to erosion and this problem is related to soil structure, and second, it affects plant growth (Yadav et al. 2011). Since sodic soils have high amounts of exchangeable sodium, individual sand, silt, and clay particles may be separated rather than grouped into larger particles. As a result of this dispersion, the surface becomes compact and impenetrable, enabling less air, precipitation, or runoff water to enter (Gawel 2006). Thus, the plant may not receive enough moisture and oxygen to develop. Salts may remain on the soil surface, so they cannot leach out of the root zone. Plants can be harmed by the effects of salt or toxicity (El-Maarouf-Bouteau and Bailly 2008). High amounts of soluble salts in salty and saline-sodic soils minimize the quantity of usable water for plants to use. High amounts of sodium may contaminate certain plants. Furthermore, elevated soil pH in high-salt soils significantly alters the nutrients access to the plants (Qadir et al. 2007). Because of the elevated pH levels, certain plant nutrients lose their ionic structure and become inaccessible to the plants.

If salts in plant roots do not exceed the threshold level, then salts do not affect the plant. Soil salinity has a direct negative impact on crop production (Lambers et al. 2008). Sensitivity to salinity in agricultural crops is typically (but not limited) attributed to an accumulation of Na^+ in the soil, as additional Na^+ is detrimental to plants (Zhang et al. 2010). We believe that decreasing Na^+ absorption is a solution to overcome Na^+ accumulation in agricultural plants and an efficient method to regulate Na^+ accumulation in crop plants and improve salt tolerance. Therefore, it is of utmost importance to understand the process of Na^+ absorption by higher plant roots for regulating salt tolerance (Munns et al. 2012). Hence, the aim of this chapter is to highlight and discuss recent advances in our understanding of the mechanisms of Na^+ uptake by plant roots at both physiological and molecular levels (Apse and Blumwald 2007). We summarize that further research is needed to understand method of root Na^+ uptake in higher plants; specifically low-affinity Na^+ intake is needed because it is the mechanism through which sodium affects plants' growth in saline soil conditions.

Osmotic Tolerance

It is imperative to test newly inserted genes at the field level to check their efficiency in improving crop productivity and salinity resistance. Just like salinity has several varied impacts on a plant, there are often several different pathways for plants to survive this stress. (Penumarthi et al. 2020). These processes can be divided into three main groups: first, osmotic tolerance controlled by long-range signals, which restricts shoot development; second: ion exclusion, where Na^+ and Cl^- transport mechanisms are decreasing toxic level of Na^+ and Cl^- accumulation in the leaves; and lastly, tissue tolerance, where high levels of Na^+ and Cl^- occur in leaves but are divided into the cellular and intracellular levels (particularly in vacuole) (She et al. 2016). Third, the osmotic resistance/tolerance, which is associated with reducing shoot growth, affects and is associated with the unknown sensing and signalling pathways. Osmotic shock, also known as osmotic stress, is a physiologic dysfunction induced by a drastic shift in the concentration of solutes around a cell, which triggers a dramatic change in water flow through the cell membrane. Osmotic stress is a vital abiotic issue that has affected the evolution of the plants due to drought, salinity, or cold stress. Drought and salinity stress stimulated the lowering of water potential for plant cells caused by osmotic stress (Upadhyaya et al. 2013). The dehydration produced by drought, salinity, and freeze is clearly referred to as osmotic stresses; chilling and hypoxia trigger osmotic stresses that indirectly impact the water absorption loss. Therefore, the long-term aim of agricultural biotechnology is to improve seed tolerance to osmotic stress (Wu et al. 2012). Plants display an extensive range of responses under osmotic stress ranging from the responses at the entire plant to the level of metabolism. Morphological and developmental variations in life cycles, shooting inhibition, and increasing root growth are all plant reactions. Under extreme osmotic pressure, salt stress has a much more harmful effect on plants as Cl^- and Na^+ toxicity increases (Öktem et al. 2006). Nowadays, studying salt tolerance mechanisms of the plant and resistance of the plant to salinity has become a hot debate among researchers (Acosta-Motos et al. 2017). By examining the process of salt tolerance in plants based on molecular and biochemical responses to salt stress in plants, we may be able to better understand how plants respond to salt stress. As proteomic technology develops, we can now combine it with knowledge of the majority of plants' genomic sequences.

Proteomics Study in Plant Responses and Tolerance to Salt Stress

Proteomics has many uses; proteomic-based approaches have been extensively used in numerous crop species to explain alterations in cellular processes at the protein level in response to salt stress. Proteins are essential macromolecules that play a variety of roles in plant stress resistance/tolerance (Afroz et al. 2011). To comprehend intricate plant salt stress signalling systems. The limitations of the aforementioned technology have been overcome with the development of non-gel-based quantitative proteomics methods in recent years (Agarwal and Khurana 2020). The mass spectrometry proteomics technique (isobaric tags for relative and absolute quantification) may be used to assess cell metabolic variations (Aghaei and Komatsu 2013). iTRAQ, on the other hand, is extensively used in plant quantitative proteomics (Eldakak et al. 2013). In addition, this technology is also used to explain the functional variation of *Brassica napus* guard cells and mesophyll cells (Zhu et al. 2009). Moreover, Peng et al. (2015) effectively studied the protein profile of plant responses to mineral nutrient deficiencies or excesses, such as the response of *Citrus sinensis* roots to boron deficiency. Many Distributed Engineering Plants (DEPs) in tomatoes are identified using iTRAQ protein profile review (Zhu et al. 2018) and in *Arabidopsis thaliana* and *Brassica juncea* (Alvarez et al. 2009) subjected to alkali stress and salt stress. Finally, iTRAQ's molecular process of plant reaction to abiotic stress can be extensively utilized in the future.

The proteomics research provides a modern approach to identifying and studying proteins and biochemical processes for the physiological responses to crops and stress (Ghosh and Xu 2014). Therefore, the analysis of plants at proteomic levels can help to understand the mechanisms of stress tolerance. In addition, the knowledge of the primary metabolic proteins that participated in stress resistance can be applied in biotechnological applications (transgenic/recombinant) formation (Kosová et al. 2011). Furthermore, research into established metabolic pathways eventually aids in the advancement of anti-stress techniques (Kerins and Ooi 2018). The proteomic technique is very useful because it assists plant physiologists in studying plants more accurately in understanding what is going on in the cell as a result of an external stimulus (Mittler 2006). In recent years, mass spectrometry-based proteomics has gained popularity worldwide because it allowed scientists to research plants more accurately than ever (Bendixen 2005). For example, a variety of strategies have been used to separate and identify/characterize different proteins in various plant species, including polyacrylamide gel electrophoresis (PAGE), two-dimensional liquid chromatography (2D-LC), pro-Q Diamond stain, sodium dodecyl sulphate (SDS)-PAGE, mass spectrometry, MALDI-TOF, fluorescence, Coomassie brilliant blue (CBB)-stained 2-DE, 2D gel electrophoresis ion-exchange chromatography (IEC), 2D PAGE, 2D difference GE (2D-DIGE), and non-gel-based LC-MS. All these methods have shown sound results in characterization of proteins.

Ion Uptake/Homeostasis

Understanding proteins, specifically root proteins, is critical in salt tolerance mechanisms (Zhang and Webster 2009). Na^+ reaches the plant roots primarily by apoplastic or symplastic pathways, which involve several Na^+ transport transmembrane proteins like HKT and Na^+/H^+ antiporters (Keisham et al. 2018). Various proteins released under stress or non-stress conditions are genotype specific, whereas others are affected by the amount and length of salinity stress (Soda et al. 2013). For example, a previous study of Zhao et al. (2013) was conducted to investigate the plasma membrane of rice under salt stress conditions; 24 distinct proteins were identified that were correlated with the association and signalling of protein-protein (reorine and 14-3-3 protein), and these proteins were engaged in regulating potassium (K) ion channel (Hashiguchi et al. 2010). Another research was performed by Budak et al. (2013) to test the leaf and root proteomes of two wheat cultivars under saline stress conditions. They discovered that most of the proteins released during stress were cultivar specific, whereas others were stress sensitive. They proposed that increased salinity resistance in wheat was linked to increased efflux of toxic by-products and also ionic/osmotic homeostasis.

Role of Phytohormones for Abiotic Stress Tolerance

Phytohormones are naturally occurring chemical molecules or substances that elicit physiological reactions in plants at a very low level (Khan et al. 2012). Not only conventional phytohormones like auxins, gibberellins, cytokinins, ethylene, and ABA play an integral role during abiotic stress tolerance but also many other compounds that have hormone-like properties including brassinosteroids, methyl jasmonates, salicylic acid, and strigolactones also play a vital role in response to abiotic stress tolerance (Kumar et al. 2015). Plants are vulnerable to various abiotic stresses during their life cycle including high temperature, low-temperature salt, drought, flooding, allelochemicals, oxidative, light, radiation, wind, and heavy metal stresses (Vardhini 2016, Roychoudhury and Tripathi 2019). Among the different abiotic stresses in plants, salt stress is a significant factor limiting agricultural production (Devi et al. 2017). The function of the aforementioned phytohormones in alleviating salt stress is gradually unfolded. Phytohormones are highly specialized molecular signalling compounds that play an important role in plant defense mechanisms against various abiotic stresses (Miller et al. 2010). Plants are often subjected to biotic and abiotic stresses that limit their growth and productivity. Brassinosteroids are considered as the sixth class of plant hormones, they have a pleiotropic effect, and they can protect the plant from salinity stress (Zhu et al. 2018). Plant hormones are chemical substances that spontaneously synthesize in plants, control growth and various physiological mechanisms away from the synthesis site, and show their activity even in minute concentrations.

In general, salt in soil water impedes metabolism by reducing the plant's capacity to absorb water. The high concentration of salt enters the plant through the transpiration phase, causing cell wounding and disrupting transpiration, resulting in osmotic stress. Crop plants respond to adverse external stress caused by environmental conditions by suppressing their native biological processes, causing their growth, production, and productivity to decline. Plant hormones are small chemical signalling molecules. Their complex hormone signalling mechanisms and crosstalk capabilities make them ideal candidates for initiating defensive responses (Peleg and Blumwald 2011). Plants have evolved different mechanisms that detect stress signal, enhance optimal growth response, and respond to unfavourable environmental conditions, and these mechanisms play critical roles in assisting plants to acclimate to salinity stress. Plant hormones are organic compounds that are generated from one part of the plant and transferred to the other, where they induce physiological reaction at very low concentrations. Plant hormones are naturally occurring substances; they are referred to as plant growth regulators when chemically synthesized.

Plants are normally susceptible to environmental conditions including drought or excessive soil and water salinity. Plant development can be reduced in saline conditions due to the influence of specific ions on metabolism or due to unfavourable water relations.

To optimize plant growth under saline conditions, multiple approaches are being used. One approach is to develop salt-tolerant genotypes of various crops. Attempting to increase salt tolerance through traditional plant breeding is relatively time consuming and complex and depends on current genetic resources.

Additionally, further research has been conducted to alleviate this problem, such as better management and the application of endogenously generated plant growth regulators. In this framework, previously Arzani (2008) discussed the role of different levels of phytohormones, including ABA, gibberellic acid (GA), cytokinins (CK), jasmonates (JA), IAA, and triazoles to alleviate salinity problem. Recently, phytohormones and osmolytes have been implicated in reducing the negative effects of salinity. Under salinity stress, osmolyte like proline retains cellular homeostasis through osmotic regulation and favourably induces physiological processes (Sharma et al. 2019; Dhiman et al. 2021). Under salinity tension, phytohormones play a vital role in attenuating physiological responses that ultimately result in plants adjusting to harsh environment. Although enormous research has been conducted on individual role of both proline and phytohormones, the systematic analysis of phytohormones' relationship with proline under salinity tension is missing (Javid et al. 2011).

Interaction Between Na^+ and K^+ in Plants

When compared to genotypes that are vulnerable to salt stress, they maintain a robust K^+/Na^+ ratio in the cytosol (Almeida et al. 2017). Various plant genotypes exhibit different salt tolerance methods; they all depend on the regulation and function of K^+ and Na^+ transporters and H^+ pumps, which produce the driving force for K^+ and Na^+ transport. Salt stress is a big problem for modern agriculture all over the world. However, progress in the production of salt-tolerant crops has been plodding. There are numerous causes for this delay, but the reason is that salt tolerance is dependent on the coordinated regulation of hundreds of genes that may be the most significant. To overcome salt stress, plants have developed unique mechanisms to control K^+ and Na^+ tissue and cellular homeostasis (Almeida et al. 2017). Furthermore, these mechanisms depend on H^+, K^+, and Na^+ transporters. Numerous metabolic studies in recent years have established a clear link between SOS1, HKTs, and NHXs transporters in K^+ and Na^+ homeostasis and salt tolerance. Modification of some of these genes in model crop plants resulted in higher yield under controlled environment (Brini and Masmoudi 2012); however, its application in field condition is minimal. The development of salt-tolerant crops is unlikely to be fruitful only if new technological solutions that permit for the fine-tuning and regulation of multiple genes, especially in a tissue-specific manner, are developed. Furthermore, certain ion transporters are engaged in the important cellular mechanism, and overexpression of certain genes may cause substantial perturbations of associated cellular and physiological mechanisms, restricting the improvement of the salt stress response.

Thus, future research initiatives should be conducted towards a deeper understanding of the molecular pathway, such as epigenetic modifications, post-translation modifications, and transcription factors, underlying the regulation of those transporters to provide critical insights for the growth of salt-tolerant plants (Reguera et al. 2012).

Interactions Between Na^+ and Mg^{2+} in Plants

Higher concentration of Na^+-induced accumulation of Ca^{2+} and Mg^{2+} in the plant. The addition of salt raised the concentration of Na^+, thus inhibiting the accumulation of Ca^{2+} and Mg^{2+}. Drought impaired Na^+ deposition while accelerating Ca^{2+} and Mg^{2+} accumulation (Khorshidi et al. 2009). Significant Mg-Na associations have been identified to control potassium and calcium intake through cotton. In contrast to commonly accepted assumptions about cation antagonists, it was discovered under certain circumstances that sodium raised the potassium and calcium contents of plants (Gopalapillai et al. 2013). When Mg^{2+} was limited, the impact of Na^+ was more prominent in plants, including phenotypic and physiological modifications, cell Mg^{2+} homeostasis regulated by Mg^{2+} transporters, monogalactosyl diacylglycerol (MGD) signalling, interactions between Mg^{2+} and other ions, and roles of Mg^{2+} in plant secondary metabolism. Mg^{2+} also affects secondary metabolism in plants. MGD upsurges cardenolide production in callus cultures of *Digitalis davisiana Heywood*, *D. lamarckii Ivanina*, *D. trojana Ivanina*, and *D. cariensis*. Storing of phenolic compounds and putrescine in plant cells is a universal response to stress in plants. Mg^{2+} treatment increases anthocyanins in red grape cell suspension culture while decreasing catabolism, as it does in other ornamental plants by spraying the foliage or drenching pot plants with Mg^{2+} solution (Abdel-Kader et al. 2012). Mg^{2+} treatment of Aster plants improved pigment concentrations without raising the activity of main enzymes in the anthocyanin biosynthetic pathway in flower buds (Nissim-Levi et al. 2007). Flowers from Anigozanthos, Limonium, Gypsophila, and Aconitum all had a similar influence (Miller et al. 2011). However, little is known regarding Mg's function in secondary metabolism in plants.

Interactions Between Na^+ and Ca^{2+} in Plants

Ca^{2+} regulates Na^+ transportation through Na^+, K^+-ATPase. Ca^{2+} concentrations in the physiological range of 0.08–5 M inhibit Na^+, K^+-ATPase activity (Lepik et al. 2004). Cell excitation triggers depolarization, which raises the cytosolic Ca^{2+} concentration above the threshold level and, as a result, inhibits some isoforms of the Na^+, K^+-ATPase in a concentration- and affinity-dependent manner (Keisham et al. 2018). However, a higher concentration of Ca^{2+} impedes the activity of Na^+, K^+-ATPase irrespective of the isomer. The enzyme's α subunit is linked to Ca^{2+} inhibition, with the $\alpha 2$ isomer possessing a stronger affinity for Ca^{2+} at physiological amounts of 0.08–5 M and at higher concentrations of 10 mM (Cereijido et al. 2016). Ca^{2+} contends with Mg^{2+} for ATP at a concentration of 10^{-5} M and reduces the concentration of Mg^{2+}-ATP, which is a rate-limiting substrate for the enzyme. Calmodulin and calnaktin, two Ca^{2+} binding proteins, are reported to influence the Ca^{2+} inhibition of this enzyme (Peinelt and Apell 2002). The Ca^{2+} inhibition mechanism connected with the enzyme subunit is non-competitive, and it does not contend with Na^+ for its binding site (Meldrum and Rogawski 2007).

Conclusion

Abiotic stress, including cold, drought, salt, and heavy metals, primarily affects plant growth and ultimately restrains crop productivity. Due to the constant shifts of climate change and environmental degradation induced by human activities, abiotic stress has become a significant challenge to food security. Plants may react and adapt to abiotic stress by initiating a variety of genetic, cellular, and physiological alterations. A greater understanding of plants' susceptibility to abiotic stress can help conventional and current breeding applications to increase stress tolerance. Studies on particular indigenous plants with greater stress tolerance often add significantly to our knowledge of stress tolerance. Plants are frequently subjected to biotic and abiotic stresses throughout their life cycle. Drought, extreme temperature, salinity, nutrition (excess, deficiency), and flooding are the most common abiotic stresses that cause a reduction in crop growth and yield. According to projections, abiotic stresses caused 50% of yield losses.

Nevertheless, abiotic stresses induce many alterations in the hormonal, molecular, and biochemical mechanisms that occur in plants. It has been broadly documented that many proteins react to these stresses at the pre- and post-translational stages. It would be simple to explain the mechanisms of stress resistance in plants if we understood the function of these stress-inducible proteins. It is essential to further explain the processes involved in plant stress responses using modern biological technology, particularly stress responses of certain wild plant species with very high-stress tolerances, which will eventually be used in the development of stress-tolerant plants.

References

Abd El-Samad, H.M. (2016). The potential role of osmotic pressure to exogenous application of phytohormones on crop plants grown under different osmotic stress. *Am. J. Plant Sci.* 7: 937.

Abdel-Kader, M.M., El-Mougy, N.S., Aly, M.D.E., and Lashin, S.M. (2012). Integration of biological and fungicidal alternatives for controlling foliar diseases of vegetables under greenhouse conditions. *Int. J. Agric. For.* 2: 38–48.

Acosta-Motos, J.R., Ortuño, M.F., Bernal-Vicente, A. et al. (2017). Plant responses to salt stress: adaptive mechanisms. *Agronomy* 7: 18.

Adem, G.D., Roy, S.J., Zhou, M. et al. (2014). Evaluating contribution of ionic, osmotic and oxidative stress components towards salinity tolerance in barley. *BMC Plant Biol.* 14: 1–13.

Afroz, A., Ali, G.M., Mir, A., and Komatsu, S. (2011). Application of proteomics to investigate stress-induced proteins for improvement in crop protection. *Plant Cell Rep.* 30: 745–763.

Agarwal, P. and Khurana, P. (2020). TaZnF, a C3HC4 type RING zinc finger protein from Triticum aestivum is involved in dehydration and salinity stress. *J. Plant Biochem. Biotechnol.* 29: 395–406.

Aghaei, K. and Komatsu, S. (2013). Crop and medicinal plants proteomics in response to salt stress. *Front. Plant Sci.* 4: 8.
Ahmad, R. and Anjum, M.A. (2020). Physiological and molecular basis of salinity tolerance in fruit crops. In: *Fruit Crops*, 445–464. Elsevier.
Allam, N.G., Kinany, R., El-Refai, E., and Ali, W.Y. (2018). Potential use of beneficial salt tolerant bacteria for improving wheat productivity grown in salinized soil. *J Microbiol Res* 8: 43–53.
Almeida, D.M., Oliveira, M.M., and Saibo, N.J.M. (2017). Regulation of Na+ and K+ homeostasis in plants: towards improved salt stress tolerance in crop plants. *Genet. Mol. Biol.* 40: 326–345.
Alvarez, S., Berla, B.M., Sheffield, J. et al. (2009). Comprehensive analysis of the Brassica juncea root proteome in response to cadmium exposure by complementary proteomic approaches. *Proteomics* 9: 2419–2431.
Apse, M.P. and Blumwald, E. (2007). Na+ transport in plants. *FEBS Lett.* 581: 2247–2254.
Arzani, A. (2008). Improving salinity tolerance in crop plants: a biotechnological view. *Vitr. Cell. Dev. Biol.* 44: 373–383.
Ashraf, M. (2009). Biotechnological approach of improving plant salt tolerance using antioxidants as markers. *Biotechnol. Adv.* 27: 84–93.
Assaha, D.V.M., Ueda, A., Saneoka, H. et al. (2017). The role of Na+ and K+ transporters in salt stress adaptation in glycophytes. *Front. Physiol.* 8: 509.
Aydin, A., Kant, C., and Turan, M. (2012). Humic acid application alleviate salinity stress of bean (Phaseolus vulgaris L.) plants decreasing membrane leakage. *African J. Agric. Res.* 7: 1073–1086.
Basak, N., Chaudhari, S.K., and Sharma, D.K. (2015). Impact of varying Ca/Mg waters on ionic balance, dispersion, and clay flocculation of salt-affected soils. *Commun. Soil Sci. Plant Anal.* 46: 827–844.
Beck, D.A., Johnson, G.R., and Spolek, G.A. (2011). Amending greenroof soil with biochar to affect runoff water quantity and quality. *Environ. Pollut.* 159: 2111–2118.
Bendixen, E. (2005). The use of proteomics in meat science. *Meat Sci.* 71: 138–149.
Benková, E. and Hejátko, J. (2009). Hormone interactions at the root apical meristem. *Plant Mol. Biol.* 69: 383.
Bhatnagar-Mathur, P., Vadez, V., and Sharma, K.K. (2008). Transgenic approaches for abiotic stress tolerance in plants: retrospect and prospects. *Plant Cell Rep.* 27: 411–424.
Bhattacharyya, A.J. and Maier, J. (2004). Second phase effects on the conductivity of non-aqueous salt solutions:"soggy sand electrolytes". *Adv. Mater.* 16: 811–814.
Bodale, I. and Filipov, F. (2019). The negative effect of snow melting water infiltration in ploughed layer at the edges of polytunnel. *Int. Multidiscip. Sci. GeoConference SGEM* 19: 539–546.
Brini, F. and Masmoudi, K. (2012). Ion transporters and abiotic stress tolerance in plants. *Int. Sch. Res. Not.* 2012.
Budak, H., Akpinar, B.A., Unver, T., and Turktas, M. (2013). Proteome changes in wild and modern wheat leaves upon drought stress by two-dimensional electrophoresis and nanoLC-ESI–MS/MS. *Plant Mol. Biol.* 83: 89–103.
Carillo, P., Annunziata, M.G., Pontecorvo, G. et al. (2011). Salinity stress and salt tolerance. *Abiotic Stress plants–Mechanisms Adapt.* 1: 21–38.
Cereijido, M., Larre, I., Paez, O., Shoshani, L., and Ponce, A. (2016). "Na+/K+-ATPase drives most asymmetric transports and modulates the phenotype of epithelial cells," in Hamilton,

K., Devor, D. (eds) *Ion Channels and Transporters of Epithelia in Health and Disease* (Springer, New York, NY), 351–374. https://doi.org/10.1007/978-1-4939-3366-2_10

Chanwala, J., Satpati, S., Dixit, A. et al. (2020). Genome-wide identification and expression analysis of WRKY transcription factors in pearl millet (Pennisetum glaucum) under dehydration and salinity stress. *BMC Genomics* 21: 1–16.

Choudhary, O. P. (2017). "Use of amendments in ameliorating soil and water sodicity," in Arora, S., Singh, A., Singh, Y. (eds) *Bioremediation of Salt Affected Soils: An Indian Perspective* (Springer, Cham), 195–210. https://doi.org/10.1007/978-3-319-48257-6_10

Christmann, A., Grill, E., and Huang, J. (2013). Hydraulic signals in long-distance signaling. *Curr. Opin. Plant Biol.* 16: 293–300.

Colmenero-Flores, J.M., Arbona, V., Morillon, R., and Gómez-Cadenas, A. (2020). Salinity and water deficit. In: *The Genus Citrus* (ed. M. Talon, M. Caruso and F.G. Gmitter), 291–309. Woodhead Publishing.

Dawood, M.A.O., Koshio, S., Fadl, S.E. et al. (2020). The modulatory effect of mannanoligosaccharide on oxidative status, selected immune parameters and tolerance against low salinity stress in red sea bream (Pagrus major). *Aquac. Reports* 16: 100278.

Deinlein, U., Stephan, A.B., Horie, T. et al. (2014). Plant salt-tolerance mechanisms. *Trends Plant Sci.* 19: 371–379.

Devi, E.L., Kumar, S., Singh, T.B. et al. (2017). Adaptation strategies and defence mechanisms of plants during environmental stress. In: *Medicinal Plants and Environmental Challenges* (ed. M. Ghorbanpour and A. Varma), 359–413. Cham: Springer. https://doi.org/10.1007/978-3-319-68717-9_2.

Dhiman, P., Rajora, N., Bhardwaj, S. et al. (2021). Fascinating role of silicon to combat salinity stress in plants: an updated overview. *Plant Physiol. Biochem.* 162: 110–123.

Dobrev, P.I. and Vankova, R. (2012). Quantification of abscisic acid, cytokinin, and auxin content in salt-stressed plant tissues. In: *Plant Salt Tolerance. Methods in Molecular Biology*, vol. 913 (ed. S. Shabala and T. Cuin), 251–261. Totowa, NJ: Humana Press https://doi.org/10.1007/978-1-61779-986-0_17.

Douglas, A.E. (2015). Multiorganismal insects: diversity and function of resident microorganisms. *Annu. Rev. Entomol.* 60: 17.

Dudley, L.M., Ben-Gal, A., and Shani, U. (2008). Influence of plant, soil, and water on the leaching fraction. *Vadose Zo. J.* 7: 420–425.

Egamberdieva, D. (2009). Alleviation of salt stress by plant growth regulators and IAA producing bacteria in wheat. *Acta Physiol. Plant.* 31: 861–864.

Egamberdieva, D., Jabborova, D., and Berg, G. (2016). Synergistic interactions between Bradyrhizobium japonicum and the endophyte Stenotrophomonas rhizophila and their effects on growth, and nodulation of soybean under salt stress. *Plant Soil* 405: 35–45.

Egamberdieva, D., Wirth, S., and Abd_Allah, E.F. (2018). Plant hormones as key regulators in plant-microbe interactions under salt stress. In: *Plant Microbiome: Stress Response. Microorganisms for Sustainability* (ed. D. Egamberdieva and P. Ahmad), 165–182. Singapore: Springer https://doi.org/10.1007/978-981-10-5514-0_7.

Ejaz, M., Ali, M., Hasan, H. et al. (2020). Chapter 5 – Salt-regulating genes in wheat. In: *Climate Change and Food Security with Emphasis on Wheat* (ed. M. Ozturk and A. Gul), 77–91. Academic Press.

Eldakak, M., Milad, S.I., Nawar, A.I., and Rohila, J.S. (2013). Proteomics: a biotechnology tool for crop improvement. *Front. Plant Sci.* 4: 35.

El-Esawi, M.A., Sinha, R.P., Chauhan, D.K. et al. (2020). Chapter 35 – Role of ionomics in plant abiotic stress tolerance. In: *Plant Life Under Changing Environment* (ed. D.K. Tripathi, V. Pratap Singh, D.K. Chauhan, et al.), 835–860. Academic Press.

El-Maarouf-Bouteau, H. and Bailly, C. (2008). Oxidative signaling in seed germination and dormancy. *Plant Signal. Behav.* 3: 175–182.

Esitken, A. (2011). Use of plant growth promoting rhizobacteria in horticultural crops. In: *Bacteria in Agrobiology: Crop Ecosystems* (ed. D. Maheshwari), 189–235. Berlin, Heidelberg: Springer https://doi.org/10.1007/978-3-642-18357-7_8.

Ezlit, Y. D., Smith, R. J., and Raine, S. R. (2010). A Review of Salinity and Sodicity in Irrigation. Project Report. University of Southern Queensland, Toowoomba, Australia.

Fahad, S., Hussain, S., Bano, A. et al. (2015). Potential role of phytohormones and plant growth-promoting rhizobacteria in abiotic stresses: consequences for changing environment. *Environ. Sci. Pollut. Res.* 22: 4907–4921.

Farley, H.M. and Smith, Z.A. (2020). *Sustainability: If It's Everything, Is It Nothing?* Routledge.

Farooqi, Z.U.R., Ayub, M.A., ur Rehman, M.Z. et al. (2020). Regulation of drought stress in plants. In: *Plant Life Under Changing Environment* (ed. D.K. Tripathi, V. Pratap Singh, D.K. Chauhan, et al.), 77–104. Academic Press.

Flowers, T.J., Glenn, E.P., and Volkov, V. (2019). Could vesicular transport of Na+ and Cl–be a feature of salt tolerance in halophytes? *Ann. Bot.* 123: 1–18.

Frukh, A., Ahmad, A., and Siddiqi, T.O. (2019). Chapter 31 – Proteomics insights into salt stress signaling in plants. In: *Plant Signaling Molecules*, 479–497. Woodhead Publishing.

Galvani, A. (2006). "The challenge of the food sufficiency through salt tolerant crops," in Amils, R., Ellis-Evans, C., Hinghofer-Szalkay, H. (eds) *Life in Extreme Environments* (Springer, Dordrecht), 437–450. https://doi.org/10.1007/978-1-4020-6285-8_28

Gao, F., Dai, C., and Liu, X. (2010). Mechanisms of fungal endophytes in plant protection against pathogens. *African J. Microbiol. Res.* 4: 1346–1351.

García, J.E., Maroniche, G., Creus, C. et al. (2017). in vitro PGPR properties and osmotic tolerance of different Azospirillum native strains and their effects on growth of maize under drought stress. *Microbiol. Res.* 202: 21–29.

Gawel, L.J. (2006). *A Guide for Remediation of Salt/Hydrocarbon Impacted Soil*. Bismarck, ND: North Dakota Industrial Commission, Department of Mineral Resources.

Geise, G.M., Falcon, L.P., Freeman, B.D., and Paul, D.R. (2012). Sodium chloride sorption in sulfonated polymers for membrane applications. *J. Memb. Sci.* 423: 195–208.

Ghosh, D. and Xu, J. (2014). Abiotic stress responses in plant roots: a proteomics perspective. *Front. Plant Sci.* 5: 6.

Gopalapillai, Y., Hale, B., and Vigneault, B. (2013). Effect of major cations (Ca2+, Mg2+, Na+, K+) and anions (SO, Cl−, NO) on Ni accumulation and toxicity in aquatic plant (Lemna minor L.): Implications For Ni risk assessment. *Environ. Toxicol. Chem.* 32: 810–821.

Guerriero, G., Berni, R., Muñoz-Sanchez, J.A. et al. (2018). Production of plant secondary metabolites: examples, tips and suggestions for biotechnologists. *Genes (Basel)* 9: 309.

Harb, A., Krishnan, A., Ambavaram, M.M.R., and Pereira, A. (2010). Molecular and physiological analysis of drought stress in Arabidopsis reveals early responses leading to acclimation in plant growth. *Plant Physiol.* 154: 1254–1271.

Hardoim, P.R., de Carvalho, T.L.G., Ballesteros, H.G.F. et al. (2019). Genome-wide transcriptome profiling provides insights into the responses of maize (Zea mays L.) to diazotrophic bacteria. *Plant Soil* 451 (1): 121–143.

Hasanuzzaman, M., Nahar, K., Alam, M. et al. (2013). Physiological, biochemical, and molecular mechanisms of heat stress tolerance in plants. *Int. J. Mol. Sci.* 14: 9643–9684.

Hasegawa, P.M. (2013). Sodium (Na+) homeostasis and salt tolerance of plants. *Environ. Exp. Bot.* 92: 19–31.

Hashiguchi, A., Ahsan, N., and Komatsu, S. (2010). Proteomics application of crops in the context of climatic changes. *Food Res. Int.* 43: 1803–1813.

Hayat, R., Ali, S., Amara, U. et al. (2010). Soil beneficial bacteria and their role in plant growth promotion: a review. *Ann. Microbiol.* 60: 579–598.

Iqbal, A., Fahad, S., Iqbal, M. et al. (2020). Special adaptive features of plant species in response to drought. In: *Salt and Drought Stress Tolerance in Plants. Signaling and Communication in Plants* (ed. M. Hasanuzzaman and M. Tanveer), 77–118. Cham: Springer. https://doi.org/10.1007/978-3-030-40277-8_4.

Isayenkov, S.V. and Maathuis, F.J.M. (2019). Plant salinity stress: many unanswered questions remain. *Front. Plant Sci.* 10: 80.

Jamil, A., Riaz, S., Ashraf, M., and Foolad, M.R. (2011). Gene expression profiling of plants under salt stress. *CRC. Crit. Rev. Plant Sci.* 30: 435–458.

Javid, M.G., Sorooshzadeh, A., Moradi, F. et al. (2011). The role of phytohormones in alleviating salt stress in crop plants. *Aust. J. Crop Sci.* 5: 726–734.

Jull, L.G. (2009). Winter salt injury and salt-tolerant landscape plants. *Univ. Wisconsin Coop. Ext A* 3877.

Kamran, M., Parveen, A., Ahmar, S. et al. (2020). An overview of hazardous impacts of soil salinity in crops, tolerance mechanisms, and amelioration through selenium supplementation. *Int. J. Mol. Sci.* 21: 148.

Kaya, C., Higgs, D., Ashraf, M. et al. (2020). Integrative roles of nitric oxide and hydrogen sulfide in melatonin-induced tolerance of pepper (Capsicum annuum L.) plants to iron deficiency and salt stress alone or in combination. *Physiol. Plant.* 168: 256–277.

Keisham, M., Mukherjee, S., and Bhatla, S.C. (2018). Mechanisms of sodium transport in plants—progresses and challenges. *Int. J. Mol. Sci.* 19: 647.

Kerins, M.J. and Ooi, A. (2018). The roles of NRF2 in modulating cellular iron homeostasis. *Antioxid. Redox Signal.* 29: 1756–1773.

Khaled, H. and Fawy, H.A. (2011). Effect of different levels of humic acids on the nutrient content, plant growth, and soil properties under conditions of salinity. *Soil Water Res.* 6: 21–29.

Khan, N.A., Nazar, R., Iqbal, N., and Anjum, N.A. (2012). *Phytohormones and Abiotic Stress Tolerance in Plants.* Springer Science & Business Media.

Khan, M.M.A., Khatun, A., and Islam, M.T. (2016). *Promotion of Plant Growth by Phytohormone Producing Bacteria.* New York, NY: *Microbes Action; Nov. Sci. Publ.*

Khorshidi, M.B., Yarnia, M., and Hassanpanah, D. (2009). Salinity effect on nutrients accumulation in alfalfa shoots in hydroponic condition. *J. Food Agric. Env.* 7: 787–790.

Kim, S.K., Son, T.K., Park, S.Y. et al. (2006). Influences of gibberellin and auxin on endogenous plant hormone and starch mobilization during rice seed germination under salt stress. *J. Environ. Biol.* 27: 181.

Kosová, K., Vítámvás, P., Prášil, I.T., and Renaut, J. (2011). Plant proteome changes under abiotic stress—contribution of proteomics studies to understanding plant stress response. *J. Proteomics* 74: 1301–1322.

Kumar, I.S., Rao, S.R., and Vardhini, B.V. (2015). Role of phytohormones during salt stress tolerance in plants. *Curr. Trends Biotechnol. Pharm.* 9: 334–343.

Lambers, H., Chapin, F.S. III, and Pons, T.L. (2008). *Plant Physiological Ecology*. Springer Science & Business Media.

Läuchli, A. and Grattan, S.R. (2007). Plant growth and development under salinity stress. In: *Advances in Molecular Breeding Toward Drought and Salt Tolerant Crops* (ed. M.A. Jenks, P.M. Hasegawa and S.M. Jain), 1–32. Dordrecht: Springer https://doi.org/10.1007/978-1-4020-5578-2_1.

Lepik, M., Liira, J., and Zobel, K. (2004). The space-use strategy of plants with different growth forms, in a field experiment with manipulated nutrients and light. *Folia Geobot.* 39: 113–127.

Lewis, R. (2002). Using transgenesis to create salt-tolerant plants: the hope is to grow crops where none could thrive before. *Sci.* 16: 26–28.

Lipiec, J., Doussan, C., Nosalewicz, A., and Kondracka, K. (2013). Effect of drought and heat stresses on plant growth and yield: a review. *Int. Agrophysics* 27: 463–477.

Maathuis, F.J.M., Ahmad, I., and Patishtan, J. (2014). Regulation of Na+ fluxes in plants. *Front. Plant Sci.* 5: 467.

Meldrum, B.S. and Rogawski, M.A. (2007). Molecular targets for antiepileptic drug development. *Neurotherapeutics* 4: 18–61.

Mensah, J.K. and Ihenyen, J. (2009). Effects of salinity on germination, seedling establishment and yield of three genotypes of mung bean (Vigna mungo L. Hepper) in Edo State, Nigeria. *Nig Ann Nat. Sci* 8: 17–24.

Mercado-Blanco, J., Abrantes, I., Barra Caracciolo, A. et al. (2018). Belowground microbiota and the health of tree crops. *Front. Microbiol.* 9: 1006.

Miller, G.A.D., Suzuki, N., Ciftci-Yilmaz, S., and Mittler, R.O.N. (2010). Reactive oxygen species homeostasis and signalling during drought and salinity stresses. *Plant. Cell Environ.* 33: 453–467.

Miller, R., Owens, S.J., and Rørslett, B. (2011). Plants and colour: flowers and pollination. *Opt. Laser Technol.* 43: 282–294.

Minhas, P.S., Ramos, T.B., Ben-Gal, A., and Pereira, L.S. (2020). Coping with salinity in irrigated agriculture: Crop evapotranspiration and water management issues. *Agric. Water Manag.* 227: 105832.

Mishra, J., Fatima, T., and Arora, N.K. (2018). Role of secondary metabolites from plant growth-promoting rhizobacteria in combating salinity stress. In: *Plant Microbiome: Stress Response. Microorganisms for Sustainability*, vol. 5 (ed. D. Egamberdieva and P. Ahmad), 127–163. Singapore: Springer https://doi.org/10.1007/978-981-10-5514-0_6.

Mittler, R. (2006). Abiotic stress, the field environment and stress combination. *Trends Plant Sci.* 11: 15–19.

Munns, R., Husain, S., Rivelli, A.R., et al. (2002). Avenues for increasing salt tolerance of crops, and the role of physiologically based selection traits, in Walter J. Horst (eds), *Progress in Plant Nutrition: Plenary Lectures of the XIV International Plant Nutrition Colloquium. Developments in Plant and Soil Sciences*, vol 98 (Springer, Hannover, Germany), 93–105. https://doi.org/10.1007/978-94-017-2789-1_7.

Munns, R., James, R.A., Xu, B. et al. (2012). Wheat grain yield on saline soils is improved by an ancestral Na+ transporter gene. *Nat. Biotechnol.* 30: 360–364.

Munns, R., Day, D.A., Fricke, W. et al. (2020). Energy costs of salt tolerance in crop plants. *New Phytol.* 225: 1072–1090.

Nissim-Levi, A., Ovadia, R., Forer, I., and Oren-Shamir, M. (2007). Increased anthocyanin accumulation in ornamental plants due to magnesium treatment. *J. Hortic. Sci. Biotechnol.* 82: 481–487.

Öktem, H.A., Eyidogan, F., Selçuk, F. et al. (2006). Osmotic stress tolerance in plants: transgenic strategies. In: *Floriculture, Ornamental and Plant Biotechnology*, vol. 3, 194–208. Global Science Books UK.

Peinelt, C. and Apell, H.-J. (2002). Kinetics of the Ca2+, H+, and Mg2+ interaction with the ion-binding sites of the SR Ca-ATPase. *Biophys. J.* 82: 170–181.

Peleg, Z. and Blumwald, E. (2011). Hormone balance and abiotic stress tolerance in crop plants. *Curr. Opin. Plant Biol.* 14: 290–295.

Peng, H.-Y., Qi, Y.-P., Lee, J. et al. (2015). Proteomic analysis of Citrus sinensis roots and leaves in response to long-term magnesium-deficiency. *BMC Genomics* 16: 1–24.

Penumarthi, A., Basak, P., Smooker, P., and Shukla, R. (2020). Hitching a ride: enhancing nucleic acid delivery into target cells through nanoparticles. In: *Nanoscience in Medicine Vol 1. Environmental Chemistry for a Sustainable World*, vol. *39* (ed. H. Daima, N. PN, S. Ranjan, et al.), 373–457. Cham: Springer https://doi.org/10.1007/978-3-030-29207-2_11.

Pérez-Alfocea, F., Ghanem, M.E., Gómez-Cadenas, A., and Dodd, I.C. (2011). Omics of root-to-shoot signaling under salt stress and water deficit. *Omi. A J. Integr. Biol.* 15: 893–901.

Qadir, M., Oster, J.D., Schubert, S. et al. (2007). Phytoremediation of sodic and saline-sodic soils. *Adv. Agron.* 96: 197–247.

Raney, W.A. (1960). The dominant role of nitrogen in leaching losses from soils in humid regions. *Agron. J.* 52: 563–566.

Reddy, I.N.B.L., Kim, B.-K., Yoon, I.-S. et al. (2017). Salt tolerance in rice: focus on mechanisms and approaches. *Rice Sci.* 24: 123–144.

Reguera, M., Peleg, Z., and Blumwald, E. (2012). Targeting metabolic pathways for genetic engineering abiotic stress-tolerance in crops. *Biochim. Biophys. Acta (BBA)-Gene Regul. Mech.* 1819: 186–194.

Reynolds, M. and Tuberosa, R. (2008). Translational research impacting on crop productivity in drought-prone environments. *Curr. Opin. Plant Biol.* 11: 171–179.

Rolston, H. (2020). New environmental ethics: the next millennium for life on Earth, A-(2012) and 2nd edition.

Roychoudhury, A. and Tripathi, D.K. (2019). *Molecular Plant Abiotic Stress: Biology and Biotechnology*, vol. 1, 1–480. John Wiley & Sons.

Saharan, B.S. and Nehra, V. (2011). Plant growth promoting rhizobacteria: a critical review. *Life Sci. Med. Res. 21* (1): 30.

Sairam, R.K. and Tyagi, A. (2004). Physiology and molecular biology of salinity stress tolerance in plants. *Curr. Sci.* 86 (3): 407–421. http://www.jstor.org/stable/24108735.

Salehi-Lisar, S.Y. and Bakhshayeshan-Agdam, H. (2020). Agronomic crop responses and tolerance to drought stress. In: *Agronomic Crops* (ed. M. Hasanuzzaman), 63–91. Springer.

Samaddar, S., Chatterjee, P., Choudhury, A.R. et al. (2019). Interactions between Pseudomonas spp. and their role in improving the red pepper plant growth under salinity stress. *Microbiol. Res.* 219: 66–73.

Satisha, J., Laxman, R.H., Upreti, K.K. et al. (2020). Mechanisms of abiotic stress tolerance and their management strategies in fruit crops. In: *New Frontiers in Stress Management for Durable Agriculture* (ed. A. Rakshit, H.B. Singh, A.K. Singh, et al.), 579–607. Springer.

Sharma, A., Shahzad, B., Kumar, V. et al. (2019). Phytohormones regulate accumulation of osmolytes under abiotic stress. *Biomolecules* 9: 285.

She, Q., Wang, R., Fane, A.G., and Tang, C.Y. (2016). Membrane fouling in osmotically driven membrane processes: a review. *J. Memb. Sci.* 499: 201–233.

Shrivastava, P. and Kumar, R. (2015). Soil salinity: A serious environmental issue and plant growth promoting bacteria as one of the tools for its alleviation. *Saudi J. Biol. Sci.* 22: 123–131.

da Silva, E.C., Nogueira, R.J.M.C., de Araújo, F.P. et al. (2008). Physiological responses to salt stress in young umbu plants. *Environ. Exp. Bot.* 63: 147–157.

Soda, N., Kushwaha, H.R., Soni, P. et al. (2013). A suite of new genes defining salinity stress tolerance in seedlings of contrasting rice genotypes. *Funct. Integr. Genomics* 13: 351–365.

Tester, M. and Davenport, R. (2003). Na+ tolerance and Na+ transport in higher plants. *Ann. Bot.* 91: 503–527.

Upadhyaya, H., Sahoo, L., and Panda, S.K. (2013). Molecular physiology of osmotic stress in plants. In: *Molecular Stress Physiology of Plants* (ed. G. Rout and A. Das), 179–192. India: Springer https://doi.org/10.1007/978-81-322-0807-5_7.

Vardhini, B.V. (2016). Brassinosteroids are potential ameliorators of heavy metal stresses in plants. In: *Plant Metal Interaction* (ed. P. Ahmad), 209–237. Elsevier.

Wahid, A., Arshad, M., and Farooq, M. (2009). Cadmium phytotoxicity: responses, mechanisms and mitigation strategies: a review. In: *Organic Farming, Pest Control and Remediation of Soil Pollutants. Sustainable Agriculture Reviews*, vol. 1 (ed. E. Lichtfouse), 371–403. Dordrecht: Springer https://doi.org/10.1007/978-1-4020-9654-9_17.

Wu, S.G., Huang, L., Head, J. et al. (2012). Phytotoxicity of metal oxide nanoparticles is related to both dissolved metals ions and adsorption of particles on seed surfaces. *J Pet Env. Biotechnol* 3: 126.

Xiong, L. and Zhu, J. (2002). Molecular and genetic aspects of plant responses to osmotic stress. *Plant. Cell Environ.* 25: 131–139.

Xiong, Y.-W., Li, X.-W., Wang, T.-T. et al. (2020). Root exudates-driven rhizosphere recruitment of the plant growth-promoting rhizobacterium Bacillus flexus KLBMP 4941 and its growth-promoting effect on the coastal halophyte Limonium sinense under salt stress. *Ecotoxicol. Environ. Saf.* 194: 110374.

Yadav, S., Irfan, M., Ahmad, A., and Hayat, S. (2011). Causes of salinity and plant manifestations to salt stress: a review. *J. Environ. Biol.* 32: 667.

Yang, Z.-B., Rao, I.M., and Horst, W.J. (2013). Interaction of aluminium and drought stress on root growth and crop yield on acid soils. *Plant Soil* 372: 3–25.

Zhang, L. and Webster, T.J. (2009). Nanotechnology and nanomaterials: promises for improved tissue regeneration. *Nano Today* 4: 66–80.

Zhang, J.-L., Flowers, T.J., and Wang, S.-M. (2010). Mechanisms of sodium uptake by roots of higher plants. *Plant Soil* 326: 45–60.

Zhao, Q., Nakashima, J., Chen, F. et al. (2013). Laccase is necessary and nonredundant with peroxidase for lignin polymerization during vascular development in Arabidopsis. *Plant Cell* 25: 3976–3987.

Zhu, M., Dai, S., McClung, S. et al. (2009). Functional differentiation of Brassica napus guard cells and mesophyll cells revealed by comparative proteomics. *Mol. Cell. Proteomics* 8: 752–766.

Zhu, Z., Zhang, Y., Liu, J. et al. (2018). Exploring the effects of selenium treatment on the nutritional quality of tomato fruit. *Food Chem.* 252: 9–15.

Zörb, C., Geilfus, C., and Dietz, K. (2019). Salinity and crop yield. *Plant Biol.* 21: 31–38.

14

Selenium Species in Plant Life: Uptake, Transport, Metabolism, and Biochemistry

Zaid Ulhassan[1], Ali Raza Khan[1], Wardah Azhar[1], Yasir Hamid[2], Durgesh Kumar Tripathi[3], and Weijun Zhou[1]

[1] *Institute of Crop Science, Ministry of Agriculture and Rural Affairs Key Laboratory of Spectroscopy Sensing, Zhejiang University, Hangzhou, China*
[2] *Ministry of Education (MOE) Key Lab of Environ. Remediation and Ecol. Health, College of Environmental and Resources Science, Zhejiang University, Hangzhou, China*
[3] *The Amity Institute of Organic Agriculture, Amity University Uttar Pradesh, Noida, India*

Selenium Speciation in the Soil-Plant System

In plants, selenium (Se) exists in different forms such as nano-selenium (Se°), elemental Se [selenide (Se^{-2})], inorganic Se [selenite (Se^{+4}) and selenate (Se^{+6})], and organic Se [selenomethionine (SeMet), selenocysteine (SeCys), methylselenol, and Se-methylselenocysteine (MetSeCys)] (Wu et al. 2015). The distribution, accumulation, and transport of these Se species within plant tissues are adjusted by particular plant (Yin et al. 2019). In soils, lower adsorption capacity of Se^{+6} leads to higher water solubility as compared to Se^{+4} (Mayland et al. 1991). In this way, Se^{+4} is dominant species in acidic soils or reducing conditions and alkaline soils or oxidizing conditions favour Se^{+6}. Therefore, there is higher bioavailability/mobility in oxidizing conditions and lower mobility in reducing conditions. The distribution of geochemical or residual fractions such as exchangeable ions, carbonates, and iron or manganese oxides in soils depend on the difference between bioavailability and mobility of Se species (Shaheen et al. 2018; Qin et al. 2017), ultimately affecting the Se-transport in soil-plant system.

Accumulation and Uptake of Selenium Species by Plants

The availability of Se in soil and target plant control the accumulation of inorganic/organic Se species. Based on the accumulation capacity, plants are categorized into hyperaccumulators (>1000 μg Se g^{-1} DW) such as *Astragalus* (Fabaceae), secondary accumulators (≤1000 μg Se g^{-1} DW) such as *Brassica juncea*, and non-hyperaccumulators (<100 μg Se g^{-1} DW)

Beneficial Chemical Elements of Plants: Recent Developments and Future Prospects, First Edition. Edited by Sangeeta Pandey, Durgesh Kumar Tripathi, Vijay Pratap Singh, Shivesh Sharma, and Devendra Kumar Chauhan.

(Terry et al. 2000; Galeas et al. 2007; Pilon-Smits and LeDuc 2009). The difference in the accumulation capacity of Se levels among hyperaccumulators and non-hyperaccumulators may be due to diverse kinetic characteristics, transporters regulation, and active sites (El Mehdawi et al. 2018). Hyperaccumulators greatly convert Se into organic forms and gain more organic Se forms in grains or reproduction systems, while non-hyperaccumulators contain more inorganic Se forms in shoots or seeds (Schiavon and Pilon-Smits 2017a) and may not cause toxic effects up to certain Se dose.

Mostly, plants uptake Se via inorganic Se such as Se^{+4} and Se^{+6} and organic Se such as SeMet and SeCys species (Gupta and Gupta 2016). Within plant species, these Se speciation may differ in absorption capacity from soils and passage pathways. Plants exploit small amount of Se from soils and diverse solubility levels for different Se species. Elemental Se forms are not taken by plants due to their water insolubilities, while organic forms have relatively higher solubilities and are easily assimilated by plants (Natasha et al. 2018). Thus, organic Se forms can be utilized in fertilizer applications. The greater affinity of plants for Se^{+6} resulted in greater assimilation rate as compared to Se^{+4}. Mainly, Se^{+6} uptake by plants from soils is higher as compared to Se^{+4} and organic Se forms (White et al. 2004). Thus, Se uptake from soil to plant system depends upon the type of Se form and transfer rate.

Inorganic Se species (Se^{+4} and Se^{+6}) are readily available to plants as compared to other forms. Se^{+4} competes for phosphorus ions for their uptake by plant roots; thus it transports via phosphate transporters and utilizes passive transport for absorption (Li et al. 2008; Wan et al. 2016; Ulhassan et al. 2019b). Earlier studies reported the involvement of phosphate transporters such as *OsPT2* (Zhang et al. 2014a) in rice, (Ulhassan et al. 2019b) in brassica napus and *OsPT8* (Song et al. 2017) in tobacco. In addition, a silicon influx transporter *OsNIP2;1* was found to enhance the uptake of Se^{+4} by rice under acidic soil conditions (Zhao et al. 2010). Studies concerning the exact uptake or transport mechanisms of inorganic Se species are still needed for further investigations in different plant species under varied environmental conditions.

Se^{+6} carried out absorption via active transport system and competes with sulphate (S) for its uptake by plants that suggested the involvement of sulphate transporters in its transport mechanism (Gupta and Gupta 2016). While, S deficiency enhances the Se^{+6} uptake by plants that proposed that S deficiency aggravated the S-transporter transcripts that boosts the Se^{+6} transport (Li et al. 2008). Target plants decide the transport selectivity between Se^{+6} and S as both compete with each other for transportation (White 2016). First report confirmed the participation of high-affinity sulphate transport (SULTR1; 2) in facilitating the Se^{+6} uptake via *Arabidopsis thaliana* roots (Shibagaki et al. 2002). Later, SULTR1; 1 was also involved in the transport of Se^{+6} into plant parts (Barberon et al. 2008). Because these S-transporters are transported through xylem, higher Se^{+6} levels were noticed in xylem sap. Then, Se^{+6} is converted to organic Se forms and transported to reproductive organs, but transporters for organic Se forms are not identified yet.

S-transporters related to influx (SULTR3; 1) and efflux (SULTR4; 1, SULTR4; 2) are involved in the transport of Se^{+6} into plastids and extrusion from vacuoles (Cao et al. 2013; White 2016). These findings suggested the possible involvement of S-transporters in the transport mechanisms of all plant species (probably). Though, mechanistic approaches by which SULTR transporters choose/prefer particular ions for their transport are still lacking. Overall, Se^{+6} and Se^{+4} uptake by plants is improved by low concentrations of S and P in soil, facilitating plants to enhance the sulphate and phosphorus transporter genes and higher nutrient availability. Eventually, bioavailability of Se^{+6} and Se^{+4} would be maximum in the absence of S and P (Li et al. 2008).

Transport Mechanisms of Selenium Species within Plants

The transportation of inorganic Se species (Se^{+6} and Se^{+4}) from root to aerial parts is carried out by xylem. Root symplastic plasma membrane takes part in the translocation of Se^{+6} to cytoplasm and then to selenide, while Se^{+4} translocation in internal organelles is carried out by apoplast-xylem transport mechanisms and remained mostly in roots without chemical transformation (Pyrzynska 2009). This suggested the abundance of Se^{+6} in xylem and higher transport potential as compared to Se^{+4}.

Organic Se forms are mostly transported by xylem and phloem (Carey et al. 2012). Usually, Se^{+6} remains in roots and is converted into organic Se forms such as SeMet and SeCys or related volatile compounds via unknown mechanisms. The exploration of transporters for organic Se forms is obligatory to fully understand the translocation mechanisms for Se^{+6} in different plants. Only few studies proposed that organic Se form may transport via amino acid transporters (Kikkert and Berkelaar 2013), but no confirmed reports are documented. Recent study revealed the participation of *NRT1.1B*, rice peptide transporter (PTR) family, in the transport of organic Se forms to aerial plant parts that ultimately improves the grains yield (Zhang et al. 2019). But the molecular mechanisms for the translocation of organic Se forms in different plants and varied growth stages are still lacking.

Earlier reports also confirmed the greater accumulation of Se^{+4} in root tissues (Li et al. 2019; Kaur and Sharma 2018; Ulhassan et al. 2018, 2019a, b, c), and higher accumulation of Se^{+6} or organic Se forms was observed in grains or aerial parts (Deng et al. 2017; Gong et al. 2018). The uptake and transport mechanisms involved in the uptake of Se species via different transporters in different plant organs are explained in Figure 14.1. These findings may alter with the type of Se form, plant species, Se^{+6} to organic Se conversion rate, or environmental stress conditions. The pattern and abundance of inorganic/organic Se transportation may vary with respect to different organs at different growth stages. These outcomes proposed that type of Se form greatly affects the distribution of Se species in plant parts and alters with the target plant species.

Selenium Metabolism in Plants

Four steps are generally involved in Se metabolism by plants.

Step 1: Conversion of Selenate into Selenite and Selenide

The reduction of selenate to elemental Se is carried out with the contribution of ATP sulphurylase (APS) and phosphosulphatereductase (APR) enzymes (Gupta and Gupta 2017).

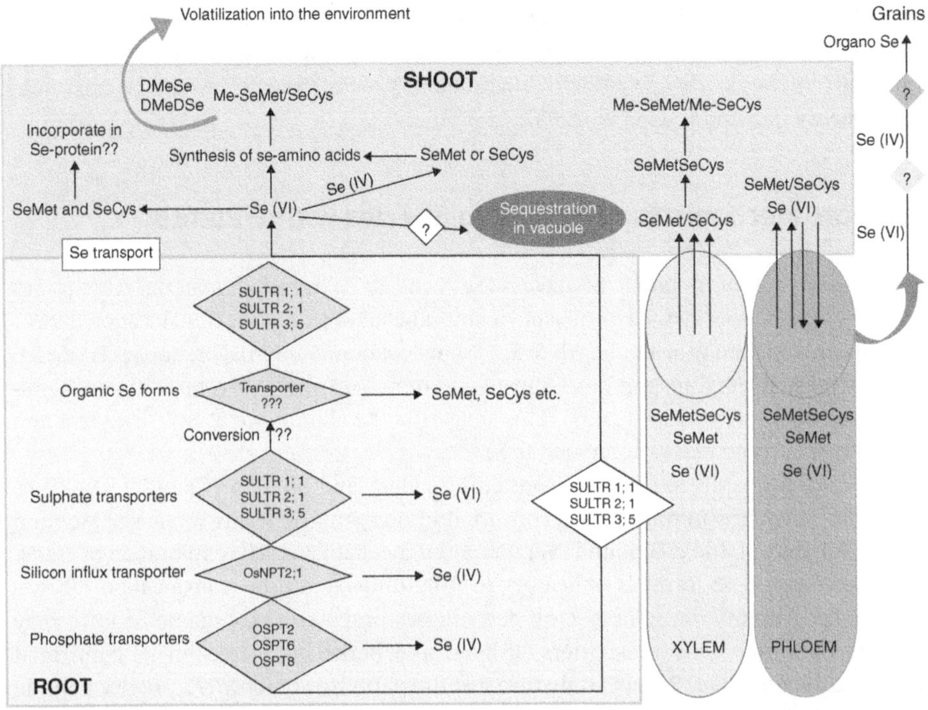

Figure 14.1 Diagram displays the involvement of transporters in the uptake and transport of Se species in different plant organs via xylem and phloem. Phosphate transporters (OSPT2, OSPT6, OSPT8) and silicon influx transporter (OsNPT 2;1) directs Se (IV) in plant tissue. Sulphate transporters (SULTR 1;1, SULTR 2;1, SULTR 3;5) are most probably involved in the uptake of Se (VI) into aerial plant parts where it is converted into organic forms and then volatilized (DMeSe/DMeDSe) into the environment. The identification of exact transporters for organic Se forms are still not explored.

Initially, ATP is formed with the combination of selenate and APS, and then adenosine phosphoselenate (APSe) is formed (Sors et al. 2005a). Then, reduction of APSe to selenate with the help of APR takes place (Pilon-Smits and Quinn 2010). In parallel, APSe is reduced to phosphoadenosine phosphoselenate (PAPSe) and facilitates in the production of secondary metabolites (Pilon-Smits 2015). The reduction process for selenate is carried out in chloroplasts, while these contributing enzymes exist in chloroplast and cytosols. Later on, selenate to selenide conversion is done with the successive contribution of sulphite reductase (SiR) or enzymes glutathione (GSH)/glutaredoxins (GRXs) (Wallenberg et al. 2010).

Step 2: Selenide to Selenocysteine (SeCys) Transformation

The products from serine acetyl transferase (SAT) and *O*-acetylserinethiollyase (OAS-TL) produce cysteine synthase (CS) enzyme, which is involved in the conversion of selenide to selenocysteine (SeCys) (Pilon-Smits and Womg 2012). Selenide to SeCys conversion is catalyzed by CS through the contribution of OAS. At the same time, SeCys is formed as the result of uninterrupted actions of selenomethyltransferase (SMT) (Pilon-Smits and Quinn 2010). The conversion of selenite to SeCys is carried out with the formation of selenopersulphide/glutathionyl-selenol (GS-SeH) with the help of NADPH as a reducing agent and eventually converted GS-SeH to SeCys (Terry et al. 2000).

Step 3: Transformation of Selenocysteine (SeCys) into Elemental Se^0 and Alanine (Ala)

The transformation of SeCys to Se^0 and Ala via SeCys-lyase (SL) or methyl-SeCys (MeSeCys) is carried out with SeCys lyase or SMT along with translation to SeMet (Gupta and Gupta 2017). Whereas these enzymatic transformations depend upon the target plant species and involvement of environmental factors. One metabolic pathway of SeCys is the methylation to produce MeSeCys via SeCys methyltransferase (SeCysMT) (Natasha et al. 2018). Then, MeSeCys is converted to methylselenoglutathione (S-MeSeGS) through methionine-γ-lyase. Additional metabolic activity performed by MeSeCys is the production of γ-glutamylmethyl-SeCys (γ-GluMeSeCys) coupled with glutamate (Freeman et al. 2007). Furthermore, MeSeCys is converted to volatile Se compounds such as dimethyl diselenide (DMDSe). SeCys metabolism also involves conjugation with *O*-phosphohomoserine catalyzed by γ-cystathionine synthase and then forms Se-homocysteine (SeHCys) catalyzed by cystathionine-β-lyase. The coupling of SeHCys and OS produced selenohomolan (SeHLan). Eventually, SeCys is carried to cytosol and transformed into SeMet and MeSeMet volatile compounds via the involvement of different enzymes (Pilon-Smits and Quinn 2010).

Step 4: Metabolic Pathways of Methyl Selenomethionine (MeSeMet)

For metabolic activity, two paths are adapted by MeSeMet. In first pathway, MeSeMet is transformed into dimethylselenopropionate (DMSeP) and dimethylselenide (DMSe) by the responses from decarboxylation and transamination. Another path includes the hydrolysis of MeMet into DMSe (Pilon-Smits and Quinn 2010). Precursors of SeCys were converted into non-protein thiols such as MeSeCys, SeCysMet, or γ-GMeSeCys that minimized the production of SeMet (Natasha et al. 2018). Usually, plants either reduce or avoid misincorporation of SeCys or SeMet into unintended selenoproteins, which may impair the protein structure and cause Se phytotoxicities. While hyperaccumulators adapted particular mechanistic approaches to handle SeCys-induced toxicities. They converted toxic SeCys into less toxic organic Se forms/compounds (Freeman et al. 2007). In cytosol, SMT is involved in stimulating the SeCys-induced MeSeCys synthesis and then is converted to methyl-SeCys-selenooxide (MeSeCysSeO). At the end, MeSeCysSeO is dispersed into methaneseleniol (MSe) via the contribution of Cys sulphoxide lyase and finally volatilized into DMDSe and DMSe in hyperaccumulators and non-hyperaccumulators, respectively (Pilon-Smits and Quinn 2010; Figure 14.2).

Biochemistry of Selenium

Is Selenium an Essential Trace Element for Plants?

There are increasing evidences that Se is essential for microalgae, prokaryotes, and animals including mammals (Novoselov et al. 2002; Rayman 2012; Schiavon and Pilon-Smits 2017a, b). Regarding the Se biochemistry for higher plants, its essentiality is still

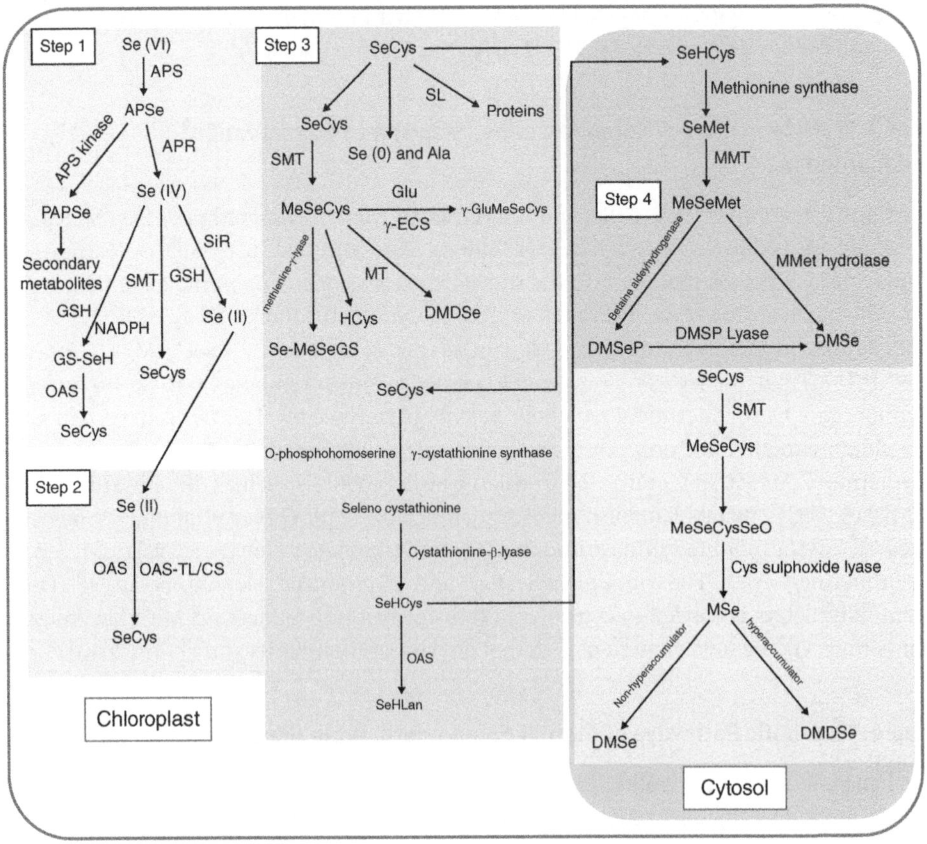

Figure 14.2 Possible steps involved in selenium metabolism by plants. Adapted from Pilon-Smits and Quinn, 2010.

controversial and debatable. Although Se has been regarded as an essential element for plants having higher Se-accumulation capacity and easily assimilates due to chemical similarity with sulphur (Terry et al. 2000). A number of indications suggested that Se is required for Se accumulators that are prevalent in Se-rich soils (Gupta and Gupta 2017; Terry et al. 2000). There is no evidence that Se is required for non-hyperaccumulators (Both et al. 2020). To explore the essentiality of Se in higher plants, many studies have been made to find out whether plants enhance the availability or production of key selenoproteins as previously reported in animals. Following are the important biochemical approaches utilized by plant scientists to uncover the essential Se species and improve the tolerance ability of plants.

Conversion of Inorganic to Organic Selenium Forms (The First Step of the Se-Assimilation Pathway)

The conversion of inorganic to organic Se forms is carried out by S-assimilation pathways due to the chemical similarity with S (Bulteau and Chavatte 2015; Schiavon and

Pilon-Smits 2017a). This rate-limiting step in Se assimilation into plants is carried out by an enzyme ATP sulphurylase, which pairs selenate/sulphate to ATP and formed adenosine 5′-phosphoselenate (APSe) or 5′-phosphosulphate (APS). ATP sulphurylase is responsible for the conversion of Se^{+6} into Se-amino acids (once Se^{+6} is entered by plant roots) (Pilon-Smits and LeDuc 2009; Schiavon et al. 2015). In *A. thaliana*, the localization of four isoforms of ATP sulphurylase such as APS (1, 3, 4) were identified in plastids (Anjum et al. 2015) and APS2 was localized in both plastids and cytosols (Bohrer et al. 2015). Surprisingly, the elevated expressions of APS2 (over 120 fold) in the roots (and 2–4 fold in leaves) of *S. pinnata* (Se hyperaccumulator) as compared to *Stanleya elata* (Se non-hyperaccumulator) proposed that the overexpression of APS2 may enhance the tolerance or accumulation potential of *S. pinnata* (Wang et al. 2018; Schiavon et al. 2015). It is still unclear whether APS2 prefers selenate or sulphate. This suggested that APS can be effectively utilized in genetic engineering with the aim to improve plants' Se uptake, accumulation, and tolerance potential for biofortification/phytoremediation.

Previously, transgenic studies illustrated that the overexpression of *A. thaliana* APS1 in transgenic *B. juncea* enhanced the Se^{+6} reduction with greater accumulation of organic Se compounds in shoots and roots as compared to higher accumulation of Se^{+6} in the organs of wild-type plants (Pilon-Smits et al. 1999). In another study, overexpression of APS in transgenic *B. juncea* resulted in three times higher Se accumulation as compared to wild types (Huysen et al. 2004). Further field experiments on seleniferous soils confirmed the greater Se accumulation in transgenic *B. juncea* than wild types (Bañuelos et al. 2005). As Se-assimilation pathway proceeds, an enzyme APS reductase (APR) converts APSe to selenite. This reaction occurs in plastids and rate-limiting step in Se^{+6} assimilation (Suter et al. 2000; Sors et al. 2005a).

Earlier transgenic experiments on *A. thaliana* evidenced in backing APR. In *apr2-1* mutant, enhanced levels of Se^{+6} and less concentration of Se^{+4} implicated the importance of APR2 in converting APSe into Se^{+4} (Grant et al. 2011; Chao et al. 2014). In addition, lower S flux from sulphate to reduced S compounds (Chao et al. 2014) confirmed the lower Se^{+6} assimilation. These *apr2-1* mutants also had lowered Se^{+6} tolerance due to decreased glutathione levels (Grant et al. 2011). While overexpression of APR in plants enhanced the Se flux and ultimately higher Se^{+6} assimilation rate to amino acids (Sors et al. 2005a). These results suggested the roles of APR in Se^{+6} assimilation pathways. The catalytic activity of APR2 varies among different *A. thaliana* species and links with the differences in S and Se metabolisms (Chao et al. 2014). No correlation between Se hyperaccumulation and APR transcripts were found among eight *Astragalus* species (Sors et al. 2005b). This proposes that APR corresponds (in terms of Se-assimilation rate) more to non-hyperaccumulators than hyperaccumulators, although thorough investigations are needed.

In next step, Se^{+4} is converted to Se^{2-} proceeded by their incorporation into Se-amino acids. This conversion (Se^{+4} to Se^{2-}) is proposed to occur in the presence of sulphite reductase (SiR) (White 2016; Yarmolinsky et al. 2012) or glutathione reduction via selenodiglutathione (GSSeSG) and selenopersulphide (GSSeH) (Terry et al. 2000). At the end, Se^{2-} is converted to SeCys with the help of cysteine synthase that stimulates the conversion of *O*-acetylserine (OAS) and Se^{2-} to SeCys (Sors et al. 2005a; Terry et al. 2000). While the conversion of SeCys to SeMet is implied by selenocystathione and selenohomocysteine (SeHCys) as intermediates and catalyzed by cystathionine γ-synthase (CGS), *O*-phosphohoroserine

(OPH), and SeCys (Van Huysen et al. 2003). And, it is the rate-limiting step in the conversion of SeCys to volatile Se compounds such as DMSe (Sors et al. 2009; Van Huysen et al. 2003).

Adaptive Mechanisms by Plants to Evade Selenium Toxicity Participation of Se-Amino Acids

Noticeably, Se caused phytotoxicities as the result of misincorporation of Se-amino acids into proteins (Pilon-Smits and Womg 2012; Van Hoewyk et al. 2008). Thus, prevention of misincorporation into proteins is crucial to avoid Se phytotoxicity. Furthermore, inorganic Se anions may cause oxidative stress by depleting the GSH pool and extra generation of superoxide radical that impairs the iron-sulphur clusters, mitochondria, and chloroplast iron-sulphur proteins (Fisher et al. 2016). There is another possibility that Se may misincorporate into iron-sulphur (Fe-S) clusters as the involved enzyme form Fe-S clusters by releasing elemental S from SeCys and thus may utilize SeCys as substrate (Van Hoewyk et al. 2005). One of the tolerance mechanisms exploited by hyperaccumulators is the restricting of misincorporation of Se-amino acids into proteins (Brown and Shrift 1982). Another mechanism used by hyperaccumulators is the methylation of SeCys to methyl SeCys (MeSeCys) through SeCys methyltransferase (SMT) (Pilon-Smits and LeDuc 2009) and ultimately SeCys replaces Cys in proteins and is effectively reduced. SMT is chloroplast localized mainly found in hyperaccumulators and non-hyperaccumulators of genus *Astragalus*. While isoforms of SMT in hyperaccumulators produce MeSeCys and display priority for SeCys methylation over Cys (Neuhierl et al. 1999; Sors et al. 2009). This elucidates the reason behind the higher accumulation of MeSeCys in their organs by Se hyperaccumulators such as *Astragalus bisulcatus* and *S. pinnata* than non-hyperaccumulators containing inorganic Se forms (mainly selenate) (Sors et al. 2005b; Freeman et al. 2006, 2010; Neuhierl et al. 1999; Pickering et al. 2003). Although SMTs (from *A. bisulcatus*) greatly accumulate in hyperaccumulators, but they can also be induced by Se (mainly as MeSeCys and γ-glutamyl-MeSeCys) in non-hyperaccumulators such as *A. thaliana* (LeDuc et al. 2004; Ellis et al. 2004).

Volatilization of Selenium Organic Compounds

Se phytovolatilization is the process by which plants metabolize inorganic/organic Se forms into gaseous forms (e.g. DMSe), and is a key factor to remove Se from contaminated soils (Tagmount et al. 2002). Additional mechanism by which hyperaccumulators tolerate extra accumulation of Se in their tissues is the conversion (by oxidation/methylation) of MeSeCys into volatile compounds such as dimethyldiselenide (DMDSe) (Pilon-Smits and LeDuc 2009). Initially, MeSeCys is converted to methylselenocysteineselenideoxide (MeSeCysSeO) and then transferred into methaneselenol (CH_3SeH) with the help of Cys sulphoxide lyase (Chin and Lindsay 1994; Ellis and Salt 2003). The production of DMDSe mainly occurred in leaves (Ellis and Salt 2003). The enzymatic conversion of SeCys to the formation of SeMet is potential target for the phytoremediation potential of transgenic plants. At first, SeCys is converted to Se-cystathionine catalyzed by cystathionine-γ-synthase (CGS) (Pilon-Smits and

Womg 2012) via condensation of OPH and SeCys (Van Huysen et al. 2003; Sors et al. 2005a, b). The conversion of SeCys to volatile DMSe is mainly governed by the CGS (rate-limiting enzyme) (Van Huysen et al. 2003). The overexpression of CGS in transgenic *B. juncea* enhanced the Se phytovolatilization by three fold and simultaneous lower shoot or root Se levels when compared with wild types (Huysen et al. 2003), indicating the importance of this strategy in Se phytoremediation.

In non-hyperaccumulators, potential mechanism involved in avoiding Se phytotoxicities is the volatilization of SeMet into DMSe (Terry et al. 2000; Schiavon and Pilon-Smits 2017a, b). DMSe is synthesized by the methylation of SeMet to produce Se-methyl Se-Met (SeMM) in a reaction catalyzed by *S*-adenosyl-L-Met:Met-*S*-methyl transferase (MMT) (Tagmount et al. 2002), and then SeMM carries the reaction intermediates to form molecule 3-dimethylselenoniopropionate (DMSeP) or directly via methylmethionine hydrolase (Chin and Lindsay 1994; Kocsis 1998; Tagmount et al. 2002). Besides minimizing the toxic Se-amino acids into less toxic volatile compounds, the production of volatile Se compounds (e.g. DMDSe and DMSe) in plants might play a defensive role against herbivores (Schiavon and Pilon-Smits 2017a, b; Quinn et al. 2010). Primarily, DMDSe is produced in hyperaccumulators and DMSe is abundantly produced in non-hyperaccumulators (Pilon-Smits and LeDuc 2009). The production of these volatile compounds may be the result of different biochemical reactions within non-hyperaccumulator plants. While the release of volatile Se forms in hyperaccumulators may result from the corresponding actions between Se and S. Therefore, evaporation of volatile Se compounds and involved assimilation pathways should be focused before or during field research.

Involvement of Selenocysteine Lyase

Aside from the production of phytovolatilization, selenocysteine lyase (SL) is thought to be involved in preventing Se phytotoxicity. The breakdown of SeCys into elemental Se and alanine is carried out with the help of SL and ultimately reduced the SeCys, which is required for the production of malformed proteins (Van Hoewyk et al. 2005). In plants, SL exists in two isoforms: one isoform localizes in cytosol (Kushnir et al. 2001) and other in mitochondria and plastids (Pilon-Smits et al. 2002). NifS-like cysteine desulphurase proteins (analogous to SL) were characterized in *A. thaliana* (Ye et al. 2005). The function of NifS is to separate S from Cys and form FeS clusters (Pilon-Smits et al. 2002). The overexpression of AtCpNifS (chloroplast-localized NifS protein) from *Arabidopsis* enhanced the Se accumulation and tolerance by 2.2 and 1.9 folds, respectively (Van Hoewyk 2013). Additional reports documented that expression of SL (from mouse) enhanced the Se accumulation by 2 and 1.5 folds in Indian mustard (Garifullina et al. 2003) and *Arabidopsis* (Bañuelos et al. 2007), respectively. These findings suggested the potential of SL enzyme in exploiting phytoremediation purpose, besides its role in reducing Se phytotoxicity.

Sequestration of Selenium Organic Compounds

Another potential mechanism adapted by hyperaccumulators to tolerate higher Se levels is the sequestration of organic compounds (mainly MeSeCys) with selenocystathionine in minor fractions into specific compartments (Cappa et al. 2015). In *S. pinnata*, Se is predominantly stored in the vacuoles of epidermal cells along the peripheral area of leaves (Cappa et al. 2015). In non-hyperaccumulators, Se was mainly limited to vascular tissues (Cappa et al. 2014; Freeman et al. 2006). Higher accumulation of MeSeCys was noticed in the flowers of hyperaccumulator *S. pinnata* (Quinn et al. 2011). This suggested that organic Se forms targeted specific sequestration patterns critical to Se tolerance in plants.

Antioxidant Defense Mechanisms

To handle the environmental stresses, plants boost their endogenous defense system (Ali et al. 2018a, b, c; Hassan et al. 2017; Mwamba et al. 2020). Generally, hyperaccumulators had stimulating antioxidant defence systems to cope with the extra accumulation of reactive oxygen species (ROS) caused by excessive Se levels in cellular organs. Excessive Se can desynchronize the cellular redox status by overproducing the ROS that impairs the protein structures, lipid peroxidation, and oxidative injuries (Dai et al. 2020; Gupta and Gupta 2017; Ulhassan et al. 2018, 2019a, b; Mostofa et al. 2017, 2020). While low Se levels can enhance the plants' defense system by minimizing the abiotic stress-induced oxidative stress (Ulhassan et al. 2019c, 2022; Feng et al. 2013; Hasanuzzaman et al. 2012; Malik et al. 2012). Previous studies documented that Se is involved in antioxidant defense metabolism in hyperaccumulators (Freeman et al. 2010; Ulhassan et al. 2019a, c; Hasanuzzaman et al. 2012; Malik et al. 2012). It was noticed that hyperaccumulator plants such as *S. pinnata* had 1.5-fold higher antioxidant potential then non-hyperaccumulators such as *Stanleya albescens*. The concentrations of oxidative markers (O_2^- and H_2O_2) in *S. pinnata* were lowered as compared to non-hyperaccumulator *S. albescens*, after the exposure of 20 μM selenate for 10 weeks. Notably, under the same conditions, hyperaccumulators had 1.3-fold higher total glutathione levels (1.4-fold GSH + 1.2-fold GSSG) than non-hyperaccumulator plants (Freeman et al. 2010).

Among non-enzymatic antioxidants, GSH, *S*-comprised metabolite, constitutes foremost defense system in response to oxidative stressors and mainly formed from the amino acids glutamate (Glu) and glycine (Gly). GSH, with the help of ascorbate peroxidase (APX), participate in glutathione-ascorbate (GSH-AsA) cycle (Foyer and Noctor 2012; Nazir et al. 2020). Exogenous applications of inorganic Se forms (selenite and selenate) have been reported to enhance the plant defense by activating the GSH-Px and biosynthesis of GSH, PCs, proline, ascorbate, flavonoids, carotenoids, and alkaloids. Likewise, Se can stimulate the dismutation of superoxide radical into H_2O_2 (Feng et al. 2013). Ultimately, Se-induced increase in antioxidant activities led to reduction in lipid peroxidation (due to lower ROS accumulation) under environmental stress conditions (Feng et al. 2013).

Involvement of Phytohormones or Signalling Molecules

Besides antioxidants, the applications of phytohormones such as melatonin (MEL) (Ulhassan et al. 2019b), salicylic acid (SA) (Mostofa et al. 2020), ethylene (EA), jasmonic acid (JA) (Tamaoki et al. 2008), and nitric oxide (NO) (Dai et al. 2020) enhance the Se tolerance in hyperaccumulators/non-hyperaccumulators. Furthermore, RNA seq study confirmed the higher biosynthesis of JA in hyperaccumulator *S. pinnata* (greater Se tolerance) as compared to non-hyperaccumulator *S. albescens* (Wang et al. 2018). Recent investigations suggested that overexpression of ethylene-responsive ERF genes enhanced the Se tolerance in *A. thaliana* (Jiang et al. 2020). However, the molecular mechanisms involved in the accumulation and tolerance capacity of hyperaccumulators/non-hyperaccumulators are still not clear. In addition, the interaction of inorganic/organic Se species with other defence systems is yet to be further explained.

General Conclusions and Future Prospects

Depending upon the Se form and target plant species, Se is generally considered an essential element for plants. Still, its essentiality or phytotoxicity (particularly inorganic Se forms) for higher plants required further investigations. Previous studies evidenced that the beneficial effects of Se forms on overall plant growth can be related to the SeCys or SeMet comprising selenoproteins especially in higher plants. Se essentiality is manifested in stimulating antioxidant defence system, minimize external environmental stresses, and enhance plant growth and provide protection against herbivores. In current investigations, the accumulation, transport, metabolism, and biochemistry of Se species in plants is mostly related to sulphur metabolism (although Se and S behave differently in plants). Different plant species (hyperaccumulators and non-hyperaccumulators) utilize diverse approaches to uptake and accumulate Se and directly/indirectly related to the differences in metabolic pathways. In-depth knowledge of Se or S assimilation pathways can help to develop strategies for the biofortification of Se-deficient areas and phytoremediation of seleniferous soils. The excessive Se levels in soil-plant system and their induced phytotoxic effects can be scavenged by the participation of phytohormones, antioxidants, and other biochemical approaches (as explained in preceding sections).

Further identification of potential genes in Se uptake, assimilation, and Se-metabolic pathways can help plant scientists to develop crop varieties and boost our understandings concerning the Se accumulation and tolerance mechanisms. Ultimately, these plants can be further utilized for Se-biofortification (fulfil Se deficiency) and phytoremediation (handle excess Se) purposes. Molecular approaches confirmed that overexpression of transporter proteins or enzymes contribution in the uptake and assimilation was directly/indirectly correlated with Se metabolism and volatilization mechanisms. The activation of antioxidant/non-enzymatic antioxidants and phytohormones associated with signalling pathways can enhance Se acquisition and directly/indirectly affect the primary/secondary metabolism. Furthermore, the utilization of bioremediation agents such as bacteria, fungi, and algae should be focused in upcoming research to validate their roles in Se uptake, volatilization, and remediation of seleniferous soils. Better understandings of Se metabolism in plants may help in improving human health via its accumulation in food chain.

Abbreviations: Se (IV); selenite, Se (VI); selenate, SeMet; selenomethionine, SeCys; selenocysteine, SeMetSeCys; selenomethyl selenocysteine, DMeSe; dimethyl selenide, DMDSe; dimethyl diselenide.

Acknowledgments

This work was supported by the Science and Technology Department of Zhejiang Province (2021C02064-2), Collaborative Innovation Center for Modern Crop Production co-sponsored by Province and Ministry (CIC-MCP), and the Agriculture and Rural Affairs Department of Zhejiang Province (2021XTTGLY0202). We thank Rui Sun from Agricultural Experiment Station of Zhejiang University for his assistance.

References

Ali, S., Jin, R., Gill, R.A. et al. (2018a). Beryllium stress-induced modifications in antioxidant machinery and plant ultrastructure in the seedlings of black and yellow seeded oilseed rape. *Biomedical Research International* 2018: 1–14. https://doi.org/10.1155/2018/1615968.

Ali, S., Gill, R.A., Mwamba, T.M. et al. (2018b). Differential cobalt-induced effects on plant growth, ultrastructural modifications, and antioxidative response among four *Brassica napus* L. cultivars. *International Journal of Environmental Science and Technology* 15: 1–16. https://doi.org/10.1007/s13762-017-1629-z.

Ali, S., Gill, R.A., Ulhassan, Z. et al. (2018c). Insights on the responses of Brassica napus cultivars against the cobalt-stress as revealed by carbon assimilation, anatomical changes and secondary metabolites. *Environmental and Experimental Botany* 156: 183–196.

Anjum, N.A., Gill, R., Kaushik, M. et al. (2015). ATP-sulfurylase, sulfur-compounds, and plant stress tolerance. *Frontiers in Plant Science* 6: 210.

Bañuelos, G., Terry, N., Leduc, D. et al. (2005). Field trial of transgenic indian mustard plants shows enhanced phytoremediation of selenium-contaminated sediment. *Environmental Science and Technology* 39: 1771–1777.

Bañuelos, G., Leduc, D., Pilon-Smits, E.A.H., and Terry, N. (2007). Transgenic indian mustard overexpressing selenocysteine lyase or selenocysteine methyltransferase exhibit enhanced potential for selenium phytoremediation under field conditions. *Environmental Science and Technology* 41: 599–605.

Barberon, M., Berthomieu, P., Clairotte, M. et al. (2008). Unequal functional redundancy between the two Arabidopsis thaliana high-affinity sulphate transporters SULTR1; 1 and SULTR1; 2. *New Phytologist* 180 (3): 608–619.

Bohrer, A.S., Yoshimoto, N., Sekiguchi, A. et al. (2015). Alternative translational initiation of ATP sulfurylase underlying dual localization of sulfate assimilation pathways in plastids and cytosol in Arabidopsis thaliana. *Frontiers in Plant Science* 5: 750.

Both, E.B., Stonehouse, G.C., Lima, L.W. et al. (2020). Selenium tolerance, accumulation, localization and speciation in a Cardamine hyperaccumulator and a non-hyperaccumulator. *Science of the Total Environment* 703: 135041.

Brown, T.A. and Shrift, A. (1982). Selenium: toxicity and tolerance in higher plants. *Biological Reviews* 57 (1): 59–84.

Bulteau, A.L. and Chavatte, L. (2015). Update on selenoprotein biosynthesis. *Antioxidants & Redox Signaling* 23 (10): 775–794.

Cao, M.J., Wang, Z., Wirtz, M. et al. (2013). SULTR 3; 1 is a chloroplast-localized sulfate transporter in *Arabidopsis thaliana*. *The Plant Journal* 73 (4): 607–616.

Cappa, J.J., Cappa, P.J., El Mehdawi, A.F. et al. (2014). Characterization of selenium and sulfur accumulation across the genus Stanleya (Brassicaceae): A field survey and common-garden experiment. *American Journal of Botany* 101 (5): 830–839.

Cappa, J.J., Yetter, C., Fakra, S. et al. (2015). Evolution of selenium hyperaccumulation in Stanleya (Brassicaceae) as inferred from phylogeny, physiology and X-ray microprobe analysis. *New Phytologist* 205 (2): 583–595.

Carey, A.M., Scheckel, K.G., Lombi, E. et al. (2012). Grain accumulation of selenium species in rice (Oryza sativa L.). *Environmental Science & Technology* 46 (10): 5557–5564.

Chao, D.Y., Baraniecka, P., Danku, J. et al. (2014). Variation in sulfur and selenium accumulation is controlled by naturally occurring isoforms of the key sulfur assimilation enzyme ADENOSINE 5′-PHOSPHOSULFATE REDUCTASE2 across the Arabidopsis species range. *Plant Physiology* 166 (3): 1593–1608.

Chin, H.W. and Lindsay, R.C. (1994). Mechanisms of formation of volatile sulfur compounds following the action of cysteine sulfoxide lyases. *Journal of Agricultural and Food Chemistry* 42 (7): 1529–1536.

Dai, Z., Rizwan, M., Gao, F. et al. (2020). Nitric oxide alleviates selenium toxicity in rice by regulating antioxidation, selenium uptake, speciation and gene expression. *Environmental Pollution* 257: 113540.

Deng, X., Liu, K., Li, M. et al. (2017). Difference of selenium uptake and distribution in the plant and selenium form in the grains of rice with foliar spray of selenite or selenate at different stages. *Field Crops Research* 211: 165–171.

El Mehdawi, A.F., Jiang, Y., Guignardi, Z.S. et al. (2018). Influence of sulfate supply on selenium uptake dynamics and expression of sulfate/selenate transporters in selenium hyperaccumulator and non-hyperaccumulator Brassicaceae. *New Phytologist* 217 (1): 194–205.

Ellis, D.R. and Salt, D.E. (2003). Plants, selenium and human health. *Current Opinion in Plant Biology* 6 (3): 273–279.

Ellis, D.R., Sors, T.G., Brunk, D.G. et al. (2004). Production of Se-methylselenocysteine in transgenic plants expressing selenocysteine methyltransferase. *BMC Plant Biology* 4 (1): 1471–2229.

Feng, R., Wei, C., and Tu, S. (2013). The roles of selenium in protecting plants against abiotic stresses. *Environmental and Experimental Botany* 87: 58–68.

Fisher, B., Yarmolinsky, D., Abdel-Ghany, S. et al. (2016). Superoxide generated from the glutathione-mediated reduction of selenite damages the iron-sulfur cluster of chloroplastic ferredoxin. *Plant Physiology and Biochemistry* 106: 228–235.

Foyer, C.H. and Noctor, G. (2012). Managing the cellular redox hub in photosynthetic organisms. *Plant, Cell & Environment* 35 (2): 199–201.

Freeman, J.L., Quinn, C.F., Marcus, M.A. et al. (2006). Selenium-tolerant diamondback moth disarms hyperaccumulator plant defense. *Current Biology* 16 (22): 2181–2192.

Freeman, J.L., Lindblom, S.D., Quinn, C.F. et al. (2007). Selenium accumulation protects plants from herbivory by Orthoptera via toxicity and deterrence. *New Phytologist* 175 (3): 490–500.

Freeman, J.L., Tamaoki, M., Stushnoff, C. et al. (2010). Molecular mechanisms of selenium tolerance and hyperaccumulation in Stanleya pinnata. *Plant Physiology* 153 (4): 1630–1652.

Galeas, M.L., Zhang, L.H., Freeman, J.L. et al. (2007). Seasonal fluctuations of selenium and sulfur accumulation in selenium hyperaccumulators and related nonaccumulators. *New Phytologist* 173 (3): 517–525.

Garifullina, G.F., Owen, J.D., Lindblom, S.D. et al. (2003). Expression of a mouse selenocysteine lyase in Brassica juncea chloroplasts affects selenium tolerance and accumulation. *Physiol Plantarum* 118: 538–544.

Gong, R., Ai, C., Zhang, B., and Cheng, X. (2018). Effect of selenite on organic selenium speciation and selenium bioaccessibility in rice grains of two Se-enriched rice cultivars. *Food Chemistry* 264: 443–448.

Grant, K., Carey, N.M., Mendoza, M. et al. (2011). Adenosine 5′-phosphosulfate reductase (APR2) mutation in Arabidopsis implicates glutathione deficiency in selenate toxicity. *Biochemical Journal* 438 (2): 325–335.

Gupta, S. and Gupta, M. (2016). Alleviation of selenium toxicity in *Brassica juncea* L.: salicylic acid-mediated modulation in toxicity indicators, stress modulators, and sulfur-related gene transcripts. *Protoplasma* 253 (6): 1515–1528.

Gupta, M. and Gupta, S. (2017). An overview of selenium uptake, metabolism, and toxicity in plants. *Front. Plant Sci.* 7: 2074.

Hasanuzzaman, M., Hossain, M.A., and Fujita, M. (2012). Exogenous selenium pretreatment protects rapeseed seedlings from cadmium-induced oxidative stress by upregulating antioxidant defense and methylglyoxal detoxification systems. *Biological Trace Element Research* 149 (2): 248–261.

Hassan, Z., Ali, S., Ahmad, R. et al. (2017). Biochemical and molecular responses of oilseed crops to heavy metal stress. In: *Oilseed Crops: Yield and Adaptations Under Environmental Stress*, 236–248. John Wiley & Sons https://doi.org/10.1002/9781119048800.ch13.

Huysen, T.V., Abdel-Ghany, S., Hale, K. et al. (2003). Overexpression of cystathionine-γ-synthase enhances selenium volatilization in Brassica juncea. *Planta* 218 (1): 71–78.

Huysen, T., Terry, N., and Pilon-Smits, E.A.H. (2004). Exploring the selenium phytoremediation potential of transgenic Indian mustard overexpressing ATP sulfurylase or cystathionine-γ-synthase. *International Journal of Phytoremediation* 6 (2): 111–118.

Jiang, L., Yang, J., Liu, C. et al. (2020). Overexpression of ethylene response factor *ERF96* gene enhances selenium tolerance in Arabidopsis. *Plant Physiology and Biochemistry* 149: 294–300.

Kaur, M. and Sharma, S. (2018). Influence of selenite and selenate on growth, leaf physiology and antioxidant defense system in wheat (Triticum aestivum L.). *Journal of the Science of Food and Agriculture* 98 (15): 5700–5710.

Kikkert, J. and Berkelaar, E. (2013). Plant uptake and translocation of inorganic and organic forms of selenium. *Archives Environmental Contamination and Toxicology* 65: 458–465.

Kocsis, M.G. (1998). Dimethylsulfoniopropionate biosynthesis in *Spartina alterniflora* (L.) Evidence that S-methylmethionine and dimethylsulfoniopropylamine are intermediates. *Plant Physiology* 117: 273–281.

Kushnir, S., Babiychuk, E., Storozhenko, S. et al. (2001). A mutation of the mitochondrial ABC transporter Sta1 leads to dwarfism and chlorosis in the Arabidopsis mutant starik. *Plant Cell* 13: 89–100.

Li, H.F., McGrath, S.P., and Zhao, F.J. (2008). Selenium uptake, translocation and speciation in wheat supplied with selenate or selenite. *New Phytologist* 178: 92–102.

Li, Y., Zhu, N., Liang, X. et al. (2019). A comparative study on the accumulation, translocation and transformation of selenite, selenate, and SeNPs in a hydroponic-plant system. *Ecotoxicology and Environmental Safety* 189: 109955.

LeDuc, D.L., Tarun, A.S., Montes-Bayon, M. et al. (2004). Overexpression of selenocysteine methyltransferase in Arabidopsis and Indian mustard increases selenium tolerance and accumulation. *Plant Physiology* 135 (1): 377–383.

Tamaoki, M., Freeman, J.L., and Pilon-Smits, E.A. (2008). Cooperative ethylene and jasmonic acid signaling regulates selenite resistance in Arabidopsis. *Plant Physiology* 146 (3): 1219–1230.

Malik, J.A., Goel, S., Kaur, N. et al. (2012). Selenium antagonises the toxic effects of arsenic on mungbean (Phaseolus aureus Roxb.) plants by restricting its uptake and enhancing the antioxidative and detoxification mechanisms. *Environmental and Experimental Botany* 77: 242–248.

Mayland, H.F., Gough, L.P., and Stewart, K.C. (1991). Chapter E: Selenium mobility in soils and its absorption, translocation, and metabolism in plants. *Proceedings of the 1990 Billings Land Reclamation Symposium on Selenium in Arid and Semiarid Environments*, western United States. U. S. Geological Survey Circular No. 1064, 55–64.

Mostofa, M.G., Hossain, M.A., Siddiqui, M.N. et al. (2017). Phenotypical, physiological and biochemical analyses provide insight into selenium-induced phytotoxicity in rice plants. *Chemosphere* 178: 212–223.

Mostofa, M.G., Rahman, M.M., Siddiqui, M.N. et al. (2020). Salicylic acid antagonizes selenium phytotoxicity in rice: selenium homeostasis, oxidative stress metabolism and methylglyoxal detoxification. *Journal of Hazardous Materials* 394: 122572.

Mwamba, T.M., Islam, F., Ali, B. et al. (2020). Comparative metabolomic responses of low- and high-cadmium accumulating genotypes reveal the cadmium adaptive mechanism in Brassica napus. *Chemosphere* 250: 126308. doi: https://doi.org/10.1016/j.chemosphere.2020.126308.

Natasha, S.M., Niazi, N.K., Khalid, S. et al. (2018). A critical review of selenium biogeochemical behavior in soil-plant system with an inference to human health. *Environmental Pollution* 234: 915–934.

Nazir, M.M., Ulhassan, Z., Zeeshan, M. et al. (2020). Toxic metals/metalloids accumulation, tolerance, and homeostasis in *Brassica* oilseed species. In: *The Plant Family Brassicaceae* (ed. M. Hasanuzzaman). Singapore: Springer https://doi.org/10.1007/978-981-15-6345-4_13.

Neuhierl, B., Thanbichler, M., Lottspeich, F. et al. (1999). A family of S-Methylmethionine-dependent Thiol/Selenol Methyltransferases role in selenium tolerance and evolutionary relation. *Journal of Biological Chemistry* 274 (9): 5407–5414.

Novoselov, S.V., Rao, M., Onoshko, N.V. et al. (2002). Selenoproteins and selenocysteine insertion system in the model plant cell system, Chlamydomonas reinhardtii. *The EMBO Journal* 21 (14): 3681–3693.

Pickering, I.J., Wright, C., Bubner, B. et al. (2003). Chemical form and distribution of selenium and sulfur in the selenium hyperaccumulator Astragalus bisulcatus. *Plant Physiology* 131 (3): 1460–1467.

Pilon-Smits, E.A.H., De Souza, M.P., Hong, G. et al. (1999). Selenium volatilization and accumulation by twenty aquatic plant species. *Journal of Environmental Quality* 28: 1011–1018.

Pilon-Smits, E.A.H., Garifullina, G.F., Abdel-Ghany, S.E. et al. (2002). Characterization of a NifS-like chloroplast protein from Arabidopsis. Implications for its role in sulfur and selenium metabolism. *Plant Physiology* 130: 1309–1318.

Pilon-Smits, E.A. and LeDuc, D.L. (2009). Phytoremediation of selenium using transgenic plants. *Current Opinion in Biotechnology* 20 (2): 207–212.

Pilon-Smits, E.A.H. and Quinn, C.F. (2010). Selenium metabolism in plants. In: *Cell Biology of Metal and Nutrients* (ed. R. Hell and R. Mendel), 225–241. Berlin: Springer.

Pilon-Smits, E.A. and Womg, M.H. (2012). Plant selenium metabolism–genetic manipulation, phytotechnological applications, and ecological implications. In: *Environmental Contamination: Health Risks and Ecological Restoration* (ed. M.H. Womg), 293–311. Boca Raton, FL: CRC Press.

Pilon-Smits, E.A. (2015). Selenium in plants. In: *Progress in Botany (Genetics-Physiology-Systematics-Ecology)*, vol. 76 (ed. U. Lüttge and W. Beyschlag), 93–107. Cham, Switzerland: Springer.

Pyrzynska, K. (2009). Selenium speciation in enriched vegetables. *Food Chemistry* 114: 1183–1191.

Qin, H.B., Zhu, J.M., Lin, Z.Q. et al. (2017). Selenium speciation in seleniferous agricultural soils under different cropping systems using sequential extraction and X-ray absorption spectroscopy. *Environmental Pollution*, 225, 361–369. doi:https://doi.org/10.1016/j.envpol.2017.02.062

Quinn, C.F., Freeman, J.L., Reynolds, R.J. et al. (2010). Selenium hyperaccumulation offers protection from cell disruptor herbivores. *BMC Ecol.* 10 (1): 1–11.

Quinn, C.F., Prins, C.N., Freeman, J.L. et al. (2011). Selenium accumulation in flowers and its effects on pollination. *New Phytologist*, 192 (3), 727–737.

Rayman, M.P. (2012). Selenium and human health. *The Lancet* 379 (9822): 1256–1268.

Shaheen, S.M., Ali, R.A., Abowaly, M.E. et al. (2018). Assessing the mobilization of As, Cr, Mo, and Se in Egyptian lacustrine and calcareous soils using sequential extraction and biogeochemical microcosm techniques. *Journal of Geochemical Exploration*, 191: 28–42. doi:https://doi.org/10.1016/j.gexplo.2018.05.003

Schiavon, M. and Pilon-Smits, E.A.H. (2017a). Selenium biofortification and phytoremediation phytotechnologies: A review. *Journal of Environment Quality* 46 (1), 10–19. doi: https://doi.org/10.2134/jeq2016.09.0342

Schiavon, M. and Pilon-Smits, E.A.H. (2017b). The fascinating facets of plant selenium accumulation – biochemistry, physiology, evolution and ecology. *New Phytologist*, 213 (4), 1582–1596. doi:https://doi.org/10.1111/nph.14378

Schiavon, M., Pilon, M., Malagoli, M. et al. (2015). Exploring the importance of sulfate transporters and ATP sulphurylases for selenium hyperaccumulation—a comparison of Stanleya pinnata and Brassica juncea (Brassicaceae). *Frontiers in Plant Science* 6: 2.

Shibagaki, N., Rose, A., McDermott, J.P. et al. (2002). Selenate-resistant mutants of Arabidopsis thaliana identify Sultr1; 2, a sulfate transporter required for efficient transport of sulfate into roots. *The Plant Journal* 29 (4): 475–486.

Song, Z., Shao, H., Huang, H. et al. (2017). Overexpression of the phosphate transporter gene OsPT8 improves the Pi and selenium contents in Nicotiana tabacum. *Environmental and Experimental Botany* 137: 158–165.

Sors, T., Ellis, D., and Salt, D. (2005a). Selenium uptake, translocation, assimilation and metabolic fate in plants. *Photosynth. Res.* 86: 373–389.

Sors, T.G., Ellis, D.R., Na, G.N. et al. (2005b). Analysis of sulfur and selenium assimilation in *Astragalus* plants with varying capacities to accumulate selenium. *The Plant Journal* 42: 785–797. doi: https://doi.org/10.1111/j.1365-313X.2005.02413.x

Sors, T.G., Martin, C.P., and Salt, D.E. (2009). Characterization of selenocysteine methyltransferases from Astragalus species with contrasting selenium accumulation capacity. *The Plant Journal* 59 (1): 110–122.

Suter, M., Von Ballmoos, P., Kopriva, S. et al. (2000). Adenosine 5′-phosphosulfate sulfotransferase and adenosine 5′-phosphosulfate reductase are identical enzymes. *Journal of Biological Chemistry* 275: 930–936.

Tagmount, A., Berken, A., and Terry, N. (2002). An essential role of S-adenosyl-l-methionine: l-methionine S-methyl transferase in selenium volatilization by plants. Methylation of selenomethionine to selenium-methyl-l-selenium-methionine, the precursor of volatile selenium. *Plant Physiology* 130 (2): 847–856.

Terry, N., Zayed, A.M., De Souza, M.P. et al. (2000). Selenium in higher plants. *Annual Review of Plant Biology* 51: 401–432.

Ulhassan, Z., Ali, S., Gill, R.A. et al. (2018). Comparative orchestrating response of four oilseed rape (*Brassica napus*) cultivars against the selenium stress as revealed by physio-chemical, ultrastructural and molecular profiling. *Ecotoxicology and Environmental Safety* 161: 634–647.

Ulhassan, Z., Gill, R.A., Ali, S. et al. (2019a). Dual behavior of selenium: Insights into physio-biochemical, anatomical and molecular analyses of four Brassica napus cultivars. *Chemosphere* 225: 329–341.

Ulhassan, Z., Huang, Q., Gill, R.A. et al. (2019b). Protective mechanisms of melatonin against selenium toxicity in Brassica napus: insights into physiological traits, thiol biosynthesis, and antioxidant machinery. *BMC Plant Biology.* 19, 507 doi: https://doi.org/10.1186/s12870-019-2110-6

Ulhassan, Z., Gill, R.A., Huang, H. et al. (2019c). Selenium mitigates the chromium toxicity in Brassicca napus L. by ameliorating nutrients uptake, amino acids metabolism and antioxidant defense system. *Plant Physiology and Biochemistry* 145: 142–152. https://doi.org/10.1016/j.plaphy.2019.10.035.

Ulhassan, Z., Khan, A.R., Sheteiwy, M.S. et al. (2022). Selenium-Mediated Regulation of Antioxidant Defense System and Improved Heavy Metals Tolerance in Plants. In: *Antioxidant Defense in Plants* (eds. T. Aftab and K.R Hakeem). Singapore: Springer, 369–382. https://doi.org/10.1007/978-981-16-7981-0_16.

Van Hoewyk, D., Garifullina, G.F., Ackley, A.R. et al. (2005). Overexpression of AtCpNifS enhances selenium tolerance and accumulation in Arabidopsis. *Plant Physiology* 139 (3): 1518–1528.

Van Hoewyk, D., Takahashi, H., Inoue, E. et al. (2008). Transcriptome analyses give insights into selenium-stress responses and selenium tolerance mechanisms in *Arabidopsis. Physiologia Plantarum* 132 (2): 236–253.

Van Hoewyk, D. (2013). A tale of two toxicities: malformed selenoproteins and oxidative stress both contribute to selenium stress in plants. *Annals of Botany* 112: 965–972.

Van Huysen, T., Abdel-Ghany, S., Hale, K.L. et al. (2003). Overexpression of cystathionine-γ-synthase enhances selenium volatilization in Brassica juncea. *Planta* 218 (1): 71–78.

Wallenberg, M., Olm, E., Hebert, C. et al. (2010). Selenium compounds are substrates for glutaredoxins: a novel pathway for selenium metabolism and a potential mechanism for selenium-mediated cytotoxicity. *The Biochemical Journal* 429: 85–93. doi: https://doi.org/10.1042/BJ20100368

Wan, Y., Yu, Y., Wang, Q. et al. (2016). Cadmium uptake dynamics and translocation in rice seedling: influence of different forms of selenium. *Ecotoxicology Environmental Safety* 133: 127e134.

Wang, J., Cappa, J.J., Harris, J.P. et al. (2018). Transcriptome-wide comparison of selenium hyperaccumulator and nonaccumulator Stanleya species provides new insight into key processes mediating the hyperaccumulation syndrome. *Plant Biotechnology Journal* 16 (9): 1582–1594.

White, P.J., Bowen, H.C., Parmaguru, P. et al. (2004). Interactions between selenium and sulphur nutrition in *Arabidopsis thaliana*. *Journal of Experimental Botany* 55: 1927–1937. doi: https://doi.org/10.1093/jxb/erh192

White, P.J. (2016). Selenium accumulation by plants. *Annals of botany* 117 (2): 217–235.

Wu, Z., Bañuelos, G.S., Lin, Z.Q. et al. (2015). Biofortification and phytoremediation of selenium in China. *Frontiers in Plant Science* 6: 136. doi: https://doi.org/10.3389/fpls.2015. 00136

Ye, H., Garifullina, G.F., Abdel-Ghany, S.E. et al. (2005). The chloroplast NifS-like protein of Arabidopsis thaliana is required for iron – sulfur cluster formation in ferredoxin. *Planta* 220: 602–608.

Yin, H., Qi, Z., Li, M. et al. (2019). Selenium forms and methods of application differentially modulate plant growth, photosynthesis, stress tolerance, selenium content and speciation in Oryza sativa L. *Ecotoxicology and Environmental Safety* 169: 911–917.

Yarmolinsky, D., Brychkova, G., Fluhr, R., and Sagi, M. (2012). Sulfite reductase protects plants against sulfite toxicity. *Plant Physiology* 161 (2): 725–743.

Zhang, L., Hu, B., Li, W. et al. (2014a). OsPT2, a phosphate transporter, is involved in the active uptake of selenite in rice. *New Phytologist* 201: 1183e1191.

Zhao, X.Q., Mitani, N., Yamaji, N. et al. (2010). Involvement of silicon influx transporter OsNIP2;1 in selenite uptake in rice. *Plant Physiology* 153: 1871–1877.

Zhang, L., Hu, B., Deng, K. et al. (2019). NRT1. 1B improves selenium concentrations in rice grains by facilitating selenomethinone translocation. *Plant Biotechnology Journal* 17 (6): 1058–1068.

15

Lanthanides as Beneficial Elements for Plants

Fernando C. Gómez-Merino[1], Libia F. Gómez-Trejo[2], Rubén Ruvalcaba-Ramírez[2], and Libia I. Trejo-Téllez[1]

[1] *College of Postgraduates in Agricultural Sciences, Laboratory of Plant Nutrition, Texcoco, State of Mexico, Mexico*
[2] *Department of Plant Protection, Chapingo Autonomous University, Texcoco, State of Mexico, Mexico*

Introduction

The lanthanides or lanthanoids (Ln) comprise 15 metallic chemical elements with atomic numbers 57–71 from lanthanum (La) through lutetium (Lu). Thus, apart from La and Lu, this series includes the following elements listed in alphabetical order: cerium (Ce), dysprosium (Dy), erbium (Er), europium (Eu), gadolinium (Gd), holmium (Ho), neodymium (Nd), praseodymium (Pr), promethium (Pm), samarium (Sm), terbium (Tb) thulium (Tm), and ytterbium (Yb). The lanthanides, together with scandium (Sc) and yttrium (Y), integrate the rare earth elements (REE) group of 17 chemically similar elements (Figure 15.1).

Interestingly, the technical term REE refers to the laborious and complicated methods of separation of these elements from reasonably rare minerals than with the scarcity of any of them. Indeed, the REE can be more abundant on Earth than other commonly used metals. For instance, Ce has been ranked number 26 in terms of average crustal abundance of chemical elements, being even more bountiful than copper (Cu), while Nd is more abundant than gold (Au). In fact, the four most abundant REE, namely Ce, Y, La, and Nd, resemble average crustal abundance of industrial metals like chromium (Cr), nickel (Ni), molybdenum (Mo), or zinc (Zn) (Panichev 2015; Van Gosen et al. 2017). Importantly, lanthanides are subdivided into light lanthanides (La to Gd) and heavy lanthanides (Tb to Lu). A detailed estimated distribution of these elements is presented in Figure 15.2. The light lanthanides are more bountiful in the Earth's crust than the heavy ones. It is noteworthy mentioning that Pm is extremely rare and unstable in nature. Pm endures relatively prompt radioactive decay (half-life is 2.62 years) (Hinderer 2010), and in natural environments it is practically non-existent (0.0018% of the Earth's crust content) (Aide 2018). In nature, carbonate or phosphate minerals such as xenotime,

Beneficial Chemical Elements of Plants: Recent Developments and Future Prospects, First Edition. Edited by Sangeeta Pandey, Durgesh Kumar Tripathi, Vijay Pratap Singh, Shivesh Sharma, and Devendra Kumar Chauhan.

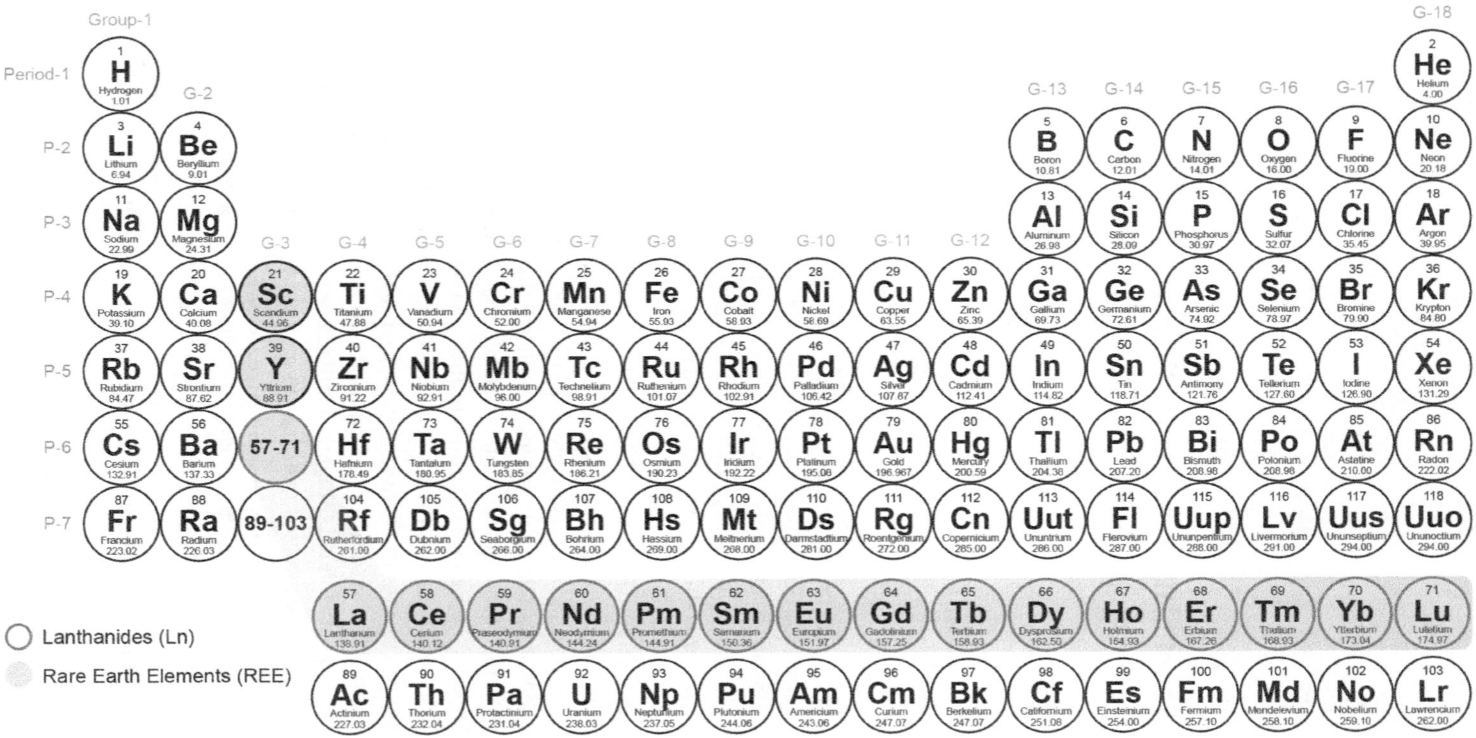

Figure 15.1 Periodic table depicting the lanthanide or lanthanoid (Ln) series of metallic elements and the rare earth elements (REE). Lanthanides comprises the 15 elements with atomic numbers 57–71, from lanthanum through lutetium. These elements, along with scandium and yttrium, are often collectively known as REE. Scandium: Sc; yttrium: Y; lanthanum: La; cerium: Ce; praseodymium: Pr; neodymium: Nd; promethium: Pm; samarium: Sm; europium: Eu; gadolinium: Gd; terbium: Tb; dysprosium: Dy; holmium: Ho; erbium: Er; thulium: Tm; ytterbium: Yb; and lutetium: Lu.

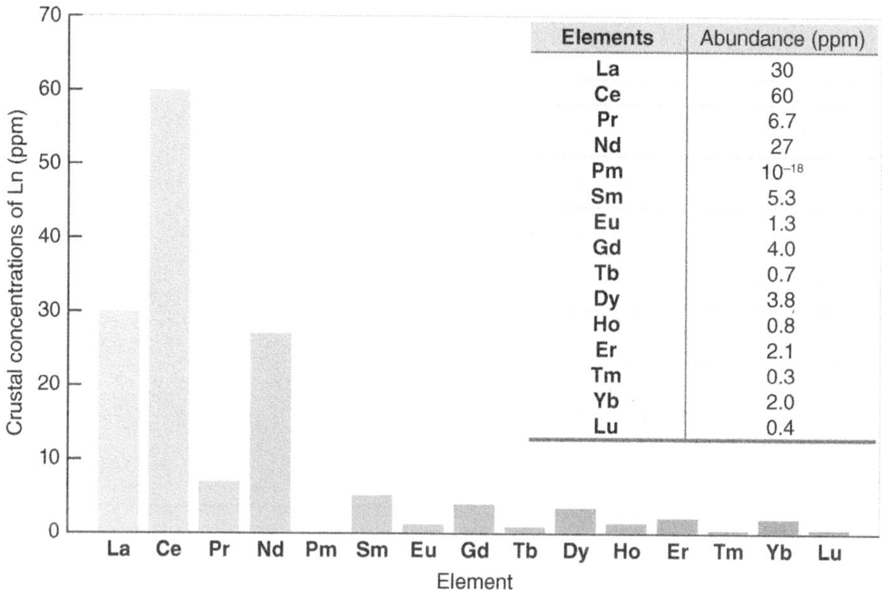

Figure 15.2 Crustal concentrations of the 15 lanthanide chemical elements found in natural ores. Lanthanum: La; cerium: Ce; praseodymium: Pr; neodymium: Nd; promethium: Pm; samarium: Sm; europium: Eu; gadolinium: Gd; terbium: Tb; dysprosium: Dy; holmium: Ho; erbium: Er; thulium: Tm; ytterbium: Yb; and lutetium: Lu.

monazite, and bastnäsite are the primary sources of lanthanides and REE in general (Long et al. 2012; Xie et al. 2014). In these minerals multiple REE often co-occur (Cotruvo 2019).

Though informally, the symbol 'Ln' is often used to represent the lanthanides. This series of chemical elements is located in the f-block, within the period six in the periodic table, and they all form trivalent cations, denoted as 'Ln^{3+}'. In these elements, chemical properties are highly dependent on the ionic radius, which diminishes continuously from 103.2 pm in La^{3+} to 86.1 pm in Lu^{3+} for a coordination number of 6 (CN 6), and from 116.0 pm in La^{3+} to 97.7 pm in Lu^{3+} for a coordination number of 8 (CN 8) (Aide 2018). This decrease is a phenomenon known as 'Lanthanide Contraction', in which electrons fill the f-subshell as a result of the reduction in atomic size of the elements. The f subshell is not shielded, thus the atomic size decreases as the nuclear charge still augments (Seitz et al. 2007) (Figure 15.3).

Lanthanides are considered transition metals, but their unique properties let them being classified apart from the rest of the elements in the periodic table. In particular, lanthanides as a group share very similar physical and chemical properties, which can be attributed to their electron shell structure: with each increase in atomic number across the group, an electron is added to an inner incomplete subshell (4f) rather than to an outer level (Van Gosen et al. 2017). Thus, the lanthanide contraction phenomenon is produced by selectively filling inner subshells, which results in a progressive decrease in the ionic radius of these cations. As a result of this contraction, the fractionation of the Ln^{3+} is high, leading to unusual and useful chemical and physical properties, including unique magnetic and optical properties (Van Gosen et al. 2017). Thus, in the lanthanides, radius

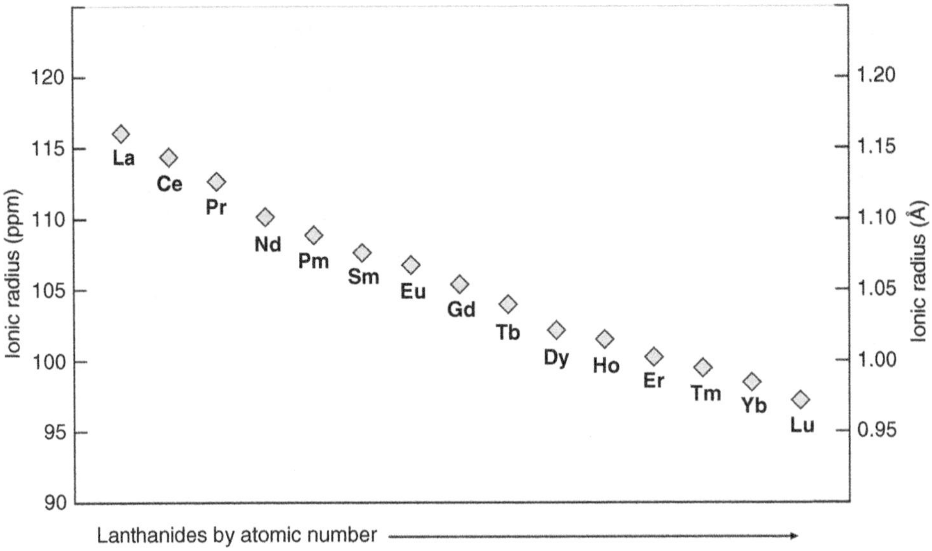

Figure 15.3 Ionic radii (in picometres on the left and in Ångström on the right) of the trivalent lanthanide (Ln) ions for a coordination number of 8 (CN 8). A systematic decrease in the radius of the trivalent Ln ions with raising atomic number is evident, a phenomenon referred to as 'Lanthanide Contraction'. Lanthanum: La; cerium: Ce; praseodymium: Pr; neodymium: Nd; promethium: Pm; samarium: Sm; europium: Eu; gadolinium: Gd; terbium: Tb; dysprosium: Dy; holmium: Ho; erbium: Er; thulium: Tm; ytterbium: Yb; and lutetium: Lu.

of the anionic species, atomic number (Z), oxidation state, and coordination number (CN) significantly affect the ionic radius (r_{ion}) of each particular element (Aide 2018). This phenomenon happens because the 6s and 5p orbitals enter the 4f subshell, since the 4f orbital is not shielded from the raising nuclear change, which reduces the atomic radius of the atom across the series. Consequently, there is a decrease in atomic size throughout the series. This fact, together with the unique spectroscopic hallmark of the 4f electrons, has led to the extensive use of lanthanides in biochemical tests, while their ionic character favors hard, carboxylate ligands in biological systems and versatile coordination geometries (Cotruvo 2019). Furthermore, the high charge observed in Ln^{3+} confers them strong electropositivity, enabling them to become hard Lewis acids capable of polarizing bonds upon coordination and consequently modify the electrophilic power of compounds (Aspinall 2018).

Adjacent lanthanides have a similar ionic radius, which complicates the separation processes of each element in naturally occurring ores. Therefore, in order to separate them, ion-exchange chromatography or countercurrent exchange methods are currently used (Krättli et al. 2013; Johnson and Nash 2015; Baldwin et al. 2018). Once separated, the purified metals and their compounds exhibit significant variation amongst themselves in magnetic, optical, electrical, and electronic properties, which make them useful in industry, life science, and medicine, especially if they are in alloys with other metals (Cotton and Harrowfield 2012; Charalampides et al. 2015; Teo et al. 2016). Table 15.1 summarizes some of the most salient physical and chemical properties of lanthanides.

Table 15.1 Main useful chemical and physical hallmarks of the lanthanide series of chemical elements.

Chemical element	Atomic number	Density ($g\ cm^{-3}$)	Metal radius (pm)	Ion ($^{+3}$)	Ionic radius (pm) CN 6	Ionic radius (pm) CN 8	Unpaired electrons	Color
Lanthanum (La)	57	6.16	162.0	La^{3+}	103.2	116.0	0	Colorless
Cerium (Ce)	58	6.77	181.8	Ce^{3+}	101.0	114.3	1	Colorless
Praseodymium (Pr)	59	6.77	182.4	Pr^{3+}	99.0	112.6	2	Green
Neodymium (Nd)	60	7.01	181.4	Nd^{3+}	98.3	110.9	3	Reddish purple
Promethium (Pm)	61	7.26	183.4	Pm^{3+}	97.0	109.3	4	Pink
Samarium (Sm)	62	7.52	180.4	Sm^{3+}	95.8	107.9	5	Pale yellow
Europium (Eu)	63	5.24	208.4	Eu^{3+}	94.7	106.6	6	Colorless to pale pink
Gadolinium (Gd)	64	7.90	180.4	Gd^{3+}	93.8	105.3	7	Colorless
Terbium (Tb)	65	8.23	177.3	Tb^{3+}	92.3	104.0	6	Pale yellow
Dysprosium (Dy)	66	8.54	178.1	Dy^{3+}	91.2	102.7	5	Yellow green
Holmium (Ho)	67	8.79	176.2	Ho^{3+}	90.1	101.5	4	Tannish yellow
Erbium (Er)	68	9.06	176.1	Er^{3+}	89.0	100.4	3	Rose pink
Thulium (Tm)	69	9.32	175.9	Tm^{3+}	88.0	99.4	2	Pale green
Ytterbium (Yb)	70	6.90	193.3	Yb^{3+}	86.8	98.5	1	Colorless
Lutetium (Lu)	71	9.84	173.8	Lu^{3+}	86.1	97.7	0	Colorless

Lanthanides in Biological Systems

Although deemed non-essential elements for life, the lanthanides are indeed biologically active and have numerous applications of importance in biological systems (Cotton and Harrowfield 2012). At the level of biomolecules, Ln extensively interacts with sugars (Yang et al. 2003) and amino acids (Zheng 2012). Furthermore, Ln^{3+} species may replace calcium bound to proteins and block Ca^{2+} permeable cation channels (Halaszovich et al. 2000), since these chemical species have similar size and a higher charge as compared to Ca^{2+} (Cotton and Harrowfield 2012). Furthermore, they may play a role as antagonists of other metals such as zinc and as catalytic centres of metalloenzymes (Ascenzi et al. 2020).

In living organisms, it was not until 2011 that Hibi et al. (2011) reported for the first time the key role of lanthanides in the activation of a methanol dehydrogenase in methylotrophic bacteria. This seminal discovery afforded a plausible explanation for the biological impact of lanthanides in living organisms and has opened up new areas of biology and biochemistry (Good et al. 2018). Concomitantly, interest in the bioinorganic chemistry of this series of elements is growing at great speed as increasing number of Ln-dependent bacteria are being unveiled (Lumpe et al. 2018). In particular, methylotrophic bacteria tend to prefer the light lanthanides (La to Gd) as cofactors in their alcohol dehydrogenase enzymes (Fitriyanto et al. 2011; Nakagawa et al. 2012; Good et al. 2016; Lv and Tani 2018;

Ochsner et al. 2019). It has been demonstrated that proline significantly contributes to lanthanide selectivity by a putative conformational modification of Ca^{2+} binding (Cotruvo et al. 2018). Importantly, La-dependent alcohol dehydrogenases display a broader distribution in biological systems as compared to the Ca-dependent types, since the first are predicted to have arisen prior to the Ca-dependent methanol dehydrogenases in the evolution of living organisms on the Earth (Cotruvo 2019). Since nearly 350 thousand prokaryotic genomes have so far been sequenced and deposited in the GeneBank (https://www.ncbi.nlm.nih.gov/genome/browse/#!/overview/; April 2021), a plethora of data resources are available for study and research purposes. Indeed, challenging approaches taking advantage of diverse bioinformatics platforms have been proposed in recent years (i.e. Gómez-Merino et al. 2015; Land et al. 2015; Ellis 2020; Mahadevakumar and Sridhar 2020; Hugenholtz et al. 2021; Li et al. 2021).

In addition to the advances in the characterization of the specific biological functions of lanthanides in bacteria, these elements have also been reported to accumulate in plants (Markert 1987; Tyler 2004; Haneklaus et al. 2015). Importantly, lanthanides have been proposed to break the normal evolution of plants to initiate leaf cell endocytosis, which ultimately modified the whole metabolism of the organism (Yang et al. 2018). Furthermore, an important biostimulation role of lanthanides in plant biology has been increasingly reported (García-Jiménez et al. 2017; Gómez-Merino and Trejo-Téllez 2018; Ramírez-Olvera et al. 2018; Gómez-Merino et al. 2020a, 2020b, Deshmukh et al. 2020).

In animals, Ce may reduce oxidative damage and palliate motor and cognitive disorders derived from traumatic brain impairments (Pałasz et al. 2019). Nevertheless, La^{3+} and Gd^{3+} have been demonstrated to block different Ca^{2+}, K^+, and Na^+ channels in human and animal cells, while Dy^{3+} and La^{3+} hinder Ca^{2+} and Mg^{2+} ATPases and Eu^{3+} and Tb^{3+} impair the normal activity of calcineurin (Pałasz and Czekaj 2000). Lanthanides also modulate neurotransmitter metabolism and function, chemical synapse, ionotropic receptors (i.e. GABA, NMDA, and kainite), cell proliferation, and apoptosis, thus altering vital physiological processes in both animals and humans (Pałasz and Czekaj 2000; Pałasz et al. 2019). The specific mechanisms by which lanthanides trigger such effects deserve further research, though some proposals have been made. It is believed that this series of elements might exert their action locally within the gastrointestinal tract, especially affecting the bacterial micro-flora and consequently the processes of nutrient uptake, digestibility, and utilization (Redling 2006). Furthermore, the baicalin complexes containing La^{3+}, Y^{3+}, and Nd^{3+} exhibit antioxidative and anti-inflammatory activities in mice RAW 264.7 macrophages (Zeng and Xu 2015). In medicinal chemistry, Ln-based molecules and nanoparticles have been tested in bioimaging, biosensing, molecular delivery, photodynamic and radiation therapy, among others (Teo et al. 2016). This emerging field in Ln applications has tremendous potential for use in numerous industries including medicine, food, animal husbandry, and crop production, to cite just some examples. Indeed, Ln have been widely exploited in food authentication and traceability studies (Aceto et al. 2017; Aceto et al. 2019). This is based on the fact that plants can absorb and transport Ln from soil and water and distribute them to all plant tissues and organs (i.e. roots, leaves, stems, flowers, fruits, and seeds) (Liang et al. 2008). In plant tissues, such distribution follows a concentration gradient tightly related to the concentration profile present in a given environment, which can be used as an effective geographical fingerprint marker (Ascenzi et al. 2020).

Lanthanides in Plants

With more than 44 million metric tons in REE reserves as of 2019 (Statistica 2020), China is unsurprisingly the world's leading producer and user of these elements. After China, important REE reserves are found in Brazil (22 million metric tons), Vietnam (21 million), Russia (12 million), India (7 million), Australia (3.3 million), the United States (1.4 million), Tanzania (890 thousand), and South Africa (790 thousand metric tons) (Statistica 2020).

As a consequence of the abundance and early interest in these elements in China, since the early 1970s, lanthanides have been the subject of extensive research and applications in different industries, including agriculture (Guo 1993; Pang et al. 2002). Indeed, some of the pioneering studies testing lanthanides to improve crop productivity were performed in China (Ji and Cui 1988; Hu et al. 2004). Other major contributors to the development of Ln-related science and technology, especially Ln in plants, are the United States, France, Japan, and Germany (Ramos et al. 2016, Tripathi et al. 2021).

According to Brown et al. (1990), the uptake and translocation of Ln in plants may be influenced by soil chelates, while increasing applications of nitrogen (N) and potassium (K) fertilization, and decreasing phosphorus (P) supply may stimulate La absorption by root cells. Furthermore, the Casparian strip in the primary roots prevent Ln acquisition, whereas significant interactions among different Ln have been observed (Hu et al. 2004).

Lanthanide availability in soils is largely dependent on the water soluble and exchangeable fractions of these elements, and thus physical and chemical soil properties including clay content, pH, and cation exchange capacity (CEC) are key factors controlling Ln uptake by root cells (El-Ramady 2009). In general, a higher availability of lanthanides in the growth media increases their uptake by plants, as has been documented in various studies (Wyttenbach et al. 1994; Ramírez-Martínez et al. 2012; Basu et al. 2015; de Oliveira et al. 2015; Ramírez-Olvera et al. 2018, Deshmukh et al. 2020).

According to El-Ramady (2009), the concentrations of lanthanides in plant tissues are typically found as follows: root > leaf > stem > flower > fruit. Nonetheless, this concentration pattern may vary according to factors such as application method, concentration of the elements in growth media, plant tissue, and genotypes. For instance, in rice plants treated with 100 μM $CeCl_3$ in the nutrient solution in hydroponics, root tissues had approximately 5000 mg Ce kg^{-1}, while in shoots the concentration found was c. 50 mg Ce kg^{-1} on a dry basis (Ramírez-Olvera et al. 2018). In corn and mung bean, the application of 5 μM La or Ce produced approximately 4000 mg kg^{-1} of each element in roots and nearly 30 mg Ce kg^{-1} in shoots (Diatloff et al. 2008).

Once inside the plants, lanthanides can form chelates and combine with molecules such as proteins, nucleic acids, amino acids, nucleotide acids, pigments, and cellulose (Hu et al. 2004).

Patterns of accumulation and partitioning of lanthanides in plant tissues are substantially affected by the formation of Ln complexes in the soil solution (El-Ramady 2009). In the root rhizosphere, organic exudates may control soil-normalized Ln fractionation patterns (Wyttenbach et al. 1998), though exogenous and intrinsic ligands may also affect this process. In soils containing abundant light lanthanides, organic acids acting as ligands such as citrate, malate, and oxalate are usually found in the rhizosphere (Jones 1998; Wei et al. 2001), while heavy lanthanide abundance may result from the formation of Ln-complexes with intrinsic chelators in the xylem cells (Ding et al. 2003). In wheat, light and heavy Ln enrichments were evident in stems and leaves, respectively (Ding et al. 2005).

Beneficial Effects of Lanthanides in Plants

Some lanthanides have been regarded as inorganic biostimulants in various crop species, since they can enhance attributes related to biomass production, productivity traits, yield quality, and responses to several environmental stressors. Hence, these elements are being widely applied as supplemental fertilizers in agriculture (Tyler 2004; Gómez-Merino and Trejo-Téllez 2018; Cotruvo 2019, Tripathi et al. 2021).

Although the lanthanide series of chemical elements encompasses 15 elements (from La to Lu in the periodic table), most of the studies and reports have been focused on La and Ce. Here we highlight some of the most salient beneficial effects of these elements when applied at low or medium doses. Since these elements trigger hormesis in plants, suitable application concentrations and frequencies would significantly contribute to avoiding negative effects.

Lanthanum is the first chemical element in the list of lanthanides. It has atomic number 57, atomic mass 138.90547 u, electronic configuration [Xe] $5d^1 6s^2$, and often oxidation state La^{3+}. La has been ranked number 28 in terms of average crustal abundance, being approximately three times as copious as lead (Pb), comprising c. 25% of the total Ln content in the carbonate-fluoride mineral bastnäsite and the phosphate mineral monazite (Xu et al. 2012).

Beneficial effects triggered by La can be explained at physiological, biochemical, and molecular levels. Shoot and root biomass weight and N concentration were enhanced in coconut palm in response to the application of 1 g earth chloride containing 22–24% La_2O_3 (Wahid et al. 2000). In *Arabidopsis*, La promoted floral initiation and reproductive growth (He and Loh 2000). In hydroponically grown tobacco seedlings, enhanced production of dry matter together with higher contents of chlorophylls and accelerated photosynthetic light reactions were observed when plants were treated with 5–20 mg $LaCl_3$ L^{-1} (Chen et al. 2001). In cucumber, lanthanum applications displayed better stimulation responses at an optimal concentration of 0.02 mM La^{3+}, since it enhanced growth and increased Mn and Fe concentrations (Zeng et al. 2000). Also in cucumber plants, La may facilitate absorption, transport, and distribution of beneficial elements such as Se, Co, V, and Tc (Huang et al. 2003). Moreover, stimulated growth parameters, increased chlorophyll and carotenoid contents, and higher activity of antioxidant enzymes including catalase (CAT), peroxidase (POX), and superoxide dismutase (SOD) were observed when plants were exposed to 2.0–20.0 μM $LaCl_3$ (Shi et al. 2005). In this same crop species exposed to NO_3^- stress, the enzymatic activity of Mg^{2+}-ATPase and Ca^{2+}-ATPase, chlorophyll and carotenoids contents, as well as leaf gas exchange were enhanced after application of 20 μM $LaCl_3$ (Gao et al. 2009). Decreased cell membrane permeability as well as reduced malondialdehyde (MDA), hydrogen peroxide (H_2O_2), and proline contents were observed in soya bean exposed to UV-B stress when treated with 20 mg La^{3+} L^{-1} (Yan et al. 2007). In addition, chlorophyll content, photosynthetic rate, incidence of binucleated cells, as well as root and shoot biomass were increased in soya bean after the application of 5–10 μM La (de Oliveira et al. 2015). In maize and mung bean, trivalent lanthanum positively affected growth when applied at >0.2 μM La (Diatloff et al. 2008). Interestingly, maize plant roots can accumulate approximately 60% more La than the shoots, while in the range of 25–100 μM, La increased chlorophyll index and shoot biomass (Duarte et al. 2018). Nonetheless, in durum wheat exposed to ≤100 μM La seed germination and seedling growth were inhibited (d'Aquino et al. 2009). In tulip, flower stem diameter and length were higher when plants received 10 μM La through a fertigation system (Ramírez-Martínez et al. 2009), while a significant accumulation of Ca, K, and La was observed in flower stems treated with 10 and 20 μM La in the vase solution (Ramírez-Martínez et al. 2012). In rice plants exposed to 0.05 mmol La^{3+} L^{-1}, significant increase of root growth was observed, though when applying 1.0 and 1.5 mmol La^{3+} L^{-1} a drastic decrease in this parameter was registered, whereas the accumulation of macronutrients (i.e. K, Ca, and Mg), micronutrients (i.e. Fe, Zn, Mn, Cu, and Mo), and sodium in the roots was also affected by La^{3+} treatments, thus pointing to a hormetic role of this element in rice (Liu et al. 2012a, 2013). Furthermore, Liu et al. (2012b) showed that La^{3+} is involved in signal transduction networks mediated by calmoduline (CaM), entering into the plant cells via intracellular Ca^{2+} channels in the plasma membrane. Recently, Wang et al. (2019) reported a mechanism for clathrin-mediated endocytosis of extracellular La^{3+} cargoes, which requires extracellular arabinogalactan proteins that are anchored on the outer face of the plasma membrane in *Arabidopsis*.

In maize, mung bean, and black bean, the application of 5–50 μM La substantially improved attributes related to seed germination, root and shoot growth, as well as fresh and dry biomass weight (Chaturvedi et al. 2014). In citrus plants, the application of 50 mg $LaCl_3 \times 7H_2O$ also increased biomass weight and growth parameters (Turra et al. 2015).

Increased chlorophyll and protein contents were observed after the application of La in lemon grass (an REE accumulator), which can accumulate 20.72 mg La g^{-1} and 25.62 mg La g^{-1} on a dry basis in roots and shoots, respectively (Basu et al. 2015). In sweet pepper, the application of 10 μM significantly increased seedling growth and the biosynthesis of chlorophylls, soluble sugars, and soluble protein concentration (García-Jiménez et al. 2017). In barley plants established in soils having approximately 200 mg La kg^{-1}, both leaves and stems readily accumulate La, while biomass weight steadily increases nearly 14% compared to the control, with increased concentrations of Ca in leaves and Si in stems (Fastovets et al. 2017). Indeed, La has been regarded as a beneficial element capable of stimulating plant responses in a hormetic manner, with tremendous potential to elicit beneficial effects when applied at low doses (Agathokleous et al. 2018). With 703 La-induced hormetic responses studied in different plant species, the maximum biological response to low La concentrations was determined to be 150% below the control response. The corresponding geometric mean of 142% as compared to the untreated control was set at 56 μM, which is usually affected by the pH value of the growth media, concentrations below the no-observed-adverse-effect-level (NOAEL, estimated in ~250 μM La), and time window (Agathokleous et al. 2019). Accordingly, the application of 150 mg $La(NO_3)_3 \times 6H_2O$ L^{-1} increased the level of the antioxidant protective enzymes SOD and POX, while significantly decreasing MDA contents and thus reducing photosynthesis damage in P-deficient adzuki bean seedlings (Lian et al. 2019). In switchgrass seedlings, La and abscisic acid coregulate chlorophyll biosynthesis and subsequent structural modification of this molecule (He et al. 2020). Furthermore, the application of 40 μM La (either as $La(NO_3)_3$ or $LaCl_3$) was demonstrated to extend vase life of cut tulip flowers by enhancing some physiological (i.e. water consumption) and biochemical variables (i.e. concentrations of chlorophylls, proteins, and sugars) (Gómez-Merino et al. 2020a). Complementarily, La also retards and enhances postharvest quality attributes in this ornamental species (Gómez-Merino et al. 2020b).

Cerium (Ce) is the second element of this series, and the most abundant among all lanthanides in the Earth's crust. This element has the atomic number 58, atomic mass 140.116 u, electronic configuration [Xe] $4f^1 5d^1 6s^2$, and is most often found as Ce^{3+}. Importantly, Ce is the 26th most abundant element, found in concentrations ≤66 ppm of the Earth's crust, a value that is just behind that of copper (Cu; 68 ppm), and half as much as chlorine (Cl; 145 ppm) and five times as much as Pb (14 ppm). It is mainly found in minerals such as monazites and bastnäsites (Xu et al. 2012).

Like all other lanthanides, which are not considered essential for plants, but rather categorized as beneficial elements, Ce has shown diverse effects depending on the dose applied. Positive effects have been observed in: rice seed germination (Fashui et al. 2002; Ramírez-Olvera et al. 2018); plant height of *Arabidopsis thaliana* (He and Loh 2000); root growth and volume in maize and mung bean (Diatloff et al. 2008), air potato or yam (Wang et al. 2010), and diverse medicinal plants (Zhang et al. 2013); biomass weight in rice (Fashui et al. 2002), maize (Gong et al. 2011), and lettuce (Barbieri et al. 2013); yield of Chinese cabbage (Ma et al. 2014); chlorophyll concentration in *A. thaliana* (Wang et al. 2012); total sugars in Chinese cabbage (Ma et al. 2014) and pak choi (Hu et al. 2015); and the nutrimental status of several species (Diatloff et al. 2008; Liu et al. 2013; Shtangeeva 2014; Chen et al. 2015a; Chen et al. 2015b). In Mg-deficient spinach plants, the application of $CeCl_3$ stimulated the enzymatic activity of glutamic-pyruvic transaminase (GPT), glutamate dehydrogenase (GDH), glutamate synthase (GltS), nitrate reductase (NR), nitrite reductase (NiR), and urease, all of them implicated in N metabolism and thus enhancing plant growth (Yin et al. 2009). Coincidently, Ce^{3+} applications stimulated antioxidant metabolism and nutrient status in roots and shoots of rice plants in a hormetic manner (Liu et al. 2012c). Nevertheless, in wheat, Ce applications negatively affected root length, dry matter weight of shoots and roots, as well as the nutrient status (Hu et al. 2002). Radish and tomato seeds decreased their germination percentage when exposed to Ce in an acidic soil with pH 4.08, a condition that triggered a higher mobility and availability of Ce in the soil solution (Thomas et al. 2014). The absorption of nutrients can also be affected in Ce-treated plants, and such an effect can be either negative or positive (Diatloff et al. 2008; Liu et al. 2012c). In spinach, Ce may in part substitute Mg functions (Yin et al. 2009), while in maize N assimilation and photosystem II (PSII) activity can be improved in response to Ce (Zhao et al. 2012). In duckweed, plant growth was stimulated with 100 μM Ce, while drastic decreases were observed when plants were exposed to 500 and 1000 μM Ce, with evident chlorosis, deficient chlorophyll and carotenoid concentrations, and enhanced antioxidant activity (Zicari et al. 2018). Ultimately, the effect of conventional Ce on plant physiology and metabolism is largely dependent on the applied dose, frequency of application, time of exposure, agronomic management, developmental stage, and genotype tested.

However, Ce has also been tested in the form of nanoparticles, which have both stimulatory and toxic effects on plants and soil biology. Microbial communities interacting in the soil system, soil properties and dynamics, genotypic variation of plants evaluated, as well as nanofertilizer properties and characteristics are among the factors that most affect the effects of nanoceria on plants (El-Ramady et al. 2018). In rice, cerium oxide nanoparticles ($nCeO_2$) at 500 mg kg^{-1} soil altered the nutritional value of grains by reducing the concentrations of mineral nutrients (i.e. Fe and S), as well as those of starch, lauric and valeric acids, and glutelin and prolamin (Rico et al. 2013). Instead, the application of 125–500 mg $nCeO_2$ kg^{-1} to the soil in wheat crops improved plant growth, shoot biomass, and grain yield, but when applied at 500 mg kg^{-1} it decreased the linoleic acid contents (Rico et al. 2014). In barley, kernels are negatively affected by $nCeO_2$, with the potential of negatively impacting malt and feed production (Poščić et al. 2016). In lettuce, ryegrass, and tomato, the application of 250 or 500 μg $nCeO_2$ mL did not affect seed germination but significantly decreased root length (Andersen et al. 2016). In rapeseed exposed to salt stress (50 mM NaCl), the application of 500 mg $nCeO_2$ kg^{-1} shortened the root apoplastic barriers that allowed more Na^+ transport to shoots and less accumulation of Na^+ in roots and thus better physiological plant performance (Rossi et al. 2017). In soya bean, the application of 100 mg $nCeO_2$ kg^{-1} positively impacted photosynthesis at the moisture content above 70% of field capacity (θfc), but when it was reduced to 55% θfc, the positive effects of Ce were no longer evident (Cao et al. 2018). In pea and wheat, exposure to uncoated and glucose-, levan-, and pullulan-coated $nCeO_2$ did not affect seed germination rate, but significantly stimulated total phenolic content as well as root and shoot elongation (Milenković et al. 2019). Using stevia leaf extracts, nickel-doped cerium oxide nanoparticles green synthesized protected plants against harmful ultraviolet rays, making them a suitable choice for use in sunscreens (Khatami et al. 2019).

Other lanthanides, including Nd^{3+} applied at low concentrations (1, 3, or 5 mg L^{-1}), promoted seed germination of sicklepod, whereas when applied at doses ≥5 mg L^{-1}, they inhibited germination (Yao et al. 2008). In milkvetch, the application of 6 mg Nd^{3+} L^{-1} had a more positive effect on seed germination as compared to burdock oligosaccharide or NaCl (Sun et al. 2008). In seeds of dong quai or female ginseng (*Angelica sinensis*), the application of 80 mg Nd^{3+} L^{-1} caused the highest germination rate (57%), which was 22% higher than that of the control (Chu et al. 2010). In rice, Nd^{3+} did not significantly impact mitochondria at low concentrations (<200 μM), whereas at high concentrations (≥ 200–500 μM) it induced severe damages to membranes and ultrastructures of mitochondria, with a concomitant mitochondria permeability transition pore (mPTP) opening (Xia et al. 2015). Both Eu and Y inhibited the nitrification activity of the bacteria *Nitrosomonas europaea* and *Nitrobacter winogradskyi*, which have a key role in nitrogen metabolism for plants (Fujita et al. 2015). In common milkweed, showy ticktrefoil, switchgrass, radish, and tomato, exposure to Pr, Nd, Sm, Tb, Dy, and E in the soil did not affect speed of germination at low levels, though radish in Nd soils and tomato in Er soils decreased this speed at the highest dosage evaluated (Carpenter et al. 2015). Both Gn and Nd have shown to significantly increase chlorophyll contents by up to 21% relative to the control culture, while they prevented the incorporation of heavy isotopes of magnesium and maintained efficient photosynthetic rates (Řezanka et al. 2016). In sunflower, the application of low doses of La and Nd (≤100 mg kg^{-1}) had stimulatory effects on shoot and root, respectively, whereas Nd enhanced the number of leaves and chlorophyll content of pak choi (Rezaee et al. 2018).

Summing up, lanthanides may differentially affect seed germination, root growth, biomass accumulation, primary and secondary metabolism, as well as absorption of both essential and non-essential nutrients in plants. Since they have dose-response hormetic effects, they may trigger stimulatory responses when applied at low to moderate concentrations, whereas negative or toxic effects are observed when applied at higher doses. To date, only a few lanthanides (i.e. La, Ce, Nd and Eu) have been studied in plants. Therefore, further research in this area should be aimed at deciphering the impact of other elements of this series on the yields and quality of more cultivated plants.

Conclusions and Future Research Needs

It is evident that lanthanides have an enormous impact in relevant living systems' fields, including plant biology and agriculture. Applying timely, small-amount doses of these elements may improve yields and crop quality, while ameliorating some effects of major environmental stressors. However, there are critical knowledge gaps thwarting an in-depth understanding of their role in both natural and artificial systems that, if achieved, could help lead to innovative solutions in different fields related to crop production, biopharming, and food and beverage authentication studies, mainly regarding geographical origin, as well as environmental impacts. Special attention should be paid to Ln-microbiological interactions; hormetic dose-response curves; novel models based on morphological, physiological, biochemical, and molecular mechanisms explaining the hormetic effects observed, if any; and their total environmental impacts. Understanding such mechanisms and processes may enable approaches that are more accurate and thus improve crop production and productivity, while reducing the potential of toxic impacts on water bodies, soils, wildlife, vegetation, and human life. Since lanthanides have been shown to ameliorate some human disorders, a more in-depth understanding of their potential mechanistic functions in the course of a disease and as therapeutic agents needs further attention. Indeed, since plant organs and tissues can accumulate these elements, human consumption of such products could be a way of prescribing these elements with therapeutic and diagnostic applications in modern medicine. Though we consider research on lanthanides to be still in its infancy, new approaches will undoubtedly increase our knowledge of these elements and lead to greater use of their unique properties in terms of the development of innovations in the fields of plant sciences and agriculture, as well as in other industries including medicine.

References

Aceto, M., Musso, D., Calà, E. et al. (2017). Role of lanthanides in the traceability of the milk production chain. *Journal of Agricultural and Food Chemistry* 65: 4200–4208. https://doi.org/10.1021/acs.jafc.7b00916.

Aceto, M., Calà, E., Musso, D. et al. (2019). A preliminary study on the authentication and traceability of extra virgin olive oil made from Taggiasca olives by means of trace and ultra-trace elements distribution. *Food Chemistry* 298: 125047. https://doi.org/10.1016/j.foodchem.2019.125047.

Agathokleous, E., Kitao, M., and Calabrese, E.J. (2018). The rare earth element (REE) lanthanum (La) induces hormesis in plants. *Environmental Pollution* 238: 1044–1047. https://doi.org/10.1016/j.envpol.2018.02.068.

Agathokleous, E., Kitao, M., and Calabrese, E.J. (2019). Hormetic dose responses induced by lanthanum in plants. *Environmental Pollution* 244: 332–341. https://doi.org/10.1016/j.envpol.2018.10.007.

Aide, M. (2018). Lanthanide soil chemistry and its importance in understanding soil pathways: Mobility, plant uptake, and soil health. In: *Lanthanides* (ed. N.S. Awwad and A.T. Mubarak), 49–65. London: IntechOpen https://doi.org/10.5772/intechopen.79238.

Andersen, C.P., King, G., Plocher, M. et al. (2016). Germination and early plant development of ten plant species exposed to titanium dioxide and cerium oxide nanoparticles. *Environmental Toxicology and Chemistry* 35: 2223–2229. https://doi.org/10.1002/etc.3374.

Ascenzi, P., Bettinelli, M., Bof, A. et al. (2020). Rare earth elements (REE) in biology and medicine. *Rendiconti Lincei. Scienze Fisiche e Naturali* https://doi.org/10.1007/s12210-020-00930-w.

Aspinall, H.C. (2018). *Chemistry of the F-Block Elements. eBook*. London: Routledge https://doi.org/10.1201/9781315139258.

Baldwin, A.G., Ivanov, A.S., Williams, N.J. et al. (2018). Outer-sphere water clusters tune the lanthanide selectivity of diglycolamides. *ACS Central Science* 4: 739–747. https://doi.org/10.1021/acscentsci.8b00223.

Barbieri, A.P.P., Espindola, M.C.G., Menezes, N.L., and Henrique, D.F.S. (2013). Lettuce seeds treated with cerium and lanthanum aqueous solutions. *Pesquisa Agropecuária Tropical* 43: 104–109. https://doi.org/10.1590/S1983-40632013000100013.

Basu, A., Panda, S.S., and Dhal, N.K. (2015). Effect and accumulation of lanthanum on the growth and physiological activities of *Cymbopogon flexuosus* (Nees Ex Steud.) W. Watson. *Current World Environment* 10: 951–956. https://doi.org/10.12944/CWE.10.3.26.

Brown, P.H., Rathjien, A.H., Graham, R.D., and Tribe, D.E. (1990). Rare earth elements in biological systems. In: *Handbook on the Physics and Chemistry of Rare Earths* (ed. K.A. Geschneider and J.L. Eyring), 423–452. New York: Elsevier https://doi.org/10.1016/S0168-1273(05)80135-7.

Cao, Z., Rossi, L., Stowers, C. et al. (2018). The impact of cerium oxide nanoparticles on the physiology of soybean (*Glycine max* (L.) Merr.) under different soil moisture conditions. *Environmental Science and Pollution Research International* 25: 930–939. https://doi.org/10.1007/s11356-017-0501-5.

Carpenter, D., Boutin, C., Allison, J.E. et al. (2015). Uptake and effects of six rare earth elements (REEs) on selected native and crop species growing in contaminated soils. *PLoS One* 10: e0129936. https://doi.org/10.1371/journal.pone.0129936.

Charalampides, G., Vatalis, K.I., Apostoplos, B., and Ploutarch-Nikolas, B. (2015). Rare Earth Elements: Industrial applications and economic dependency of Europe. *Procedia Economics and Finance* 24: 126–135. https://doi.org/10.1016/S2212-5671(15)00630-9.

Chaturvedi, N., Gannavarapu, R., and Dhal, N.K. (2014). Effect of lanthanum on the growth and physiological activities of *Zea mays*, *Vigna radiata* and *Vigna mungo*. *International Journal of Environmental Science* 4: 653–659. https://doi.org/10.6088/ijes.2014040404505.

Chen, W.J., Tao, Y., Gu, Y.H., and Zhao, G.W. (2001). Effect of lanthanide chloride on photosynthesis and dry matter accumulation in tobacco seedlings. *Biological Trace Element Research* 79: 169–176. https://doi.org/10.1385/733BTER:79:2:169.

Chen, S.A., Wang, X.D., Zhao, B. et al. (2015a). Regulating the cell growth and shoot induction of *Crocus sativus* embryogenic callus by rare earth elements. *Chinese Bulletin of Botany* 45: 609–614.

Chen, Y., Luo, Y., Qiu, N. et al. (2015b). Ce^{3+} induces flavonoids accumulation by regulation of pigments, ions, chlorophyll fluorescence and antioxidant enzymes in suspension cells of *Ginkgo biloba* L. *Plant Cell Tissue and Organ Culture* 123: 283–296. https://doi.org/10.1007/s11240-0150831-2.

Chu, Z., Sun, L., Liu, C. et al. (2010). Effect of Nd^{3+} and burdock oligosaccharide on the germination power of *Angelica sinensis* seeds. *Chinese Agricultural Science Bulletin* 13: 191–193.

Cotruvo, J.A.J. (2019). The chemistry of lanthanides in biology: Recent discoveries, emerging principles, and technological applications. *ACS Central Science* 5: 1496–1506. https://doi.org/10.1021/acscentsci.9b00642.

Cotruvo, J.A.J., Featherston, E.R., Mattocks, J.A. et al. (2018). Lanmodulin: A highly selective lanthanide-binding protein from a lanthanide-utilizing bacterium. *Journal of the American Chemical Society* 140: 15056–15061. https://doi.org/10.1021/jacs.8b09842.

Cotton, S.A. and Harrowfield, J.M. (2012). *Lanthanides: biological activity and medical applications. Encyclopedia of Inorganic and Bioinorganic Chemistry* 2012: https://doi.org/10.1002/9781119951438.eibc2091.

d'Aquino, L., de Pinto, M.C., Nardi, L. et al. (2009). Effect of some light rare earth elements on seed germination, seedling growth and antioxidant metabolism in *Triticum durum*. *Chemosphere* 75: 900–905. https://doi.org/10.1016/j.chemosphere.2009.01.026.

Deshmukh, R., Tripathi, D.K., and Guerriero, G. (2020). *Metalloids in Plants: Advances and Future Prospects*, vol. 1, 1–520. John Wiley & Sons.

Diatloff, E., Smith, F.W., and Asher, C.J. (2008). Effects of lanthanum and cerium on the growth and mineral nutrition of corn and mungbean. *Annals of Botany* 101: 971–982. https://doi.org/10.1093/aob/mcn021.

Ding, S.M., Liang, T., Wang, L.J., and Sun, Q. (2003). Characteristics of translocation and fractionation of rare earth elements in soil–wheat system. *Journal of Agro-Environment Science* 22: 519–523.

Ding, S.M., Liang, T., Zhang, C. et al. (2005). Accumulation and fractionation of rare earth elements (REEs) in wheat: controlled by phosphate precipitation, cell wall absorption and solution complexation. *Journal of Experimental Botany* 56: 2765–2775. https://doi.org/10.1093/jxb/eri270.

Duarte, A.C.O., de Oliveira, C., Ramos, S.J. et al. (2018). Lanthanum content and effects on growth, gas exchanges, and chlorophyll index in maize plants. *Acta Scientianum. Biological Sciences* 40: 38469. https://doi.org/10.4025/actascibiolsci.v40i1.38469.

Ellis, J.C. (2020). P finder: genomic and metagenomic annotation of RNase P RNA gene (*rnpB*). *BMC Genomics* 21: 334. https://doi.org/10.1186/s12864-020-6615-z.

El-Ramady, H.R. (2009). *Ecotoxicology of Rare Earth Elements: Ecotoxicology of Rare Earth Elements within Soil and Plant Environments*. Saarbrücken: VDM Publishing.

El-Ramady, H., Abdalla, N., Alshaal, T. et al. (2018). Plant nano-nutrition: perspectives and challenges. In: *Nanotechnology, Food Security and Water Treatment* (ed. K. Gothandam, S. Ranjan, N. Dasgupta, et al.), 129–161. Cham: Springer https://doi.org/10.1007/978-3-319-70166-0_4.

Fashui, H., Ling, W., Xiangxuan, M. et al. (2002). The effect of cerium (III) on the chlorophyll formation in spinach. *Biological Trace Element Research* 89: 263–276. https://doi.org/10.1385/BTER:89:3:263.

Fastovets, I., Kotelnikova, A., Rogova, O. et al. (2017). Effects of soil lanthanum on growth and elemental composition of plants. *Proceedings of 19th EGU General Assembly, EGU2017, 305.* 23–28 April, 2017. Vienna, Austria, https://ui.adsabs.harvard.edu/abs/2017EGUGA..19..305F/abstract.

Fitriyanto, N.A., Fushimi, M., Matsunaga, M. et al. (2011). Molecular structure and gene analysis of Ce^{3+}-induced methanol dehydrogenase of *Bradyrhizobium* sp. MAFF211645. *Journal of Bioscience Bioengineering* 111: 613–617. https://doi.org/10.1016/j.jbiosc.2011.01.015.

Fujita, Y., Barnes, J., Eslamimanesh, A. et al. (2015). Effects of simulated rare earth recycling wastewaters on biological nitrification. *Environmental Science and Technology* 49: 9460–9468. https://doi.org/10.1021/acs.est.5b01753.

Gao, Q.H., Wu, Y., Wang, X.F. et al. (2009). Alleviation effects of $LaCl_3$ on photosynthetic characteristics of cucumber seedlings under nitrate stress. *Ying Yong Sheng Tai Xue Bao* 20: 2685–2690.

García-Jiménez, A., Gómez-Merino, F.C., Tejeda-Sartorius, O., and Trejo-Téllez, L.I. (2017). Lanthanum affects bell pepper seedling quality depending on the genotype and time of exposure by differentially modifying plant height, stem diameter and concentrations of chlorophylls, sugars, amino acids, and proteins. *Frontiers in Plant Science* 8: 308. https://doi.org/10.3389/fpls.2017.00308.

Gómez-Merino, F.C. and Trejo-Téllez, L.I. (2018). The role of beneficial elements in triggering adaptive responses to environmental stressors and improving plant performance. In: *Biotic and Abiotic Stress Tolerance in Plants* (ed. S. Vats), 137–172. Singapore: Springer Nature https://doi.org/10.1007/978-981-10-9029-5_6.

Gómez-Merino, F.C., Trejo-Téllez, L.I., and Alarcón, A. (2015). Plant and microbe genomics and beyond: potential for developing a novel molecular plant nutrition approach. *Acta Physiologiae Plantarum* 37: 208. https://doi.org/10.1007/s11738-015-1952-2.

Gómez-Merino, F.C., Ramírez-Martínez, M., Castillo-González, A.M., and Trejo-Téllez, L.I. (2020a). Lanthanum prolongs vase life of cut tulip flowers by increasing water consumption and concentrations of sugars, proteins and chlorophylls. *Scientific Reports* 10: 4209. https://doi.org/10.1038/s41598-020-61200-1.

Gómez-Merino, F.C., Castillo-González, A.M., Ramírez-Martínez, M., and Trejo-Téllez, L.I. (2020b). Lanthanum delays senescence and improves postharvest quality in cut tulip (*Tulipa gesneriana* L.) flowers. *Scientific Reports* 10: 19437. https://doi.org/10.1038/s41598-020-76266-0.

Gong, X., Hong, M., Wang, Y. et al. (2011). Cerium relieves the inhibition of photosynthesis of maize caused by manganese deficiency. *Biological Trace Element Research* 141: 305–316. https://doi.org/10.1007/s12011-010-8716-z.

Good, N.M., Vu, H.N., Suriano, C.J. et al. (2016). Pyrroloquinoline quinone ethanol dehydrogenase in *Methylobacterium extorquens* AM1 extends lanthanide-dependent metabolism to multicarbon substrates. *Journal of Bacteriology* 198: 3109–3118. https://doi.org/10.1128/JB.00478-16.

Good, N.M., Walser, O.N., Moore, R.S. et al. (2018). Investigation of lanthanide-dependent methylotrophy uncovers complementary roles for alcohol dehydrogenase enzymes. *bioRxiv* 329011. https://doi.org/10.1101/329011.

Guo, B.S. (1993). The application of rare earth elements on agriculture and breeding. *Journal of the Chinese Rare Earths Society* 15: 37–43.

Halaszovich, C.R., Zitt, C., Jungling, E., and Luckhoff, A. (2000). Inhibition of TRP3 channels by lanthanides. Block from the cytosolic side of the plasma membrane. *Journal of Biological Chemistry* 275: 37423–37428. https://doi.org/10.1074/jbc.M007010200.

Haneklaus, S.H., Schnug, E., Lottermoser, B.G., and Hu, Z. (2015). Lanthanides. In: *Handbook of Plant Nutrition* (ed. A.V. Barker and D.J. Pilbeam), 625–650. Boca Raton: CRC Press.

He, Y. and Loh, C. (2000). Cerium and lanthanum promote floral initiation and reproductive growth of *Arabidopsis thaliana*. *Plant Science* 159: 117–124. https://doi.org/10.1016/s0168-9452(00)00338-1.

He, X., You, P., and Sun, Y. (2020). Lanthanum and abscisic acid coregulate chlorophyll production of seedling in switchgrass. *PLoS One* 15: e0232750. https://doi.org/10.1371/journal.pone.0232750.

Hibi, Y., Asai, K., Arafuka, H. et al. (2011). Molecular structure of La^{3+}-induced methanol dehydrogenase-like protein in *Methylobacterium radiotolerans*. *Journal of Bioscience Bioengineering* 111: 547–549. https://doi.org/10.1016/j.jbiosc.2010.12.017.

Hinderer, J.H. (2010). Radioisotopic impurities in promethium-147 produced at the ORNL high flux isotope reactor. Master's Thesis, University of Tennessee, 2010. https://trace.tennessee.edu/utk_gradthes/717.

Hu, X., Ding, Z., Chen, Y. et al. (2002). Bioaccumulation of lanthanum and cerium and their effects on the growth of wheat (*Triticum aestivum* L.) seedlings. *Chemosphere* 48: 621–629. https://doi.org/10.1016/S0045-6535(02)00109-1.

Hu, Z., Richter, H., Sparovek, G., and Schnug, E. (2004). Physiological and biochemical effects of Rare Earth Elements on plants and their agricultural significance: A review. *Journal of Plant Nutrition* 27: 183–220. https://doi.org/10.1081/PLN-120027555.

Hu, Y.F., Yuan, C.M., Jiang, S.L. et al. (2015). Direct applications of cerium and yttrium on vegetable pak choi. *Fresenius Environmental Bulletin* 24: 4173–4178.

Huang, Z., Chen, G., and Du, J. (2003). Influence of lanthanum on the uptake of trace elements in cucumber plant. *Biological Trace Element Research* 95: 185–195. https://doi.org/10.1385/BTER:95:2:185.

Hugenholtz, P., Chuvochina, M., Oren, A. et al. (2021). Prokaryotic taxonomy and nomenclature in the age of big sequence data. *The ISME Journal* https://doi.org/10.1038/s41396-021-00941-x.

Ji, Y.J. and Cui, M.Z. (1988). Toxicological studies on safety of rare earths used in agriculture. *Biomedical and Environmental Science* 1: 270–276.

Johnson, A.T. and Nash, K.L. (2015). Mixed monofunctional extractants for trivalent actinide/lanthanide separations: TALSPEAK-MME. *Solvent Extraction and Ion Exchange* 33: 642–655. https://doi.org/10.1080/07366299.2015.1085225.

Jones, D.L. (1998). Organic acids in the rhizosphere. A critical review. *Plant and Soil* 205: 25–44. https://doi.org/10.1023/A:1004356007312.

Khatami, M., Sarani, M., Mosazadeh, F. et al. (2019). Nickel-doped cerium oxide nanoparticles: green synthesis using stevia and protective effect against harmful ultraviolet rays. *Molecules* 24: 4424. https://doi.org/10.3390/molecules24244424.

Krättli, M., Müller-Späth, T., Ulmer, N. et al. (2013). Separation of lanthanides by continuous chromatography. *Industrial and Engineering Chemistry Research* 52: 8880–8886. https://doi.org/10.1021/ie3031482.

Land, M., Hauser, L., Jun, S.R. et al. (2015). Insights from 20 years of bacterial genome sequencing. *Functional and Integrative Genomics* 15: 141–161. https://doi.org/10.1007/s10142-015-0433-4.

Li, W., O'Neill, K.R., Haft, D.H. et al. (2021). RefSeq: expanding the Prokaryotic Genome Annotation Pipeline reach with protein family model curation. *Nucleic Acids Research* 9 (D1): D1020–D1028. https://doi.org/10.1093/nar/gkaa1105.

Lian, H., Qin, C., Zhang, L. et al. (2019). Lanthanum nitrate improves phosphorus-use efficiency and tolerance to phosphorus-deficiency stress in *Vigna angularis* seedlings. *Protoplasma* 256: 383–392. https://doi.org/10.1007/s00709-018-1304-3.

Liang, T., Ding, S., Song, W. et al. (2008). A review of fractionations of rare earth elements in plants. *Journal of Rare Earths* 26: 7–15. https://doi.org/10.1016/S1002-0721(08)60027-7.

Liu, D., Lin, Y., and Wang, X. (2012a). Effects of lanthanum on growth, element uptake, and oxidative stress in rice seedlings. *Journal of Plant Nutrition and Soil Science* 175: 907–911. https://doi.org/10.1002/jpln. 201200016841.

Liu, D., Wang, X., Chen, X. et al. (2012b). Effects of lanthanum on the change of calcium level in the root cells of rice. *Communications in Soil Science and Plant Analysis* 43: 1994–2003. https://doi.org/10.1080/843 00103624.2012.693231.

Liu, D., Wang, X., Lin, Y. et al. (2012c). The effects of cerium on the growth and some antioxidant metabolisms in rice seedlings. *Environmental Science and Pollution Research* 19: 3282–3291. https://doi.org/10.1007/s11356-012-0844-x.

Liu, D., Wang, X., Zhang, X., and Gao, Z. (2013). Effect of lanthanum on growth and accumulation of rice seedlings. *Plant, Soil and Environment* 59: 196–200. https://doi.org/10.17221/760/2012-PSE.

Long, K.R., Van Gosen, B.S., Foley, N.K., and Cordier, D. (2012). The principal rare earth elements deposits of the United States: A summary of domestic deposits and a global perspective. In: *Non-Renewable Resource Issues* (ed. R. Sinding-Larsen and F.W. Wellmer), 131–155. Dordrecht: Springer https://doi.org/10.1007/978-90-481-8679-2_7.

Lumpe, H., Pol, A., Op den Camp, H.J.M., and Daumann, L.J. (2018). Impact of the lanthanide contraction on the activity of a lanthanide-dependent methanol dehydrogenase. A kinetic and DFT study. *Dalton Transactions* 47: 10463. https://doi.org/10.1039/c8dt01238e.

Lv, H. and Tani, A. (2018). Genomic characterization of methylotrophy of *Oharaeibacter diazotrophicus* strain SM30^T. *Journal of Bioscience Bioengineering* 126: 667–675. https://doi.org/10.1016/j.jbiosc.2018.05.023.

Ma, J.J., Ren, Y.J., and Yan, L.Y. (2014). Effects of spray application of lanthanum and cerium on yield and quality of Chinese cabbage (*Brassica chinensis* L.) based on different seasons. *Biological Trace Element Research* 160: 427–432. https://doi.org/10.1007/s12011-014-0062-0.

Mahadevakumar, S. and Sridhar, K.R. (2020). Plant-microbe interaction: Current developments and future challenges. In: *Advances in Plant Microbiome and Sustainable Agriculture* (ed. A.N. Yadav, A.A. Rastegari, N. Yadav and D. Kour), 1–38. Singapore: Springer https://doi.org/10.1007/978-981-15-3208-5_1.

Markert, B. (1987). The pattern of distribution of lanthanide elements in soils and plants. *Phytochemistry* 26: 3167–3170. https://doi.org/10.1016/S0031-9422(00)82463-2.

Milenković, I., Mitrović, A., Algarra, M. et al. (2019). Interaction of carbohydrate coated cerium-oxide nanoparticles with wheat and pea: Stress induction potential and effect on development. *Plants (Basel)* 8: 478. https://doi.org/10.3390/plants8110478.

Nakagawa, T., Mitsui, R., Tani, A. et al. (2012). A catalytic role of XoxF1 as La^{3+}-dependent methanol dehydrogenase in *Methylobacterium extorquens* strain AM1. *PLoS One* 7: e50480. https://doi.org/10.1371/journal.pone.0050480.

Ochsner, A.M., Hemmerle, L., Vonderach, T. et al. (2019). Use of rare-earth elements in the phyllosphere colonizer *Methylobacterium extorquens* PA1. *Molecular Microbiology* 111: 1152–1166. https://doi.org/10.1111/mmi.14208.

de Oliveira, C., Ramos, S.J., Siqueira, J.O. et al. (2015). Bioaccumulation and effects of lanthanum on growth and mitotic index in soybean plants. *Ecotoxicology and Environmental Safety* 122: 136–144. https://doi.org/10.1016/j.ecoenv.2015.07.020.

Pałasz, A. and Czekaj, P. (2000). Toxicological and cytophysiological aspects of lanthanides action. *Acta Biochimica Polonica* 47: 1107–1114. https://doi.org/10.18388/abp.2000_3963.

Pałasz, A., Segovia, H.Y., Skowronek, R., and Worthington, J.J. (2019). Molecular neurochemistry of the lanthanides. *Synapse* 73: e22119. https://doi.org/10.1002/syn.22119.

Pang, X., Li, D., and Peng, A. (2002). Application of rare-earth elements in the agriculture of China and its environmental behavior in soil. *Environmental Science and Pollution Research* 9: 143–148. https://doi.org/10.1007/BF02987462.

Panichev, A.M. (2015). Rare Earth Elements: Review of medical and biological properties and their abundance in the rock materials and mineralized spring waters in the context of animal and human geophagia reasons evaluation. *Achievements in the Life Sciences* 9: 95–103. https://doi.org/10.1016/j.als.2015.12.001.

Poščić, P., Mattiello, A., Fellet, G. et al. (2016). Effects of cerium and titanium oxide nanoparticles in soil on the nutrient composition of barley (*Hordeum vulgare* L.) kernels. *International Journal of Environmental Research and Public Health* 13: 577. https://doi.org/10.3390/ijerph13060577.

Ramírez-Martínez, M., Gómez-Merino, F.C., Trejo-Téllez, L.I. et al. (2009). Effect of lanthane on quality of tulip flower 'Ile de France'. *Acta Horticulturae* 847: 295–300. https://doi.org/10.17660/ActaHortic.9032009.847.39904.

Ramírez-Martínez, M., Trejo-Téllez, L.I., Gómez-Merino, F.C. et al. (2012). Bioaccumulation of potassium, calcium and lanthanum in tulip treated with lanthanum. *Terra Latinoamericana* 30: 229–238.

Ramírez-Olvera, S.M., Trejo-Téllez, L.I., García-Morales, S. et al. (2018). Cerium enhances germination and shoot growth, and alters mineral nutrient concentration in rice. *PLoS One* 13: e0194691. https://doi.org/10.1371/journal.pone.0194691.

Ramos, S.J., Dinali, G.S., Oliveira, C. et al. (2016). Rare earth elements in the soil environment. *Current Pollution Reports* 2: 28–50. https://doi.org/10.1007/s40726-016-0026-4.

Redling, K. (2006). Rare Earth Elements in agriculture with emphasis on animal husbandry. Doctoral Thesis, Ludwig Maximilian University of Munich. Munich, Germany https://edoc.ub.uni-muenchen.de/5936/1/Redling_Kerstin.pdf.

Rezaee, A., Hale, B., Santos, R.M. et al. (2018). Accumulation and toxicity of lanthanum and neodymium in horticultural plants (*Brassica chinensis* L. and *Helianthus annuus* L.). *The Canadian Journal of Chemical Engineering* 96: 2263–2272. https://doi.org/10.1002/cjce.23152.

Řezanka, T., Kaineder, K., Mezricky, D. et al. (2016). The effect of lanthanides on photosynthesis, growth, and chlorophyll profile of the green alga *Desmodesmus quadricauda*. *Photosynthesis Research* 130: 335–346. https://doi.org/10.1007/s11120-016-0263-9.

Rico, C.M., Morales, M.I., Barrios, A.C. et al. (2013). Effect of cerium oxide nanoparticles on the quality of rice (*Oryza sativa* L.) grains. *Journal of Agricultural and Food Chemistry* 61: 11278–11285. https://doi.org/10.1021/jf404046v.

Rico, C.M., Lee, S.C., Rubenecia, R. et al. (2014). Cerium oxide nanoparticles impact yield and modify nutritional parameters in wheat (*Triticum aestivum* L.). *Journal of Agricultural and Food Chemistry* 62: 9669–9675. https://doi.org/10.1021/jf503526r.

Rossi, L., Zhang, W., and Ma, X. (2017). Cerium oxide nanoparticles alter the salt stress tolerance of *Brassica napus* L. by modifying the formation of root apoplastic barriers. *Environmental Pollution* 229: 132–138. https://doi.org/10.1016/j.envpol.2017.05.083.

Seitz, M., Oliver, A.G., and Raymond, K.N. (2007). The lanthanide contraction revisited. *Journal of the American Chemical Society* 129: 11153–11160. https://doi.org/10.1021/ja072750f.

Shi, P., Chen, G.C., and Huang, Z.W. (2005). Effects of La^{3+} on the active oxygen-scavenging enzyme activities in cucumber seedling leaves. *Russian Journal of Plant Physiology* 52: 294–297. https://doi.org/10.1007/s11183-005-0044-3.

Shtangeeva, I. (2014). Europium and cerium accumulation in wheat and rye seedlings. *Water, Air, and Soil Pollution* 225: 1964. https://doi.org/10.1007/s11270-014-1964-3.

Statistica (2020). Rare earth reserves worldwide as of 2019 by country. https://www.statista.com/statistics/277268/rare-earth-reserves-by-country/

Sun, L.J., Wu, S.Y., Zhang, Y.L. et al. (2008). Effect of rare earth Nd, burdock oligosaccharide and NaCl on seed germination of Radix Astragali. *China Seed Industry* 1: 41–42.

Teo, R.D., Termini, J., and Gray, H.B. (2016). Lanthanides: applications in cancer diagnosis and therapy. *Journal of Medicinal Chemistry* 59: 6012–6024. https://doi.org/10.1021/acs.jmedchem.5b01975.

Thomas, P.J., Carpenter, D., Boutin, C., and Allison, J.E. (2014). Rare Earth Elements (REEs): effects on germination and growth of selected crop and native plant species. *Chemosphere* 96: 57–66. https://doi.org/10.1016/j.chemosphere.2013.07.020.

Tripathi, D.K., Singh, V.P., Ahmad, P. et al. (2021). Metalloids in plant biology: new avenues in their research. *Journal of Hazardous Materials* 422: 126738.

Turra, C., Fernandes, E.A.N., Bacchi, M.A. et al. (2015). Effects of lanthanum on citrus plant. *International Journal of New Technology and Research* 1: 48–50.

Tyler, G. (2004). Rare Earth Elements in soil and plant systems. A review. *Plant and Soil* 267: 191–206. https://doi.org/10.1007/s11104-005-4888-2.

Van Gosen, B.S., Verplanck, P.L., Seal, R.R. et al. (2017). Rare-earth elements. Critical mineral resources of the United States. Economic and environmental geology and prospects for future supply. U.S. Geological Survey Professional Paper 1802: O1–O31. https://doi.org/10.3133/pp1802O.

Wahid, P.A., Valiathan, M.S., Kamalam, N.V. et al. (2000). Effect of rare earth elements on growth and nutrition of coconut palm and root competition for these elements between the palm and *Calotropis gigantean*. *Journal of Plant Nutrition* 23: 329–338. https://doi.org/10.1080/01904160009382019.

Wang, J.N., Sha, S., Li, Y.Z. et al. (2010). Effect of $Ce(NO_3)_3$ on callus growth and adventitious root of stem from *Dioscorea zingberensis*. *Journal of Nanjing Normal University (Natural Science Edition)* 33: 94–97.

Wang, X., Lin, Y., Liu, D. et al. (2012). Cerium toxicity, uptake and translocation in *Arabidopsis thaliana* seedlings. *Journal of Rare Earths* 30: 579–585. https://doi.org/10.1016/S1002-0721(12)60094-5.

Wang, L., Cheng, M., Yang, Q. et al. (2019). Arabinogalactan protein–rare earth element complexes activate plant endocytosis. *Proceedings of the National Academy of Sciences USA* 116: 14349–14357. https://doi.org/10.1073/pnas.1902532116.

Wei, Z.G., Yin, M., Zhang, X. et al. (2001). Rare earth elements in naturally grown fern *Dicranopteris linearis* in relation to their variation in South-Jiangxi region (Southern China). *Environmental Pollution* 114: 345–355. https://doi.org/10.1016/S0269-7491(00)00240-2.

Wyttenbach, A., Schleppi, P., Bucher, J. et al. (1994). The accumulation of the rare earth elements and of scandium in successive needle age classes of Norway spruce. *Biological Trace Element Research* 41: 13–29. https://doi.org/10.1007/BF02917214.

Wyttenbach, A., Furrer, V., Schleppi, P., and Tobler, L. (1998). Rare Earth Elements in soil and in soil-grown plants. *Plant and Soil* 199: 267–273. https://doi.org/10.1023/A:1004331826160.

Xia, C.F., Lv, L., Chen, X.Y. et al. (2015). Nd(III)-induced rice mitochondrial dysfunction investigated by spectroscopic and microscopic methods. *The Journal of Membrane Biology* 248: 319–326. https://doi.org/10.1007/s00232-015-9773-1.

Xie, F., Zhang, T.A., Dreisinger, D., and Doyle, F. (2014). A critical review on solvent extraction of rare earths from aqueous solutions. *Minerals Engineering* 56: 10–28. https://doi.org/10.1016/j.mineng.2013.10.021.

Xu, Y., Liu, H., Meng, Z. et al. (2012). Decomposition of bastnasite and monazite mixed rare earth minerals calcined by alkali liquid. *Journal or Rare Earths* 30: 155–158. https://doi.org/10.1016/S1002-0721(12)60014-3.

Yan, S., Huang, X., and Zhou, Q. (2007). Effect of lanthanum (III) on reactive oxygen metabolism of soybean seedlings under supplemental UV-B irradiation. *Journal of Rare Earths* 25: 352–358. https://doi.org/1008org/10.1016/S1002-0721(07)60435-9.

Yang, L., Su, Y., Xu, Y. et al. (2003). Interactions between metal ions and carbohydrates. Coordination behavior of neutral erythritol to Ca(II) and lanthanide ions. *Inorganic Chemistry* 42: 5844–5856. https://doi.org/10.1021/ic0300464.

Yang, Q., Wang, L., He, J. et al. (2018). Direct imaging of how lanthanides break the normal evolution of plants. *Journal of Inorganic Biochemistry* 182: 158–169. https://doi.org/10.1016/j.jinorgbio.2018.01.020.

Yao, J., Liu, C.Y., Qin, G.Z. et al. (2008). The effect of Nd^{3+} and burdock oligosaccharide on the germination of *Cassia obtusifolia* seeds. *Chinese Agricultural Science Bulletin* 24: 69–72.

Yin, S., Ze, Y., Liu, C. et al. (2009). Cerium relieves the inhibition of nitrogen metabolism of spinach caused by magnesium deficiency. *Biological Trace Element Research* 132: 247–258. https://doi.org/10.1007/1011s12011-009-8392-z.

Zeng, Y. and Xu, T. (2015). Anti-oxidative and anti-inflammatory activity of baicalin complexes with trivalent lanthanum, yttrium and neodymium. *Latin American Journal of Pharmacy* 35 (5A): 1228–1233.

Zeng, F.L., Shi, P., Zhang, M.F., and Deng, R.W. (2000). Effect of lanthanum on ion absorption in cucumber seedling leaves. *Biological Trace Element Research* 78: 265–270. https://doi.org/10.1385/BTER:78:1-3:265.

Zhang, C., Li, Q., Zhang, M. et al. (2013). Effects of rare earth elements on growth and metabolism of medicinal plants. *Acta Pharmaceutica Sinica B* 3: 20–24. https://doi.org/10.1016/j.apsb.2012.12.005.

Zhao, H., Zhou, Q., Zhou, M. et al. (2012). Magnesium deficiency results in damage of nitrogen and carbon cross-talk of maize and improvement by cerium addition. *Biological Trace Element Research* 148: 102–109. https://doi.org/10.1007/s12011-012-9340-x.

Zheng, Z. (2012). *Lanthanides: Amino Acid Compounds*. Encyclopedia of Inorganic and Bioinorganic Chemistry https://doi.org/10.1002/9781119951438.eibc2063.

Zicari, M.A., d'Aquino, L., Paradiso, A. et al. (2018). Effect of cerium on growth and antioxidant metabolism of Lemna minor L. *Ecotoxicology and Environmental Safety* 163: 536–543. https://doi.org/10.1016/j.ecoenv.2018.07.113.

Index

Beneficial Chemical Elements of Plants: Recent Developments and Future Prospects, First Edition. Edited by Sangeeta Pandey, Durgesh Kumar Tripathi, Vijay Pratap Singh, Shivesh Sharma, and Devendra Kumar Chauhan.